Communications and Control Engineering

Series Editors

E.D. Sontag · M. Thoma · A. Isidori · J.H. van Schuppen

Published titles include:

Stability and Stabilization of Infinite Dimensional Systems with Applications
Zheng-Hua Luo, Bao-Zhu Guo and Omer Morgul

Nonsmooth Mechanics (Second edition)
Bernard Brogliato

Nonlinear Control Systems II
Alberto Isidori

L_2-Gain and Passivity Techniques in Nonlinear Control
Arjan van der Schaft

Control of Linear Systems with Regulation and Input Constraints
Ali Saberi, Anton A. Stoorvogel and Peddapullaiah Sannuti

Robust and H_∞ Control
Ben M. Chen

Computer Controlled Systems
Efim N. Rosenwasser and Bernhard P. Lampe

Control of Complex and Uncertain Systems
Stanislav V. Emelyanov and Sergey K. Korovin

Robust Control Design Using H_∞ Methods
Ian R. Petersen, Valery A. Ugrinovski and Andrey V. Savkin

Model Reduction for Control System Design
Goro Obinata and Brian D.O. Anderson

Control Theory for Linear Systems
Harry L. Trentelman, Anton Stoorvogel and Malo Hautus

Functional Adaptive Control
Simon G. Fabri and Visakan Kadirkamanathan

Positive 1D and 2D Systems
Tadeusz Kaczorek

Identification and Control Using Volterra Models
Francis J. Doyle III, Ronald K. Pearson and Bobatunde A. Ogunnaike

Non-linear Control for Underactuated Mechanical Systems
Isabelle Fantoni and Rogelio Lozano

Robust Control (Second edition)
Jürgen Ackermann

Flow Control by Feedback
Ole Morten Aamo and Miroslav Krstić

Learning and Generalization (Second edition)
Mathukumalli Vidyasagar

Constrained Control and Estimation
Graham C. Goodwin, María M. Seron and José A. De Doná

Randomized Algorithms for Analysis and Control of Uncertain Systems
Roberto Tempo, Giuseppe Calafiore and Fabrizio Dabbene

Switched Linear Systems
Zhendong Sun and Shuzhi S. Ge

Subspace Methods for System Identification
Tohru Katayama

Digital Control Systems
Ioan D. Landau and Gianluca Zito

Multivariable Computer-controlled Systems
Efim N. Rosenwasser and Bernhard P. Lampe

Dissipative Systems Analysis and Control (Second edition)
Bernard Brogliato, Rogelio Lozano, Bernhard Maschke and Olav Egeland

Algebraic Methods for Nonlinear Control Systems (Second edition)
Giuseppe Conte, Claude H. Moog and Anna Maria Perdon

Polynomial and Rational Matrices
Tadeusz Kaczorek

Simulation-based Algorithms for Markov Decision Processes
Hyeong Soo Chang, Michael C. Fu, Jiaqiao Hu and Steven I. Marcus

Iterative Learning Control
Hyo-Sung Ahn, Kevin L. Moore and YangQuan Chen

Distributed Consensus in Multi-vehicle Cooperative Control
Wei Ren and Randal W. Beard

Control of Singular Systems with Random Abrupt Changes
El-Kébir Boukas

Nonlinear and Adaptive Control with Applications
Alessandro Astolfi, Dimitrios Karagiannis and Romeo Ortega

Felix L. Chernousko · Igor M. Ananievski ·
Sergey A. Reshmin

Control of Nonlinear Dynamical Systems

Methods and Applications

Springer

F.L. Chernousko
Russian Academy of Sciences
Institute for Problems in Mechanics
Vernadsky Ave. 101-1
Moscow
Russia 119526
chern@ipmnet.ru

I.M. Ananievski
Russian Academy of Sciences
Institute for Problems in Mechanics
Vernadsky Ave. 101-1
Moscow
Russia 119526
anan@ipmnet.ru

S.A. Reshmin
Russian Academy of Sciences
Institute for Problems in Mechanics
Vernadsky Ave. 101-1
Moscow
Russia 119526
reshmin@ipmnet.ru

ISBN: 978-3-642-08970-1 e-ISBN: 978-3-540-70784-4

DOI: 10.1007/978-3-540-70784-4

Communications and Control Engineering ISSN: 0178-5354

Cover design: Integra Software Services Pvt. Ltd.

Printed on acid-free paper

9 8 7 6 5 4 3 2 1

springer.com

Preface

This book is devoted to new methods of control for complex dynamical systems and deals with nonlinear control systems having several degrees of freedom, subjected to unknown disturbances, and containing uncertain parameters. Various constraints are imposed on control inputs and state variables or their combinations.

The book contains an introduction to the theory of optimal control and the theory of stability of motion, and also a description of some known methods based on these theories.

Major attention is given to new methods of control developed by the authors over the last 15 years. Mechanical and electromechanical systems described by nonlinear Lagrange's equations are considered. General methods are proposed for an effective construction of the required control, often in an explicit form. The book contains various techniques including the decomposition of nonlinear control systems with many degrees of freedom, piecewise linear feedback control based on Lyapunov's functions, methods which elaborate and extend the approaches of the conventional control theory, optimal control, differential games, and the theory of stability.

The distinctive feature of the methods developed in the book is that the controls obtained satisfy the imposed constraints and steer the dynamical system to a prescribed terminal state in finite time. Explicit upper estimates for the time of the process are given. In all cases, the control algorithms and the estimates obtained are strictly proven.

The methods are illustrated by a number of control problems for various engineering systems: robotic manipulators, pendular systems, electromechanical systems, electric motors, multibody systems with dry friction, etc. The efficiency of the proposed approaches is demonstrated by computer simulations.

The authors hope that the monograph will be a useful contribution to the scientific literature on the theory and methods of control for dynamical systems. The book could be of interest for scientists and engineers in the field of applied mathematics, mechanics, theory of control and its applications, and also for students and postgraduates.

Moscow,
April 2008

Felix L. Chernousko
Igor M. Ananievski
Sergey A. Reshmin

Contents

Introduction

There exist numerous methods for the design of control for dynamical systems.

The classical methods of the theory of automatic control are meant for linear systems and represent the control in the form of a linear operator applied to the current phase state of the system. Shortcomings of this approach are obvious both in the vicinity of the prescribed terminal state as well as far from it. Near the terminal state, the magnitude of the control becomes small, so that control possibilities are not fully realized. As a result, the time of the control process occurs to be, strictly speaking, infinite, and the phase state can only tend asymptotically to the terminal state as time goes to infinity. On the other hand, far from the terminal state, the control magnitude becomes large and can violate the constraints usually imposed on the control. That is why it is difficult and often impossible to take account of the constraints imposed when the linear methods are used. Moreover, the classical methods based on linear models are usually inapplicable to nonlinear systems; at least, their applicability should be justified thoroughly.

In principle, the methods of the theory of optimal control can be applied to nonlinear systems. These methods take account of various constraints imposed on the control and, though with considerable complications, on the state variables. The methods of optimal control bring a dynamical system to a prescribed terminal state in an optimal (in a certain sense) way; for example, in a minimum time. However, to construct the optimal control for a nonlinear system is a very complicated problem, and its explicit solution is seldom available. Especially difficult is the construction of a feedback optimal control for a nonlinear system, even for a system with a small number of degrees of freedom and even with the help of modern computers.

There exist a number of other general methods of control: the method of systems with variable structure [123, 116, 115], the method of feedback linearization [70, 71, 91], and their various generalizations. However, these methods usually do not take into account constraints imposed on the control and state variables. Moreover, being very general, these methods do not take account of specific properties of mechanical systems such as conservation laws or the structure of basic equations of motions that can be presented in the Lagrangian or the Hamiltonian forms. Some other control methods applicable to nonlinear mechanical systems were developed in [61, 62, 94, 95, 59, 51, 52, 85, 90, 118].

In this book, some methods of control for nonlinear mechanical systems subjected to perturbations and uncertainties are proposed. These methods are applicable in the presence of various constraints on control and state variables. By taking into account some specific properties inherent in the equations of mechanical systems, these methods yield more efficient control algorithms compared with the methods developed for general systems of differential equations.

The authors' objective was to develop control methods having the following features.

1. Methods are applicable to nonlinear mechanical systems described by the Lagrange equations.

2. Methods are applicable to systems with many degrees of freedom.

3. Methods take into account the constraints imposed on the control, and, in a number of cases, also on the state variables as well as on both the control and state variables.

4. Methods bring the control system to the prescribed terminal state in finite time, and an efficient upper estimate is available for this time.

5. Methods are applicable in the presence of uncertain but bounded external perturbations and uncertain parameters of the system. Thus, the methods are robust.

6. There exist efficient algorithms for the construction of the desired feedback control.

7. Efficient sufficient controllability conditions are stated for the control methods proposed.

8. In all cases, a rigorous mathematical justification of the proposed methods is given.

It is clear that the above requirements are very useful and important from the standpoint of the control theory as well as various practical applications.

Several methods are proposed and developed in the book, and not all of them possess all of the features 1–8 listed above. Properties 3, 4, 7, and 8 are always fulfilled, whereas other features are inherent in some of the methods and not present in others.

The book consists of 10 chapters.

Chapters 2, 3, 5, and 6 deal with nonlinear mechanical systems with many degrees of freedom governed by Lagrange's equations and subjected to control and perturbation forces.

These equations are taken in the form:

$$\frac{d}{dt}\frac{\partial T}{\partial \dot{q}_i} - \frac{\partial T}{\partial q_i} = U_i + Q_i, \quad i = 1, \dots, n. \tag{0.1}$$

Here, t is time, the dots denote time derivatives, q_i are the generalized coordinates, $\dot{q}_i$ are the generalized velocities, U_i are the generalized control forces, Q_i are all other generalized forces including uncertain perturbations, n is the number of degrees of freedom, and $T(q, \dot{q})$ is the kinetic energy of the system. The kinetic energy is a symmetric positive definite quadratic form of the generalized velocities $\dot{q}_i$:

$$T(q,\dot{q}) = \frac{1}{2}\langle A(q)\dot{q},\dot{q}\rangle = \frac{1}{2}\sum_{j,k=1}^{n} a_{jk}(q)\dot{q}_j\dot{q}_k. \tag{0.2}$$

Here, q and $\dot{q}$ are the n-vectors of the generalized coordinates and velocities, respectively, and the brackets $\langle\cdot,\cdot\rangle$ denote the scalar product of vectors.

The quadratic form (0.2) satisfies the conditions

$$m|\dot{q}|^2 \le \langle A(q)\dot{q},\dot{q}\rangle \le M|\dot{q}|^2 \tag{0.3}$$

for any $q \in R^n$ and $\dot{q} \in R^n$, where m and M are positive constants such that $M > m$. Condition (0.3) implies that all eigenvalues of the matrix $A(q)$, for all $q \in R^n$, belong to the interval $[m,M]$.

In Chapters 2 and 3, the coefficients of the quadratic form (0.2) are supposed to be known functions of the coordinates: $a_{jk} = a_{jk}(q)$. In Chapters 5 and 6, the functions $a_{jk}(q)$ may be unknown but the constants m and M in (0.3) are given. Also, the case of rheonomic systems for which $T = T(q,\dot{q},t)$ is considered in Chapter 5.

We suppose that the control forces are subjected to the geometric constraints at any time instant:

$$|U_i| \le U_i^0, \qquad i = 1,\dots,n, \tag{0.4}$$

where U_i^0 are given constants.

The generalized forces Q_i may be more or less arbitrary functions of the coordinates, velocities, and time; these functions may be unknown but are assumed bounded by the inequality

$$|Q_i(q,\dot{q},t)| \le Q_i^0, \qquad i = 1,\dots,n, \tag{0.5}$$

The constants Q_i^0 are supposed to be known, and certain upper bounds are imposed on Q_i^0 in order to achieve the control objective.

The control problem is formulated as follows:

Problem 0.1. It is required to construct the feedback control $U_i(q,\dot{q})$ that brings system (0.1) subject to constraints (0.3)–(0.5) from the given initial state

$$q(t_0) = q^0, \qquad \dot{q}(t_0) = \dot{q}^0 \tag{0.6}$$

at a given initial time instant $t = t_0$ to the prescribed terminal state with zero terminal generalized velocities

$$q(t_*) = q^*, \qquad \dot{q}(t_*) = 0 \tag{0.7}$$

in finite time. The time instant t_* is not prescribed but an upper estimate on it should be obtained.

In some sections of Chapter 3, the case of nonzero terminal velocities $\dot{q}_i(t_*) \neq 0$ is also considered.

In many practical applications, it is desirable to bring the system from the state (0.6) to the state (0.7) as soon as possible, i.e., to minimize t_*. However, to construct the exact solution of this time-optimal control problem for the nonlinear system is a very difficult problem, especially, if one desires to obtain the feedback control. The methods proposed in Chapters 2, 3, and 5–8 do not provide the time-optimal control but include certain procedures of optimization of the time t_*. Therefore, these methods may sometimes be called suboptimal.

The main difficulties arising in the construction of the control for system (0.1) are due to its nonlinearity and its high order. The complex nonlinear dynamical interaction of different degrees of freedom of the system is characterized by the elements $a_{jk}(q)$ of the matrix $A(q)$ of the kinetic energy. Another property that complicates the construction of the control is the fact that the dimension n of the control vector is two times less than the order of system (0.1).

Manipulation robots can be regarded as typical examples of mechanical or electromechanical systems described by equations (0.1). Being an essential part of automated manufacturing systems, these robots can serve for various technological operations. A manipulation robot is a controlled mechanical system that consist of one or several manipulators, a control system, drives (actuators), and grippers. A manipulator can perform a wide range of three-dimensional motions and bring objects (instruments and/or workpieces) to a prescribed position and orientation in space. Various types of drives, namely, electric, hydraulic, pneumatic, and other, are employed in robotic manipulators, the electric drives being the most widespread.

The manipulator is a multibody system that consists of several links connected by joints. The drives are usually located at the joints or inside links adjacent to the joints. Relative angular or linear displacements of neighboring links are usually chosen as the generalized coordinates q_i of the manipulator. The kinetic energy $T(q,\dot{q})$ of the manipulator consists of the kinetic energy of its links and also, if the drives are taken into account, the kinetic energy of electric drives and gears. The Lagrange equations (0.1) of the manipulator involve the generalized forces Q_i due to the weight and resistance forces; the latter are often not known exactly and may change during operations. Moreover, parameters of the manipulator may also change in an unpredictable way. Therefore, some of the forces Q_i should be regarded as uncertain perturbations. The control forces U_i are forces and/or torques produced by the drives.

Since the manipulator is a nonlinear multibody system subject to uncertain perturbations, it is quite natural to consider the problem of control for the manipulator as a nonlinear control problem formulated above as Problem 0.1.

Let us outline briefly the contents of Chapters 1–10.

Chapter 1 gives an introduction to the theory of optimal control. Basic concepts and results of this theory, especially the Pontryagin maximum principle, are often used throughout the book. The maximum principle is formulated and illustrated by several examples. The feedback optimal controls obtained for these examples are often referred to in the following chapters.

In Chapters 2 and 3, the methods of decomposition for Problem 0.1 are proposed and developed. The essence of these methods is a transformation of the original nonlinear system (0.1) with n degrees of freedom to the set of n independent linear subsystems

$$\ddot{x}_i = u_i + v_i, \quad i = 1, \dots, n. \tag{0.8}$$

Here, x_i are the new (transformed) generalized coordinates, u_i are the new controls, and forces v_i include the generalized forces Q_i, as well as the nonlinear terms that describe the interaction of different degrees of freedom in system (0.1). The perturbations v_i in system (0.8) are treated as uncertain but bounded forces; they can also be regarded as the controls of another player that counteract the controls u_i.

The original constraints (0.3)–(0.5) imposed on the kinetic energy and generalized forces of system (0.1) are, under certain conditions, reduced to the following normalized constraints on controls u_i and disturbances v_i:

$$|u_i| \le 1, \quad |v_i| \le \rho_i, \quad \rho_i < 1, \quad i = 1, \dots, n. \tag{0.9}$$

By applying the approach of differential games [69, 79] to system (0.8) subject to constraints (0.9), we obtain the feedback control $u_i(x_i, \dot{x}_i)$ that solves the control problem for the ith subsystem, if $\rho_i < 1$.

Besides the game-theoretical technique, a simpler approach to the control construction is also considered, where the perturbations in system (0.8) are completely ignored. As the control $u_i(x_i, \dot{x}_i)$ of the ith subsystem (0.8) we choose the time-optimal feedback control for the system

$$\ddot{x}_i = u_i, \quad i = 1, \dots, n.$$

It is shown that this simplified approach is effective, i.e., brings the ith subsystem (0.8) to the prescribed terminal state, if and only if the number ρ_i in (0.9) does not exceed the golden section ratio:

$$\rho_i < \rho^* = \frac{1}{2}(\sqrt{5} - 1) \approx 0.618.$$

In other words, uncertain but bounded perturbations can be neglected while constructing the feedback control, if and only if their magnitude divided by the magnitude of the control does not exceed the golden section ratio ρ^*.

Two versions of the decomposition method presented in Chapters 2 and 3 differ both by the assumptions made and the results obtained.

The assumptions of the second version (Chapter 3) are less restrictive; on the other hand, the time of the control process is usually less for the first version (Chapter 2).

As a result of each decomposition method, explicit feedback control laws for the original system (0.1) are obtained. These control laws $U_i = U_i(q, \dot{q})$, $i = 1, \dots, n$, satisfy the imposed constraints (0.4) and bring the system to the terminal state (0.7) under any admissible perturbations $Q_i(q, \dot{q}, t)$ subject to conditions (0.5). Sufficient controllability conditions are derived for the methods proposed. The time of control t_* is finite, and explicit upper bounds on t_* are given.

Certain generalizations and modifications of the decomposition methods are presented in Chapters 2 and 3. The original system (0.1) with n degrees of freedom can be reduced to sets of subsystems more complicated than (0.8); these subsystems can be either linear or nonlinear, and these two cases are examined. The decomposition method is extended to the case of nonzero prescribed terminal velocity $\dot{q}_i(t_*)$ in (0.7), and also to the problem of tracking the prescribed trajectory of motion.

Control problems for the manipulation robots with several degrees of freedom are considered as examples illustrating the methods proposed. Purely mechanical models of robots as well as electromechanical models that take account of processes in electric circuits are considered.

Chapter 4 briefly presents basic concepts and results of the theory of stability. Here, the notion of the Lyapunov function plays the central role, and the corresponding theorems using this notion are formulated. The Lyapunov functions are widely used in the following Chapters 5 and 6.

In these chapters, the method of control based on the piecewise linear feedback for system (0.1)–(0.7) is presented. The required control vector U is sought in the form

$$U = -\beta(q - q^*) - \alpha\dot{q}, \quad U = (U_1, \ldots, U_n), \tag{0.10}$$

where α and β are scalar coefficients.

During the motion, the coefficients increase in accordance with a certain algorithm and may tend to infinity as the system approaches the terminal state (0.7), i.e., $t \to t_*$. However, the control forces (0.10) stay bounded and satisfy the imposed constraints (0.4).

In Chapter 5, the coefficients α and β are piecewise constant functions of time. These coefficients change when the system reaches certain prescribed ellipsoidal surfaces in $2n$-dimensional phase space. In Chapter 6, the coefficients α and β are continuous functions of time.

In both Chapters 5 and 6, the proposed algorithms are rigorously justified with the help of the second Lyapunov method. It is proven that this control technique brings the system (0.1) to the prescribed terminal state (0.7) in finite time. An explicit upper bound for this time is obtained.

The methods of Chapters 5 and 6 are applicable not only in the case of uncertain perturbations satisfying (0.5), but also if the matrix A of the kinetic energy (0.2) is uncertain. It is only necessary that restrictions (0.3) hold and the constants m and M be known.

The approach based on the feedback control (0.10) is extended also to rheonomic systems whose kinetic energy is a second-order polynomial of the generalized velocities with coefficients depending explicitly on the generalized coordinates and time (Chapter 5). The coefficients of the kinetic energy are assumed unknown, and the system is acted upon by uncertain perturbations. The control algorithm is given that brings the rheonomic system to the prescribed terminal state by a bounded control force.

Several examples of controlled multibody systems are considered in Chapters 5 and 6. Some parameters of the systems, namely, masses, coefficients of stiffness and friction, are assumed unknown, and uncertain perturbations are also taken into account. It is shown that the methods proposed in Chapters 5 and 6 can control such systems and bring them to the terminal state; moreover, the methods are efficient even if the sufficient controllability conditions derived in Chapters 5 and 6 are not satisfied.

Note that, together with the methods discussed in the book, there are other approaches that ensure asymptotic stability of a given state of the system, i.e., bring the system to this state as $t \to \infty$. In practice, one needs to bring the system to the vicinity of the prescribed state; therefore, the algorithms based on the asymptotic stability practically solve the control problem in finite time. However, as the required vicinity of the terminal state decreases and tends to zero, the time of motion for the control methods ensuring the asymptotic stability increases and tends to infinity. By contrast, the methods proposed in this book ensure that the time of motion is finite, and explicit upper bounds for this time are given in Chapters 2, 3, 5, and 6.

In Chapters 1–6, systems with finitely many degrees of freedom are considered; these are described by systems of ordinary differential equations. A number of books and papers (see, for example, [25, 122, 86, 113, 117, 87]) are devoted to control problems for systems with distributed parameters that are described by partial differential equations. The methods of decomposition proposed in Chapters 2 and 3 can also be applied to systems with distributed parameters.

In Chapter 7, control systems with distributed parameters are considered. These systems are described by linear partial differential equations resolved with respect to the first or the second time derivative. The first case corresponds, for example, to the heat equation, and the second to the wave equation. The control is supposed to be distributed and bounded; it is described by the corresponding terms in the right-hand side of the equation. The control problem is to bring the system to the zero terminal state in finite time. The proposed control method is based on the decomposition of the original system into subsystems with the help of the Fourier method. After that, the time-optimal feedback control is applied to each mode. A peculiarity of this control problem is that there is an infinite (countable) number of modes.

Sufficient controllability conditions are derived. The required feedback control is obtained, together with upper estimates for the time of the control process. These results are illustrated by examples.

In Chapters 8–10, we return to control systems governed by ordinary differential equations.

In Chapter 8, we consider linear systems subject to various constraints. Control and phase constraints, as well as mixed constraints imposed on both the control and the state variables are considered. Integral constraints on control and state variables are also taken into account. Though the original systems are linear, the presence of complex constraints makes the control problem essentially nonlinear and rather complicated.

Note that various constraints on control and state variables are often encountered in applications. For example, if the system includes an electric drive, it is usually necessary to take account of constraints on the angular velocity of the shaft, the control torque, and also on their combination. Integral constraints typically occur, if there are energy restrictions.

The approach developed in Chapter 8 is a generalization of the well-known Kalman's method [72, 73]. This method, originally proposed for the control of linear systems in the absence of constraints, is based on the representation of the control as a linear combination of the eigenmodes of motion. In Chapter 8, this method is extended to some cases with different constraints. Explicit control laws are obtained for various oscillatory systems, in particular, a system of many oscillators controlled by one bounded control. For certain systems of the second order, the controls obtained are compared with time-optimal controls. The method is applied also to systems of the fourth (and higher) order with mixed constraints. The models considered here correspond to mechanical and electromechanical systems containing an oscillator and an electric motor. Sufficient controllability conditions derived in Chapter 8 ensure that the control obtained brings the system to the prescribed state in finite time, and all mixed constraints are satisfied.

Chapter 9 is devoted to several control problems for a simple dynamical system with one degree of freedom described by the second Newton's law and subject to different constraints that model real constraints typical for actuators. The system is to be brought to the origin of the coordinate system in the phase plane.

First, the time-optimal control problem is considered in the presence of mixed constraints imposed on the control and state variables. The time-optimal feedback control is obtained. As an example, a control problem for the electric drive is examined.

Next, a constraint is imposed on the rate of change of the control force. Such a constraint is often inherent in various drives. The resultant equations are reduced to a third-order system. The time-optimal control problem for this system is solved, and the required control is obtained in the open-loop as well as in the feedback form. The solution of this problem is based on a group-invariant approach that reduces the number of the essential phase variables from three to two.

At the end of Chapter 9, it is supposed that the absolute value of the control force can grow only gradually, with a bounded rate, whereas this force can be switched off instantly. Under these assumptions, which model real drives, we find the control that brings the system to a prescribed state and has the simplest possible structure.

In Chapter 10, two time-optimal control problems for the nonlinear pendulum are solved. The pendulum is a classical nonlinear system that often serves as a test model in nonlinear dynamics and control theory. We assume that the bounded control torque is applied to the axis of the pendulum. The terminal state is either the upper unstable or the lower stable equilibrium position of the pendulum; thus, we study the time-optimal swing-up and damping control problems, respectively. The peculiarity of these problems is that the pendulum has a cylindrical phase space and an infinite number of equivalent equilibrium positions which differ by 2π. The feedback controls for both the swing-up and the dumping cases have a very complicated structure, which is obtained numerically for a wide range of the system parameters.

Thus, a number of new methods for the control of nonlinear dynamical systems are presented in the book. The control algorithms are described, their rigorous mathematical proof is given, and a number of specific control problems are analyzed and solved by these methods.

This book is mostly based on the results obtained by the authors during the last two decades.

Chapter 1
Optimal control

In the following chapters of the book we will often use the approach and concepts of the optimal control theory. Also, some of the proposed methods of control utilize certain results obtained for particular optimal control problems and use these results as integral parts of our control algorithms. Thus, it would be useful to recall the basic concepts of the optimal control theory and describe the solution of several typical problems.

1.1 Statement of the optimal control problem

We consider a general dynamical system subjected to control and described by the following nonlinear *differential equation*

$$\dot{x} = f(x,u,t). \tag{1.1.1}$$

Here, $x = (x_1,\dots,x_n)$ is the n-dimensional vector of state and $u = (u_1,\dots,u_m)$ is the m-dimensional vector of control; these vectors are functions of time t: $x = x(t)$, $u = u(t)$. The dot $\cdot$ denotes differentiation with respect to time. The n-dimensional vector $f(x,u,t)$ is a given function of its arguments. Equation (1.1.1) is sometimes called *equation of motion*.

Control systems can also be described by more general classes of equations: differential algebraic equations (DAE), integro-differential equations, functional differential equations, etc. In this book, we mostly restrict ourselves to differential equations (1.1.1).

To formulate the optimal control problem, we should, in addition to (1.1.1), impose *boundary conditions, constraints*, and an *optimality criterion*, or a *cost functional*. The control process is considered on the time interval $t \in [t_0, T]$, the ends t_0 and T of this interval may be fixed or free.

In general, the *boundary conditions* can be stated as follows:

$$(t_0, x(t_0)) \in X_0, \qquad (T, x(T)) \in X_T, \tag{1.1.2}$$

where X_0 and X_T are given sets in the $(n+1)$-dimensional (t,x)-space.

Let us restrict ourselves to the case mostly considered in this book, where the initial data are fixed so that the set X_0 in (1.1.2) is a given point (t_0, x^0) in the (t,x)-space. Hence, we have the initial condition

$$x(t_0) = x^0. \tag{1.1.3}$$

Here, the time instant t_0 and the vector x^0 are fixed.

We assume also that the set X_T in (1.1.2) is defined by r equations in the x-space

$$X_T = \{t = T, x : g_i(x) = 0\}, \quad i = 1, \ldots, r \leq n, \tag{1.1.4}$$

whereas the terminal time T may be either fixed or free. Here, $g_i(x)$ are given scalar functions of x such that the Jacobian matrix

$$G = \left(\frac{\partial g_i}{\partial x_j}\right), \quad i = 1, \ldots, r, \quad j = 1, \ldots, n, \tag{1.1.5}$$

has the maximal possible rank r on the set defined by (1.1.4).

The simplest case of the conditions (1.1.2) often referred to in this book is the so-called *two-point boundary conditions* where both vectors $x(t_0)$ and $x(T)$ are fixed. In this case, in addition to (1.1.3) we have

$$x(T) = x^1, \tag{1.1.6}$$

where x^1 is a given vector. The terminal time T may be fixed or free. Note that in the case of (1.1.6) we have

$$g_i = x_i - x_i^1, \quad i = 1, \ldots, n, \quad r = n, \quad G = I,$$

in (1.1.4) and (1.1.5), where I is the identity matrix.

Constraints may be imposed on the control u, the state x, or both. *Control constraints* are often expressed in the form

$$u(t) \in U, \qquad t \in [t_0, T], \tag{1.1.7}$$

where U is a given closed set in the m-dimensional u-space.

State constraints can be expressed in a similar way

$$x(t) \in V, \qquad t \in [t_0, T], \tag{1.1.8}$$

where V is a given closed set in the n-dimensional x-space.

Both sets in (1.1.7) and (1.1.8) may depend on time so that we have $U = U(t)$ and $V = V(t)$. Note that the boundary conditions (1.1.2) can be formally considered as a particular case of the state constraints imposed at two specific time instants $t = t_0$ and $t = T$.

In more general [than (1.1.7) and (1.1.8)] case of *mixed constraints*, we have

$$u(t) \in U(x(t),t), \quad t \in [t_0,T], \tag{1.1.9}$$

where $U(x,t)$ is, for all x and $t \in [t_0,T]$, a closed set in the m-dimensional u-space; this set depends on x and t. The constraint (1.1.9) can be also expressed as follows:

$$(u(t),x(t)) \in W(t), \qquad t \in [t_0,T]. \tag{1.1.10}$$

Here, $W(t)$ is, for any $t \in [t_0,T]$, a closed set in the $(m+n)$-dimensional (u,x)-space.

All constraints described by (1.1.7)–(1.1.10) are sometimes called *geometric*; they are imposed on the values of the control and state at any given instant t.

Another class of constraints are *integral constraints* that can be imposed on control and state variables. These constraints can be either of equality or inequality type, and the integrals can be taken over either a fixed or variable time interval.

Integral constraints can be often reduced to the boundary conditions and geometric state constraints. As an example, let us consider two integral constraints: an equality type constraint with a fixed interval of integration and an inequality type constraint with a variable integration interval. We have

$$\begin{gathered} \int_{t_0}^{T} \varphi_1(x(t),u(t),t)dt = c_1, \\ \int_{t_0}^{t} \varphi_2(x(\tau),u(\tau),\tau)d\tau \ge c_2(t), \quad t \in [t_0,T], \end{gathered} \tag{1.1.11}$$

where φ_1 and φ_2 are given functions, c_1 is a constant, and c_2 is a given function of t.

We introduce additional state variables x_{n+i} defined by the following equations and boundary conditions

$$\dot{x}_{n+i} = \varphi_i(x,u,t), \quad x_{n+i}(t_0) = 0, \quad i = 1,2.$$

Then our integral constraints (1.1.11) can be rewritten as follows:

$$x_{n+1}(T) = c_1, \quad x_{n+2}(t) \ge c_2(t), \quad t \in [t_0,T].$$

Thus our integral constraints (1.1.11) are reduced to the boundary condition for $x_{n+1}(T)$ and the state constraint imposed on $x_{n+2}(t)$.

The *cost functional*, or the *optimality criterion*, is mostly given as a function depending on the terminal values of the state variables and time

$$J = F(x(T),T) \tag{1.1.12}$$

or as an integral functional

$$J = \int_{t_0}^{T} f_0(x(t), u(t), t)dt. \tag{1.1.13}$$

Here, $F(x,t)$ and $f_0(x,u,t)$ are given functions of their arguments. Each type of the functionals (1.1.12) and (1.1.13) can be reduced to the other one.

If the original functional is given in the terminal form (1.1.12), we introduce the function

$$f_0(x,u,t) = \frac{\partial F}{\partial t} + \left\langle \frac{\partial F}{\partial x}, f(x,u,t) \right\rangle. \tag{1.1.14}$$

Here, $\partial/\partial x$ denotes the vector of gradient, and brackets $\langle .,. \rangle$ denote the scalar product of vectors.

Then, taking into account (1.1.1), (1.1.3), and (1.1.14), we reduce the terminal functional (1.1.12) to the integral one

$$J = \int_{t_0}^{T} f_0(x,u,t)dt + \text{const}.$$

Vice versa, if we have an integral functional (1.1.13), we introduce an additional state variable by the following equation and initial condition

$$\dot{x}_0 = f_0(x,u,t), \qquad x_0(t_0) = 0$$

and express our functional (1.1.13) in the terminal form

$$J = x_0(T).$$

Also, combinations of terminal and integral functionals can be considered as the optimality criteria; these combinations can be also reduced to one of the basic types (1.1.12) or (1.1.13).

More complicated example of the cost functional is the minimum (or maximum) of some given function $\psi(x,t)$ over the time interval $[t_0, T]$, i.e.,

$$J = \min_t \psi(x(t),t), \quad t \in [t_0, T]. \tag{1.1.15}$$

In general, this kind of the functional cannot be reduced to the conventional types (1.1.12) and (1.1.13). However, this reduction is possible, if the derivative

$$\frac{d\psi}{dt} = \frac{\partial \psi}{\partial t} + \left\langle \frac{\partial \psi}{\partial x}, f(x,u,t) \right\rangle = g(x,t)$$

does not depend on u, and the function $\psi(x(t),t)$ has only one minimum with respect to $t \in [t_0, T]$. Then our functional (1.1.15) can be expressed as follows:

$$J = \psi(x(\tau), \tau),$$

where τ is the new terminal time, and the following terminal boundary condition

$$g(x(\tau),\tau) = 0$$

should be imposed on τ and $x(\tau)$.

In this book, we will not deal with cost functionals of the type presented by (1.1.15).

Now we can formulate the optimal control problem in general terms.

For the given system, find the *control* $u(t)$ and the corresponding *state trajectory* $x(t)$ such that they provide the minimal possible value of the *optimality criterion* J under the imposed *boundary conditions* and *constraints*.

We will always take the system in the form (1.1.1) and impose initial and terminal conditions in one of the forms (1.1.2)–(1.1.4) or (1.1.6). The constraints will be imposed in one of the forms (1.1.7)–(1.1.10), whereas the cost functional J will be given by (1.1.12) or (1.1.13).

Without loss of generality, we will always deal with the minimization of the cost functional J. If we are interested in the maximization of the functional J, it is sufficient just to change the sign of the functional and minimize $(-J)$.

The control $u(t)$ and state $x(t)$ as the functions of time that correspond to the minimal possible value of the functional J are called the *optimal control* and *optimal state trajectory*, respectively.

The problem of optimal control formulated above is very important for numerous practical applications. In particular, this problem arises naturally in control of such mechanical and electromechanical systems as various vehicles (aircraft and spacecraft, rockets, automobiles, ships, and other transport systems), industrial and mobile robots, motors, machines, machine tools, etc. In these applications, the state variables x_i, $i = 1,\dots,n$, are usually generalized coordinates and velocities of the mechanical part of the system under consideration as well as electric currents in the electric part of the system. The variables u_i, $i = 1,\dots,m$, denoted usually control forces and torques, electric voltages, and other controls acting upon the system. The boundary conditions and constraints reflect real limitations and bounds imposed upon the system under consideration.

For example, if u is the thrust acting upon the aircraft, the constraint (1.1.7) expresses the bounds upon the magnitude and direction of the thrust. Since these bounds may depend on the altitude and velocity of the aircraft, we come to the condition (1.1.9), where the set U at any instant t depends on the current state $x(t)$. Integral constraints (1.1.11) may reflect bounds imposed upon the energy or fuel expenditure. The cost functional (1.1.12) can be related to the desired position and velocity at the terminal state of the aircraft, whereas the integral functional (1.1.13) can be the measure of the fuel or energy consumption.

An important particular case of the optimality criterion J is the time of the control process. This case can be considered, if we put $F = T$ in (1.1.12) or $f_0 = 1$ in (1.1.13). The optimal control with this cost functional is called *time-optimal*.

1.2 The maximum principle

In this section, we will formulate the maximum principle for a class of optimal control problems. This principle was first stated and proved by Pontryagin and his colleagues [93]. Comprehensive information and complete proofs of the maximum principle can be found in numerous books; see, for example, [19, 24, 84, 83].

We consider the system (1.1.1) under the initial condition (1.1.3), terminal condition (1.1.4), control constraint (1.1.7), and the cost functional (1.1.13). Thus, our optimal control problem is determined by the following set of equations and conditions

$$\dot{x} = f(x,u,t), \quad x(t_0) = x^0, \quad t \in [t_0, T],$$

$$g_i(x(T)) = 0, \quad i = 1,\dots,r \leq n, \quad u(t) \in U, \tag{1.2.1}$$

$$J = \int_{t_0}^{T} f_0(x(t),u(t),t) \to \min.$$

The functions f and f_0 as well as their first derivatives with respect to x_i, $i = 1,\dots,n$, are assumed to satisfy the Lipschitz conditions with respect to x and u. The functions g_i, $i = 1,\dots,r$, are smooth, and the rank of the matrix G from (1.1.5) at the set defined by (1.1.4) is equal to r. The set U in (1.2.1) is a closed set in R^m. We will consider two cases: the terminal time T is either fixed or free.

The control $u(t)$ will be called *admissible*, if it is a piecewise continuous function of t for $t \in [t_0, T]$ and satisfies the constraint $u(t) \in U$ for all $t \in [t_0, T]$.

The admissible control $u(t)$ is called *optimal*, if it corresponds to the minimal possible value of the cost functional J among all admissible controls.

Suppose the optimal control exists. If we substitute it into our equation of motion and integrate this equation subject to the given initial condition [see (1.2.1)], we obtain the *optimal state trajectory* $x(t)$.

Let us introduce the additional state variable x_0 defined by the following differential equation and initial condition

$$\dot{x}_0 = f_0(x,u,t), \qquad x_0(t_0) = 0. \tag{1.2.2}$$

Then our functional J from (1.2.1) can be expressed as $J = x_0(T)$.

We introduce now the $(n+1)$-dimensional *adjoint*, or *conjugate*, vector with the components

$$\bar{p}(t) = (p_0, p_1, \dots, p_n) \tag{1.2.3}$$

and the *Hamiltonian* H defined by

$$H(\bar{p},x,u,t) = \sum_{i=0}^{n} p_i f_i(x,u,t) = p_0 f_0 + \langle p, f \rangle. \tag{1.2.4}$$

Here, $p = (p_1, \dots, p_n)$.

Note that our equation of motion from (1.2.1) together with (1.2.2) can be rewritten in terms of the Hamiltonian (1.2.4) as follows:

$$\dot{x}_i = \frac{\partial H}{\partial p_i}, \qquad i = 0, 1, \dots, n. \tag{1.2.5}$$

The components of the adjoint vector (1.2.3) will obey the following differential equations

$$\dot{p}_i = -\frac{\partial H}{\partial x_i}, \qquad i = 0, 1, \dots, n. \tag{1.2.6}$$

Hence, the variables x_i and p_i, $i = 0, 1, \dots, n$, satisfy the system of Hamiltonian equations (1.2.5) and (1.2.6). The components x_i of the state vector and the adjoint variables p_i play the role of the coordinates and impulses, respectively, of the Hamiltonian system.

Now we can formulate the maximum principle that is a necessary optimality condition. We consider both cases: of the fixed or free terminal time T.

Theorem 1.1. *Let $u(t)$ be an optimal control for the problem defined by (1.2.1) and $x(t)$ be the corresponding optimal trajectory. Then there exists a nonzero adjoint vector $\bar{p}(t)$ satisfying the adjoint system (1.2.6) and such that*

1)

$$H(\bar{p}(t), x(t), u(t), t) = \sup_{u \in U} H(\bar{p}(t), x(t), u, t) \tag{1.2.7}$$

for $t \in [t_0, T]$;

2)

$$p_0 = \text{const} \leq 0; \tag{1.2.8}$$

3) the following boundary conditions hold

$$p_i(T) = \sum_{j=1}^{r} \lambda_j \frac{\partial g_j(x(T))}{\partial x_i}, \quad i = 1, \dots, n, \tag{1.2.9}$$

where λ_j are constants;

4) if the terminal time T is free, then

$$H(\bar{p}(T), x(T), u(T), T) = 0. \tag{1.2.10}$$

The proof of Theorem 1.1 can be found in books [93, 19, 24, 84, 83]. We will restrict ourselves only with comments on this theorem.

Conditions (1.2.9) are called the *transversality conditions*. They are absent in the case of the two-point problem where the boundary conditions are given by (1.1.3) and (1.1.6). In this case, we have $r = n$, and the number of unknown constants λ_i is equal to the number of equations (1.2.9), so that these equations do not provide any additional conditions.

If there are no boundary conditions (1.1.4), we have $r = 0$, and the transversality conditions (1.2.9) become $p_i(T) = 0$, $i = 1, \dots, n$.

If the closed set U is bounded, then the upper bound of H over $u \in U$ in (1.2.7) is attained, and this equation implies

$$u(t) = \arg\max_{v \in U} H(\bar{p}(t), x(t), v, t). \tag{1.2.11}$$

Of course, the control u providing the maximum in (1.2.11) may be not unique.

Suppose it is unique so that we can express u as a single valued function of $\bar{p}, x, t$ by means of (1.2.11). Then we have

$$u = V(\bar{p}, x, t), \tag{1.2.12}$$

where V is a given function of its arguments.

By substituting u from (1.2.12) into the Hamiltonian system formed by (1.2.5) and (1.2.6) for $i = 1, \ldots, n$, we obtain a system of $2n$ differential equations for $2n$ variables x_i and p_i, $i = 1, \ldots, n$.

Let us consider the boundary conditions related to this system. We have n initial (at $t = t_0$) and r terminal (at $t = T$) conditions from (1.2.1) as well as n transversality conditions (1.2.9). The latter conditions contain r unknown parameters $\lambda_1, \ldots, \lambda_r$ that can be excluded from (1.2.9), since the corresponding matrix G defined by (1.1.5) has the rank r. After such elimination of λ_j from (1.2.9), the transversality conditions will consist of $n - r$ equalities imposed on $x(T)$ and $p(T)$. Thus, the total number of boundary conditions for our system will be equal to the number $2n$ of the variables x_i and p_i, $i = 1, \ldots, n$.

If the terminal time T is not fixed, we have an additional unknown parameter T and an additional boundary condition given by (1.2.10).

Note that the Hamiltonian (1.2.4) and, therefore, the right-hand sides of (1.2.11) and (1.2.12) depend also on p_0. Since the Hamiltonian (1.2.4) does not depend on x_0, we have $\partial H / \partial x_0 = 0$. Hence, by virtue of (1.2.6), p_0 is constant. Consider the following linear transformation of the adjoint vector

$$p_i \to \mu p_i, \quad i = 0, 1, \ldots, n, \quad \mu > 0, \tag{1.2.13}$$

where μ is an arbitrary positive constant.

Under transformation (1.2.13), the Hamiltonian (1.2.4) is transformed similarly: $H \to \mu H$, whereas the equations (1.2.5) and (1.2.6) stay invariant. As a result, the function V from (1.2.12) will also stay invariant, and the constants λ_j in (1.2.9) will be simply multiplied by μ.

Thus, without loss of generality, we can restrict ourselves to two cases in (1.2.8). If $p_0 = \text{const} < 0$, we normalize the adjoint vector and set $p_0 = -1$; otherwise, we have $p_0 = 0$. The first case is called regular, or normal, and takes place usually in well-posed problems. The second case is singular (abnormal) and occurs in ill-posed problems.

Therefore, in both cases of the fixed and free terminal time T and under the assumptions made, the maximum principle formally reduces the optimal control problem to the two-point boundary value problem for the nonlinear system of differential equations of the $2n$th order.

Consider now some particular cases of the general problem defined by (1.2.1).

Let the system (1.1.1) be autonomous, i.e., its right-hand side does not depend explicitly on t. Instead of (1.1.1), we have

$$\dot{x} = f(x,u). \tag{1.2.14}$$

The following theorem holds for this case.

Theorem 1.2. *For the optimal control problem defined by (1.2.1), where the equation of motion (1.1.1) is replaced by the autonomous equation (1.2.14), Theorem 1.1 holds and, besides,*

$$H(\bar{p}(t),x(t),u(t)) \equiv \text{const} \tag{1.2.15}$$

for the optimal control. If the terminal time T is free, then the constant in (1.2.15) is zero, i.e.,

$$H(\bar{p}(t),x(t),u(t)) \equiv 0. \tag{1.2.16}$$

Thus, for the autonomous system, the Hamiltonian is the first integral of the Hamiltonian system (1.2.5), (1.2.6) along the optimal trajectory. This fact can be used in order to check the correctness of the obtained optimal solution. This is especially useful for computational methods.

In case of free time T, we need an additional boundary condition. On the strength of (1.2.16), we have

$$H(\bar{p}(t_0),x(t_0),u(t_0)) = 0, \quad H(\bar{p}(T),x(T),u(T)) = 0. \tag{1.2.17}$$

By virtue of the first integral (1.2.15), only one of the conditions (1.2.17) is an independent one, and one of them follows from the other. Hence, any of these conditions and only one of them can be imposed as an additional boundary condition in case of free time T.

Let us consider an important case of time-optimal control for the autonomous system (1.2.14). In terms of (1.2.1), we have $f_0 = 1$, and the Hamiltonian (1.2.4) can be presented as follows:

$$H = p_0 + H_1, \qquad H_1(p,x,u) = \langle p, f(x,u)\rangle. \tag{1.2.18}$$

Substituting (1.2.18) into conditions of Theorems 1.1 and 1.2 for the autonomous system (1.2.14), we come to the following assertion.

Theorem 1.3. *Let $u(t)$ be an optimal control for the time-optimal control problem defined by (1.2.14), (1.1.3), (1.1.4), and (1.1.7), and $x(t)$ be the corresponding optimal trajectory. Then there exists a nonzero adjoint vector $p(t)$ satisfying the adjoint system*

$$\dot{p}_i = -\frac{\partial H_1}{\partial x_i}, \qquad i = 1,\dots,n, \tag{1.2.19}$$

where the truncated Hamiltonian H_1 is defined by (1.2.18), and such that

1)

$$H_1(p(t),x(t),u(t)) = \sup_{u\in U} H_1(p(t),x(t),u), \quad t\in[t_0,T]; \tag{1.2.20}$$

2) the transversality conditions

$$p_i(T) = \sum_{j=1}^{r} \lambda_j \frac{\partial g_j(x(T))}{\partial x_i}, \quad i=1,\dots,n, \tag{1.2.21}$$

hold, where λ_j are constant;

3)

$$H_1(p(t),x(t),u(t)) \equiv \text{const} \geq 0. \tag{1.2.22}$$

It follows from (1.2.22) that $H_1 = \text{const}$ is the first integral of our Hamiltonian system defined by (1.2.14) and (1.2.19). One can normalize the adjoint variables using the transformation (1.2.13) and set $H_1 = 1$ or $H_1 = 0$. If $H_1 = 1$, we have the regular case, whereas $H_1 = 0$ corresponds to the abnormal one. The value of H_1 can be fixed at one and only one time instant: for example, in the regular case we can impose the following boundary condition

$$H_1(p(T),x(T),u(T)) = 1. \tag{1.2.23}$$

This additional boundary condition is necessary because of an additional unknown, namely, terminal time T.

The approach to the optimal control briefly described above and based upon the maximum principle meets many difficulties and open questions. Let us mention some of them considering the general case treated in Theorem 1.1.

1) The maximum principle is only a necessary optimality condition; if it is satisfied, the corresponding control can still be not optimal.

2) The maximal value of the Hamiltonian over $u \in U$ in (1.2.11) can be reached at many values of the control. Hence, it may be impossible to express u as a single valued function given by (1.2.12).

3) The constant p_0 may be equal to either -1 or 0, that is, the problem may be abnormal.

4) Equations (1.2.5) and (1.2.6) are strongly nonlinear, even if the system (1.1.1) is linear.

5) The boundary value problem for (1.2.5) and (1.2.6) is usually very complicated. It can have many solutions or no solution at all.

The maximum principle is, together with the dynamic programming [21], one of the basic cornerstones lying in the foundation of the mathematical theory of optimal control. This principle was generalized and applied to various classes of optimal control problems.

It was proved that for some of these classes, including time-optimal control for linear systems and so-called linear-quadratic problems, the maximum principle provides not only necessary but also sufficient optimality conditions, and exact optimal solutions can be obtained by means of this principle [93, 19, 24, 84, 83]. On the other hand, a number of efficient computational methods based on the maximum principle [24, 84, 89, 33] were developed. Even if these methods do not always overcome the difficulties listed above and cannot provide the rigorous proof of the control optimality, they make it possible to find reasonable and close to optimal solutions for many practical problems. In all cases, a preliminary knowledge of the specific properties of the engineering, technological, or economical problem under consideration helps to choose the relevant computational algorithm and an initial approximation and, as a result, to find a good approximation to the desired optimal solution.

1.3 Open-loop and feedback control

Let us discuss the important notions of open-loop and closed-loop, or feedback, control.

The optimal control $u(t)$ considered above is a function of time. But since it corresponds to a certain initial condition (1.1.3), it depends also on the initial data t_0 and x^0. Hence, it can be presented as

$$u = \tilde{u}(t;t_0,x^0). \tag{1.3.1}$$

This form of control is called *open-loop*, or *program*, control.

In practical problems, the engineers are interested usually in another form of control called *closed-loop*, or *feedback control*. It defines the control as a function of the current state and, maybe, also of time. The feedback control can be represented as the following function

$$u = \bar{u}(x,t). \tag{1.3.2}$$

The feedback optimal control (1.3.2) is sometimes called also as the *synthesis of optimal control.*

The open-loop control (1.3.1) does not use measurement results, it does not take into account possible disturbances and errors. As a result, the open-loop control can be applied only in the ideal situation where the motion of the system is precisely determined by the equation of motion (1.1.1), the chosen control $u(t)$, and initial condition (1.1.3).

In practical problems, the system (1.1.1) is usually subjected to unknown disturbances. Moreover, there are inaccuracies in the mathematical model of the system and parametric errors. Hence, it is quite natural to prefer the feedback control that is based on the current measurements of state. The feedback form of control is widely used in applications as an on-line control.

A natural question arises: what is the relation between the open-loop (1.3.1) and feedback (1.3.2) forms of optimal control?

To establish this relation, we consider first the case where the feedback optimal control given by (1.3.2) is known, and we wish to find the open-loop control for the given initial condition (1.1.3).

Let us substitute the control (1.3.2) into (1.1.1) and integrate this equation under the given initial condition

$$\dot{x} = f(x, \bar{u}(x,t), t), \qquad x(t_0) = x^0. \tag{1.3.3}$$

Denote the solution of the initial value problem (1.3.3) by

$$x = x(t; t_0, x^0). \tag{1.3.4}$$

Now we substitute the obtained optimal trajectory (1.3.4) into the feedback control (1.3.2).

Thus, we obtain the desired open-loop optimal control for the given initial condition

$$\bar{u}(x(t; t_0, x^0), t) = \tilde{u}(t; t_0, x^0). \tag{1.3.5}$$

Consider now the inverse situation and suppose the open-loop control (1.3.1) is known for all t, t_0, and x^0. Let us set the initial time t_0 equal to the current time t and regard the initial state x^0 as the current state x in the open-loop control. In other words, we consider that each current time instant is an initial one. Then the open-loop control coincides with the feedback and we have

$$\tilde{u}(t; t, x) = \bar{u}(x, t). \tag{1.3.6}$$

Thus, the relation between the open-loop and feedback optimal control is determined by (1.3.5) and (1.3.6).

In case of the autonomous system (1.2.14), this relation is simplified.

The open-loop control for the system (1.2.14) depends on the time difference $t - t_0$, and we have instead of (1.3.1)

$$u = \tilde{u}(t - t_0; x^0). \tag{1.3.7}$$

The feedback control (1.3.2) here does not depend on time

$$u = \bar{u}(x). \tag{1.3.8}$$

The initial value problem (1.3.3) takes the form

$$\dot{x} = f(x, \bar{u}(x)), \qquad x(t_0) = x^0$$

and its solution can be expressed as follows:

$$x = x(t - t_0; x^0). \tag{1.3.9}$$

By substituting the optimal trajectory (1.3.9) into the feedback control (1.3.8), we obtain

$$\bar{u}(x(t - t_0; x^0)) = \tilde{u}(t - t_0; x^0). \tag{1.3.10}$$

Similarly, instead of the relationship (1.3.6), we have

$$\tilde{u}(0; x) = \bar{u}(x). \tag{1.3.11}$$

The open-loop and feedback controls in the autonomous case are given by the respective equations (1.3.7) and (1.3.8) and satisfy (1.3.10) and (1.3.11).

The approach of the maximum principle described in the previous section is aimed at the obtaining the optimal control for a given initial condition. Thus, the maximum principle can provide the open-loop control in the form given by (1.3.1) or (1.3.7), for autonomous systems.

The transformation of the open-loop control into the feedback one described by (1.3.6) and (1.3.11) is realizable, only if the open-loop optimal control can be found for *all* possible initial data. This situation occurs in rather rare and simple cases. In the next section, we consider two such cases, where the feedback optimal control is obtained by means of the maximum principle.

Note that the feedback optimal control can be, in principle, obtained by the method of dynamic programming [21]. However, this method requires a great amount of computation and needs a huge memory. Up till now, the approach of dynamic programming is used for optimal control problems of low dimension ($n \leq 3$). Moreover, this method represents the feedback control only numerically, not in an explicit form.

1.4 Examples

We consider below two examples of linear dynamical systems for which the feedback control will be obtained by means of the maximum principle [93].

Example 1

Consider a mechanical system with one degree of freedom controlled by a bounded force. Let us take the equation of motion and the control constraint as follows

$$\ddot{x} = u, \qquad |u| \leq 1. \tag{1.4.1}$$

Without loss in generality, we assume that the mass of the system and the maximal admissible forces are equal to unity. Note that (1.4.1) describes not only the progressive motion of a body along the x-axis but also its rotation by the angle x about the fixed axis; the control u denotes the force or torque, respectively.

Let us rewrite (1.4.1) as a system

$$\dot{x}_1 = x_2, \qquad \dot{x}_2 = u, \qquad |u| \leq 1. \tag{1.4.2}$$

We impose arbitrary initial conditions

$$x_1(t_0) = x_1^0, \qquad x_2(t_0) = x_2^0 \tag{1.4.3}$$

and zero terminal conditions

$$x_1(T) = 0, \qquad x_2(T) = 0. \tag{1.4.4}$$

We will find the time-optimal control for the problem defined by (1.4.2)–(1.4.4).

Using Theorem 1.3, we introduce the truncated Hamiltonian defined by (1.2.18) as follows:

$$H_1(p,x,u) = p_1 x_2 + p_2 u. \tag{1.4.5}$$

Its maximum with respect to u under the constraint $|u| \leq 1$ is attained at

$$u = \operatorname{sign} p_2. \tag{1.4.6}$$

The adjoint system given by (1.2.19) for the Hamiltonian (1.4.5) has the form

$$\dot{p}_1 = 0, \qquad \dot{p}_2 = -p_1. \tag{1.4.7}$$

Integrating system (1.4.7), we get

$$p_1(t) = c_1, \qquad p_2(t) = -c_1 t + c_2, \tag{1.4.8}$$

where c_1 and c_2 are constants. By substituting p_2 from (1.4.8) into (1.4.6), we have

$$u(t) = \operatorname{sign}(c_2 - c_1 t). \tag{1.4.9}$$

The linear function of time $(c_2 - c_1 t)$ can change its sign not more than once over the time interval $[t_0, T]$. Hence, the control $u(t)$ is equal to ± 1 for all $t \in [t_0, T]$, and the switch, i.e., the change of the control sign, can happen not more than once over the time interval $t \in [t_0, T]$.

The control that takes only the maximal and minimal admissible values is called the *bang-bang* control.

Let us consider arcs of phase trajectories in the (x_1, x_2)-plane corresponding to $u = 1$ and $u = -1$. By substituting $u = 1$ into (1.4.2), we obtain

$$\frac{dx_1}{dx_2} = x_2.$$

Integrating this equation, we get

$$x_1 = \frac{1}{2} x_2^2 + A, \tag{1.4.10}$$

where A is an arbitrary constant. Similarly, for the control $u = -1$, we obtain from (1.4.2)

$$x_1 = -\frac{1}{2} x_2^2 + B, \tag{1.4.11}$$

where B is an arbitrary constant.

The families of parabolas corresponding to (1.4.10) and (1.4.11) are shown in Fig. 1.1 by solid and dashed (broken) lines, respectively. According to (1.4.2), x_2 grows (decreases) for $u = 1$ ($u = -1$). Hence, we can determine the direction of motion, or the direction of the time growth, along the phase trajectories; this direction is shown by arrows in Fig. 1.1.

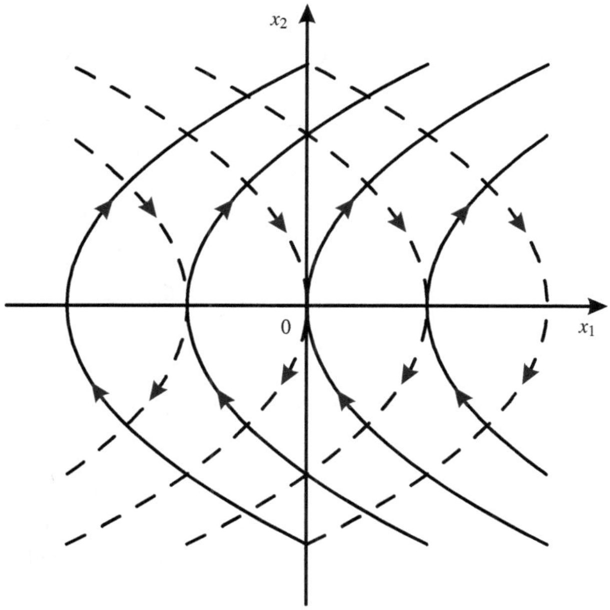

Fig. 1.1 Phase trajectories for Example 1

Let us now construct the optimal phase trajectory starting at the given initial point (1.4.3) and ending at the zero terminal point (1.4.4). We can see from Fig. 1.1 that this trajectory can reach the terminal point only along the one of two parabolas given (1.4.10) and (1.4.11) for $A = 0$ and $B = 0$, respectively, in the directions shown by arrows. Since our bang-bang control has no more than one switch, the phase trajectory consists of no more than two arcs of parabolas belonging to different families (1.4.10) and (1.4.11). The last part of the trajectory is an arc reaching the zero point and corresponding to $A = 0$ in (1.4.10) or $B = 0$ in (1.4.11). Hence, the first part of the trajectory should belong to the other family of parabolas. This part starts at the initial point (1.4.3) and intersects the parabola of the other family reaching the zero point.

Thus, we obtain the field of optimal phase trajectories shown in Fig. 1.2 by thin lines.

The locus of states where the bang-bang control changes its sign is called the *switching curve*. In our example, the switching curve consists of two semi-parabolas of different families that reach the zero point. These semi-parabolas are optimal phase trajectories themselves. The switching curve is shown by the thick line in Fig. 1.2.

Since the families of optimal trajectories given by (1.4.10) and (1.4.11) correspond to $u = 1$ and $u = -1$, respectively, the feedback optimal control can be easily

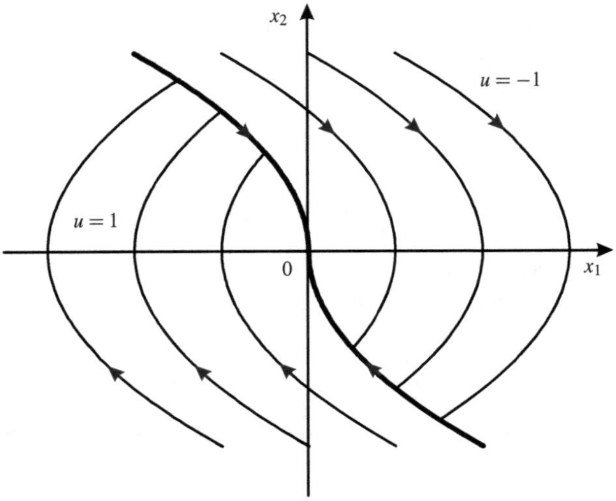

Fig. 1.2 Optimal phase trajectories for Example 1

determined by Fig. 1.2. This control $u(x_1,x_2)$ is equal to 1 and -1 to the left and right of the switching curve, respectively. Along the switching curve, the control is determined as $u = \operatorname{sign} x_1$ or $u = -\operatorname{sign} x_2$. Thus, the optimal feedback control in our example is determined as follows:

$$u = 1 \quad \text{if} \quad x_1 < -\frac{1}{2}x_2|x_2|,$$

$$u = -1 \quad \text{if} \quad x_1 > -\frac{1}{2}x_1|x_2|,$$

$$u = \operatorname{sign} x_1 = -\operatorname{sign} x_2 \quad \text{if} \quad x_1 = -\frac{1}{2}x_1|x_2|.$$

By introducing the switching function

$$\psi(x_1,x_2) = -x_1 - \frac{1}{2}x_2|x_2|, \tag{1.4.12}$$

we can express our feedback control in the form

$$\begin{aligned} u(x_1,x_2) &= \operatorname{sign} \psi(x_1,x_2) \quad \text{if} \quad \psi \neq 0, \\ u(x_1,x_2) &= \operatorname{sign} x_1 = -\operatorname{sign} x_2 \quad \text{if} \quad \psi = 0. \end{aligned} \tag{1.4.13}$$

Thus, we have obtained the optimal feedback control for our system (1.4.2). Since this system is autonomous, the feedback control (1.4.13) does not depend on time explicitly, in accordance with (1.3.8).

Example 2

Consider now the time-optimal control problem for a linear oscillator. The equation of motion and the control constraint are given by

$$\ddot{x} + x = u, \qquad |u| \leq 1$$

or, in the form of a system similar to (1.4.2), by the equations

$$\dot{x}_1 = x_2, \qquad \dot{x}_2 = -x_1 + u, \qquad |u| \leq 1. \tag{1.4.14}$$

The initial and terminal conditions are defined again by (1.4.3) and (1.4.4).

The truncated Hamiltonian H_1 for our system (1.4.14) is given by

$$H_1(p,x,u) = p_1 x_2 + p_2(-x_1 + u). \tag{1.4.15}$$

Its maximum with respect to u is attained again at u given by (1.4.6). The adjoint system for the Hamiltonian (1.4.15) is

$$\dot{p}_1 = p_2, \qquad \dot{p}_2 = -p_1.$$

Integrating this system, we obtain

$$p_2(t) = C\sin(t+\alpha), \quad p_1(t) = -C\cos(t+\alpha),$$

where C and α are arbitrary constants. Here, we can set $C > 0$, without loss in generality. By substituting $p_2(t)$ into (1.4.6), we have

$$u(t) = \operatorname{sign}\sin(t+\alpha). \tag{1.4.16}$$

As in Example 1, we have again the bang-bang optimal control that takes the values $u = \pm 1$. By contrast to (1.4.9), the control defined by (1.4.16) can have many switches.

To determine the arcs of phase trajectories corresponding to $u = 1$ and $u = -1$, we substitute these values of control into (1.4.14) and obtain

$$\frac{dx_1}{dx_2} = \frac{x_2}{-x_1 + u} \quad \text{for} \quad u = \pm 1.$$

By integrating these equation, we get

$$\begin{aligned} (x_1 - 1)^2 + x_2^2 &= r_+^2 \quad \text{for} \quad u = 1, \\ (x_1 + 1)^2 + x_2^2 &= r_-^2 \quad \text{for} \quad u = -1. \end{aligned} \tag{1.4.17}$$

These curves are families of circles of arbitrary radii r_+ and r_- with centers at the points $(1,0)$ and $(-1,0)$ for the controls $u = 1$ and -1, respectively. The families of circles defined by (1.4.17) are shown in Fig. 1.3 by solid and dashed lines for $u = 1$ and $u = -1$, respectively.

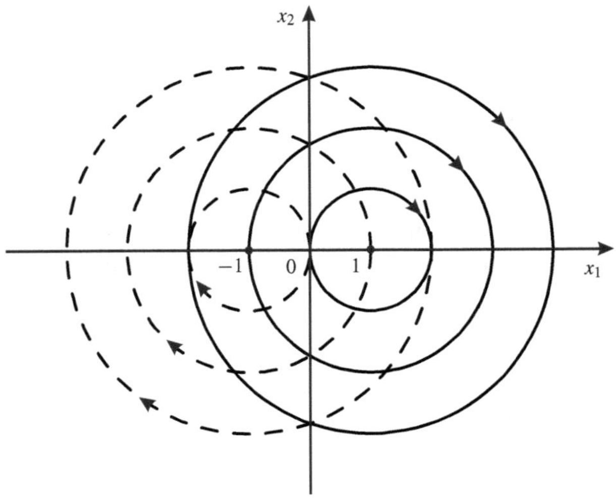

Fig. 1.3 Phase trajectories for Example 2

On the strength of (1.4.17), we can express the state coordinates x_1 and x_2 as follows:

$$\begin{aligned} x_1 - 1 &= r_+ \cos\varphi_+, \quad x_2 = r_+ \sin\varphi_+ \quad \text{for} \quad u = 1, \\ x_1 + 1 &= r_- \cos\varphi_-, \quad x_2 = r_- \sin\varphi_- \quad \text{for} \quad u = -1, \end{aligned} \tag{1.4.18}$$

where φ_+ and φ_- are angles of rotation about the centers of the respective circles. Substituting (1.4.18) into (1.4.14), we obtain

$$\dot{\varphi}_+ = -1, \qquad \dot{\varphi}_- = -1.$$

Thus, the direction of rotation of phase trajectories along the circles given by (1.4.18) is clockwise. This direction is shown by arrows in Fig. 1.3. The angles of rotation φ_+ and φ_- change linearly with time t with a unit angular velocity, so that one revolution occurs in the time interval equal to 2π.

It follows from (1.4.16) that the control $u(t)$ changes its sign at time intervals equal to π. Thus, the time intervals between the control switches are equal to π, and the phase trajectory always travels along a half-circle between the neighbouring switches.

Taking into account the above considerations, we come to the field of optimal phase trajectories depicted in Fig. 1.4. Here, thin lines correspond to the trajectories, and the switching curve is shown by a thick line. This curve consists of semi-circles of the unit radii given by (1.4.17) for $r_+ = r_- = 1$; the semi-circles of the first family (1.4.7) lie in the quadrant $x_1 \geq 0$, $x_2 \leq 0$, and the semi-circles of the second family (1.4.17) lie in the quadrant $x_1 \leq 0$, $x_2 \geq 0$. The whole picture is centrally symmetric with respect to the origin of coordinates. Only two semi-circles of the switching curve adjacent to the zero point are arcs of the optimal trajectories; all other semi-circles are intersected by the trajectories.

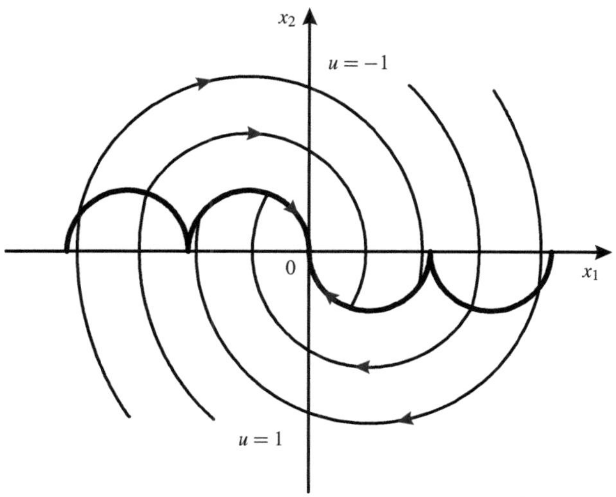

Fig. 1.4 Optimal phase trajectories for Example 2

The feedback optimal control is equal to 1 (-1) below (above) the switching curve.

Analytically, the feedback optimal control can be presented in the same form (1.4.13) as in Example 1, where the switching function $\psi(x_1,x_2)$ is given by

$$\begin{aligned} \psi(x_1,x_2) &= (-x_1^2 - 2x_1)^{1/2} - x_2 \quad \text{if} \quad -2 \leq x_1 \leq 0, \\ \psi(x_1,x_2) &= \psi(x_1+2,x_2) \quad \text{if} \quad x_1 < -2, \\ \psi(x_1,x_2) &= -\psi(-x_1,-x_2) \quad \text{if} \quad x_1 > 0. \end{aligned} \tag{1.4.19}$$

Here, the first expression defines the semi-circle of the second family (1.4.17) belonging to the switching curve and adjacent to the origin, and the other expressions (1.4.19) define all other semi-circles of this curve.

The optimal feedback controls obtained above for Examples 1 and 2 will be used below as integral parts of the non-optimal feedback controls for more general nonlinear systems.

Chapter 2
Method of decomposition (the first approach)

In this chapter, we present the method of decomposition for the control design in nonlinear dynamical systems. We suppose the system is described by Lagrangian ordinary differential equations and subjected to controls, the number of independent control forces being equal to the number of degrees of freedom. The material of the chapter is based on papers [27, 28, 29, 34, 98, 102, 42, 43, 44].

2.1 Problem statement and game approach

2.1.1 Controlled mechanical system

Consider a nonlinear control system whose dynamics is described by Lagrange's equations

$$\frac{d}{dt}\frac{\partial T}{\partial \dot{q}_i} - \frac{\partial T}{\partial q_i} = U_i + Q_i, \quad i = 1,\ldots,n. \tag{2.1.1}$$

Here, $q = (q_1,\ldots,q_n)$ denotes the generalized coordinates of the system, n is the number of its degrees of freedom, and a dot over a letter denotes the derivative with respect to the time t. The generalized forces consist of the control forces U_i to be determined and terms Q_i representing all other external and internal forces, including the uncontrolled perturbations.

The kinetic energy of the system T is given by a quadratic form

$$T(q,\dot{q}) = \frac{1}{2}\sum_{j,k=1}^{n} a_{jk}(q)\dot{q}_j\dot{q}_k, \tag{2.1.2}$$

where a_{jk} are elements of a symmetric positive-definite matrix $A(q)$ of order $n \times n$. Substituting (2.1.2) into (2.1.1), we write the equations of motion in the form

$$A(q)\ddot{q} = U + S(q,\dot{q},t). \tag{2.1.3}$$

Here, $U = (U_1, \ldots, U_n)$ is the vector of the control forces and $S = (S_1, \ldots, S_n)$ is the vector function

$$S(q,\dot{q},t) = Q(q,\dot{q},t) - \sum_{j,k=1}^{n} \Gamma_{jk}\dot{q}_j\dot{q}_k, \tag{2.1.4}$$

where $\Gamma_{jk} = (\Gamma_{1jk}, \ldots, \Gamma_{njk})$ are n-dimensional vectors with components

$$\Gamma_{ijk} = \frac{\partial a_{ij}}{\partial q_k} - \frac{1}{2}\frac{\partial a_{jk}}{\partial q_i}. \tag{2.1.5}$$

We impose the constraints

$$|U_i| \le U_i^0, \quad i = 1, \ldots, n, \tag{2.1.6}$$

on the control forces, where $U_i^0 > 0$ are given positive constants.

The initial conditions for system (2.1.3)

$$q(t_0) = q^0, \qquad \dot{q}(t_0) = \dot{q}^0 \tag{2.1.7}$$

lie in a given domain Ω in $2n$-dimensional phase space: $\{q, \dot{q}\} \in \Omega$.

Let us formulate the control problem:

Problem 2.1. Find a feedback control $U = U(q, \dot{q})$ satisfying constraint (2.1.6) and bringing system (2.1.3) from an arbitrary initial state (2.1.7) in domain Ω to a given state with zero velocities

$$q(t_*) = q^*, \qquad \dot{q}(t_*) = 0 \tag{2.1.8}$$

in finite time (instant $t_* > t_0$ is not fixed).

2.1.2 Simplifying assumptions

Problem 2.1 will be solved under certain simplifying assumptions (conditions), which are formulated below.

We represent the matrix $A(q)$ in the form

$$A(q) = B(q)A_*,$$
$$B(q) = E + [A(q) - A_*]A_*^{-1} \equiv A(q)A_*^{-1}, \tag{2.1.9}$$

where A_* is some constant symmetric positive-definite $n \times n$ matrix and E is the $n \times n$ identity matrix. Matrix $B(q)$ is nonsingular; hence, the inverse matrix $B^{-1}(q)$ exists. We multiply both sides of (2.1.3) by $B^{-1}(q)$ and, using the relationships (2.1.9), transform (2.1.3) to the form

$$A_*\ddot{q} = U + V(q, \dot{q}, t, U). \tag{2.1.10}$$

Here, we use the notation

$$V = V' + V'', \quad V' = B^{-1}(q)S(q,\dot{q},t),$$
$$V'' = [B^{-1}(q) - E]U. \tag{2.1.11}$$

By virtue of this notation, (2.1.10) is equivalent to the original equation (2.1.3). Let us suppose that the conditions

$$V_i' = -\lambda_i (A_* \dot{q})_i + V_i^*,$$
$$|V_i^* + V_i''| \le V_i^0 < U_i^0, \quad i = 1, \ldots, n, \tag{2.1.12}$$

hold for all $t \ge t_0$, all $\{q,\dot{q}\} \in \Omega$, and all U satisfying (2.1.6). Here, $V_i^0 > 0$ and $\lambda_i > 0$ are constants. If all λ_i are equal to zero, conditions (2.1.12) become more simple

$$|V_i| \le V_i^0 < U_i^0, \quad i = 1, \ldots, n, \tag{2.1.13}$$

for all $t \ge t_0$, all $\{q,\dot{q}\} \in \Omega$, and all U satisfying (2.1.6).

The following lemma allows to check whether condition (2.1.13) is satisfied.

Lemma 2.1. *Suppose that, for any n-dimensional vector z, the following conditions are satisfied for all* $t \ge t_0$ *and all* $\{q,\dot{q}\} \in \Omega$*:*

$$|A_* z| \ge \mu_* |z|, \quad |[A(q) - A_*]z| \le \mu |z|,$$
$$|S_i(q,\dot{q},t)| \le \vartheta U_i^0, \quad i = 1, \ldots, n, \tag{2.1.14}$$
$$0 < \mu < \mu_*, \quad \vartheta > 0,$$

where μ_**,* μ*, and* ϑ *are constants. Then, for all* $t \ge t_0$*, all* $\{q,\dot{q}\} \in \Omega$*, and all U meeting (2.1.6), the components of vector V in (2.1.11) satisfy the estimates*

$$|V_i| \le \vartheta U_i^0 + \mu(\mu_* - \mu)^{-1}(1+\vartheta)|U^0|, \qquad i = 1, \ldots, n,$$
$$U^0 = (U_1^0, \ldots, U_n^0). \tag{2.1.15}$$

We note that, since A_* is a positive-definite matrix, we can take as μ_* any positive number not exceeding its smallest eigenvalue.

Proof. From the first of inequalities (2.1.14), we have

$$|A_*^{-1} z| \le \mu_*^{-1} |z|. \tag{2.1.16}$$

Here and below, z denotes any n-dimensional vector. We define

$$L = [A(q) - A_*]A_*^{-1}. \tag{2.1.17}$$

It follows from (2.1.16) and the second of inequalities (2.1.14) that

$$|Lz| \le \mu \mu_*^{-1} |z|. \tag{2.1.18}$$

By virtue of (2.1.17), we can rewrite relationship (2.1.9) for B in the form

$$Bz = z + Lz. \tag{2.1.19}$$

With the aid of (2.1.19) and (2.1.18), we obtain the estimate

$$|Bz| \ge |z| - |Lz| \ge (1 - \mu \mu_*^{-1})|z|. \tag{2.1.20}$$

It follows from condition (2.1.14) of the lemma that $(1 - \mu \mu_*^{-1}) > 0$. Setting $z = B^{-1}z'$ in (2.1.20), we obtain

$$|B^{-1}z'| \le (1 - \mu \mu_*^{-1})^{-1} |z'|. \tag{2.1.21}$$

Inequalities (2.1.18) and (2.1.21) yield

$$|LB^{-1}z| \le \mu (\mu_* - \mu)^{-1} |z|. \tag{2.1.22}$$

Let us set $z = B^{-1}z'$ in (2.1.19):

$$B^{-1}z' = z' - LB^{-1}z'. \tag{2.1.23}$$

Using (2.1.11) for V' and (2.1.23) with $z' = S$, we represent the components V_i of the vector V' in the form

$$V_i' = (B^{-1}S)_i = S_i - (LB^{-1}S)_i, \quad i = 1, \dots, n. \tag{2.1.24}$$

The subscripts denote the components of the vectors. By virtue of the third of conditions (2.1.14) and inequality (2.1.22), we obtain from (2.1.24)

$$\begin{aligned} |V_i'| &\le |S_i| + |(LB^{-1}S)_i| \le \vartheta U_i^0 + \mu(\mu_* - \mu)^{-1}|S| \\ &\le \vartheta U_i^0 + \mu(\mu_* - \mu)^{-1} \vartheta |U^0|, \qquad i = 1, \dots, n. \end{aligned} \tag{2.1.25}$$

Here, we used notation (2.1.15) for U^0. We substitute (2.1.23) with $z' = U$ into (2.1.11) for vector V'':

$$V_i'' = (B^{-1}U - U)_i = -(LB^{-1}U)_i, \quad i = 1, \dots, n. \tag{2.1.26}$$

From this equation, using inequalities (2.1.22) and (2.1.6), we obtain

$$\begin{aligned} |V_i''| &\le |(LB^{-1}U)_i| \le |LB^{-1}U| \le \mu(\mu_* - \mu)^{-1}|U| \\ &\le \mu(\mu_* - \mu)^{-1}|U^0|, \qquad i = 1, \dots, n. \end{aligned} \tag{2.1.27}$$

This and inequality (2.1.25) imply (2.1.15). This completes the proof of the lemma. □

Corollary 2.1. *If, under the conditions of Lemma 2.1, $\vartheta < 1$ and μ is sufficiently small, then condition (2.1.13) is satisfied.*

Remark 2.1. We should take for matrix A_* some "average" value of the matrix $A(q)$ for domain Ω. In particular, we can choose for A_* the matrix $A(q^*)$ for some value of vector q^*, for example, $A(q^*)$, $A(q^0)$, or $A((q^0+q^*)/2)$. Then, if the domain Ω is sufficiently small, the matrix $A(q)$ will differ only slightly from A_* for all the motions considered, and the number μ will, under conditions (2.1.14) of Lemma 2.1, be sufficiently small. Thus, by virtue of Corollary 2.1, condition (2.1.13) can be ensured for a given nonlinear system (2.1.3) if, first, the possibilities of control are increased [that is, if the constants U_i^0 in (2.1.6) are increased so that the condition $\vartheta < 1$ holds] and, second, the domain Ω is decreased, so that $A(q)$ is close to A_* (that is, the number μ is decreased).

We shall show in Sect. 2.5 that formulation of Problem 2.1 and condition (2.1.12) are natural often and are satisfied for manipulation robots with electromechanical drives.

2.1.3 Decomposition

Let us turn to the solution of Problem 2.1 with condition (2.1.12) satisfied. We assume that all motions of system (2.1.3) considered lie in the domain Ω.

If condition (2.1.12) is satisfied, system (2.1.3) can, by virtue of (2.1.10)–(2.1.12), be represented in the form

$$(A_*\ddot{q})_i + \lambda_i (A_*\dot{q})_i = U_i + \tilde{V}_i, \quad \tilde{V}_i = V_i^* + V_i'', \quad i = 1,\dots,n. \tag{2.1.28}$$

In system (2.1.28), we make the change of variables

$$A_*(q - q^*) = y, \tag{2.1.29}$$

where q^* was introduced in (2.1.8). We obtain

$$\ddot{y}_i + \lambda_i \dot{y}_i = U_i + \tilde{V}_i, \quad i = 1,\dots,n. \tag{2.1.30}$$

For the terms on the right-hand sides of (2.1.30), we have, by virtue of (2.1.6), (2.1.28), and (2.1.12), the constraints

$$|U_i| \le U_i^0, \quad |V_i| \le V_i^0 < U_i^0, \quad i = 1,\dots,n. \tag{2.1.31}$$

After the change of variables (2.1.29), the initial conditions (2.1.7) and the boundary conditions (2.1.8) take the forms

$$y(t_0) = A_*(q^0 - q^*), \quad \dot{y}(t_0) = A_*\dot{q}^0, \tag{2.1.32}$$

$$y(t_*) = \dot{y}(t_*) = 0. \tag{2.1.33}$$

Thus, Problem 2.1 reduces to the construction of a control $U(y, \dot{y})$ that brings system (2.1.30) from an arbitrary initial state (2.1.32) to state (2.1.33) under constraint (2.1.31). System (2.1.30) consists of n subsystems, each with a single degree of freedom. Each of the subsystems has its own scalar control U_i that satisfies constraint (2.1.31). In this subsystem, we treat the function V_i as a perturbation subject to constraint (2.1.31), but otherwise arbitrary. Then the result obtained can be summed up in the form of the following assertion.

Theorem 2.1. *Suppose that condition (2.1.12) is satisfied and that all the motions of system (2.1.3) that are being considered lie in the domain Ω. Then, to solve Problem 2.1, it is sufficient to solve n control problems for the linear subsystems (2.1.30) with a single degree of freedom. In each of these problems, it is necessary to construct a scalar control $U_i(y_i, \dot{y}_i)$ satisfying constraint (2.1.31) and taking the ith subsystem (2.1.30) from an arbitrary initial state (2.1.32) to the coordinate origin (2.1.33) in finite time for arbitrary admissible perturbations V_i satisfying constraint (2.1.31).*

The described approach to the control decomposition has been first suggested in [27] for the case $\lambda = 0$ and in [29] for the general case $\lambda \geq 0$.

2.1.4 Game problem

Let us consider the ith subsystem (2.1.30) and, in it, let us set

$$y_i = U_i^0 x, \quad U_i = U_i^0 u, \quad \tilde{V}_i = U_i^0 v. \tag{2.1.34}$$

Then, this system combined with constraints (2.1.31) takes the form

$$\ddot{x} + \lambda \dot{x} = u + v, \quad |u| \leq 1, \quad |v| \leq \rho < 1, \tag{2.1.35}$$

and the boundary conditions (2.1.32) and (2.1.33) become

$$x(0) = \xi, \quad \dot{x}(0) = \eta, \quad x(\tau) = \dot{x}(\tau) = 0. \tag{2.1.36}$$

In (2.1.35) and (2.1.36), we used the notation

$$\begin{gathered} \rho = \frac{V_i^0}{U_i^0} < 1, \quad \xi = (U_i^0)^{-1} y_i(t_0) = (U_i^0)^{-1} [A_*(q^0 - q^*)]_i, \\ \eta = (U_i^0)^{-1} \dot{y}_i(t_0) = (U_i^0)^{-1} (A_* \dot{q}^0)_i, \quad \lambda = \lambda_i, \\ \tau = t_* - t_0, \quad i = 1, \ldots, n. \end{gathered} \tag{2.1.37}$$

Without loss of generality, the initial instant of time is taken equal to zero.

Let us consider the problem of bringing system (2.1.35) to the coordinate origin in the shortest time, that is, for minimum τ in (2.1.36). We treat this problem as a differential game, in which one of the players (the controlling side) chooses a control u, and the second player (the opponent) chooses a perturbation v. We will use the approach of the theory of differential games [79] and construct a feedback control $u(x,\dot{x})$ that brings system (2.1.35) to the coordinate origin in the shortest guaranteed time τ for an arbitrary admissible perturbation v. We note that this differential game (2.1.35) and (2.1.36) is a linear differential game of similar objects.

Its solution reduces [79] to the solution of the time-optimal control problem for the system

$$\ddot{x}+\lambda\dot{x}=(1-\rho)u, \quad |u|\le 1, \quad \tau\to\min, \tag{2.1.38}$$

with the boundary conditions (2.1.36). The sought control $u(x,\dot{x})$ and the minimum guaranteed time τ in the game problem (2.1.35) and (2.1.36) coincide with the synthesis of the optimal control and optimal time for problem (2.1.38) and (2.1.36). We note that system (2.1.38) is obtained from (2.1.36) for a perturbation equal to $v=-\rho u$ that is the optimal control of the opponent choosing the perturbation v. In other words, the worst perturbation in this problem can be taken in the form $v=-\rho u$.

Thus, as a result of the decomposition, the solution of Problem 2.1 is reduced to the construction of the time-optimal feedback control for system (2.1.38) and (2.1.36).

2.2 Control of the subsystem and feedback control design

2.2.1 Optimal control for the subsystem

Let us rewrite the time-optimal control problem (2.1.38) and (2.1.36) in the form

$$\dot{x}_1=x_2, \quad \dot{x}_2=-\lambda x_2+w, \quad w=(1-\rho)u, \quad |u|\le 1, \tag{2.2.1}$$

$$x_1(0)=\xi, \quad x_2(0)=\eta, \quad x_1(\tau)=x_2(\tau)=0, \tag{2.2.2}$$

$$0\le\rho<1, \quad \lambda\ge 0, \quad \tau\to\min, \quad (x_1=x, \quad x_2=\dot{x}).$$

This problem is easily solved by means of the maximum principle, see Sect. 1.2. Here, we present the necessary relationships.

Hamilton's function for system (2.2.1) is equal to

$$H=p_1x_2+p_2[(1-\rho)u-\lambda x_2], \quad |u|\le 1,$$

where p_1 and p_2 are conjugate variables. From this we obtain, by virtue of the maximum principle,

$$u=\operatorname{sign} p_2=\pm 1. \tag{2.2.3}$$

The adjoint system has the form

$$\dot{p}_1 = 0, \quad \dot{p}_2 = -p_1 + \lambda p_2.$$

Integrating it, we obtain

$$p_2 = C_1 + C_2 e^{\lambda t} \quad \text{for} \quad \lambda > 0,$$

$$p_2 = C_1 + C_2 t \quad \text{for} \quad \lambda = 0,$$

where C_1 and C_2 are arbitrary constants. It follows that $p_2(t)$ is a monotone function for $\lambda \geq 0$. Therefore, control (2.2.3) has no more than one switching point.

For a constant $w = \text{const}$, the general solution of system (2.2.1) has the form

$$\begin{aligned} x_1 &= B_1 + \lambda^{-1} w(t-\tau) - \lambda^{-1}(B_2 - \lambda^{-1} w)[e^{-\lambda(t-\tau)} - 1], \\ x_2 &= \lambda^{-1} w + (B_2 - \lambda^{-1} w) e^{-\lambda(t-\tau)} \qquad \text{for} \qquad \lambda > 0, \end{aligned} \tag{2.2.4}$$

$$\begin{aligned} x_1 &= B_1 + B_2(t-\tau) + \frac{1}{2} w (t-\tau)^2, \\ x_2 &= B_2 + w(t-\tau) \qquad \text{for} \qquad \lambda = 0. \end{aligned} \tag{2.2.5}$$

Here and in what follows, all the relationships are given separately for the cases $\lambda > 0$ and $\lambda = 0$. We note that the case $\lambda = 0$ can be obtained by taking the limit as $\lambda \to +0$. The arbitrary constants B_1 and B_2 in (2.2.4) and (2.2.5) are chosen in such a way that, for $B_1 = B_2 = 0$, the zero boundary conditions (2.2.2) hold for $t = \tau$. Eliminating $t - \tau$ from (2.2.4) and (2.2.5), we obtain the equations for the phase trajectories

$$x_1 = B' - \lambda^{-1} x_2 - \lambda^{-2} w \log|1 - \lambda w^{-1} x_2| \qquad \text{for} \qquad \lambda > 0, \tag{2.2.6}$$

$$x_1 = B' + (2w)^{-1} x_2^2 \qquad \text{for} \qquad \lambda = 0. \tag{2.2.7}$$

Here, B' is a new constant expressed in terms of B_1 and B_2. In deriving (2.2.6), we assume that $\lambda B_2 \neq w$. If $\lambda B_2 = w$, we obtain from (2.2.4) the equation for the phase trajectory in the form

$$x_2 = \lambda^{-1} w. \tag{2.2.8}$$

The phase trajectories (2.2.7) for $\lambda = 0$ are parabolas that are symmetric about the x_1-axis. They can be obtained successively, one from another, by a parallel translation along the x_1-axis.

Let us consider trajectory (2.2.6) for $\lambda > 0$, $u = 1$, and $B' = 0$. Using (2.2.1) with $w = 1 - \rho$, we obtain the following properties of the curve $x_1(x_2)$:

- As x_2 increases from $-\infty$ to 0, x_1 decreases from ∞ to 0 and attains a zero minimum at $x_2 = 0$;
- In the interval $x_2 \in (0, \lambda^{-1}(1-\rho))$, the value of x_1 increases from 0 to ∞;
- In the interval $(\lambda^{-1}(1-\rho), \infty)$, the value of x_1 decreases from ∞ to $-\infty$.

Thus, the curve $x_1(x_2)$ consists of two branches which approach the asymptote $x_2 = \lambda^{-1}(1-\rho)$. By (2.2.8), this asymptote is itself also a phase trajectory for $u = 1$. Dependence $x_1(x_2)$ is shown for $u = 1$ and $B' = 0$ in Fig. 2.1, where the arrows indicate the direction of increasing t. The phase trajectories corresponding to $u = 1$ and arbitrary B' in (2.2.6) are obtained from the curve described above by a translation along the x_1 axis.

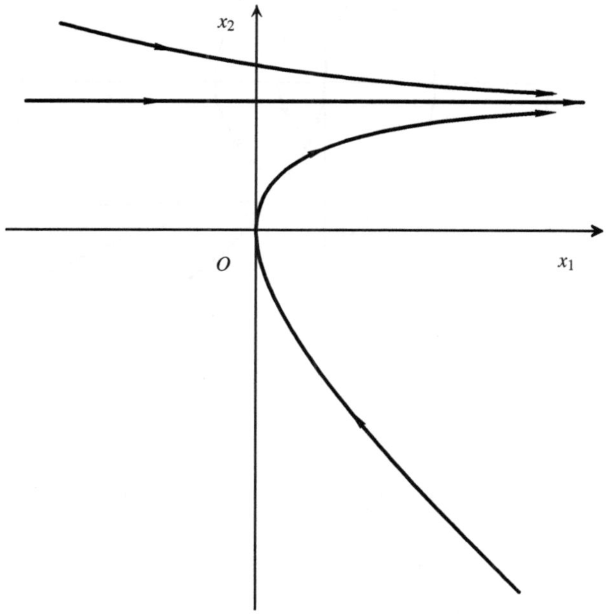

Fig. 2.1 Phase trajectories for $w = \text{const}$ and $\lambda > 0$

If we simultaneously change the signs of x_1, x_2, u, B_1, B_2, and B' in (2.2.4)–(2.2.8), the resulting relationships remain valid. Consequently, the phase trajectories corresponding to $u = -1$ are obtained by means of the central symmetry from the trajectories described above and corresponding to $u = 1$.

The only phase trajectories that reach the coordinate origin as t increases are curves (2.2.6) and (2.2.7) with $B' = 0$ and $u = \pm 1$. Motions along these curves are described by (2.2.4) and (2.2.5) with $B_1 = B_2 = 0$ and $u = \pm 1$. These two semitrajectories [(2.2.4) for $\lambda > 0$ and (2.2.5) for $\lambda = 0$] constitute the switching curve of the optimal control: the only possible change of sign of the control u along each trajectory takes place on this curve. As a result, we arrive at the field of optimal phase trajectories that is shown in Fig. 2.2 for $\lambda > 0$. Here, the bold curves represent the switching lines and the arrows indicate the direction of increasing t. For the field of optimal trajectories for $\lambda = 0$, see Fig. 1.2, Sect. 1.4.

The feedback optimal control can be represented in the form [see (1.4.13)]

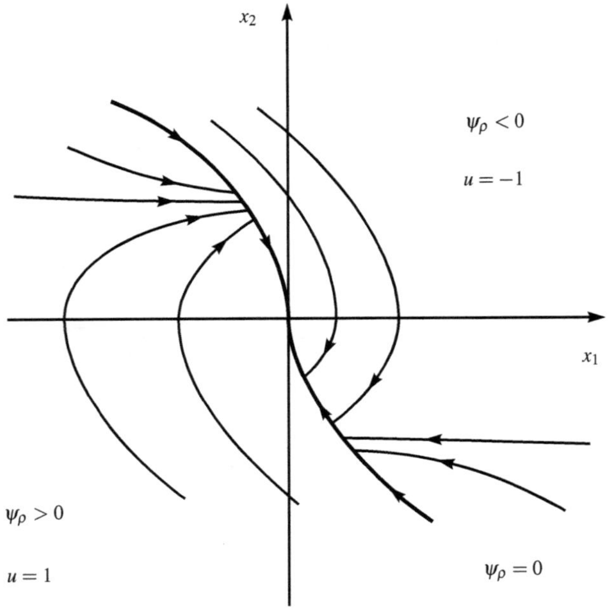

Fig. 2.2 Optimal phase trajectories

$$
\begin{aligned}
u(x_1,x_2) &= \operatorname{sign}\psi_\rho(x_1,x_2) \qquad \text{for} \qquad \psi_\rho \neq 0,\\
u(x_1,x_2) &= \operatorname{sign}x_1 = -\operatorname{sign}x_2 \qquad \text{for} \qquad \psi_\rho = 0,
\end{aligned} \tag{2.2.9}
$$

where $\psi_\rho(x_1,x_2)$ is the switching function. It is equal to

$$
\begin{aligned}
\psi_\rho(x_1,x_2) &= -x_1 - \lambda^{-1}x_2\\
&+\lambda^{-2}(1-\rho)\log[1+\lambda(1-\rho)^{-1}|x_2|]\operatorname{sign}x_2 \qquad \text{for} \qquad \lambda > 0,\\
\psi_\rho(x_1,x_2) &= -x_1 - x_2|x_2|[2(1-\rho)]^{-1} \qquad \text{for} \qquad \lambda = 0.
\end{aligned} \tag{2.2.10}
$$

Let us also determine the time necessary to reach the coordinate origin for the optimal trajectory starting from arbitrary initial conditions (2.2.2). To be definite, suppose that the initial point lies in the region $\psi_\rho \leq 0$ and that the only possible switching takes place at an instant $s \in [0,\tau]$. The point $(x_1(s),x_2(s))$ lies, on the one hand, on the phase trajectory corresponding to $u = -1$ that passes through the initial point and, on the other hand, on the switching curve for $u = 1$. Equating the corresponding expressions (2.2.5) (we have $B_1 = B_2 = 0$ on the switching curve), we obtain, for $\lambda = 0$,

$$x_1(s) = B_1 - B_2\theta - \frac{1}{2}(1-\rho)\theta^2 = \frac{1}{2}(1-\rho)\theta^2,$$
$$x_2(s) = B_2 + (1-\rho)\theta = -(1-\rho)\theta, \quad \theta = \tau - s > 0. \tag{2.2.11}$$

Let us also write down condition (2.2.2) determining that the phase trajectory (2.2.5) with $u = -1$ passes through the initial point:

$$\xi = B_1 - B_2\tau - \frac{1}{2}(1-\rho)\tau^2,$$
$$\eta = B_2 + (1-\rho)\tau. \tag{2.2.12}$$

Eliminating constants B_1 and B_2 from (2.2.11) and (2.2.12), we obtain two equations for θ and τ. Solving them, we find

$$\tau(\xi,\eta) = \frac{1}{1-\rho}\left\{2\left[\frac{1}{2}\eta^2 - (1-\rho)\xi\gamma\right]^{1/2} - \eta\gamma\right\},$$
$$\gamma = \operatorname{sign}\psi_\rho, \quad (\lambda = 0). \tag{2.2.13}$$

Here, we take into account the symmetry of the phase trajectories. Function ψ_ρ is defined in (2.2.10). On the switching curve, that is, for $\psi_\rho = 0$, we can take for γ in (2.2.13) any of the numbers $\gamma = \pm 1$: the value of $\tau(\xi,\eta)$ is the same in both cases.

The optimal time for $\lambda > 0$ is obtained in an analogous manner. Here, instead of (2.2.5), we use formulas (2.2.4). We have finally (see [2])

$$\tau(\xi,\eta) = 2\lambda^{-1}\log\{M^{1/2} + [M - 1 + \lambda\eta\gamma(1-\rho)^{-1}]^{1/2}\},$$
$$M = \exp[-(\lambda\eta + \lambda^2\xi)\gamma(1-\rho)^{-1}], \tag{2.2.14}$$
$$\gamma = \operatorname{sign}\psi_\rho, \quad (\lambda > 0),$$

where ψ_ρ is given by (2.2.10) for $\lambda > 0$.

Equations (2.2.9), (2.2.10) and (2.2.13), (2.2.14) determine the synthesis of the optimal control and the minimum guaranteed time τ in the game problem (2.1.35) and (2.1.36). If the perturbation v is different from the worst one ($v \neq -\rho u$), then the phase trajectories will differ from the optimal ones. However, the time needed to get the system to the coordinate origin will not exceed τ given by (2.2.13) and (2.2.14). We note that, when the motion has arrived on the switching curve, it will, for arbitrary admissible perturbations, proceed along that curve until it gets to the coordinate origin. If $v \neq -\rho u$, a sliding regime of motion along the switching curves is realized. Thus, if $v = 0$ on the switching curve, the control assumes the values $u = \pm 1$ with infinitely many changes of sign, so that we have "on the average" $u = 1-\rho$ or $u = -(1-\rho)$ on the corresponding branches of the switching curve.

2.2.2 Simplified control for the subsystem

With the method of control proposed in Sect. 2.2.1, we did not assume the perturbation, that is, the function v in system (2.1.35), to be known. However, we did assume its maximum possible value [ρ with constraint (2.1.35)] to be known, and the control synthesis given by (2.2.9) and (2.2.10) depends on this maximum value.

There is another possible approach to determine the control in a system with perturbations. In it, the perturbations are completely ignored at the stage of the control design and are taken into account only in the modelling and processing of the control. This approach, which is completely natural in the case of small perturbations, we shall call the simplified approach.

Below, we compare the two approaches, and determine to what extent ignoring the perturbations in the control design is justified.

By Theorem 2.1, if condition (2.1.12) holds, the system in question in the form (2.1.3), (2.1.10), or (2.1.30) is broken down into n subsystems of the form (2.1.35). Therefore, a comparison of the two approaches needs to be made only for system (2.1.35).

If we neglect perturbation v in system (2.1.35), it takes the form

$$\ddot{x} + \lambda\dot{x} = u, \quad |u| \le 1. \tag{2.2.15}$$

Let us write down the time-optimal control for system (2.2.15) with the boundary conditions (2.1.36). Since system (2.2.15) coincides with system (2.1.38) at $\rho = 0$, the desired control is determined by (2.2.9) and (2.2.10) in which we set $\rho = 0$. We obtain

$$\begin{aligned}
u(x_1,x_2) &= \operatorname{sign}\psi_0(x_1,x_2) \quad \text{for} \quad \psi_0 \neq 0,\\
u(x_1,x_2) &= \operatorname{sign}x_1 = -\operatorname{sign}x_2 \quad \text{for} \quad \psi_0 = 0,\\
\psi_0(x_1,x_2) &= -x_1 - \lambda^{-1}x_2 + \lambda^{-2}\log[1+\lambda|x_2|]\operatorname{sign}x_2, \quad (\lambda > 0),\\
\psi_0(x_1,x_2) &= -x_1 - \frac{1}{2}x_2|x_2|, \quad (\lambda = 0).
\end{aligned} \tag{2.2.16}$$

The switching curve $\psi_0 = 0$ for the feedback control (2.2.16) is the dashed curve in Fig. 2.3. For comparison, the solid curve in Fig. 2.3 shows the switching curve $\psi_\rho = 0$ for control (2.2.9) and (2.2.10) with $0 < \rho < 1$. Both these curves are symmetric about the coordinate origin. The equation of the curve $\psi_0 = 0$ can be represented in the form

$$x_1 = \phi(x_2), \tag{2.2.17}$$

where $\phi(x_2)$ is a monotonically decreasing odd function of its argument.

In the case of the control law (2.2.16), system (2.1.35) takes the form

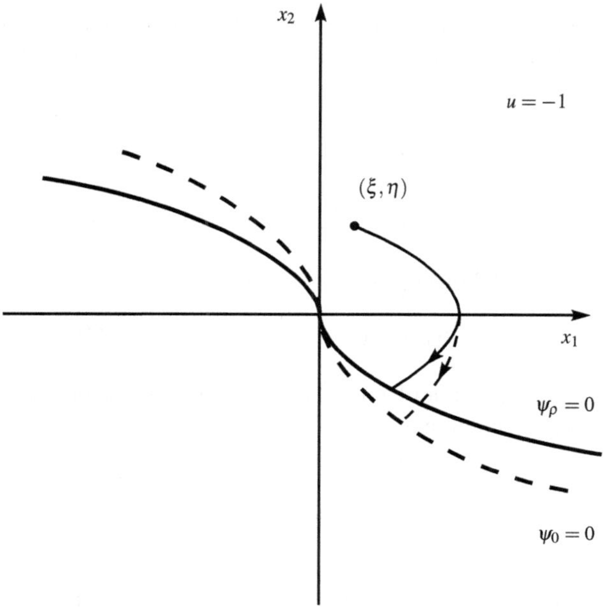

Fig. 2.3 Switching curves for $\rho>0$ and $\rho=0$

$$\dot{x}_1 = x_2,\ \dot{x}_2 = -\lambda \dot{x}_2 + u(x_1,x_2)+v, \quad \lambda \geq 0,$$
$$|v| \leq \rho < 1, \quad (x_1 = x, \quad x_2 = \dot{x}). \tag{2.2.18}$$

To estimate the possible influence of the perturbations on the motion of system (2.2.18), we pose the problem of finding the "worst" perturbation.

Problem 2.2. Find the optimal feedback control $v(x_1,x_2)$ for system (2.2.18) satisfying the constraint $|v| \leq \rho$ and having the property that the phase trajectory of that system first intersects the switching curve [$\psi_0 = 0$ or $x_1 = \phi(x_2)$, see (2.2.16) and (2.2.17)] as far as possible from the coordinate origin, that is, for as large $|x_1|$ as possible or, what amounts to the same thing, for as large $|x_2|$ as possible.

To be definite, we assume that the initial point (ξ,η) lies in the region $\psi_0<0$. Then, by (2.2.16), we have $u=-1$ on the entire trajectory in question. Then, the phase trajectory of system (2.2.18) first intersects that branch of the switching curve on which $x_1>0$ and $x_2<0$, see Fig. 2.3. As a result, Problem 2.2 is described by the following relationships:

$$\dot{x}_1 = x_2, \quad \dot{x}_2 = -\lambda x_2 - 1 + v, \quad |v| \leq \rho < 1,$$
$$\lambda \geq 0, \quad 0 \leq t \leq \tau, \quad x_1(0) = \xi, \quad x_2(0) = \eta, \tag{2.2.19}$$
$$x_1(\tau) = \phi(x_2(\tau)), \quad x_1(\tau) > 0, \quad x_2(\tau) < 0, \quad x_1(\tau) \to \max.$$

Here, τ is the terminal instant of the process that is not fixed. Function $\phi(x_2)$ in (2.2.17) and (2.2.19) is obtained from the equation $\psi_0 = 0$ [see (2.2.16)] with $\rho = 0$ and $x_2 < 0$. We obtain

$$\phi(x_2) = -\lambda^{-1}x_2 - \lambda^{-2}\log(1-\lambda x_2), \quad (\lambda > 0),$$

$$\phi(x_2) = \frac{1}{2}x_2^2, \quad (\lambda = 0).$$

It follows from these relationships that

$$\phi(x_2) \geq 0, \quad \phi'(x_2) = x_2(1-\lambda x_2)^{-1} < 0, \quad (x_2 < 0). \tag{2.2.20}$$

We note that maximization of $x_1(\tau)$ in (2.2.19) is equivalent to minimization of the following integral functional:

$$\int_0^\tau (-x_2)dt \to \min. \tag{2.2.21}$$

Let us apply the maximum principle, see Sect. 1.2, to Problem 2.2. The Hamiltonian function for problem (2.2.19) and (2.2.21) has the form

$$H = p_1x_2 + p_2(v - \lambda x_2 - 1) + x_2, \quad |v| \leq \rho, \tag{2.2.22}$$

where p_1 and p_2 are conjugate variables. They satisfy the following adjoint equations

$$\dot{p}_1 = 0, \quad \dot{p}_2 = \lambda p_2 - p_1 - 1 \tag{2.2.23}$$

and the transversality conditions

$$p_1(\tau)\phi'(x_2(\tau)) + p_2(\tau) = 0, \quad H(\tau) = 0. \tag{2.2.24}$$

We find $p_1(\tau)$ from the first of conditions (2.2.24) and substitute it into the second one, using expression (2.2.22) for the Hamiltonian H. We obtain

$$p_2[(v - \lambda x_2 - 1)\phi'(x_2) - x_2] + x_2\phi'(x_2) = 0, \quad (t = \tau). \tag{2.2.25}$$

Substituting $\phi'(x_2)$ given by (2.2.20) into (2.2.25), we get, after some simplifications,

$$x_2[p_2(v-2) + x_2] = 0, \quad (t = \tau). \tag{2.2.26}$$

Since, by (2.2.19), we have $|v| \leq \rho < 1$ and $x_2(\tau) < 0$, it follows from (2.2.26) that

$$p_2(\tau) < 0. \tag{2.2.27}$$

It follows from the maximum principle and (2.2.22) that the optimal control is expressed in the form

$$v(t) = \rho \operatorname{sign} p_2(t). \tag{2.2.28}$$

Integrating system (2.2.23), we obtain

$$p_1 = C_1, \quad p_2 = \lambda^{-1}(C_1+1) + C_2 e^{\lambda t}, \quad (\lambda > 0),$$
$$p_1 = C_1, \quad p_2 = C_2 - (C_1+1)t, \quad (\lambda = 0), \tag{2.2.29}$$

where C_1 and C_2 are constants. It follows from (2.2.29) that $p_2(t)$ is a monotone function. Consequently, the optimal control (2.2.28) has no more than one switching point.

Since system (2.2.19) is autonomous, its Hamiltonian H is constant along the optimal trajectory and, by virtue of (2.2.24), equal to zero. Then, by (2.2.22), we obtain

$$H(t) = (p_1+1)x_2 + p_2(v - \lambda x_2 - 1) \equiv 0. \tag{2.2.30}$$

At the instant of switching, we have, by (2.2.28), $p_2 = 0$. Then it follows from (2.2.30) that at that instant either $p_1 = -1$ or $x_2 = 0$.

Let us consider the first possibility. It follows from (2.2.29) that, if $p_1 = -1$, $p_2(t)$ does not change sign along the trajectory and hence a switching cannot take place for $p_1 = -1$.

The second possibility $x_2 = 0$ means that the control is switched when the trajectory crosses the line $x_2 = 0$. Since $p_2(\tau)$ is, by (2.2.27), negative, the optimal control (2.2.28) is negative for $x_2 < 0$ and positive for $x_2 > 0$. Thus, the feedback optimal control has the form

$$v(x_1, x_2) = \rho \operatorname{sign} x_2. \tag{2.2.31}$$

The optimal control in the region $\psi_0 < 0$ is constructed. We note that system (2.2.19) is, along with relationships (2.2.16), invariant with respect to the transformation $x_1 \to -x_1$, $x_2 \to -x_2$, and $v \to -v$. Consequently, the optimal feedback control $v(x_1, x_2)$ possesses the property of central symmetry, and synthesis (2.2.31) satisfies that condition. Thus, (2.2.31) provides the solution of Problem 2.2 formulated above in the whole phase plane (x_1, x_2).

2.2.3 Comparative analysis of the results

Let us use solution (2.2.31) of Problem 2.2 obtained above to analyze possible motions of system (2.2.18) for the case of the simplified control (2.2.16). We first assume that perturbation v is given by (2.2.31). For u given by (2.2.16) and v given by (2.2.31), all trajectories of system (2.2.18) consist of arcs of curves corresponding to constant $u = \pm 1$ and $v = \pm\rho$. The equations of these curves are determined by (2.2.4)–(2.2.7), in which, according to (2.2.16) and (2.2.31), we set

$$w = u + v = \operatorname{sign} \psi_0 + \rho \operatorname{sign} x_2. \tag{2.2.32}$$

One of the trajectories for the control law (2.2.16) in the case of perturbation (2.2.31) is shown in Fig. 2.3 by the thin dashed curve. The solid curve represents the optimal trajectory for control (2.2.9) and $v = -\rho u$. The arrows indicate the direction of increasing time. We note that the arcs of the optimal trajectories for the two control laws coincide in regions where the three functions ψ_0, ψ_ρ, and $(-x_2)$ have the same signs. The heavy solid and the dashed curves in Fig. 2.3 represent the switching curves $\psi_\rho = 0$ and $\psi_0 = 0$ for controls (2.2.16) and (2.2.9), respectively.

To be definite, we construct the optimal phase trajectory for control (2.2.16) that begins at the point (ξ, η) on the switching curve $\psi_0 = 0$ for $\eta \geq 0$ and ends at the point (ξ^*, η^*) on the other branch of the switching curve, that is, for $\xi^* > 0$ and $\eta^* < 0$. This trajectory lies in the region $\psi_0 < 0$ and consists of two sections that meet at $x_2 = 0$. On the first section, where $x_2 > 0$, we have, by (2.2.16) and (2.2.31), $u = -1$ and $v = \rho$. On the second section, $x_2 < 0$. Therefore, by (2.2.16) and (2.2.31), we have on it $u = -1$ and $v = -\rho$.

The first section of the trajectory passes through the initial point (ξ, η) and, on this section, we have, by (2.2.32), $w = u + v = -1 + \rho$. Consequently, its equation is, by (2.2.6) and (2.2.7), represented in the form

$$
\begin{gathered}
x_1 = B_1' - \lambda^{-1} x_2 + \lambda^{-2}(1-\rho)\log[1 + \lambda(1-\rho)^{-1} x_2], \\
B_1' = \xi + \lambda^{-1}\eta - \lambda^{-2}(1-\rho)\log[1 + \lambda(1-\rho)^{-1}\eta] \quad \text{for} \quad \lambda > 0; \\
x_1 = B_1' - [2(1-\rho)]^{-1} x_2^2, \\
B_1' = \xi + [2(1-\rho)]^{-1}\eta^2 \quad \text{for} \quad \lambda = 0; \\
0 \leq x_2 \leq \eta.
\end{gathered}
\tag{2.2.33}
$$

The second section of the trajectory passes through the final point (ξ^*, η^*), and, on it, $w = u + v = -1 - \rho$. Therefore, for the second section, we obtain from (2.2.6) and (2.2.7)

$$
\begin{gathered}
x_1 = B_2' - \lambda^{-1} x_2 + \lambda^{-2}(1+\rho)\log[1 + \lambda(1+\rho)^{-1} x_2], \\
B_2' = \xi^* + \lambda^{-1}\eta^* - \lambda^{-2}(1+\rho)\log[1 + \lambda(1+\rho)^{-1}\eta^*] \quad \text{for} \quad \lambda > 0; \\
x_1 = B_2' - [2(1+\rho)]^{-1} x_2^2, \\
B_2' = \xi^* + [2(1+\rho)]^{-1}(\eta^*)^2 \quad \text{for} \quad \lambda = 0; \\
\eta^* \leq x_2 \leq 0.
\end{gathered}
\tag{2.2.34}
$$

At the joining point of the sections, we have $x_2 = 0$, and the values of x_1 for the two sections coincide. We then obtain from (2.2.33) and (2.2.34)

$$
B_1' = B_2'. \tag{2.2.35}
$$

The points (ξ,η) and (ξ^*,η^*) belong to two branches of the switching curve $\psi_0 = 0$; also, $\eta > 0$ and $\eta^* < 0$. Consequently, on the basis of formulas (2.2.16), we obtain

$$\begin{gathered} \xi = -\lambda^{-1}\eta + \lambda^{-2}\log(1+\lambda\eta), \\ \xi^* = -\lambda^{-1}\eta^* - \lambda^{-2}\log(1-\lambda\eta^*) \quad \text{for} \quad \lambda > 0; \\ \xi = -\frac{1}{2}\eta^2, \quad \xi^* = \frac{1}{2}(\eta^*)^2 \quad \text{for} \quad \lambda = 0. \end{gathered} \tag{2.2.36}$$

We substitute into (2.2.35) expressions (2.2.33) and (2.2.34) for B_1' and B_2' and also formulas (2.2.36) expressing ξ and ξ^* in terms of η and η^*. As a result, after some simplification, we obtain the relationships

$$\begin{gathered} [1+(1-\rho)^{-1}\lambda\eta]^{1-\rho}(1+\lambda\eta)^{-1} \\ = [1+(1+\rho)^{-1}\lambda\eta^*]^{1+\rho}(1-\lambda\eta^*) \quad \text{for} \quad \lambda > 0, \\ \rho(1-\rho)^{-1}\eta^2 = (2+\rho)(1+\rho)^{-1}(\eta^*)^2 \quad \text{for} \quad \lambda = 0, \end{gathered} \tag{2.2.37}$$

where $\eta > 0$ and $\eta^* < 0$.

Equations (2.2.37) connect the values of η^* and η. Let us first consider the case $\lambda = 0$. Here, the relationship (2.2.37) takes the form

$$\begin{gathered} \left|\frac{\eta^*}{\eta}\right| = \kappa = \left[\frac{\rho(1+\rho)}{(1-\rho)(2+\rho)}\right]^{1/2}, \\ 0 \le \rho < 1. \end{gathered} \tag{2.2.38}$$

One can easily see that κ increases monotonically from 0 to ∞ as ρ increases from 0 to 1. In particular, $\kappa = 1$ for ρ equal to

$$\rho^* = \frac{1}{2}(\sqrt{5}-1) \approx 0.618. \tag{2.2.39}$$

The number ρ^* is the "golden-section" ratio. Thus, if $\lambda = 0$, then, for $\rho < \rho^*$, we have, on the basis of (2.2.38), $|\eta^*/\eta| < 1$; for $\rho = \rho^*$, we have $|\eta^*/\eta| = 1$; and for $\rho > \rho^*$, we have $|\eta^*/\eta| > 1$.

In the case $\lambda > 0$, relationship (2.2.37) defines an implicit dependence of η^* on η. To investigate this connection, we set

$$\lambda\eta = X > 0, \quad -\lambda\eta^* = Y > 0 \tag{2.2.40}$$

and represent dependence (2.2.37) in the form

$$\Phi_\rho(X) = \Psi_\rho(Y), \quad X > 0, \quad Y > 0, \quad 0 < \rho < 1,$$

$$\Phi_\rho(X) = [1 + (1-\rho)^{-1}X]^{1-\rho}(1+X)^{-1}, \tag{2.2.41}$$

$$\Psi_\rho(Y) = [1 - (1+\rho)^{-1}Y]^{1+\rho}(1+Y).$$

We note certain properties of the functions Φ_ρ and Ψ_ρ in (2.2.41). The function Φ_ρ is defined for all $X \geq 0$ and approaches zero as $X \to \infty$; the function Ψ_ρ is defined in the interval $[0, 1+\rho]$ and vanishes at $Y = 1+\rho$. Both functions are equal to unity for $X = Y = 0$. By direct differentiation of functions (2.2.41), we see that $\Phi'_\rho(X) < 0$ and $\Psi'_\rho(Y) < 0$, so that both Φ_ρ and Ψ_ρ are monotone decreasing functions. Let us also calculate the following derivative:

$$\left[\frac{\Psi_\rho(X)}{\Phi_\rho(X)}\right]' = 2[1 + (1-\rho)^{-1}X]^{\rho-2}[1 - (1+\rho)^{-1}X]^{\rho}$$
$$\times (1+X)(1-\rho^2)^{-1}X[\rho - 1 + \rho^2 - (1+\rho)X]. \tag{2.2.42}$$

We note that the expression $\rho - 1 + \rho^2$ in (2.2.42) is nonpositive for $\rho \leq \rho^*$ and positive for $\rho > \rho^*$. Consequently, for $\rho \leq \rho^*$, the ratio Ψ_ρ/Φ_ρ decreases monotonically in the interval $[0, 1+\rho]$. Therefore, $\Psi_\rho(X) < \Phi_\rho(X)$ for $0 < X \leq 1+\rho$. On the other hand, if $\rho > \rho^*$, we have $\Psi_\rho(X) > \Phi_\rho(X)$ in some interval $0 < X \leq X^* < 1+\rho$. However, $\Psi_\rho(X) < \Phi_\rho(X)$ close to $X = 1+\rho$ since

$$\Psi_\rho(1+\rho) = 0 < \Phi_\rho(1+\rho).$$

Figures 2.4 and 2.5 show graphs of the functions $\Phi_\rho(X)$ and $\Psi_\rho(Y)$ for the cases $\rho \leq \rho^*$ and $\rho > \rho^*$, respectively. These figures illustrate graphically the relationship between X and Y that is established by (2.2.41). These equations and the properties mentioned for the functions Φ_ρ and Ψ_ρ lead to the following conclusions for the case $\lambda > 0$:

- If $\rho \leq \rho^*$, we always have $Y < X$ and, by (2.2.40), $|\eta^*/\eta| \leq 1$.
- If $\rho > \rho^*$, then, for sufficiently small X, we have $Y > X$ (that is, $|\eta^*/\eta| > 1$), whereas, for sufficiently large X, we have $Y < X$ (that is, $|\eta^*/\eta| < 1$). Here, we always have $Y < 1+\rho$; that is, $|\eta^*| < (1+\rho)\lambda^{-1}$.

The trajectory that begins at an arbitrary point (ξ, η) in the phase plane can be continued indefinitely even after its intersection with the switching curve $\psi_0 = 0$ at the point (ξ^*, η^*). For this, we need to take the point (ξ^*, η^*) as the initial point and continue the motion on the basis of system (2.2.18), substituting into it control u defined by (2.2.16) and the optimal perturbation v defined by (2.2.31). The trajectory obtained in this way intersects both branches of the switching curve infinitely many times. The values of ordinates x_2 at two successive points of intersection of the switching curve $\psi_0 = 0$ are in the ratio $|\eta^*/\eta|$ that is given by formula (2.2.38) for $\lambda = 0$, and is given by (2.2.40) and (2.2.41) for $\lambda > 0$.

The nature of the motion is quite dependent on the parameters ρ and λ.

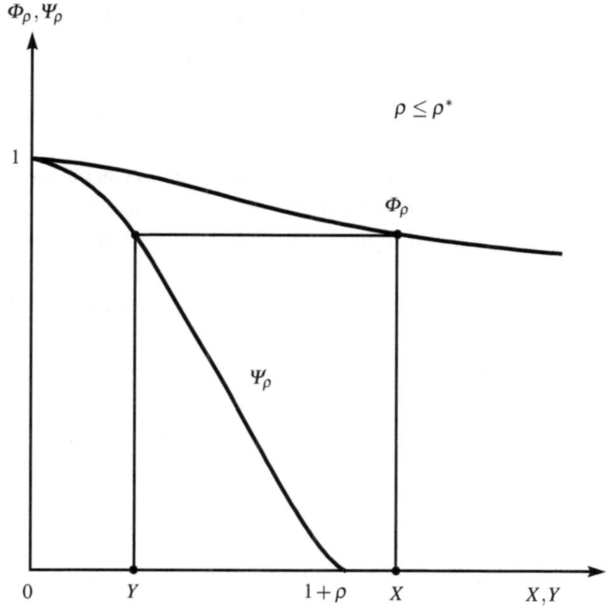

Fig. 2.4 Functions $\Phi_\rho(X)$ and $\Psi_\rho(Y)$ for the case $\rho \leq \rho^*$

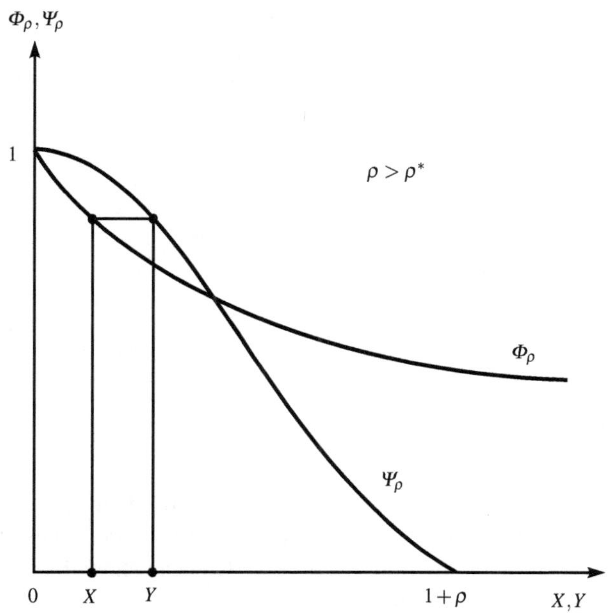

Fig. 2.5 Functions $\Phi_\rho(X)$ and $\Psi_\rho(Y)$ for the case $\rho > \rho^*$

Let us first set $\lambda = 0$. If $\rho < \rho^*$, where ρ^* is defined by (2.2.39), then $\kappa < 1$ in (2.2.38). Here, the values of $|x_2|$ at the instants of intersection of the switching curve $\psi_0 = 0$ by the trajectory decrease in a geometric progression with the denominator $\kappa < 1$. Therefore, the phase trajectory approaches the coordinate origin and reaches it in finite time, though after an infinite number of switchings.

If $\rho = \rho^*$, then $\kappa = 1$ in (2.2.38) and the phase trajectory is periodic. At equal intervals of time, it passes through the same points in the phase plane. In this case, the trajectory remains in a bounded region though it does not approach the coordinate origin.

If $\rho > \rho^*$, then $\kappa > 1$ in (2.2.38). Here, the phase trajectory moves to infinity along a spiral path.

The behavior of the phase trajectories is shown in Figs. 2.6, 2.7, and 2.8 for the cases $\rho < \rho^*$, $\rho = \rho^*$, and $\rho > \rho^*$, respectively.

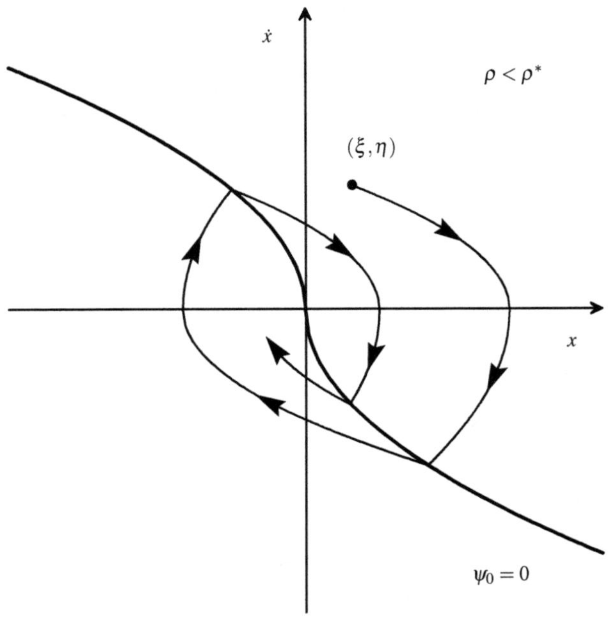

Fig. 2.6 Phase trajectory for $\rho < \rho^*$

Let us turn to the case $\lambda > 0$. If in addition $\rho \leq \rho^*$, then, by the analysis that we made, $|\eta^*/\eta| < 1$. In this case, the phase trajectory approaches the coordinate origin and reaches it in a finite time for $\rho < \rho^*$. One can show that, for $\rho = \rho^*$, the phase point approaches the coordinate origin as $t \to \infty$.

For $\rho > \rho^*$, the phase trajectory does not approach the coordinate origin but remains in a bounded region. From some instant on, we have $|x_2| \leq \lambda^{-1}(1+\rho)$ (as a consequence of the inequality $Y < 1+\rho$).

Let us now characterize the possible motions of system (2.2.18) for the control law (2.2.16) and an arbitrary perturbation $|v| \leq \rho$.

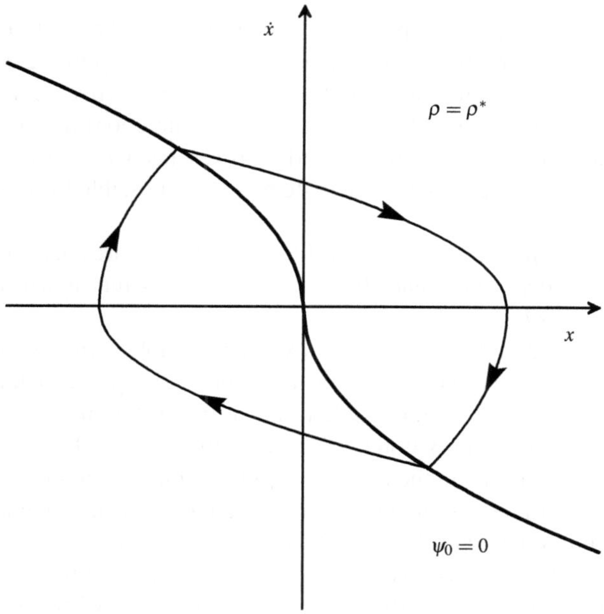

Fig. 2.7 Phase trajectory for $\rho = \rho^*$

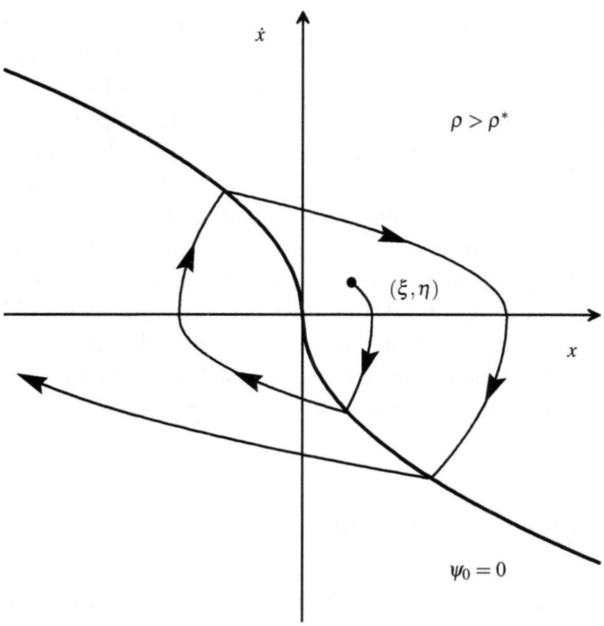

Fig. 2.8 Phase trajectory for $\rho > \rho^*$

If the ratio of the maximum possible value of the perturbation to the maximum possible value of the control is less than the "golden-section" ratio (2.2.39), namely, if $\rho < \rho^*$, then, for any admissible control and any nonnegative λ, the control law (2.2.16) ensures that system (2.2.18) will move to the coordinate origin in finite time. This follows from the fact that the origin is reached even for the "worst" perturbation (2.2.31) that tries to takes the system as far as possible from the coordinate origin.

If $\rho = \rho^*$, then, for $\lambda = 0$, the control law (2.2.16) ensures that the system will be kept in a bounded region and, for $\lambda > 0$, it ensures that it will approach the coordinate origin as $t \to \infty$.

On the other hand, if the ratio of the maximum possible perturbation to the maximum possible control exceeds the "golden-section" ratio ($\rho > \rho^*$), then there exist perturbations for which it is impossible to move the system to the coordinate origin. In the case $\lambda > 0$, perturbation (2.2.31) takes the system arbitrarily far from the coordinate origin, whereas in the case $\lambda = 0$ it takes the system out of some neighborhood of the coordinate origin though the system remains in a bounded region.

Thus, the simplified control law (2.2.16), which does not take into account the presence of perturbations, achieves the control purpose and brings the system to the coordinate origin, only if the perturbation level is sufficiently low. Specifically, the ratio of this level to the maximum level of control must not exceed the "golden-section" ratio ($\rho < \rho^*$).

In other words, one can ignore the presence of perturbations in constructing the control, only if the ratio of the maximum level of the perturbation to that of the control does not exceed the "golden-section" ratio $\rho^* \approx 0.618$.

We recall that the optimal control law (2.2.9) and (2.2.10) is based on the game approach and guarantees that system (2.1.35) will move to the coordinate origin in a finite time for any admissible perturbation, if $\rho < 1$. Thus, in comparison with the simplified control of Sect. 2.2.2, the control law described in Sect. 2.2.1 has a wider field of applicability. Furthermore, the game approach ensures the minimum guaranteed time for bringing the system to the coordinate origin because it is based on the time-optimal feedback control. However, the game approach, unlike the simplified approach, needs the knowledge of the maximal level of possible perturbations, i.e. the parameter ρ. In their structure, the two methods are similar, of a bang-bang type, and differ only in the switching curves, see Fig. 2.3.

2.2.4 Control for the initial system

Let us turn to solution of the original Problem 2.1. We obtain the feedback control in this problem on the basis of (2.1.34) and (2.1.29) in the following form:

$$U_i(q,\dot{q}) = U_i^0 u(x_1,x_2),$$

$$x_1 = x = (U_i^0)^{-1} y_i = (U_i^0)^{-1} [A_*(q-q^*)]_i, \tag{2.2.43}$$

$$x_2 = \dot{x} = (U_i^0)^{-1}\dot{y}_i = (U_i^0)^{-1}(A_*\dot{q})_i, \quad i,\dots,n.$$

Here, in the case of the optimal control of Sect. 2.2.1, function $u(x_1,x_2)$ is defined by (2.2.9), in which ψ_ρ is given by (2.2.10). Parameters λ and ρ in formulas (2.2.10) are expressed by (2.1.37); that is

$$\lambda = \lambda_i, \quad \rho = \frac{V_i^0}{U_i^0} < 1, \quad i = 1,\dots,n. \tag{2.2.44}$$

Control (2.2.43) is a bang-bang control, and it assumes the limiting admissible values: $U_i = \pm U_i^0$, $i = 1,\dots,n$. Let us describe the nature of the motion in the case of this control. We first suppose that perturbations V_i in system (2.1.28) or (2.1.30) assume at every instant the optimal ("worst") values. In terms of system (2.1.35), this means that $v = -\rho u$. In terms of system (2.1.28), we have, by virtue of (2.1.34), (2.2.43), and (2.2.44),

$$V_i = -\rho U_i^0 u = -V_i^0 (U_i^0)^{-1} U_i(q,\dot{q}), \quad i = 1,\dots,n. \tag{2.2.45}$$

In the case of perturbation (2.2.45), the motion of system (2.1.30), for the coordinate y_i, takes place along the time-optimal trajectories of system (2.1.38) or (2.2.1), that is, along the trajectories of Fig. 2.2 for $\lambda_i > 0$ or Fig. 1.2 for $\lambda = 0$. The relationship between the original coordinates q, the variables y_i, and the variables x_1 and x_2 is given by (2.1.29) and (2.2.43).

Now, if the perturbations V_i differ from the worst ones (2.2.45), as is usually the case, then the phase trajectories for each ith degree of freedom in the plane (x_1,x_2) deviate from the optimal ones. Here, the motion along the switching curves takes place in a sliding regime.

Time t_* needed to bring system (2.1.3) [or (2.1.10), (2.1.28), (2.1.30)] to the given state (2.1.8) does not exceed the maximum of the optimal times for each of subsystems (2.1.30) [or (2.1.35), (2.1.38), (2.2.1)]. We have

$$\begin{gathered} t_* \le t_0 + \max_{1\le i\le n} \tau(\xi_i,\eta_i), \quad \xi_i = (U_i^0)^{-1}[A_*(q^0 - q^*)]_i, \\ \eta_i = (U_i^0)^{-1}(A_*\dot{q}^0)_i, \quad i = 1,\dots,n. \end{gathered} \tag{2.2.46}$$

Here, we used (2.1.29) and (2.1.37) for ξ and η. The function $\tau(\xi,\eta)$ is defined by (2.2.13) for those i for which $\lambda_i = 0$ and by (2.2.14) for those i for which $\lambda_i > 0$.

We summarize the results in the form of a theorem.

Theorem 2.2. *Suppose that conditions (2.1.12) are satisfied, and that all the trajectories considered lie in domain Ω. Then, the feedback control $U(q,\dot{q})$ that solves Problem 2.1 is given by (2.2.43), where the function $u(x_1,x_2)$ is defined by (2.2.9) and (2.2.10). This control brings system (2.1.3) to the terminal state (2.1.8) no later than at time t_* defined by (2.2.46), (2.2.13), and (2.2.14). Parameters λ and ρ in these formulas are given for each degree of freedom by (2.2.44).*

The control constructed can be called suboptimal, since it is close to the time-optimal one and becomes optimal in the case of the "worst" perturbations.

Using the simplified approach described in Sect. 2.2.2, one should replace the function $u(x_1,x_2)$ in relations (2.2.43) by its expression from (2.2.16). Otherwise, the procedure for the control design for the original system remains the same as for the game approach.

2.3 Weak coupling between degrees of freedom

The first method of decomposition presented in Sect. 2.1.3 leads to the control that solves Problem 2.1. This control is given in explicit form in Sect. 2.2.4. The main assumption that made it possible to carry out the decomposition was the existence of a domain Ω in the $2n$-dimensional space $(q,\dot{q})$ in which all the motions being considered lie and in which inequalities (2.1.12) are satisfied.

By virtue of (2.1.11) for V, inequalities (2.1.12) and (2.1.13) impose constraints on the uncontrolled forces Q and the inertial terms S. We can see from (2.1.4) that the inertial terms depend quadratically on the generalized velocities $\dot{q}$. Therefore, inequalities (2.1.12) and (2.1.13) limit somehow the domain Ω with respect to $\dot{q}$, whereas the obtained control can carry the system to the region with large $|\dot{q}|$.

It is clear, on the one hand, that to solve Problem 2.1 it is necessary to impose constraints on the uncontrolled forces Q, otherwise the bounded controls U will not be able to overcome these forces Q. On the other hand, it follows from what was said above that it is expedient to impose constraints on quantities S_i during the control process. These considerations served as the basis for the modification of the first method of decomposition suggested in [98, 102].

2.3.1 Modification of the decomposition method

Let us turn again to the system described by relations (2.1.1)–(2.1.8).

The domain D in the n-dimensional q-space, where the motions of the system being considered can occur, is specified in the form of independent constraints on the coordinates q_i

$$D = \{q:\ q_i^- \le q_i \le q_i^+\}. \tag{2.3.1}$$

We will make certain simplifying assumptions concerning the kinetic energy and the generalized forces Q_i. Suppose that matrix $A(q)$ from (2.1.3) can be represented in the form

$$A(q) = J + \tilde{A}(q), \quad J = \operatorname{diag}(J_1,\dots,J_n), \quad J_i = \text{const} > 0, \tag{2.3.2}$$

where $\tilde{A}(q)$ is a symmetric matrix such that, for any n-dimensional vector z, the inequality

$$|\tilde{A}(q)z| \le \mu|z|, \qquad \mu > 0, \qquad \forall q \in D \tag{2.3.3}$$

is satisfied. Here, μ is a sufficiently small parameter, its possible values are specified below.

Furthermore, we assume that

$$\left|\frac{\partial a_{jk}}{\partial q_i}\right| \le C, \qquad C = \text{const} > 0, \qquad i,j,k = 1,\ldots,n, \tag{2.3.4}$$

and that the generalized forces Q_i can be represented in the form

$$Q_i = G_i + F_i \tag{2.3.5}$$

where $G_i(q,\dot{q},t)$ are restricted forces satisfying constraints

$$|G_i| \le G_i^0, \qquad i = 1,\ldots,n. \tag{2.3.6}$$

The magnitudes of constants G_i^0 do not exceed constants U_i^0 in constraints (2.1.6) for the permissible values of the control forces, that is

$$G_i^0 < U_i^0, \qquad i = 1,\ldots,n. \tag{2.3.7}$$

Note that, if the inequality $G_i^0 > U_i^0$ that is the inverse of (2.3.7) holds for certain i, then the system can be uncontrollable.

The forces denoted by F_i in (2.3.5) are sufficiently small at low velocities and satisfy the constraints

$$F_i = F_i(q,\dot{q},t), \qquad |F_i| \le F^0\left(|\dot{q}|\right), \qquad i = 1,\ldots,n. \tag{2.3.8}$$

Here, $F^0(\vartheta)$ is a monotonically increasing continuous function defined for $\vartheta \ge 0$ and such that $F^0(0) = 0$. The exact form of functions $G_i(q,\dot{q},t)$ and $F_i(q,\dot{q},t)$ in (2.3.5) may be unknown.

We multiply both sides of (2.1.3) by JA^{-1} [matrix J has been introduced in (2.3.2)] and obtain

$$J_i\ddot{q}_i = U_i + V_i, \tag{2.3.9}$$

$$V_i = S_i - [\tilde{A}A^{-1}(U+S)]_i. \tag{2.3.10}$$

System (2.3.9) and (2.3.10) is equivalent to the initial equation (2.1.3). Relations (2.1.4), (2.3.4)–(2.3.6), and (2.3.8) yield the constraint on the components of vector S

$$|S_i(q,\dot{q},t)| \le G_i^0 + \tilde{S}^0(\dot{q}), \qquad \tilde{S}^0(\dot{q}) = F^0\left(|\dot{q}|\right) + \frac{3}{2}C\left(\sum_{j=1}^{n}|\dot{q}_j|\right)^2. \tag{2.3.11}$$

We assume that the inequalities

$$|V_i| \le \rho_i U_i^0, \qquad \rho_i < 1 \tag{2.3.12}$$

hold, where ρ_i are constants to be specified below. We shall treat the functions V_i in (2.3.9) as independent restricted perturbations not exceeding the permissible values of the controls. In this case, the initial non-linear system is decomposed into n linear subsystems [the ith subsystem is described by the ith equation (2.3.9)] subjected to perturbations, and each subsystem has a single degree of freedom. To solve Problem 2.1, it is therefore sufficient to solve n simpler control problems for the second-order subsystems.

2.3.2 Analysis of the controlled motions

As has been done previously, we shall specify the scalar control U_i which transfers the ith subsystem (2.3.9) in finite time from an arbitrary initial state $(q_i^0, \dot{q}_i^0)$ to the terminal state $(q_i^*, 0)$ for any permissible perturbation V_i satisfying condition (2.3.12). We will again define the feedback control by (2.2.43), (2.2.9), and (2.2.10). Here, one should substitute matrix J instead of matrix A_* in (2.2.43) and set $\lambda = 0$ in (2.2.10). After these transformations, we obtain

$$\begin{gathered} U_i = -U_i^0 \operatorname{sign}(\dot{q}_i - \psi_i^*) \quad \text{for} \quad \dot{q}_i \neq \psi_i^*, \\ U_i = -U_i^0 \operatorname{sign} \dot{q}_i \quad \text{for} \quad \dot{q}_i = \psi_i^*, \\ \psi_i^*(q_i) = -\left(2X_i|q_i - q_i^*|\right)^{1/2} \operatorname{sign}(q_i - q_i^*). \end{gathered} \tag{2.3.13}$$

Here, X_i is a positive control parameter connected with constant ρ_i from (2.3.12) by the relation

$$X_i = \frac{U_i^0(1-\rho_i)}{J_i}. \tag{2.3.14}$$

In inequalities (2.3.12), we express ρ_i in terms of the control parameters X_i using (2.3.14). We obtain

$$|V_i| \leq U_i^0 - J_i X_i, \qquad i = 1, \ldots, n. \tag{2.3.15}$$

In order to specify the control law (2.3.13), it is necessary to choose the values of parameters $X_i > 0$, so that inequalities (2.3.15) are satisfied.

Remind that the above-mentioned control was obtained as the time-optimal control in a game problem in which U_i and V_i are considered as the controls of two players [79]. This control is a bang-bang control and takes its limiting permissible values:

$$U_i = \pm U_i^0.$$

The switching curve

$$\dot{q}_i = \psi_i^*(q_i)$$

consists of two parabolic branches which are symmetric about the point $(q_i^*, 0)$.

We will now specify the set Ω_i in the two-dimensional phase space of the ith subsystem (Fig. 2.9):

$$\Omega_i = \{(q_i, \dot{q}_i) : q_i^- \le q_i \le q_i^+, \quad \psi_i^- \le \dot{q}_i \le \psi_i^+\},$$
$$\psi_i^-(q_i) = \psi_i^*(q_i + q_i^* - q_i^-), \qquad \psi_i^+(q_i) = \psi_i^*(q_i + q_i^* - q_i^+). \tag{2.3.16}$$

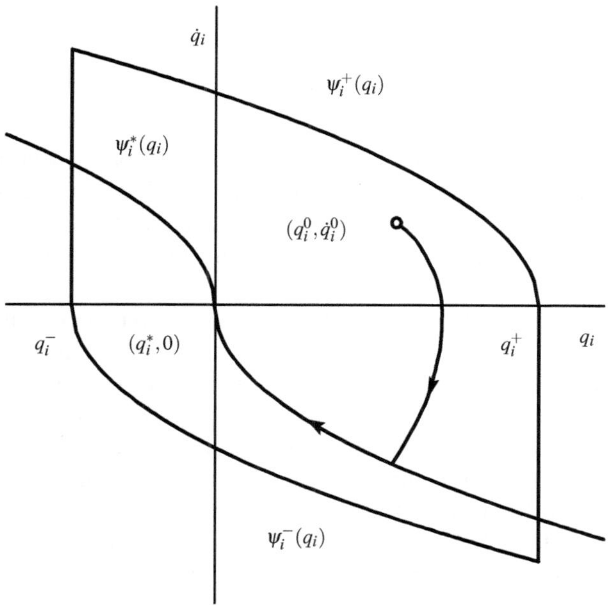

Fig. 2.9 Set Ω_i and the switching curve

We will describe the motion of subsystem (2.3.9) in the case where the control U_i is specified by (2.3.13), the disturbance V_i satisfies constraint (2.3.15), and the initial point $(q_i^0, \dot{q}_i^0)$ lies in Ω_i:

$$(q_i^0, \dot{q}_i^0) \in \Omega_i. \tag{2.3.17}$$

The control process is divided into two main stages. In the first stage, the motion is performed with a constant control until the phase point of the subsystem reaches the switching curve. To fix our ideas, we assume that $\dot{q}_i > \psi_i^*(q_i)$; then, according to (2.3.13), we have $U_i = -U_i^0$. In this case, from (2.3.9) and (2.3.15), it follows that

$$\ddot{q}_i \le -X_i. \tag{2.3.18}$$

Note that, by virtue of (2.3.16) and (2.3.13), the following equality holds

$$\frac{d\psi_i^+}{dq_i} = -\frac{X_i}{\psi_i^+}. \tag{2.3.19}$$

Along the trajectory of the subsystem in the domain Ω_i, we have, according to (2.3.16), $\dot{q}_i \le \psi_i^+$. Therefore, taking into account (2.3.18) and (2.3.19), we obtain for $\dot{q}_i > 0$

$$\frac{d\dot{q}_i}{dq_i} = \frac{\ddot{q}_i}{\dot{q}_i} \le -\frac{X_i}{\psi_i^+} = \frac{d\psi_i^+}{dq_i}, \qquad \dot{q}_i > 0. \tag{2.3.20}$$

For $\dot{q}_i < 0$, we have, in accordance with (2.3.18)

$$\frac{d\dot{q}_i}{dq_i} = \frac{\ddot{q}_i}{\dot{q}_i} > 0, \qquad \dot{q}_i < 0. \tag{2.3.21}$$

From inequalities (2.3.20) and (2.3.21) it follows that, for any perturbations, the phase trajectory of the subsystem under consideration does not intersect the curve $\dot{q}_i = \psi_i^+(q_i)$ and, due to (2.3.18), reaches the switching curve in finite time without going out beyond the limits of the domain Ω_i. This fact is proved in a similar way also for $\dot{q}_i < \psi_i^*$.

On reaching the switching curve, the phase point continues to move along it to the terminal state. The parabolic branches of the switching curve coincide with the phase trajectories of subsystem (2.3.9) in the case of the control U_i chosen in accordance with (2.3.13) and (2.3.14) and for $V_i = -\rho_i U_i$. If, however, $V_i \ne -\rho_i U_i$, then the motion occurs along a parabolic arc in a sliding mode. In this case, control U_i takes the values $\pm U_i^0$ with infinitely frequent changes of sign so that "on average" $\ddot{q}_i = X_i$ or $\ddot{q}_i = -X_i$ for the corresponding branches of the switching curve.

Hence, if conditions (2.3.16) and (2.3.17) are satisfied for all subsystems (2.3.9) at the initial instant of time, then their phase trajectories as a whole lie in the corresponding domains Ω_i. In this case, constraints (2.3.1) are satisfied, and the inequalities

$$|\dot{q}_i| \le -\psi_i^-(q_i^+) = \psi_i^+(q_i^-) = \psi_i^*(q_i^- - q_i^+) \tag{2.3.22}$$

following from (2.3.16) and (2.3.13) also hold. Introducing new notation

$$Y_i = \psi_i^*(-d_i), \qquad d_i = q_i^+ - q_i^-, \tag{2.3.23}$$

we rewrite inequality (2.3.22) in the form

$$|\dot{q}_i| \le Y_i. \tag{2.3.24}$$

Using expression (2.3.13) for ψ_i^*, we obtain

$$Y_i = (2X_i d_i)^{1/2}. \tag{2.3.25}$$

A possible phase trajectory of subsystem (2.3.9) is shown in Fig. 2.9. The direction of motion is indicated by the arrows.

It has been shown in Sect. 2.2.1 that the time for the motion of the ith subsystem (2.3.9) is maximal in the case of the "worst" perturbation $V_i = -\rho_i U_i$ and, on the strength of notation (2.3.14), is equal to [see (2.2.13)]

$$\tau_i^*(q_i^0, \dot{q}_i^0) = X_i^{-1} \left\{ 2 \left[\frac{1}{2} (\dot{q}_i^0)^2 - X_i (q_i^0 - q_i^*) \gamma_i \right]^{1/2} - \dot{q}_i^0 \gamma_i \right\},$$

$$\gamma_i = -\,\mathrm{sign}[\dot{q}_i^0 - \psi_i^*(q_i^0)], \quad \dot{q}_i^0 \neq \psi_i^*; \qquad \gamma_i = \pm 1, \quad \dot{q}_i^0 = \psi_i^*. \tag{2.3.26}$$

Since time τ required to bring the original system (2.1.1) to the terminal state (2.1.8) is defined as the greatest of the control times for each of subsystems (2.3.9), we obtain the estimate

$$\tau \leq \tau^* = \max_i(\tau_i^*), \qquad i = 1, \ldots, n. \tag{2.3.27}$$

2.3.3 Determination of the parameters

Control (2.3.13) can only be used when inequalities (2.3.15) [or (2.3.12)] are satisfied throughout the whole control process. We shall now find those control parameters X_i for which the above-mentioned relations are, in fact, satisfied.

We will first estimate the moduli of the quantities V_i from (2.3.10). Using relations (2.3.3), (2.3.11), and (2.3.24), we obtain

$$|V_i| \leq G_i^0 + \tilde{S}^0(Y) + \mu |A^{-1}(U+S)|, \qquad Y = (Y_1, \ldots, Y_n). \tag{2.3.28}$$

From (2.3.2) and (2.3.3), we have for any n-dimensional vector z

$$Az = Jz + \tilde{A}z, \quad |Az| \geq J_{\min}|z| - \mu|z| = (J_{\min} - \mu)|z|. \tag{2.3.29}$$

Here, $J_{\min}$ is the least of the quantities J_i. Let us set $z = A^{-1}z'$ in inequality (2.3.29). Then for $\mu < J_{\min}$ we obtain

$$|A^{-1}z'| \leq \frac{|z'|}{J_{\min} - \mu}. \tag{2.3.30}$$

Using (2.1.6), (2.3.11), and (2.3.24), we obtain the relations

$$|(U+S)_i| \leq |U_i| + |S_i| \leq U_i^0 + G_i^0 + \tilde{S}^0(Y) = (U^0 + G^0)_i + \tilde{S}^0(Y),$$

$$U^0 = (U_1^0, \ldots, U_n^0), \qquad G^0 = (G_1^0, \ldots, G_n^0). \tag{2.3.31}$$

Combining (2.3.28), (2.3.30), and (2.3.31), we find the final estimate for the disturbances:

$$
\begin{aligned}
|V_i| &\le G_i^0 + \tilde{S}^0(Y) + \frac{\mu}{J_{\min} - \mu}|U + S| \\
&\le G_i^0 + \left(1 + \frac{\mu n^{1/2}}{J_{\min} - \mu}\right)\tilde{S}^0(Y) + \frac{\mu}{J_{\min} - \mu}|U^0 + G^0|.
\end{aligned}
\tag{2.3.32}
$$

In inequalities (2.3.15), instead of the quantities $|V_i|$, we substitute their estimates from (2.3.32). After transformations, we obtain

$$
J_i X_i + \left(1 + \frac{\mu n^{1/2}}{J_{\min} - \mu}\right)\tilde{S}^0(Y) \le U_i^0 - G_i^0 - \frac{\mu}{J_{\min} - \mu}|U^0 + G^0|. \tag{2.3.33}
$$

System of inequalities (2.3.33) determines the permissible parameters X_i. It is a nonlinear system because the values Y_i are connected with X_i by (2.3.25).

If parameter μ is sufficiently small, such that the condition

$$
\mu < \frac{\min_i(U_i^0 - G_i^0)J_{\min}}{\min_i(U_i^0 - G_i^0) + |U^0 + G^0|} \tag{2.3.34}
$$

is satisfied, the expressions on the right-hand sides of inequalities (2.3.33) are positive. Since $\tilde{S}^0(Y) \to 0$ as $X_i \to 0$ due to (2.3.11), positive values of X_i can always be found, for which inequalities (2.3.33) and, consequently, inequalities (2.3.12) are satisfied.

We will sum up the results obtained in the form of a theorem.

Theorem 2.3. *Suppose that condition (2.3.34) is satisfied. Then the feedback control $U_i(q_i, \dot{q}_i)$ that solves Problem 2.1 in domain (2.3.16) is specified by relations (2.3.13) in which parameters X_i must be chosen in such a way that inequalities (2.3.33) are satisfied. This control transfers system (2.1.1) from the initial state (2.1.7) to the specified terminal state (2.1.8), if, at the initial instant of time, constraints (2.3.17) are satisfied. In this case, motion $q(t)$ of the system lies in domain D from (2.3.1), and the time of the control process τ does not exceed τ^* determined by expressions (2.3.26) and (2.3.27).*

We will now describe a method of selecting the permissible values of X_i. We shall seek these values in the form

$$
X_i = Z^2 d_i, \tag{2.3.35}
$$

where d_i is defined in (2.3.23), and the magnitude of Z is still unknown. We substitute (2.3.35) into inequalities (2.3.33) taking into account (2.3.25) and (2.3.11). Selecting the maximum value of Z

$$
Z_0 = \max Z \tag{2.3.36}
$$

that satisfies the inequality obtained, we calculate the control parameters X_i, using formulas (2.3.35). In this case, at least one of inequalities (2.3.33) is transformed into the equality.

Suppose, for example, constraint (2.3.8) has the form

$$|F_i| \le F^0(|\dot{q}|) = a|\dot{q}| + b|\dot{q}|^2,$$

where a and b are positive constants. Then inequalities (2.3.33) can be reduced to the form

$$Z^2 + 2g_i Z \le h_i, \tag{2.3.37}$$

where g_i and h_i are positive coefficients that can be found from (2.3.25), (2.3.11), and (2.3.33). The solution of the system of inequalities (2.3.37) can be written in the form

$$Z \le Z_0 = \min_i [(g_i^2 + h_i)^{1/2} - g_i], \quad i = 1, \ldots, n. \tag{2.3.38}$$

Conditions (2.3.33) for determining the set of permissible values X_i together with constraint (2.3.34) imposed on parameters of system (2.1.1) are sufficient conditions that differ from the necessary ones. For specific systems, it is possible to obtain more precise estimates than (2.3.32). Substituting such estimates into inequality (2.3.15) instead of V_i, one can obtain a more broad set of permissible control parameters. As a result, increasing of X_i allows one to widen domains Ω_i from (2.3.16) that bound initial velocities for the subsystems [see (2.3.17)] and, thus, significantly shorten the time of motion τ. In some instances, this allows one to relax restrictions imposed upon parameters of the system.

The control method proposed is quite simple and does not require an exact knowledge of the nonlinear terms and perturbing forces in the equations of motion. The method is not overly sensitive to slight variations in the system parameters or to additional perturbations: to take such factors into consideration, one needs only to decrease parameters X_i, leaving a sufficient "safety margin" for the controls of the corresponding subsystems.

2.3.4 Case of zero initial velocities

While solving Problem 2.1, we assumed that the initial state of the system is an arbitrary point in the domain Ω_i, see (2.3.17). Consider a special but important case of the zero initial velocities $\dot{q}^0 = 0$.

Under control (2.3.13), the coordinates q_i of all subsystems are bounded by the inequalities $\min(q_i^0, q_i^*) \le q_i(t) \le \max(q_i^0, q_i^*)$. Therefore, it is possible to restrict the domain of the possible motions, setting

$$q_i^- = \min(q_i^0, q_i^*), \qquad q_i^+ = \max(q_i^0, q_i^*) \tag{2.3.39}$$

in (2.3.1) for all $i = 1, \ldots, n$. For such presetting of the domain D, the magnitudes $d_i = q_i^+ - q_i^-$ are minimal, and hence the estimates obtained in (2.3.24) for the generalized velocities and in (2.3.32) for the disturbances are of the maximum accuracy. We suppose that the boundaries q_i^- and q_i^+ of the domain of motion are given in the form (2.3.39). Estimates (2.3.26) and (2.3.27) in this case take the form

$$\tau \leq \tau^* = \max_i(\tau_i^*), \qquad \tau_i^* = 2\sqrt{\frac{d_i}{X_i}}, \qquad i = 1,\ldots,n. \tag{2.3.40}$$

In accordance with (2.3.35), (2.3.36), and (2.3.40), we have the equal estimates on times needed for steering each of subsystems (2.3.9) to the terminal state:

$$\tau^* = \tau_i^* = \tau_0^*, \qquad \tau_0^* = 2Z_0^{-1}, \tag{2.3.41}$$

where Z_0 is defined by (2.3.36). Let us demonstrate that for control (2.3.13) with any other permissible parameters X_i, satisfying (2.3.33), but not related to each other by equalities (2.3.35) and (2.3.36), the estimate for the motion time τ^*, calculated by using (2.3.40), will be greater than τ_0^*.

Really, in order to reduce τ^*, it is required, according to (2.3.40) and (2.3.41), to increase X_i for all $i = 1,\ldots,n$. Then, due to the strict monotony of the left-hand sides of inequalities (2.3.33) with respect to X_i, all the left-hand sides will grow up, and at least one of inequalities (2.3.33), which was transformed into the equality while choosing $Z = Z_0$ according to (2.3.36), will fail. Thus, the magnitude $\tau^* = \tau_0^*$, obtained in (2.3.41), is minimal for $\dot{q}^0 = 0$ and control (2.3.13).

Modified control law

In the case $\dot{q}^0 = 0$, it is possible to modify the control law (2.3.13) so that the new (corresponding to the modified control law) estimate of the motion time will be less than the estimate obtained in (2.3.41). To this end, let us re-define functions ψ_i^* in (2.3.13) so that the switching curve $\dot{q}_i = \psi_i^*(q_i)$ (see Fig. 2.10) will consist of the parabolic arc (for $|q_i - q_i^*| \leq d_i^*$) and straight-line section (for $d_i^* < |q_i - q_i^*| \leq d_i$). Here, d_i^* is some positive constant that will be determined. We impose the only constraint upon this constant

$$d_i^* \leq \frac{1}{2} d_i, \tag{2.3.42}$$

where d_i is defined by (2.3.23) and (2.3.39). Thus, we define functions ψ_i^* in the form:

$$\begin{aligned} \psi_i^*(q_i) &= -(2X_i|q_i - q_i^*|)^{1/2}\operatorname{sign}(q_i - q_i^*) \quad \text{for} \quad |q_i - q_i^*| \leq d_i^*, \\ \psi_i^*(q_i) &= Y_i \operatorname{sign}(q_i - q_i^*) \quad \text{for} \quad d_i^* < |q_i - q_i^*| \leq d_i, \end{aligned} \tag{2.3.43}$$

Here, X_i and Y_i are positive parameters for the new control law; they are not already related to each other by (2.3.25) but subjected to the additional contraints

$$Y_i \leq (X_i d_i)^{1/2}. \tag{2.3.44}$$

Parameters d_i^* are expressed through X_i and Y_i by the following formula

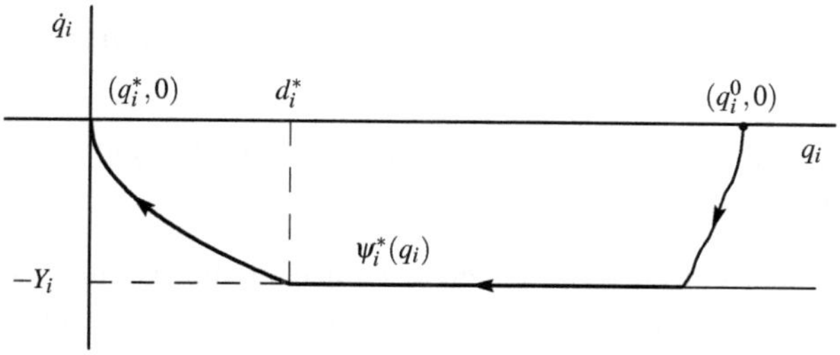

Fig. 2.10 Modified switching curve and the trajectory

$$d_i^* = \frac{Y_i^2}{2X_i}. \tag{2.3.45}$$

Due to (2.3.44) and (2.3.45), constraint (2.3.42) is satisfied.

If conditions (2.3.15) [or (2.3.12)] hold during the motion, then the control defined by (2.3.13) and (2.3.43) surely steers the system to the terminal state. Herein, the velocities of the subsystems are bounded and the inequalities $|\dot{q}_i| \leq Y_i$ hold. Hence, the estimate (2.3.32) for the maximal absolute values of the disturbances $|V_i|$ is also true. Substituting (2.3.32) into (2.3.15), we obtain inequalities that completely coincide with (2.3.33). Hence, if X_i and Y_i obey inequality (2.3.33), then the control defined by (2.3.13) and (2.3.43) solves Problem 2.1 for $\dot{q}^0 = 0$.

The following estimates of the motion time are true:

$$\tau \leq \tau^* = \max_i(\tau_i^*), \quad \tau_i^* = \frac{d_i}{Y_i} + \frac{Y_i}{X_i}, \qquad i = 1,\ldots,n, \tag{2.3.46}$$

where τ_i^* is the estimate of the motion time for the ith subsystem. Let us prove these relations.

Proof. The motion time for the ith subsystem is maximal under the worst disturbance $V_i = -\rho_i U_i$. The motion consists of three stages, see Fig. 2.10. First, according to (2.3.9), (2.3.13), (2.3.14), and (2.3.43), the motion has the constant acceleration

$$\ddot{q}_i = -X_i \operatorname{sign}(q_i^0 - q_i^*)$$

until the phase point reaches the switching curve $\dot{q}_i = \psi_i^*(q_i)$. Further, the phase point moves along the straight-line section of the switching curve with the constant velocity

$$\dot{q}_i = -Y_i \operatorname{sign}(q_i^0 - q_i^*),$$

and then, along the parabolic section of the switching curve, with the constant acceleration

$$\ddot{q}_i = X_i \operatorname{sign}(q_i^0 - q_i^*).$$

The durations of the first and the final stages are the same and equal to Y_i/X_i. The duration of the second stage is equal to $(d_i - 2d_i^*)/Y_i$. Summing up the durations of all three stages and taking into account (2.3.45), we come to the estimate (2.3.46) for τ. □

Let us demonstrate that the modified control law (2.3.43) allows one to reduce the estimate of the motion time (2.3.41).

Proof. To this end, consider the concrete choice of the parameters X_i and Y_i in (2.3.43), supposing that each Y_i depends on the corresponding X_i so that constraints (2.3.44) are trasformed into equalities

$$Y_i = (X_i d_i)^{1/2}. \tag{2.3.47}$$

Note, that (2.3.47) differs from (2.3.25) used earlier. From (2.3.45) and (2.3.47), we have

$$d_i^* = \frac{1}{2} d_i.$$

In this case, under the worst disturbances, the strait-line stage of motion is missing and relations (2.3.46) are transformed into (2.3.40). We will seek for the parameters X_i in the form of (2.3.35), as it has been done earlier, and express the parameters Y_i according to (2.3.47). Choose the maximal value of Z:

$$Z_0' = \max Z \tag{2.3.48}$$

satisfying (2.3.33). Due to the monotone dependence of the left-hand side (2.3.33) on parameters Y_i and to (2.3.47), value Z_0' is greater than Z_0 from (2.3.36). Therefore, we obtain a new upper estimate of the time of motion that is lower than the estimate given by (2.3.40) and (2.3.41):

$$\tau^* = \tau_i^* = \tau_0'^*, \qquad \tau_0'^* = 2(Z_0')^{-1}. \tag{2.3.49}$$

Thus, the modified control law really allows one to reduce the motion time estimate. □

Optimization of the control parameters

For the modified control law defined by (2.3.13) and (2.3.43), we suggest below a numerical procedure for finding the optimal permissible parameters X_i and Y_i satisfying inequalities (2.3.33) and (2.3.44) [but not connected by relations (2.3.47) and (2.3.35)], for which the motion time estimate τ^* given by (2.3.46) is minimal.

If parameters X_i and Y_i are optimal, then the quantities τ_i^* in (2.3.46) are equal, i.e.,

$$\tau^* = \tau_i^*, \qquad i = 1, \ldots, n, \tag{2.3.50}$$

and inequalities (2.3.33) are transformed into the exact equalities

$$X_i + K_i \tilde{S}^0(Y) = \Delta_i, \qquad i = 1, \ldots, n, \tag{2.3.51}$$

where

$$K_i = \left(1 + \frac{\mu n^{1/2}}{J_{\min} - \mu}\right) J_i^{-1},$$

$$\Delta_i = \left(U_i^0 - G_i^0 - \frac{\mu}{J_{\min} - \mu}|U^0 + G^0|\right) J_i^{-1}.$$

This fact is proved by the reasoning analogous to that have been used earlier and given after (2.3.41). We assume that inequality (2.3.34) holds, so that Δ_i are positive.

The procedure for finding the optimal parameters X_i and Y_i of the modified control law is as follows. In accordance with (2.3.46) and (2.3.50), we set

$$X_i = Y_i^2 (Y_i \tau^* - d_i)^{-1} \tag{2.3.52}$$

in system (2.3.33). We obtain

$$\frac{Y_i^2}{Y_i \tau^* - d_i} + K_i \tilde{S}^0(Y) \le \Delta_i. \tag{2.3.53}$$

Let us choose some initial value τ^* [for example, $\tau^* = \tau_0'^*$ from (2.3.49)] and find numerically some values of parameters Y_i satisfying (2.3.53). The set $[Y_i^-, Y_i^+]$, $i = 1, \dots, n$, in which it is possible to make this search, can be easily obtained by setting $\tilde{S}^0 = 0$ in inequalities (2.3.53). As a result, we obtain

$$Y_i^{\pm} = \frac{1}{2}\tau^* \Delta_i \pm \left[\left(\frac{1}{2}\tau^* \Delta_i\right)^2 - d_i \Delta_i\right]^{1/2}.$$

If any solution Y_i of inequalities (2.3.53) is found, then we decrease the value τ^* in (2.3.53) by some increment $\delta\tau^*$ and repeat the search of the permissible parameters Y_i corresponding to the new value of τ^*. The minimal value of τ^* for which inequalities (2.3.53) have the solution $Y_i > 0$ for all $i = 1, \dots, n$, defines together with (2.3.52) the optimal parameters X_i and Y_i.

Let us show also that the motion time estimate $\tau_0'^*$ [see (2.3.49)] obtained analytically earlier for the modified control law defined by (2.3.13) and (2.3.43) is not minimal and can be improved by using the suggested numerical procedure.

Proof. It is sufficient to show that, if $\tau_0'^*$ is chosen as the initial estimate τ^*, then, for the sufficiently low step size $\delta\tau^*$, the suggested algorithm surely finds, during the second iteration, the values of the parameters X_i and Y_i providing even less the motion time estimate

$$\tau^* = \tau_0'^* - \delta\tau^*.$$

Suppose the contrary. Let $\tau^* = \tau_0'^*$ from (2.3.49) is the minimal motion time estimate. Then, as it has been pointed above, (2.3.50) and (2.3.51) should be satisfied. Let us choose some value i ($1 \le i \le n$) and, by using (2.3.46), find the derivative $\partial\tau_i^*/\partial Y_i$, assuming that parameter X_i is connected with Y_i by the ith equality in (2.3.51):

$$\frac{\partial \tau_i^*}{\partial Y_i} = -\frac{d_i}{Y_i^2} + \frac{1}{X_i} - \frac{Y_i}{X_i^2}\frac{\partial X_i}{\partial Y_i}. \tag{2.3.54}$$

For the given i, due to the monotony of $\tilde{S}^0(Y)$ with respect to Y_i, see (2.3.11), we have from (2.3.51)

$$\frac{\partial X_i}{\partial Y_i} = -K_i \frac{\partial \tilde{S}^0(Y)}{\partial Y_i} < 0. \tag{2.3.55}$$

Values X_i and Y_i providing the estimate $\tau_i^* = \tau_0'^*$ are connected by (2.3.47). Therefore, the sign of the derivative (2.3.54) is positive

$$\frac{\partial \tau_i^*}{\partial Y_i} = -\frac{Y_i}{X_i^2}\frac{\partial X_i}{\partial Y_i} > 0. \tag{2.3.56}$$

Let us decrease the parameter Y_i for the given i by a sufficiently small value δY_i (all Y_j, $j \neq i$, are fixed) and, at the same time, increase parameters X_i, $i = 1, \ldots, n$, for all subsystems, without violating equations (2.3.51). Obviously, for such variations, the new values of the parameters will satisfy constraints (2.3.44). Herein, due to (2.3.46) and (2.3.56), the motion time estimates for all subsystems decrease. Thus, we come to the contradiction, and $\tau_0'^*$ cannot be the minimal estimate for τ^*. □

Note that control (2.3.13) with the modified switching curve (2.3.43) can be used also in the case where $\dot{q}^0 \neq 0$. The corresponding domain Ω_i that contains possible initial states for the ith subsystem is described by relations (2.3.16) and depicted in Fig. 2.11.

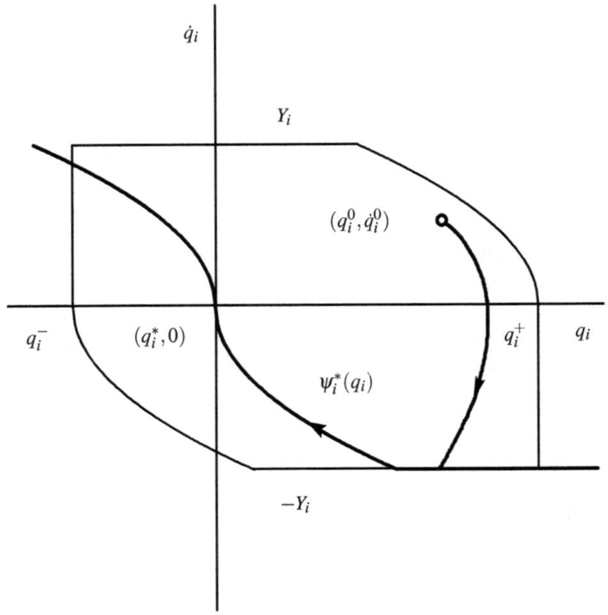

Fig. 2.11 Domain Ω_i for the modified control law

2.4 Nonlinear damping

The method of solving Problem 2.1 that we have presented in Sects. 2.1.3–2.2.4 consists of two stages: 1) the decomposition of the original nonlinear system (2.1.3) into subsystems (2.1.30) with one degree of freedom each [see (2.1.35)]; 2) the use of the game approach to construct a control for the subsystems.

Certain modifications of the suggested approach are possible at both stages. At the first stage, a system can be reduced to a collection of other subsystems, either more simple or more complicated than (2.1.35). System (2.1.35) with $\lambda = 0$ is evidently the simplest subsystem with one degree of freedom. In the case $\lambda = 0$, the imposed constraint (2.1.12) is replaced by a more simple condition (2.1.13) that can be immediately verified by means of Lemma 2.1 from Sect. 2.1.2.

At the second stage, it is not necessary to use the game approach in order to construct a control for the subsystems (see Sect. 2.2.2).

Let us consider in greater detail the modification of the described approach, in which a system with a nonlinear resistance is considered as a subsystem with one degree of freedom [28].

2.4.1 Subsystem with nonlinear damping

Let the dynamics of a system with one degree of freedom be described by the equation

$$m\ddot{q} = R(\dot{q}) + U + V(q,\dot{q},t). \tag{2.4.1}$$

Here, q is the generalized coordinate of the system, $m > 0$ is a constant inertial coefficient (the mass), $R(\dot{q})$ is the resistance, U is the control, and $V(q,\dot{q},t)$ is the disturbance.

We will assume that the resistance $R(\dot{q})$ is directed opposite to the velocity, and its magnitude increases with the velocity; it is zero in the state of rest. Also, $R(\dot{q})$ is a smooth function. Hence, we have

$$\dot{q}R(\dot{q}) < 0, \quad \frac{dR(\dot{q})}{d\dot{q}} < 0 \quad (\dot{q} \neq 0), \quad R(0) = 0. \tag{2.4.2}$$

The control and the disturbance are assumed to be bounded by geometric constraints, and the maximum disturbance is strictly less than the maximum control. We have

$$|U| \le U_0, \quad |V(q,\dot{q},t)| \le \rho U_0, \quad \rho < 1, \tag{2.4.3}$$

where $U_0 > 0$ and $\rho < 1$ are constants. In all other respects, the disturbance $V(q,\dot{q},t)$ may be an arbitrary function of its arguments.

It is required to construct a feedback control $U(q,\dot{q})$ that takes system (2.4.1) from an arbitrary initial state

$$q(t_0) = q^0, \quad \dot{q}(t_0) = \dot{q}^0 \tag{2.4.4}$$

to the prescribed terminal state with zero velocity

$$q(t_*) = q^*, \quad \dot{q}(t_*) = 0 \tag{2.4.5}$$

in finite time. Here, t_0, q^0, $\dot{q}^0$, and q^* are some given values, time t_* is not fixed.

Let $l > 0$ be a quantity of the same dimension as coordinate q. We introduce the dimensionless variables

$$\begin{gathered} x = \frac{q - q^*}{l}, \quad t' = \frac{t - t_0}{\tau_0}, \quad u = \frac{U}{U_0}, \quad f = -\frac{R}{U_0}, \\ v = \frac{V}{U_0}, \quad \tau_0 = \left(\frac{ml}{U_0}\right)^{1/2}. \end{gathered} \tag{2.4.6}$$

Making the change of variables (2.4.6) in (2.4.1), we obtain

$$\ddot{x} + f(\dot{x}) = u + v(x, \dot{x}, t). \tag{2.4.7}$$

Here and in what follows, dots denote derivatives with respect to the dimensionless time t'; in (2.4.7) and below t' is replaced by t. According to (2.4.2) and (2.4.6), the smooth function $f(z)$ has the following properties:

$$zf(z) > 0, \quad f'(z) > 0 \quad (z \neq 0), \quad f(0) = 0. \tag{2.4.8}$$

The variables u and v in (2.4.7) are constrained by [see (2.4.3) and (2.4.6)]

$$|u| \le 1, \quad |v| \le \rho, \quad \rho < 1. \tag{2.4.9}$$

After the change of variables (2.4.6), the initial conditions (2.4.4) and the final conditions (2.4.5) take the form

$$x(0) = \xi, \quad \dot{x}(0) = \eta, \tag{2.4.10}$$

$$x(\tau) = 0, \quad \dot{x}(\tau) = 0. \tag{2.4.11}$$

Here,

$$\xi = \frac{q^0 - q^*}{l}, \quad \eta = \frac{\dot{q}^0 \tau_0}{l}, \quad \tau = \frac{t_* - t_0}{\tau_0}.$$

Our control problem can now be stated in the following form.

Problem 2.3. Construct a feedback control $u(x, \dot{x})$ that satisfies constraint (2.4.9) and takes system (2.4.7) with an arbitrary disturbance v constrained by (2.4.9) from any initial state (2.4.10) to the prescribed terminal state (2.4.11) in finite time.

The formulation of the problem and the approach applied below for its solution are analogous to those described in Sects. 2.1.3–2.2.3 and are their generalization.

2.4.2 Control for the nonlinear subsystem

The game-theoretical approach

Let us consider (2.4.7) from the point of view of the differential games theory, assuming that u and v are the controls of two opponents constrained by (2.4.9). We will seek a feedback control $u(x,\dot{x})$ that takes system (2.4.7) from state (2.4.10) to state (2.4.11) in the shortest guaranteed time τ for any admissible disturbance v. Control $u(x,\dot{x})$ obtained by solving the differential game produces, as is easily seen, a solution of Problem 2.3. On the other hand, the solution of the differential game reduces [79, 80] to the solution of the time-optimal control problem for the system

$$\ddot{x}+f(\dot{x})=(1-\rho)u; \quad |u|\leq 1,\ 0\leq\rho<1,\ \tau\to\min \tag{2.4.12}$$

with the boundary conditions (2.4.10) and (2.4.11). Equation (2.4.12) is obtained from (2.4.7) for $v=-\rho u$ that corresponds to the worst (for u) opponent control: the optimal controls of the players are such that $u=\pm 1$, $v=\mp\rho$ at any instant.

Control $u(x,\dot{x})$ required in Problem 2.3 and the corresponding time τ can be found by obtaining the time-optimal control for (2.4.12) with boundary conditions (2.4.10) and (2.4.11). The corresponding time-optimal control problem is written in the form

$$\begin{gathered}\dot{x}_1=x_2, \quad \dot{x}_2=-f(x_2)+(1-\rho)u; \quad |u|\leq 1, \quad 0\leq\rho<1,\\ x_1(0)=\xi, \quad x_2(0)=\eta, \quad x_1(\tau)=x_2(\tau)=0, \quad \tau\to\min, \\ (x_1=x, \quad x_2=\dot{x}).\end{gathered} \tag{2.4.13}$$

Time-optimal control

We will solve problem (2.4.13) by the maximum principle. The Hamiltonian for problem (2.4.13) has the form

$$H=p_1x_2+p_2[(1-\rho)u-f(x_2)], \quad |u|\leq 1, \tag{2.4.14}$$

where p_1 and p_2 are conjugate variables.

The conjugate system has the form

$$\dot{p}_1=0, \quad \dot{p}_2=-p_1+f'(x_2)p_2. \tag{2.4.15}$$

Since (2.4.13) is an autonomous system, we have the first integral for our time-optimal control problem

$$H=p_1x_2+p_2[(1-\rho)u-f(x_2)]=h\geq 0, \tag{2.4.16}$$

where h is a constant.

By the maximum principle, we obtain from (2.4.14)

$$u = \operatorname{sign} p_2. \tag{2.4.17}$$

Let us consider the possibility of the existence of singular sections of the optimal trajectory on which $p_2 = 0$. On such a singular section, by the second equation in (2.4.15), we have also $p_1 = 0$. Therefore, if a singular section exists, we have $p_1 \equiv \text{const} = 0$ on the entire trajectory. But then the second equation in (2.4.15) is homogeneous with respect to p_2 on the entire trajectory, and since $p_2 = 0$ on the singular section, we have $p_2 \equiv 0$ on the entire trajectory. However, by the maximum principle, the conjugate vector does not vanish identically on the optimal trajectory. The contradiction proves that the optimal trajectory is free from singular sections. Thus, the equality $p_2 = 0$ may be observed only at isolated instants of time (switching points) and, by (2.4.17), we have $u = \pm 1$ almost everywhere.

Let us first consider the sections of the optimal trajectory where $p_2 > 0$ and $u = 1$. From (2.4.13), we obtain for these sections

$$\frac{dx_1}{dx_2} = x_2[(1-\rho)u - f(x_2)]^{-1}. \tag{2.4.18}$$

It follows from (2.4.18) that, in the (x_1, x_2)-plane, the sections of the optimal trajectory with $p_2 > 0$ are arcs of the curves

$$x_1 = \phi_\rho^+(x_2) + c^+, \tag{2.4.19}$$

where c^+ is an arbitrary constant and the function $\phi_\rho^+(x_2)$ is defined by the equality

$$\phi_\rho^+(y) = \int_0^y \frac{z\,dz}{1-\rho-f(z)}, \qquad 0 \le \rho < 1. \tag{2.4.20}$$

Let us note some properties of the function $\phi_\rho^+(y)$ that follow from (2.4.20) and (2.4.8) and are needed in the sequel. As y varies from $-\infty$ to 0, the function ϕ_ρ^+ is positive and strictly decreasing, vanishing for $y = 0$. The point $y = 0$ is the unique extremum of the function $\phi_\rho^+(y)$ (its minimum). If the transcendental equation for z^+

$$f(z^+) = 1 - \rho \tag{2.4.21}$$

is unsolvable, i.e., if $f(z) < 1 - \rho$ for all z, then the function $\phi_\rho^+(y)$ is strictly increasing for all $y \ge 0$. In this case $\phi_\rho^+(y) > 0$ for all $y \ne 0$.

If, however, z^+ is a root of (2.4.21), then this root is positive and unique by conditions (2.4.8). In this case, the function $\phi_\rho^+(y)$ is strictly increasing from 0 to ∞ in the interval $y \in (0, z^+)$ and strictly decreasing for $y > z^+$. A typical curve of the dependence (2.4.19) in the (x_1, x_2)-plane for $c^+ = 0$ is shown in Fig. 2.12 for the case where (2.4.21) has a root $z^+ > 0$. The direction in which time t increases along the trajectory according to the first equation in (2.4.13) is shown by arrows in Fig. 2.12.

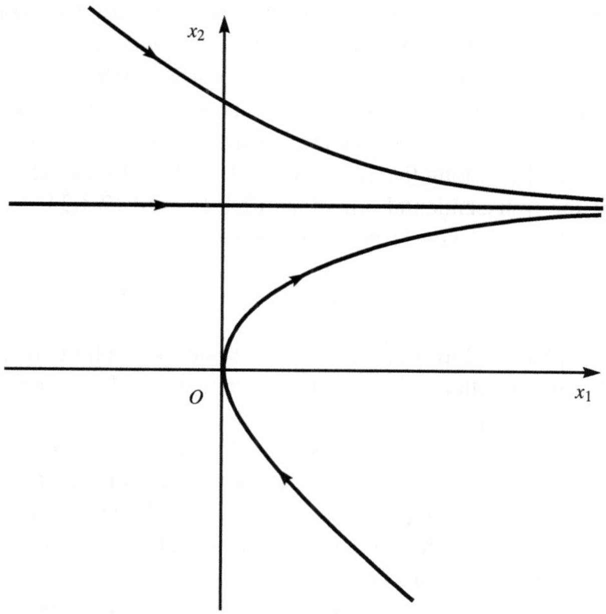

Fig. 2.12 Phase trajectory for $c^+ = 0$ and $z^+ > 0$

We similarly consider the sections of the trajectories with $p_2 < 0$. These sections are arcs of the curves

$$x_1 = \phi_\rho^-(x_2) + c^-. \tag{2.4.22}$$

Here, as in (2.4.19), c^- is an arbitrary constant and the function ϕ_ρ^- is defined by an equality similar to (2.4.20):

$$\phi_\rho^-(y) = \int_0^y \frac{zdz}{-(1-\rho) - f(z)}, \quad 0 \leq \rho < 1. \tag{2.4.23}$$

We introduce a transcendental equation for z^- similar to (2.4.21):

$$f(z^-) = -(1-\rho). \tag{2.4.24}$$

If (2.4.24) does not have a solution z^-, i.e., if $f(z) > \rho - 1$ for all z, then the function $\phi_\rho^-(y)$ from (2.4.23) is strictly increasing for $y < 0$ and strictly decreasing for $y > 0$. Here, $\phi_\rho^-(y) < 0$ for all $y \neq 0$.

If z^- is a root of (2.4.24), then it is unique and negative ($z^- < 0$) by conditions (2.4.8). In this case, the function $\phi_\rho^-(y)$ is strictly decreasing for $y \in (-\infty, z^-)$, strictly increasing for $y \in (z^-, 0)$, and again strictly decreasing for $y \in (0, \infty)$. As $y \to z^-$, this function tends to $-\infty$, and for $y = 0$ it has a local zero maximum. A typical graph of the function $\phi_\rho^-(y)$ can be obtained from the graph of the function $\phi_\rho^+(y)$ in Fig. 2.12 by a central symmetry transformation (or, equivalently, by simultaneously reversing the directions of both axes x_1 and x_2).

The curves described above are the trajectories corresponding to $p_2 > 0$ and $p_2 < 0$ that pass through the origin in the (x_1, x_2)-plane. Other curves whose arcs may be sections of the optimal trajectories are obtained from these curves by parallel translation by c^+ and c^- along the x_1 axis [see (2.4.19) and (2.4.22)].

Note that if the transcendental equations (2.4.21) and (2.4.24) have solutions, then system (2.4.13) has the corresponding solutions

$$x_2 = z^+ \ (p_2 > 0), \quad x_2 = z^- \ (p_2 < 0). \tag{2.4.25}$$

In the (x_1, x_2)-plane, solutions (2.4.25) correspond to the phase trajectories in the form of straight lines parallel to the x_1-axis. These lines are the asymptotes of curves (2.4.19) and (2.4.22), respectively (see Fig. 2.12).

Thus, the required optimal trajectories consist of sections of curves (2.4.19) and (2.4.22) with various c^+ and c^- and also, possibly, segments of the straight lines (2.4.25), if the corresponding equations (2.4.21) and (2.4.24) are solvable.

We will now show that each optimal trajectory has at most one control switching point, i.e., the function $p_2(t)$ vanishes at most once.

Suppose that this is not so, and the function $p_2(t)$ vanishes at two instants t' and t'', being positive between them. Then

$$p_2(t) > 0, \quad t \in (t', t''); \quad p_2(t') = p_2(t'') = 0. \tag{2.4.26}$$

From the first integral (2.4.16) for t' and t'', we obtain by (2.4.26)

$$p_1 x_2(t') = p_1 x_2(t'') = h \geq 0. \tag{2.4.27}$$

If $p_1 = \text{const} = 0$, then, from (2.4.15), we obtain for $p_2(t)$ a linear homogeneous equation, which with zero conditions (2.4.26) at t' and t'' has an identically zero solution $p_2(t) \equiv 0$. But this contradicts to the maximum principle that asserts the existence of a nonzero conjugate vector. Therefore, $p_1 = \text{const} \neq 0$, and, from (2.4.27), we obtain $x_2(t') = x_2(t'')$. However, on all phase trajectories except the straight lines (2.4.25) the variable x_2 is either strictly increasing or strictly decreasing as time t increases. This follows from the previous analysis of the phase trajectories and is clear from Fig. 2.12. The equality $x_2(t') = x_2(t'')$ is therefore possible only if the relevant section of the trajectory is a segment of the straight line (2.4.25), i.e.,

$$x_2(t) \equiv z^+, \quad t \in (t', t''). \tag{2.4.28}$$

Substituting (2.4.28) into the second conjugate equation (2.4.15), we obtain a linear equation with constant coefficients

$$\dot{p}_2(t) = -p_1 + kp_2, \quad k = f'(z^+) > 0,$$

where $k > 0$ by (2.4.8). The general solution of this equation has the form

$$p_2(t) = \frac{p_1}{k} + Ce^{kt}, \tag{2.4.29}$$

where C is an arbitrary constant. But solution (2.4.29) is monotone in t and cannot satisfy conditions (2.4.26) for any $p_1 \neq 0$ and C. Thus, the section of the optimal trajectory, where conditions (2.4.26) hold, cannot be a straight segment of the line (2.4.28). We have thus shown that an optimal trajectory may not include sections of the form (2.4.26).

We can similarly prove that an optimal trajectory may not include sections such that the function $p_2(t)$ is negative inside the section and vanishes at its endpoints.

Therefore, on each optimal trajectory the function $p_2(t)$ vanishes at most once, i.e., the control may have at most one switching point.

The only phase trajectories that reach the origin as the time increases are the branch of curve (2.4.19) with $c^+ = 0$ which lies in the quadrant $x_1 \geq 0$, $x_2 \leq 0$ (Fig. 2.12) and the branch of curve (2.4.22) with $c^- = 0$ which lies in the quadrant $x_1 \leq 0$, $x_2 \geq 0$. These curve branches correspond to the controls $u = 1$ and $u = -1$, respectively. The collection of these branches form the switching curve, whose equation is written as

$$x_1 = \psi_\rho(x_2). \tag{2.4.30}$$

Here, we have introduced the notation

$$\psi_\rho(y) = \phi_\rho^+(y),\ y \leq 0; \quad \psi_\rho(y) = \phi_\rho^-(y),\ y \geq 0. \tag{2.4.31}$$

By the properties of functions (2.4.20) and (2.4.23), the function $\psi_\rho(y)$ defined by (2.4.31) is strictly decreasing for all y and vanishes for $y = 0$, where it has a point of inflection.

We can now easily describe the entire field of optimal trajectories. An optimal trajectory originating from any point of the phase plane (x_1, x_2) consists of a segment of one of the families (2.4.19) or (2.4.22) and a section of switching curve (2.4.30).

The field of optimal trajectories is qualitatively shown in Fig. 2.13 for the case where (2.4.21) and (2.4.24) have roots. The thick curve in Fig. 2.13 is the switching curve (2.4.30), and the arrows indicate the direction of increase of the time t. Note that this picture of the field of optimal trajectories is similar to the picture observed for the linear resistance, see Sect. 2.2.1, Fig. 2.2.

The optimal control corresponding to this field of phase trajectories may be represented in the form

$$\begin{gathered}
u_\rho(x_1, x_2) = \operatorname{sign}[\psi_\rho(x_2) - x_1] \quad \text{for} \quad x_1 \neq \psi_\rho(x_2), \\
u_\rho(x_1, x_2) = \operatorname{sign} x_1 = -\operatorname{sign} x_2 \quad \text{for} \quad x_1 = \psi_\rho(x_2), \\
(x_1 = x, \quad x_2 = \dot{x}),
\end{gathered} \tag{2.4.32}$$

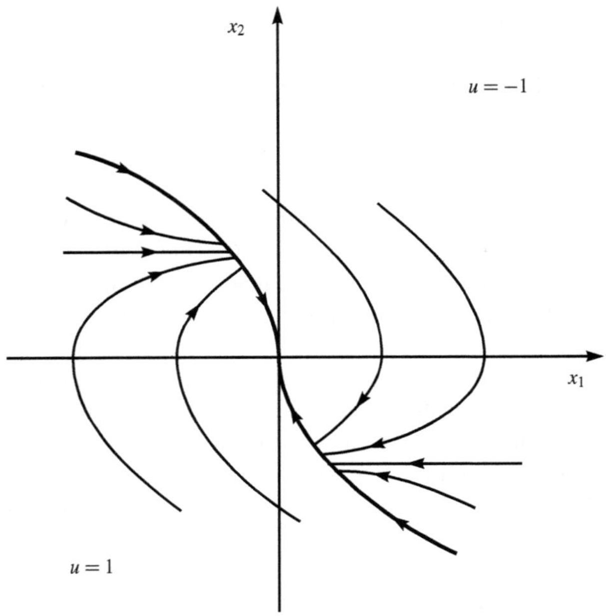

Fig. 2.13 Optimal phase trajectories

where the function ψ_ρ is defined by relationships (2.4.31), (2.4.20), and (2.4.23).

The control law (2.4.32) solves Problem 2.3. This solution may be called suboptimal, because it is time-optimal (unimprovable) when v is the "worst-case" disturbance, as assumed in the game-theoretical approach. With the worst-case disturbance $v = -\rho u$, the system moves along optimal trajectories, see Fig. 2.13. If the disturbance deviates from the worst case ($v \neq -\rho u$), which is usually so, the trajectories deviate from the optimal trajectories. The motion along the switching curve occurs in the sliding mode, and the time taken to reach the origin only diminishes.

2.4.3 Simplified control for the subsystem and comparative analysis

So far, we have assumed that the disturbance is unknown, but its maximum attainable value is known and essentially affects the feedback control. In dimensionless variables, the disturbance bound has the form $|v| \leq \rho$, see (2.4.9), and the feedback control (2.4.32) depends on the parameter ρ.

We can use a different approach to the control synthesis in the presence of disturbances, which simply ignores the disturbances (see Sect. 2.2.2).

In our case, this simplified approach means that the parameter ρ is set equal to zero during the control synthesis, and the disturbances are ignored. The control $u_0(x_1,x_2)$ obtained in this way is defined by relationships (2.4.32), (2.4.31), (2.4.20), and (2.4.23) with $\rho = 0$. The switching curve for the simplified control is given by (2.4.30) with $\rho = 0$. It is represented in Fig. 2.14 by the thick curve.

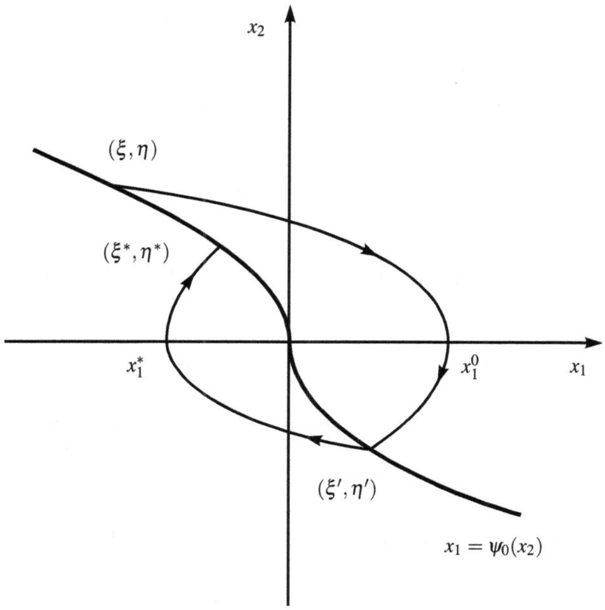

Fig. 2.14 Switching curve with $\rho = 0$ and a trajectory for simplified control

Let us compare the two control synthesis techniques — the game-theoretical and the simplified method. To this end, we will examine the dynamics of system (2.4.1) for some $\rho \in (0,1)$ under the action of the simplified control $u_0(x_1,x_2)$. We will represent this system in the form

$$\dot{x}_1 = x_2, \quad \dot{x}_2 = -f(x_2) + u_0(x_1,x_2) + v, \tag{2.4.33}$$

$$|v| \le \rho < 1, \quad (x_1 = x,\ x_2 = \dot{x}).$$

For system (2.4.33), we consider the following auxiliary problem of finding the worst-case disturbance (see analogous Problem 2.2 in Sect. 2.2.2).

Problem 2.4. Find the optimal control $v(x_1,x_2)$ for system (2.4.33) that satisfies the constraint $|v| \le \rho$ and such that the first intersection of any phase trajectory of this system with the switching curve $x_1 = \psi_0(x_2)$ lies as far as possible from the origin, i.e., at the maximum possible $|x_1|$ or, equivalently, the maximum possible $|x_2|$.

First assume that the starting point is in the region $x_1 \ge \psi_0(x_2)$. Then, by (2.4.32), we have $u_0 = -1$ for the given trajectory. The phase trajectory of system (2.4.33) first crosses that branch of the curve $x_1 = \psi_0(x_2)$, where $x_1 > 0, x_2 < 0$ [see Fig. 2.14, where it is assumed that the initial point (ξ,η) lies on the curve $x_1 = \psi_0(x_2)$]. Problem 2.4 is described by the relationships

$$\begin{gathered}\dot{x}_1 = x_2, \quad \dot{x}_2 = -f(x_2) - 1 + v, \quad |v| \leq \rho < 1, \\ x_1(0) = \xi, \quad x_2(0) = \eta, \quad \xi \geq \psi_0(\eta), \\ x_1(\tau) = \phi_0(x_2(\tau)), \quad x_1(\tau) > 0, \quad x_2(\tau) < 0, \quad x_1(\tau) \to \max.\end{gathered} \tag{2.4.34}$$

The instant τ when the process terminates is not fixed. Maximization of $x_1(\tau)$ is equivalent by (2.4.34) to the minimization of the integral functional

$$\int_0^\tau (-x_2)dt \to \min. \tag{2.4.35}$$

Applying the maximum principle to our problem defined by (2.4.34) and (2.4.35), we form the Hamiltonian

$$H = p_1 x_2 + p_2[v - 1 - f(x_2)] + x_2, \tag{2.4.36}$$

where p_1 and p_2 are the conjugate variables. They satisfy the conjugate system

$$\dot{p}_1 = 0, \quad \dot{p}_2 = f'(x_2)p_2 - p_1 - 1 \tag{2.4.37}$$

and the transversality conditions corresponding to the boundary conditions (2.4.34):

$$p_1\phi_0'(x_2) + p_2 = 0, \quad H = 0, \quad t = \tau. \tag{2.4.38}$$

From the first condition in (2.4.38), applying relationships (2.4.31) and (2.4.20) for $\rho = 0$ and noting that $x_2(\tau) < 0$ by (2.4.34), we obtain

$$p_1 = -p_2 \frac{1 - f(x_2)}{x_2}, \quad t = \tau. \tag{2.4.39}$$

Substituting (2.4.39) into (2.4.36) and using the second transversality condition in (2.4.38), we obtain after simplifications

$$H = p_2(v - 2) + x_2 = 0, \quad t = \tau.$$

Since $x_2(\tau) < 0$ and $|v| \leq \rho < 1$, we obtain from this equality

$$p_2(\tau) < 0. \tag{2.4.40}$$

We find the optimal control by maximizing H from (2.4.36) over $|v| \leq \rho$:

$$v = \rho \operatorname{sign} p_2. \tag{2.4.41}$$

Singular sections of the trajectory are ruled out. Indeed, if $p_2 \equiv 0$ in some time interval, then in this interval $p_1 = -1$ by the second equation (2.4.37). But $p_1 \equiv$ const, and therefore $p_1 = -1$ on the entire trajectory. Then the second equation in (2.4.37) becomes linear and homogeneous for p_2, and its solution with initial condition (2.4.40) does not vanish.

Thus, there are no singular sections, and (2.4.41) implies that control $v(t)$ has switching points when $p_2(t) = 0$.

Let us find the switching curve in the (x_1, x_2)-plane. Since system (2.4.34) is autonomous, its Hamiltonian (2.4.36) preserves a constant value along the optimal trajectory, and by (2.4.38) this constant value is zero:

$$H = (p_1 + 1)x_2 + p_2[v - 1 - f(x_2)] \equiv 0.$$

Hence, it follows that at the switching point, i.e., for $p_2 = 0$, we have either $p_1 = -1$ or $x_2 = 0$. But the equality $p_1 = -1$, as we have shown, implies that p_2 never vanishes. We thus have $x_2 = 0$ at the switching point, and the switching curve in this case is the ray $x_2 = 0$, $x_1 > 0$.

In order to determine the sign of the control for $x_2 < 0$ and $x_2 > 0$, it suffices to determine its sign at a single point. At the terminal time τ we have $x_2(\tau) < 0$ by (2.4.34) and $p_2(\tau) < 0$ by (2.4.40). Thus, $v = -\rho$ for $x_2 < 0$.

As a result,

$$v(x_1, x_2) = \rho \,\mathrm{sign}\, x_2. \tag{2.4.42}$$

We have obtained the optimal feedback control in the region $x_1 > \psi_0(x_2)$. To obtain the control in the region $x_1 < \psi_0(x_2)$, we note some symmetry properties. When $f(z)$ is replaced by $g(z) = -f(-z)$, we have by (2.4.20) and (2.4.23)

$$\phi_\rho^+(y) \to -\phi_\rho^-(-y), \quad \phi_\rho^-(y) \to -\phi_\rho^+(-y). \tag{2.4.43}$$

From (2.4.31) and (2.4.43), it follows that after this change

$$\psi_\rho(y) \to -\psi_\rho(-y). \tag{2.4.44}$$

Let us now make in (2.4.33) the change of variables

$$x_1 \to -x_1, \quad x_2 \to -x_2, \quad v \to -v, \quad f(z) \to -f(-z). \tag{2.4.45}$$

By (2.4.44) and (2.4.32), u_0 is changed to $-u_0$, and system (2.4.33) remains invariant. Hence, it follows that in the region $x_1 < \psi_0(x_2)$ the field of optimal trajectories and the optimal control are the same as in the region $x_1 > \psi_0(x_2)$, but with $f(z)$ replaced by $g(z) = -f(-z)$. Since synthesis (2.4.42) is independent of the specific form of function $f(z)$, it also applies in the region $x_1 < \psi_0(x_2)$. Thus, (2.4.42) defines the solution of Problem 2.2 in the entire (x_1, x_2)-plane.

Analysis of the phase trajectories

Consider the motion of system (2.4.33) under the action of the simplified control $u_0(x_1, x_2)$ defined by relationships (2.4.32), (2.4.31), (2.4.20), and (2.4.23) for $\rho = 0$ and the worst-case disturbance v from (2.4.42). Assume that the initial point ξ, η

lies on the branch of the switching curve $x_1 = \psi_0(x_2)$, where $x_1 < 0$ and $x_2 > 0$ (see Fig. 2.14). Let us investigate the phase trajectory until its next intersection with the same branch of the switching curve. This piece of the trajectory consists of four sections, each with constant u_0 and v. These sections have the following endpoints and controls (see Fig. 2.14):

$$\begin{aligned}
&1)\ (\xi,\eta) \to (x_1^0,0),\ \ u_0 = -1,\ v = \rho;\\
&2)\ (x_1^0,0) \to (\xi',\eta'),\ \ u_0 = -1,\ v = -\rho;\\
&3)\ (\xi',\eta') \to (x_1^*,0),\ \ u_0 = 1,\quad v = -\rho;\\
&4)\ (x_1^*,0) \to (\xi^*,\eta^*),\ u_0 = 1,\quad v = \rho.
\end{aligned} \tag{2.4.46}$$

The parameters of endpoints (2.4.46) satisfy relationships that reflect their position on the switching curve and on the coordinate axes (see Fig. 2.14):

$$\begin{gathered}
\xi = \psi_0(\eta), \quad \eta > 0, \quad \xi < 0; \quad x_1^0 > 0;\\
\xi' = \psi_0(\eta'), \quad \eta' < 0, \quad \xi' > 0; \quad x_1^* < 0;\\
\xi^* = \psi_0(\eta^*), \quad \eta^* > 0, \quad \xi^* < 0.
\end{gathered} \tag{2.4.47}$$

Substituting u_0 and v from (2.4.46) into (2.4.33) and integrating along the corresponding sections of the trajectory, we have

$$\xi' - \xi = \int_\eta^0 \frac{zdz}{-1+\rho-f(z)} + \int_0^{\eta'} \frac{zdz}{-1-\rho-f(z)},$$

$$\xi^* - \xi' = \int_{\eta'}^0 \frac{zdz}{1-\rho-f(z)} + \int_0^{\eta^*} \frac{zdz}{1+\rho-f(z)}.$$

Replacing ξ, ξ', and ξ^* by their expressions from (2.4.47) and using formulas (2.4.31), (2.4.20), and (2.4.23) for $\rho = 0$, we obtain

$$\begin{aligned}
\int_0^{\eta'} \frac{zdz}{1-f(z)} - \int_0^{\eta} \frac{zdz}{-1-f(z)} &= \int_0^{\eta} \frac{zdz}{1-\rho+f(z)} - \int_0^{\eta'} \frac{zdz}{1+\rho+f(z)},\\
\int_0^{\eta^*} \frac{zdz}{-1-f(z)} - \int_0^{\eta'} \frac{zdz}{1-f(z)} &= \int_0^{\eta'} \frac{zdz}{-1+\rho+f(z)} + \int_0^{\eta^*} \frac{zdz}{1+\rho-f(z)}.
\end{aligned} \tag{2.4.48}$$

Recall that $\eta' < 0$, $\eta > 0$, and $\eta^* > 0$ by (2.4.47). Set $\eta' = -\eta^0$, $\eta^0 > 0$ and transform relationships (2.4.48) so that they contain integrals only over intervals lying on the positive half-axis. Simplifying, we obtain

$$\Phi_4(\eta^0) = \kappa^2(\rho)\Phi_1(\eta), \quad \Phi_2(\eta^*) = \kappa^2(\rho)\Phi_3(\eta^0). \tag{2.4.49}$$

Here,

$$\Phi_1(y) = \Phi^+(y;f), \quad \Phi_2(y) = \Phi^-(y;f),$$

$$\Phi_3(y) = \Phi^+(y;g), \quad \Phi_4(y) = \Phi^-(y;g),$$

$$\Phi^{\pm}(y;h) = \int_0^y \frac{zdz}{(1+h)[1 \pm (1 \mp \rho)^{-1} h]}, \tag{2.4.50}$$

$$f = f(z) \geq 0, \quad g = -f(-z) \geq 0,$$

$$\kappa(\rho) = \left[\frac{\rho(1+\rho)}{(1-\rho)(2+\rho)} \right]^{1/2}.$$

Consider the transcendental equations (2.4.49) that determine η^0 and η^* for given $\eta > 0$ and $\rho \in (0,1)$. To this end, we will note some properties of functions Φ_i, $i = 1,2,3,4$, from (2.4.50). Recall that by (2.4.8) $f(z) > 0$ for $z > 0$ and $f(z) \to 0$ as $z \to 0$.

The denominators in the integrands for functions Φ_1 and Φ_3 in (2.4.50) are positive for all $z \geq 0$. Therefore, functions Φ_1 and Φ_3 are defined and bounded for all $y \geq 0$.

If the equations

$$f(z_2) = 1+\rho, \quad g(z_4) = -f(-z_4) = 1+\rho \tag{2.4.51}$$

have solutions for z_2 and z_4, then the denominators of the integrands of the corresponding functions Φ_2 and Φ_4 in (2.4.50) vanish for finite z_2 and z_4 equal to the roots of equations (2.4.51). In this case, Φ_2 and Φ_4 are monotone increasing and go to infinity at $y = z_2$ and $y = z_4$, respectively. If (2.4.51) have no solutions, then the functions Φ_2 and Φ_4 are defined for all $y > 0$. In both cases, the denominators of the integrands for the functions Φ_2 and Φ_4 have maxima over $f \geq 0$ and $g \geq 0$, which are both equal to $(2+\rho)^2(1+\rho)^{-1}/4$. We thus have the inequalities

$$\Phi_2(y) \geq \frac{1}{2}\vartheta y^2, \quad \Phi_4 \geq \frac{1}{2}\vartheta y^2, \quad \vartheta = 4(1+\rho)(2+\rho)^{-2}.$$

The functions Φ_2 and Φ_4 are thus always positive and strictly increasing, taking all values from 0 to ∞ for $y \geq 0$.

Hence, it follows that the transcendental equations (2.4.49) for any $\eta > 0$ and $\rho \in (0,1)$ have unique positive solutions $\eta^0 > 0$ and $\eta^* > 0$. These solutions are continuous and monotone functions of η.

Let us differentiate equalities (2.4.49) with respect to η. After simple reductions we obtain

$$\frac{d\eta^*}{d\eta} = \frac{\kappa^2(\rho)\Phi_3'(\eta^0)}{\Phi_2'(\eta^*)}\frac{d\eta^0}{d\eta} = \frac{\kappa^4(\rho)\Phi_3'(\eta^0)\Phi_1'(\eta)}{\Phi_2'(\eta^*)\Phi_4'(\eta^0)}. \tag{2.4.52}$$

From relationships (2.4.50) and properties (2.4.8) of the function $f(z)$, we obtain the inequalities

$$\frac{\Phi_1'(y)}{\Phi_2'(y)} < 1, \quad \frac{\Phi_3'(y)}{\Phi_4'(y)} < 1, \quad y > 0.$$

Using the second inequality, we obtain from (2.4.52)

$$\frac{d\eta^*}{d\eta} < \kappa^4(\rho)\frac{\Phi_1'(\eta)}{\Phi_2'(\eta^*)}, \quad \eta > 0. \tag{2.4.53}$$

We can verify that the function $\kappa^2(\rho)$ from (2.4.50) is strictly increasing from 0 to ∞ on $\rho \in [0,1]$, and $\kappa = 1$ for ρ equal to (see Sect. 2.2.3)

$$\rho^* = \frac{1}{2}(\sqrt{5}-1) \approx 0.618. \tag{2.4.54}$$

First assume that $\rho < \rho^*$ and therefore $\kappa^2(\rho) < \alpha$, where $\alpha < 1$ is a positive number. Then, from (2.4.53), we have

$$\frac{d\eta^*}{d\eta} < \alpha^2\frac{\Phi_1'(\eta)}{\Phi_2'(\eta^*)}, \quad \eta > 0, \tag{2.4.55}$$

and hence

$$\Phi_2(\eta^*) < \alpha^2\Phi_1(\eta), \quad \eta > 0. \tag{2.4.56}$$

We will show that $\eta^* < \eta$. Assume that this is not so, specifically $\eta^* \geq \eta$. From (2.4.50), we obtain $\Phi_2(y) > \Phi_1(y)$ for all $y > 0$. Then, by the monotonicity of the function $\Phi_2(y)$, we obtain the chain of inequalities

$$\Phi_2(\eta^*) \geq \Phi_2(\eta) > \Phi_1(\eta),$$

which leads to the contradiction with inequality (2.4.56). Thus, $\eta^* < \eta$.

Let us transform inequality (2.4.55), substituting the expressions for the derivatives Φ_1' and Φ_2' from (2.4.50) and using the positivity of the function $f(z)$:

$$\frac{d\eta^*}{d\eta} < \frac{\alpha^2\eta[1+f(\eta^*)][1-(1+\rho)^{-1}f(\eta^*)]}{\eta^*[1+f(\eta)][1+(1-\rho)^{-1}f(\eta)]} < \frac{\alpha^2\eta[1+f(\eta^*)]}{\eta^*[1+f(\eta)]}, \quad \eta > 0.$$

We can simplify the last inequality, noting that $f(\eta^*) < f(\eta)$ by the monotonicity of $f(z)$ and by the inequality $\eta^* < \eta$. We obtain

$$\frac{d\eta^*}{d\eta} < \frac{\alpha^2\eta}{\eta^*}, \quad \eta > 0.$$

Integrating this inequality with $\eta^* = 0$ when $\eta = 0$, we obtain $(\eta^*)^2 < \alpha^2\eta^2$ or $\eta^*/\eta < \alpha$.

Thus, if $\rho < \rho^*$, where ρ^* is defined in (2.4.54), then $\eta^*/\eta < \alpha$, i.e., the phase trajectory approaches the origin. The distance from the origin diminishes at a rate not slower than geometric progression. The system, therefore, reaches the prescribed state in finite time, although after infinitely many control switchings.

Suppose that the system has reached a small neighbourhood of the origin, so that η is sufficiently small. Here, η^0 and η^* are also small in view of their continuous dependence on η. Since $f(z) \to 0$ as $z \to 0$ by (2.4.8), terms $f(z)$ and $g(z)$ can be omitted in integrals (2.4.50) for small y, and we have

$$\Phi_i(y) \sim \frac{1}{2}y^2, \quad y \to 0, \quad i = 1,2,3,4.$$

The transcendental equation (2.4.49) for small η thus takes the form

$$(\eta^0)^2 = \kappa^2(\rho)\eta^2, \quad (\eta^*)^2 = \kappa^2(\rho)(\eta^0)^2.$$

Hence, we obtain

$$\frac{\eta^*}{\eta} = \kappa^2(\rho). \tag{2.4.57}$$

Let $\rho > \rho^*$ and, therefore, $\kappa^2(\rho) > 1$. Then, by (2.4.57), we obtain $\eta^* > \eta$, and the phase trajectory, even if it has reached a small neighbourhood of the origin, eventually moves away from the origin. The system does not come to the prescribed state.

Thus, with an arbitrary function $f(z)$ that satisfies condition (2.4.8), the simplified approach produces control $u_0(x_1, x_2)$ that is defined by relationships (2.4.32) for $\rho = 0$ and has the following properties:

- If $\rho < \rho^* \approx 0.618$, then, for any admissible disturbance $|v| \leq \rho$, the system reaches the origin. The time to reach the origin is finite, although the number of switchings in general is infinite.
- If $\rho > \rho^*$, there exists an admissible disturbance v defined by (2.4.42) for which the system never reaches the origin.

Therefore, simplified control guarantees a solution of Problem 2.3 only for $\rho < \rho^*$, i.e., when the ratio of the maximum allowed disturbance to the maximum allowed control does not exceed the golden section.

Specifying the form of the function $f(z)$, we can construct a more detailed picture of the phase motion. Note that the results presented here and obtained first in [28] generalize the results of [27] and [29], where the cases of zero and linear resistance, respectively, have been previously considered in detail, see Sects. 2.2.1–2.2.3.

Conclusions

The proposed control law (2.4.32) based on the game-theoretical approach takes the given system (2.4.7) to the origin in finite time for any non-linearity $f(z)$ and any uncertain disturbance, if $\rho < 1$. This control law does not require a knowledge of the disturbance; we only need to know the maximum allowed disturbance, which must not exceed the maximum control.

Let us stress the difference in the requirements imposed on functions $f(z)$ and $v(x, \dot{x}, t)$. Both these functions may be arbitrary in the framework of the corresponding conditions: (2.4.8) for $f(z)$ and (2.4.9) for v. However, the non-linear resistance function $f(z)$ should be known in order to construct the control, while the disturbance $v(x,\dot{x},t)$ is not needed.

The simplified approach to the control synthesis, which totally ignores the disturbances, is less effective. It a priori takes the system to the origin only for $\rho < \rho^* \approx 0.618$. If $\rho > \rho^*$, then there exists a disturbance for which the system never reaches the origin.

Yet both approaches have a similar structure and differ only by their switching curves.

The proposed control technique is robust with respect to various disturbances and parameter variations. These factors can be easily incorporated in the analysis, if we increase the assumed level of allowed disturbances, i.e., parameter ρ, creating a certain safety margin by this parameter.

Note that the obtained feedback control is suboptimal in the sense that it is time-optimal for the worst-case disturbance.

Our results can be applied to various dynamic systems, e.g., to control the electric motors of robotic systems. This opens up the possibility of taking into account various resistance laws that are often encountered in practice.

2.5 Applications and numerical examples

2.5.1 Application to robotics

Let us consider applications of the results obtained to problems of robot dynamics. For this, we will see that the formulation of Problem 2.1 as well as conditions (2.1.12) and (2.1.13) are typical and are often satisfied for robots.

Let us consider a manipulation robot that has n degrees of freedom and consists of n links connected by cylindrical or prismatic joints. Each link of the robot is an absolutely rigid body. The position of the ith link relative to that of the $(i-1)$st one is characterized by the relative angle of rotation (in the case of a cylindrical joint) or by a relative displacement (in the case of a prismatic joint). We take these angles and displacements as generalized coordinates $(q_1,\dots,q_n)$ determining the position of the robot. The equations of motion of the robot can be represented in the form of Lagrange's equations (2.1.1), where the kinetic energy has the form (2.1.2). The role of the generalized forces is played by the torques about the axes of cylindrical joints and by the forces in the directions of displacements in the case of prismatic joints. Here, the forces U_i in (2.1.1) are the control torques or forces caused by the motors (drives), and Q_i include all the other external and internal forces and torques: gravity, resistance, friction, various perturbations, etc.

Let us now look at the dynamics of the robot together with its drives. We suppose that each control torque or force U_i is produced by a separate direct-current electric motor, $i = 1,\dots,n$, and forces Q_i can be represented in the form (2.3.5)–

(2.3.8). The kinetic energy of the robot T is made up of the kinetic energy of its links $T^1(q,\dot{q})$ and the kinetic energy of the rotors of the electric motors $T^2(q,\dot{q},N)$. Here, $N=(N_1,\dots,N_n)$ are the gear ratios of the reduction gears, which are treated as parameters. We shall assume that $N_i \geq 1$ and neglect the inertia of the moving parts of the reduction gears. According to König's theorem, the kinetic energy of the ith rotor is equal to the sum of two terms: the kinetic energy of a point mass equal to the mass of the rotor and located at its center, and the kinetic energy of rotation of the rotor, that is

$$T_i^2(q,\dot{q},N_i) = T_i^{v}(q,\dot{q}) + T_i^{\omega}(q,\dot{q},N_i).$$

Suppose that J_i and J_i' are the moments of inertia of the ith rotor about its axis of rotation and an axis passing through the centre of inertia perpendicular to the axis of rotation. Then, if the angular velocity vector of the stator of the ith electric motor has a projection on the axis of rotation of the rotor equal to ω_i and a perpendicular component equal to ω_i', we have

$$T_i^{\omega}(q,\dot{q},N_i) = \frac{1}{2}\left[J_i(N_i\dot{q}_i+\omega_i)^2 + J_i'\omega_i'^2\right].$$

The angular velocities ω_i and ω_i' are linear combinations of the generalized velocities $\dot{q}_1,\dots,\dot{q}_n$ with coefficients depending on q. The kinetic energy of the robot can therefore be represented in the form

$$T = \frac{1}{2}\sum_{j=1}^{n} J_j(N_j\dot{q}_j)^2 + \frac{1}{2}N_{\max}\langle B\dot{q},\dot{q}\rangle, \tag{2.5.1}$$

where $B(q,N)$ is a bounded matrix such that the inequality

$$|B(q,N)z| \leq \lambda|z|, \qquad \lambda = \text{const}, \tag{2.5.2}$$

is satisfied in the case of an arbitrary vector z.

The largest and smallest of the gear ratios $N_1,\dots,N_n$ are henceforth denoted by $N_{\max}$ and $N_{\min}$, and λ is independent of N_i.

We substitute (2.5.1) into Lagrange's equations in the form of (2.1.1) and obtain

$$N_i^2 J_i\ddot{q}_i + N_{\max}[B(q,N)\ddot{q}]_i = U_i + S_i(q,\dot{q},t,N). \tag{2.5.3}$$

We divide the ith equation of (2.5.3) by N_i and make the change of variables

$$p_i = N_i q_i. \tag{2.5.4}$$

As a result, we obtain

$$J_i\ddot{p}_i + N_{\max}N_i^{-1}\sum_{j=1}^{n} B_{ij}N_j^{-1}\ddot{p}_j = N_i^{-1}(U_i+S_i). \tag{2.5.5}$$

Allowing for the fact that $N_i^{-1}U_i = M_i$, where M_i is the electromagnetic torque produced by the electric motor, we reduce system (2.5.5) to the form

$$(J+\tilde{B})\ddot{p} = M + S^*. \tag{2.5.6}$$

Here,

$$J = \mathrm{diag}(J_1,\dots,J_n), \quad \tilde{B} = N_{\max}H^{-1}BH^{-1}, \quad M = (M_1,\dots,M_n),$$
$$S^* = H^{-1}S, \qquad H = \mathrm{diag}(N_1,\dots,N_n). \tag{2.5.7}$$

Consequently, when account is taken of the change of variables (2.5.4) and notation (2.5.7), the equations of motion can be represented in the form of (2.1.3) and (2.3.2), and, by (2.5.2) and (2.5.7), we have the inequality

$$|\tilde{B}z| \le \mu|z|, \quad \mu = N_{\max}N_{\min}^{-2}\lambda, \tag{2.5.8}$$

that is analogous to constraint (2.3.3). The initial and terminal conditions can be represented in the form (2.1.7) and (2.1.8).

We will now consider different ways of formulating control problems.

1°. Suppose that the constraints

$$|M_i| \le M_i^0 \tag{2.5.9}$$

are imposed on the control torques M_i produced by the electric motors. In this case, the results obtained in the preceding sections and summarized in Theorem 2.3 can be used to construct the control. Inequality (2.3.34), rewritten in the notation of system (2.5.6), defines the permissible values of parameter μ. On substituting its value from (2.5.8) into this inequality instead of μ, we obtain a constraint on the possible values of the gear ratios of the reduction gears

$$\frac{N_{\min}^2}{N_{\max}} > \frac{\lambda}{J_{\min}}\left(1+\frac{|M^0+H^{-1}G^0|}{\min_i(M_i^0-N^{-1}G_i^0)}\right),$$
$$M^0 = (M_1^0,\dots,M_n^0), \qquad G^0 = (G_1^0,\dots,G_n^0). \tag{2.5.10}$$

Here, $J_{\min}$ is the least of the moments of inertia of the rotors $J_1,\dots,J_n$, constants G_i^0 are introduced in (2.3.5)–(2.3.7); besides, we assume that

$$G_i^0 < N_i M_i^0$$

for all $i = 1,\dots,n$.

2°. Suppose that the voltages applied to the windings of the electric motors play the role of controls. We augment the equations of motion (2.5.6) with the balance equations for the voltages in the rotor circuits and relations associating torques M_i with the currents [50]:

$$L_i \frac{dj_i}{dt} + R_i j_i + k_i^E \dot{p}_i = u_i, \qquad M_i = k_i^M j_i - b_i \dot{p}_i. \tag{2.5.11}$$

Here, L_i is the coefficient of inductance, R_i is the electrical resistance, k_i^E and k_i^M are constant coefficients, u_i is the voltage in the rotor circuit of the ith motor, term $b_i \dot{p}_i$ is the moment due to mechanical resistance, and b_i is a positive constant coefficient. The first term in the first equation (2.5.11) is usually small compared with the remaining terms. Therefore, the expression

$$M_i = k_i^M R_i^{-1}(u_i - k_i^E \dot{p}_i) - b_i \dot{p}_i$$

is obtained from (2.5.11); when this is substituted into (2.5.6), we obtain

$$(J + \tilde{B})\ddot{p} = U^* + S^{**},$$

$$S^{**} = S^* - \Lambda \dot{p}, \quad \Lambda = \mathrm{diag}(k_1^M k_1^E R_1^{-1} + b_1, \ldots, k_n^M k_n^E R_n^{-1} + b_n), \tag{2.5.12}$$
$$U^* = (k_1^M R_1^{-1} u_1, \ldots, k_n^M R_n^{-1} u_n).$$

Suppose that the constraints

$$|u_i| \le u_i^0 \tag{2.5.13}$$

are imposed on the control voltages. Constraints (2.5.13) are transformed into constraints on the components of vector U^* from (2.5.12):

$$|U_i^*| \le U_i^{*0} = k_i^M R_i^{-1} u_i^0. \tag{2.5.14}$$

The equations of motion (2.5.12) are again reduced to the form (2.1.3) and (2.3.2). Inequalities (2.5.14) are of the same form as relations (2.1.6). It is obvious that in this case we can use the method of control considered above in Sect. 2.3. By Theorem 2.3, we obtain a constraint which is analogous to (2.5.10):

$$\frac{N_{\min}^2}{N_{\max}} > \frac{\lambda}{J_{\min}} \left(1 + \frac{|U^{*0} + H^{-1}G^0|}{\min_i (U_i^{*0} - N_i^{-1} G_i^0)}\right), \quad U^{*0} = (U_1^{*0}, \ldots, U_n^{*0}). \tag{2.5.15}$$

Thus, if the gear ratios of the drives and the parameters of the robot are such that inequalities (2.5.10) and (2.5.15) are satisfied, then it is possible to construct a control which transfers the system under consideration from an initial state to a specified state in finite time. The control takes account of the existence of perturbations and structural constraints.

Remark 2.2. Considering system (2.5.3) and rewriting in its terms condition (2.3.34), one can obtain constraints imposed on parameters of the system in another form. We have

$$\frac{\min_i (N_i^2 J_i)}{N_{\max} \lambda} > 1 + \frac{|HM^0 + G^0|}{\min_i (N_i M_i^0 - G_i^0)}$$

for the case 1° of the bounded electromagnetic torques and

$$\frac{\min_i(N_i^2 J_i)}{N_{\max}\lambda} > 1 + \frac{|HU^{*0} + G^0|}{\min_i(N_i U_i^{*0} - G_i^0)}$$

for the case 2° of the bounded electric voltages. It seems that these sufficient conditions for the method of control decomposition are more efficient in the case, where moments of inertia of the rotors J_i, $i = 1, \ldots, n$ are essentially different from each other, but the difference between the effective moments of inertia $N_i^2 J_i$ is not very large.

Remark 2.3. If the elements of the matrix Λ are sufficiently large, then, in order to shorten the motion time, it is advisable to reduce system (2.5.12) to the form of (2.1.28). In this case, one should set matrix A_* equal to matrix J, and coefficients λ_i should be made equal to the corresponding elements of matrix Λ. After that, the approach described in Sects. 2.1.3–2.2.3 can be applied to the obtained subsystems with linear resistance.

3°. Recently, direct drives without gears are often used in robots. For such drives, we have $N_i = 1$ and $J_i = 0$, so that we set $J = 0$ and $H = E$ in (2.5.6) and (2.5.7). The equations of motion and the constraints are again reduced to the forms (2.1.3) and (2.1.6). However, it is no longer possible to choose matrix A_* in the form (2.3.2) and (2.3.3), since $J = 0$. This matrix must be chosen differently, for example, in the form $A_* = A(q^*)$ [see Remark 2.1 at the end of Sect. 2.1.2]. To apply the results obtained, it is necessary to verify conditions (2.1.12) or (2.1.13), and this has to be done in each specific case. In order to demonstrate this proposition, in the next subsection, the problem of the feedback control design for the two-link manipulator with direct drives is considered.

Thus, the results obtained can, under certain conditions, be used for constructing control for manipulation robots.

2.5.2 Feedback control design and modelling of motions for two-link manipulator with direct drives

The system considered in this subsection is a simplified model of a mechanical manipulation robot with two absolutely rigid links. The system can perform motions in a horizontal plane and is controlled by two torques produced by drives installed at its joints. Geometric constraints are imposed on the control torques. Using the decomposition method described in Sects. 2.3 and 2.5, we will construct the feedback control that brings the system to the prescribed terminal position.

Problem statement

Consider a mechanical two-link system (see Fig. 2.15) that consists of a stationary base G_0 and two absolutely rigid links G_1 and G_2. The elements of the system are connected by two revolute joints O_1 and O_2 such that both links can move only in horizontal plane.

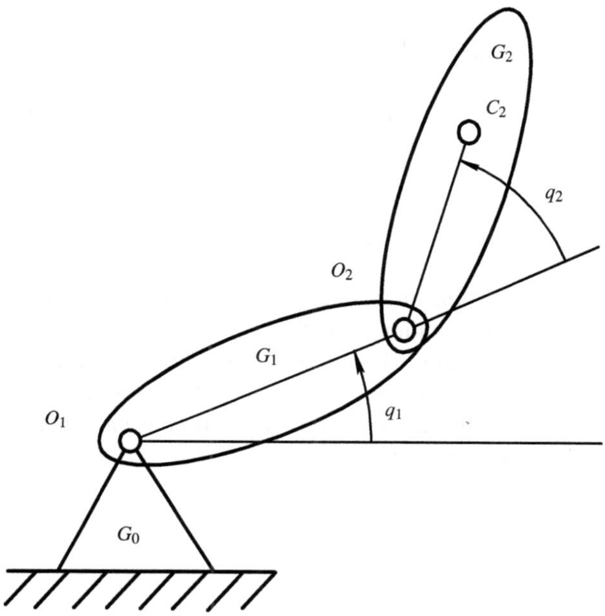

Fig. 2.15 Two-link manipulator

Lagrange's equations that describe the motion of this system are obtained are [41]:

$$\begin{aligned}&(m_2 l_1^2 + I_1 + I_2 + 2m_2 l_1 l_{g2} \cos q_2)\, \ddot{q}_1 + (I_2 + m_2 l_1 l_{g2} \cos q_2)\, \ddot{q}_2 \\ &-2m_2 l_1 l_{g2} \sin q_2\, \dot{q}_1 \dot{q}_2 - m_2 l_1 l_{g2} \sin q_2\, \dot{q}_2^2 = M_1 + Q_1, \\ &(I_2 + m_2 l_1 l_{g2} \cos q_2)\, \ddot{q}_1 + I_2\, \ddot{q}_2 + m_2 l_1 l_{g2} \sin q_2\, \dot{q}_1^2 = M_2 + Q_2.\end{aligned} \tag{2.5.16}$$

Here, q_1 is the angle of rotation of link G_1 relative to base G_0, and q_2 is the angle between the straight lines O_1O_2 and O_2C_2; C_2 is a center of mass of link G_2. Angle q_2 defines the position of link G_2 relative to link G_1; l_1 is the length of segment O_1O_2; l_{g2} is the length of segment O_2C_2; m_2 is the mass of link G_2; I_i is the moment of inertia of the ith link relative to the axis of joint O_i; and M_i and Q_i are the control torque and moment of other forces applied at the joint O_i, respectively; here and below $i = 1, 2$.

The constraints

$$|M_i| \leq M_i^0 \tag{2.5.17}$$

are imposed on the control torques. Here, M_i^0 are given constants.

Let us turn to dimensionless variables

$$t' = \left[\frac{M_2^0}{m_2 l_1 l_{g2}}\right]^{1/2} t, \quad U_i = \frac{M_i}{M_2^0}, \quad U_i^0 = \frac{M_i^0}{M_2^0},$$
$$Q_i' = \frac{Q_i}{M_2^0}, \quad \alpha = \frac{I_1 + m_2 l_1^2}{m_2 l_1 l_{g2}}, \quad \beta = \frac{I_2}{m_2 l_1 l_{g2}}. \tag{2.5.18}$$

If we now drop the primes at t' and Q_i', then relations (2.5.16) take the form

$$(\alpha + \beta + 2\cos q_2)\,\ddot{q}_1 + (\beta + \cos q_2)\,\ddot{q}_2 - (2\dot{q}_1\dot{q}_2 + \dot{q}_2^2)\sin q_2 = U_1 + Q_1,$$
$$(\beta + \cos q_2)\,\ddot{q}_1 + \beta\,\ddot{q}_2 + \dot{q}_1^2 \sin q_2 = U_2 + Q_2, \tag{2.5.19}$$

and inequalities (2.5.17) coincide with (2.1.6). Note that the conditions $\alpha\beta > 1$ and $U_2^0 = 1$ are fulfilled due to notation (2.5.18).

Next, we consider Problem 2.1 (see Sect. 2.1.2) for system (2.5.19) with constraints (2.1.6) imposed on the new controls U_i. We suppose that the domain of possible motions is given by (2.3.1). In this subsection, we assume that all external forces and disturbances are missing, i.e.,

$$Q_1 = Q_2 = 0.$$

Simplifying assumptions and decomposition of the system

Let us resolve system (2.5.19) with respect to the second derivatives $\ddot{q}_1$ and $\ddot{q}_2$ and multiply the left-hand sides of the obtained relations by some positive coefficients J_1 and J_2. Then the system takes the form (2.3.9), where the functions V_i are equal to

$$V_1 = U_1\left(J_1\frac{\beta}{\alpha\beta - \cos^2 q_2} - 1\right) - J_1 U_2 \frac{\beta + \cos q_2}{\alpha\beta - \cos^2 q_2}$$
$$+J_1\frac{\beta(\dot{q}_1 + \dot{q}_2)^2 \sin q_2 + \dot{q}_2^2 \sin q_2 \cos q_2}{\alpha\beta - \cos^2 q_2},$$
$$V_2 = U_2\left(J_2\frac{\alpha + \beta + 2\cos q_2}{\alpha\beta - \cos^2 q_2} - 1\right) - J_2 U_1 \frac{\beta + \cos q_2}{\alpha\beta - \cos^2 q_2}$$
$$-J_2\frac{(\beta + \cos q_2)(\dot{q}_1 + \dot{q}_2)^2 \sin q_2 + (\alpha + \cos q_2)\dot{q}_1^2 \sin q_2}{\alpha\beta - \cos^2 q_2}. \tag{2.5.20}$$

Suppose that the inequalities (2.3.12) hold. If we treat V_i as independent restricted disturbances, then the original nonlinear system is divided into two linear subsystems with one degree of freedom each.

Control for each of these subsystems can be given by relations (2.3.13) and (2.3.14). In what follows, we show that conditions (2.3.12) are really fulfilled under some restrictions on the parameters of the system and constants J_i.

Determination of the control parameters X_1 and X_2

Let us impose certain restrictions on the parameters of system (2.5.19) and show that there exist X_i entering control law (2.3.13) such that relations (2.3.12) are really fulfilled.

(a) Suppose the inequality

$$\beta < 1 \tag{2.5.21}$$

is true. For example, if link G_2 is a thin rod of length $l_2 < l_1$ with an arbitrary distribution of density $\rho(x)$, then

$$\beta = \frac{1}{m_2 l_1 l_{g2}} \int_0^{l_2} \rho(x) x^2 dx = \frac{1}{m_2 l_1 l_{g2}} \int_0^{l_2} xd\left(\int_0^x \rho(y) y dy\right)$$

$$= \frac{1}{m_2 l_1 l_{g2}} \left(m_2 l_2 l_{g2} - \int_0^{l_2} \int_0^x \rho(y) y dy dx \right) < 1.$$

(b) Let us require that the magnitudes of the angles q_2^- and q_2^+ in (2.3.1) are restricted:

$$-\arccos(-\beta) < q_2^-, \qquad q_2^+ < \arccos(-\beta). \tag{2.5.22}$$

From (2.5.22) and inequalities $q_2^- < q_2 < q_2^+$ that are satisfied under control (2.3.13) (see Sect. 2.3), it follows that

$$\cos q_2 > -\beta \tag{2.5.23}$$

during the whole control process.

(c) Suppose that the value U_1^0 that restricts control U_1 satisfies the inequalities

$$\frac{\beta+1}{\beta} < U_1^0 < \frac{\alpha+\beta+2}{\beta+1}. \tag{2.5.24}$$

Since

$$\frac{\alpha+\beta+2}{\beta+1} - \frac{\beta+1}{\beta} = \frac{\alpha\beta-1}{\beta(\beta+1)} > 0,$$

one can always ensure the fulfillment of relations (2.5.24) by imposing more rigid restrictions on one of torques M_i in (2.5.17). We point out that in view of (2.5.21) and (2.5.24), $U_1^0 > 2$.

(d) We choose constants J_i in system (2.3.9) so that the following inequalities hold:

$$J_1 \frac{\beta}{\alpha\beta-1} < 1, \qquad J_2 \frac{\alpha+\beta+2}{\alpha\beta-1} < 1. \tag{2.5.25}$$

Let us estimate magnitude of V_1 from (2.5.20) using assumptions (a)–(d). On the strength of inequalities (2.1.6), (2.5.23), and (2.5.25), we obtain

$$|V_1| \leq U_1^0 \left| J_1 \frac{\beta}{\alpha\beta - \cos^2 q_2} - 1 \right| + J_1 U_2^0 \frac{|\beta + \cos q_2|}{\alpha\beta - \cos^2 q_2}$$

$$+ J_1 \frac{\beta(\dot{q}_1 + \dot{q}_2)^2 + \dot{q}_2^2}{\alpha\beta - 1} = U_1^0 + J_1 \frac{\beta + \cos q_2 - U_1^0 \beta}{\alpha\beta - \cos^2 q_2}$$

$$+ J_1 \frac{\beta(\dot{q}_1 + \dot{q}_2)^2 + \dot{q}_2^2}{\alpha\beta - 1}.$$

Now, using inequalities (2.3.24) and (2.5.24), we get

$$|V_1| \leq U_1^0 + J_1 \frac{\beta + 1 - U_1^0 \beta}{\alpha\beta} + J_1 \frac{Y_2^2 + \beta\,(Y_1 + Y_2)^2}{\alpha\beta - 1}. \tag{2.5.26}$$

In a similar manner, we can obtain the following estimate on V_2

$$|V_2| \leq 1 + J_2 \frac{\beta U_1^0 + (U_1^0 - 2)\cos q_2 - \alpha - \beta}{\alpha\beta - \cos^2 q_2}$$

$$+ J_2 \frac{(\beta + 1)(\dot{q}_1 + \dot{q}_2)^2 + (\alpha + 1)\dot{q}_1^2}{\alpha\beta - 1}.$$

By virtue of relations (2.3.24), (2.5.24), and $U_1^0 > 2$, we have

$$\begin{aligned} |V_2| &\leq 1 + J_2 \frac{U_1^0(\beta + 1) - \alpha - \beta - 2}{\alpha\beta} \\ &+ J_2 \frac{(\alpha + 1)Y_1^2 + (\beta + 1)\,(Y_1 + Y_2)^2}{\alpha\beta - 1}. \end{aligned} \tag{2.5.27}$$

Let us replace values $|V_i|$ by their estimates (2.5.26) and (2.5.27) in inequalities (2.3.15). We obtain

$$\begin{aligned} X_1 + \frac{Y_2^2 + \beta\,(Y_1 + Y_2)^2}{\alpha\beta - 1} &\leq \frac{U_1^0 \beta - \beta - 1}{\alpha\beta}, \\ X_2 + \frac{(\alpha + 1)Y_1^2 + (\beta + 1)\,(Y_1 + Y_2)^2}{\alpha\beta - 1} &\leq \frac{\alpha + \beta + 2 - U_1^0(\beta + 1)}{\alpha\beta}. \end{aligned} \tag{2.5.28}$$

By virtue of (2.5.24), the expressions in the right-hand sides of inequalities (2.5.28) are positive. Let us choose Y_i according to (2.3.25). Then $Y_i \to 0$ as $X_i \to 0$. Hence, there always exist positive X_1 and X_2 that satisfy inequalities (2.5.28) and, thus, inequalities (2.3.12). Note that constants J_i do not appear in restrictions (2.5.28) directly; therefore, their specific values are not essential.

To sum up, we may state the following.

Let conditions (2.5.21), (2.5.22), and (2.5.24) be fulfilled. Then, the feedback control $U_i(q_i, \dot{q}_i)$ that solves Problem 2.1 for system (2.5.16) is given by relations

(2.3.13). In these relations, parameters X_i are chosen so that inequalities (2.5.28) should be fulfilled. This control carries system (2.5.16) from the initial position (2.1.7) to the terminal position (2.1.8), if, at the initial instant, restrictions (2.3.17) are satisfied. The trajectory of the system lies in the domain D defined by (2.3.1); time τ of the control process is not greater than value τ^* that is defined by expressions (2.3.26) and (2.3.27).

Let us show the way to choose admissible values X_i. We search for them in the form (2.3.35). In this case, inequalities (2.5.28) become

$$Z^2 \leq \frac{U_1^0\beta - \beta - 1}{\alpha\beta} \times \left[d_1 + 2\frac{d_2^2 + \beta\,(d_1 + d_2)^2}{\alpha\beta - 1}\right]^{-1},$$

$$Z^2 \leq \frac{\alpha + \beta + 2 - U_1^0(\beta + 1)}{\alpha\beta} \times \left[d_2 + 2\frac{(\alpha+1)d_1^2 + (\beta+1)\,(d_1 + d_2)^2}{\alpha\beta - 1}\right]^{-1}.$$

Now, we find the maximum value Z that satisfies the both inequalities obtained and then determine the parameters X_i by formulas (2.3.35).

Note that the set of possible values X_i may be significantly extended. For this purpose, it is necessary to obtain more precise estimates of $|V_i|$ in (2.5.26) and (2.5.27).

Computer simulation

Calculations were carried out for the following dimensional characteristics of system (2.5.16):

$$l_1 = 1\,\text{m}, \quad l_{g2} = 0.5\,\text{m}, \quad I_1 = I_2 = 3.33\,\text{kg}\cdot\text{m}^2,$$
$$m_2 = 10\,\text{kg}, \quad M_1^0 = 2.9\,\text{N}\cdot\text{m}, \quad M_2^0 = 1\,\text{N}\cdot\text{m}. \tag{2.5.29}$$

For this example, we suppose that $Q_1 = Q_2 = 0$ (see Sect. 2.5.2). The initial and terminal conditions, and also values $q_i^{\pm}$ defining the permissible domain of motion, are given by:

$$q_1^- = q_1^0 = -0.1\,\text{rad}, \quad q_2^- = q_2^0 = -0.05\,\text{rad},$$
$$\dot{q}_1^0 = \dot{q}_2^0 = q_1^* = q_2^* = q_1^+ = q_2^+ = 0. \tag{2.5.30}$$

In this case, $\alpha = 2.66$ and $\beta = 0.66 < 1$; inequalities (2.5.22) and (2.5.24) become $-2.3 < q_2^-, q_2^+ < 2.3$ and $2.5 < U_1^0 < 3.2$. Obviously, the parameters of the system (2.5.29) and (2.5.30) satisfy these restrictions. Let us choose dimensionless values for X_i that satisfy inequalities (2.5.28). For $X_1 = 1.82 \times 10^{-2}$ and $X_2 = 9.13 \times 10^{-3}$, the dimensional estimate $\tau^* = 4.68$ s of the time of control and the real time of the process $\tau = 3.64$ s are obtained. Figure 2.16 demonstrates the time histories of angular velocities $\dot{q}_1$ and $\dot{q}_2$. At the final stage, $\dot{q}_1$ and $\dot{q}_2$ vary linearly that agrees with the motion of the phase points of subsystems (2.3.9) along the parabolic sections of the switching curves. The phase trajectories of the subsystems are depicted in Figs. 2.17 and 2.18. The termination of the motion occurs at different instants for two degrees of freedom.

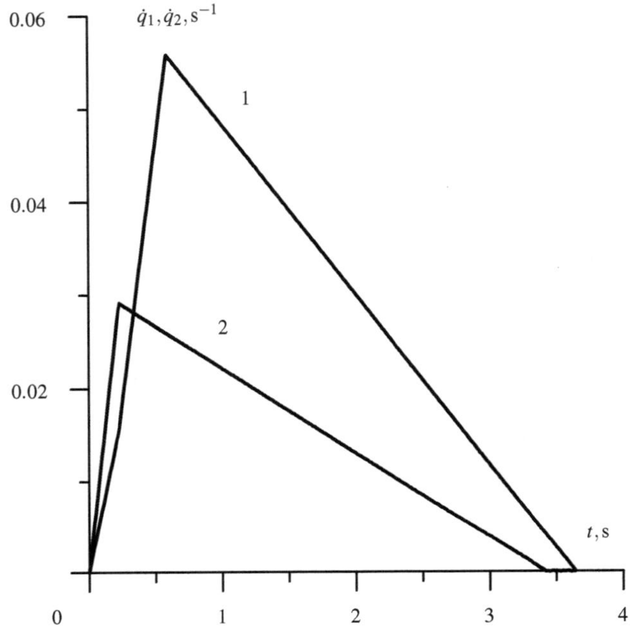

Fig. 2.16 Time history of the angular velocities

2.5.3 Modelling of motions of three-link manipulator

The three-link mechanism shown in Fig. 2.19 is chosen as an example of the control design by the method described in this chapter. This mechanism models the arm of the manipulation robot consisting of the upper and lower arms. The arm lies in the vertical plane and is connected to the vertical column supported by a fixed base.

The moment of inertia of the vertical column with respect to its axis of rotation is equal to I_1^Z. The links of the arm are the rods of the masses m_2, m_3 and lengths l_2, l_3, respectively. The centers of mass of the upper and lower arms are located exactly in the middle of the corresponding links. The principal central moments of inertia of the links with respect to the axes that are perpendicular to the rods and with respect to the longitudinal axes are equal to I_i^S and I_i^N, $i = 2, 3$, respectively.

The vertical column as well as the upper and lower arms are supplied by the actuators that include DC motors and reduction gears. For the sake of simplicity, we suppose that the axis and direction of rotation of the rotor in each electric drive coincide with the axis and direction of rotation of the corresponding joint. The masses

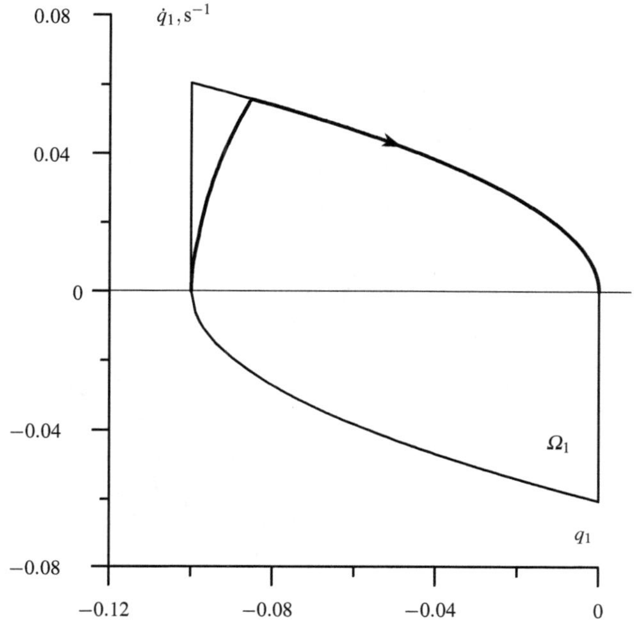

Fig. 2.17 Phase trajectory of the subsystem 1

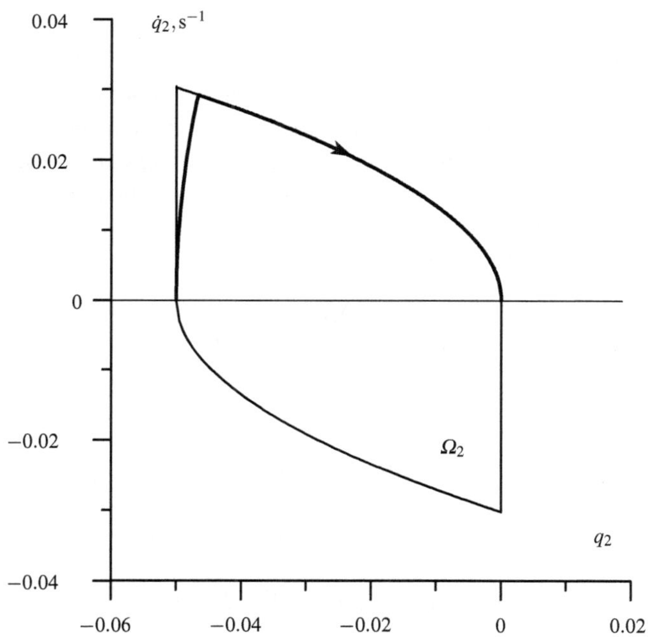

Fig. 2.18 Phase trajectory of the subsystem 2

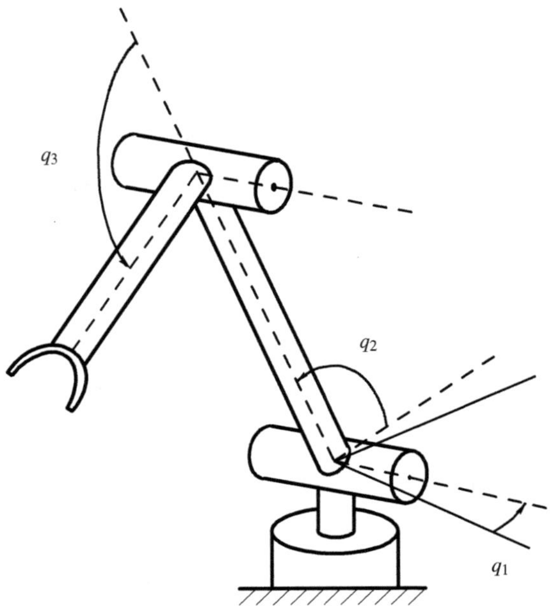

Fig. 2.19 Three-link manipulator

of the rotors of the motors are equal to m_i^R, $i = 1,2,3$. We neglect the inertia of the rotational parts of the reduction gears. The generalized coordinates q_1, q_2, and q_3 are the angles of rotation in the three cylindrical joints of the manipulator: angle q_1 of rotation of the vertical column about the vertical axis, angles q_2 and q_3 of rotation of the upper and lower arms about the corresponding horizontal axes (Fig. 2.19).

Under the assumptions made, let us obtain the elements of the matrix of the kinetic energy $A(q)$ from (2.1.2)

$$A(q) = \begin{pmatrix} a_{11} & 0 & 0 \\ 0 & a_{22} & a_{23} \\ 0 & a_{32} & a_{33} \end{pmatrix}.$$

We have

$$\begin{aligned} a_{11} &= J_1 N_1^2 + J_2' + J_3' + I_1^Z \\ &+ \frac{1}{2}\Big\{ (m_3 l_2^2 + I_2^S - I_2^N)\cos 2q_2 + (I_3^S - I_3^N)\cos 2(q_2+q_3) \\ &+ m_3 l_2 l_3 [\cos q_3 + \cos(q_3 + 2q_2)] + I_2^S + I_3^S + I_2^N + I_3^N + m_3 l_2^2 \Big\} \\ &+ \frac{1}{8}\Big\{ m_2 l_2^2 (1 + \cos 2q_2) + m_3 l_3^2 [\cos 2(q_2+q_3) + 1] \Big\}, \end{aligned}$$

$$a_{22} = J_2N_2^2 + J_3 + I_2^S + I_3^S$$
$$+ l_2^2\left(m_3^R + m_3 + \frac{1}{4}m_2\right) + m_3l_3\left(l_2\cos q_3 + \frac{1}{4}l_3\right),$$

$$a_{23} = a_{32} = J_3N_3 + I_3^S + \frac{1}{2}m_3l_3\left(l_2\cos q_3 + \frac{1}{2}l_3\right),$$
$$a_{33} = J_3N_3^2 + I_3^S + \frac{1}{4}m_3l_3^2.$$

Here, the notation for the moments of inertia J_i and J_i' of the rotors and gear ratios N_i of the reduction gears introduced in Sect. 2.5.1 is used.

As the generalized forces Q_i in (2.1.1), we will consider only torques due to the gravity in the joints (the forces of viscous and dry friction are not taken into account):

$$Q_1 = 0,$$
$$Q_2 = -9.81l_2\left(\frac{1}{2}m_2 + m_3^R + m_3\right)\cos q_2 - 9.81\cdot\frac{1}{2}m_3l_3\cos\left(q_2+q_3\right),$$
$$Q_3 = -9.81\cdot\frac{1}{2}m_3l_3\cos\left(q_2+q_3\right).$$

We will consider the case, where the constraints are imposed on the magnitude of the control electric voltages (see Sect. 2.5.1, case 2°).

Below, four variants of the simulation results (1–4) for the control of the considered system are presented. Input data for each case are shown in Tables 2.1–2.7, including the parameters of the links and electric drives, initial and terminal conditions, and domains of the possible motions. Output data are presented in Tables 2.3–2.7 and Figs. 2.20–2.24, including the control parameters, estimates of the motion time for each of three subsystems, real values of the motion time, time histories of the generalized velocities $\dot{q}_1$, $\dot{q}_2$, and $\dot{q}_3$, and phase trajectories of the subsystems. In addition, for the first set of the manipulation robot parameters, we give three variants of the simulation results (1a, 1b, and 1c) obtained by using the control method described in Sect. 2.3.4. While implementing the numerical simulations 1a, 1b, and 1c, the current states of the subsystems are determined at the discrete time instants (with finite time step). As a result, the motion along the switching curve takes place with a finite frequency of the sign change of the control; this motion approximates the sliding regime along the switching curve.

Table 2.1 Parameters of the links (variants 1–4)

i	m_i, kg	l_i, m	l_{gi}, m	I_i^S, kg×m^2	I_i^N, kg×m^2	I_i^Z, kg×m^2
1	–	–	–	–	–	0.2
2	5	0.8	0.4	0.25	0.01	–
3	5	0.8	0.4	0.25	0.01	–
1	–	–	–	–	–	0.2
2	5	0.8	0.4	0.25	0.01	–
3	4	0.64	0.32	0.20	0.01	–
1	–	–	–	–	–	0.2
2	5	0.8	0.4	0.25	0.01	–
3	4	0.64	0.32	0.17	0.086	–
1	–	–	–	–	–	0.2
2	5	0.8	0.4	0.25	0.01	–
3	4	0.74	0.37	0.18	0.009	–

Table 2.2 Parameters of the actuators (variants 1–4)

i	k_i^E, J/A	k_i^M, J/A	R_i, Ω	u_i, V	m_i^R, kg	J_i, kg×m^2	J_i', kg×m^2	N_i
1	0.04	0.04	1	27	0.5	0.00079	0.00041	160
2	0.04	0.04	1	27	0.5	0.00079	0.00041	250
3	0.04	0.04	1	27	0.5	0.00079	0.00041	150
1	0.04	0.04	0.7	27	0.4	0.00069	0.00036	120
2	0.04	0.04	0.6	27	0.25	0.00039	0.00022	180
3	0.04	0.04	0.6	27	0.25	0.00039	0.00022	150
1	0.113	0.109	0.7	42	0.4	0.00069	0.00036	150
2	0.1	0.09	0.6	36	0.25	0.00039	0.00022	250
3	0.1	0.09	0.6	36	0.25	0.00039	0.00022	200
1	0.08	0.07	0.7	27	0.4	0.00039	0.00022	120
2	0.06	0.06	0.6	27	0.25	0.00039	0.00022	180
3	0.06	0.05	0.6	27	0.25	0.00039	0.00022	150

Table 2.3 Variant 1: initial (q_i^0, $\dot{q}_i^0$) and terminal (q_i^*) conditions, domain of possible motions ($[q_i^-, q_i^+]$), control parameter (X_i), estimated (τ_i^*) and real(τ_i) motion times for the ith subsystem

i	q_i^0	$\dot{q}_i^0$, s^{-1}	q_i^*	q_i^-	q_i^+	X_i, s^{-2}	τ_i^*, s	τ_i, s
1	-1	1	0	-1	0	1.060	1.413	1.382
2	-0.3	0	0	-0.35	0.05	0.424	1.682	1.263
3	-1	0	0	-1	0	1.060	1.942	1.467

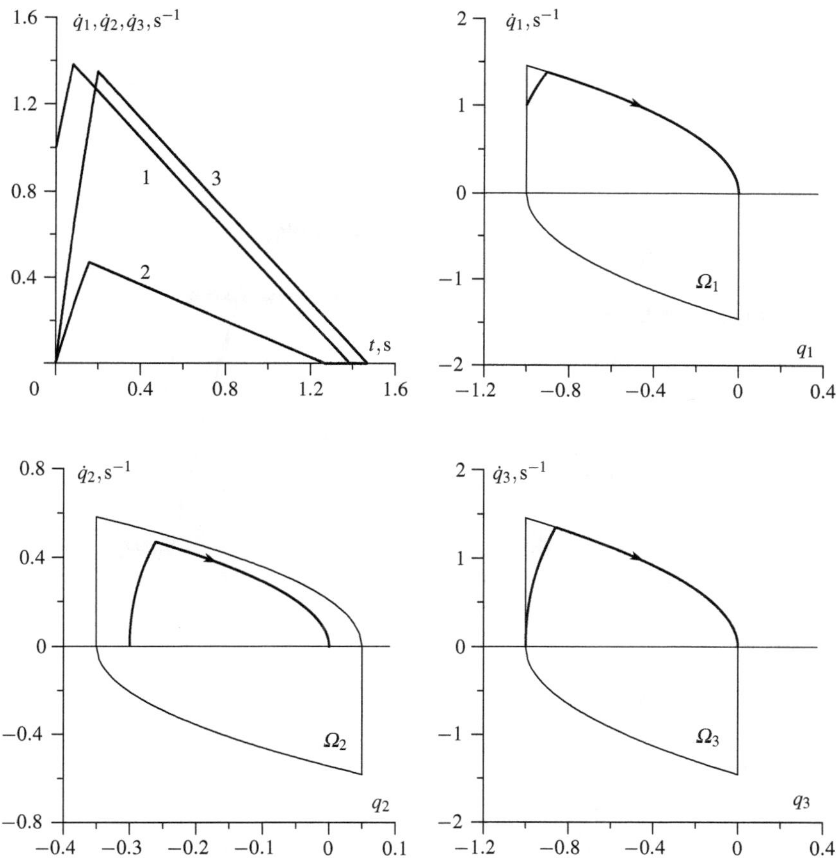

Fig. 2.20 Time histories of the generalized velocities and phase trajectories of the subsystems (variant 1)

Table 2.4 Variants 1a–1c: domain of possible motions ($[q_i^-, q_i^+]$), initial (q_i^0, $\dot{q}_i^0$) and terminal (q_i^*) conditions, and real motion time (τ_i) for the ith subsystem

i	q_i^-	q_i^+	q_i^0	$\dot{q}_i^0$, s^{-1}	q_i^*	τ_i, s
1	-1	0	-1	0	0	1.22
2	-0.3	0	-0.3	0	0	1.35
3	-0.9	0	-0.9	0	0	1.27
1	-1	0	-1	0.5	0	1.27
2	-0.35	0.05	-0.3	0	0	0.97
3	-1	0	-1	0	0	1.07
1	-1	0	-1	0.5	0	1.18
2	-0.35	0.05	-0.3	0	0	1.01
3	-1	0	1	-0.3	-0.1	1.20

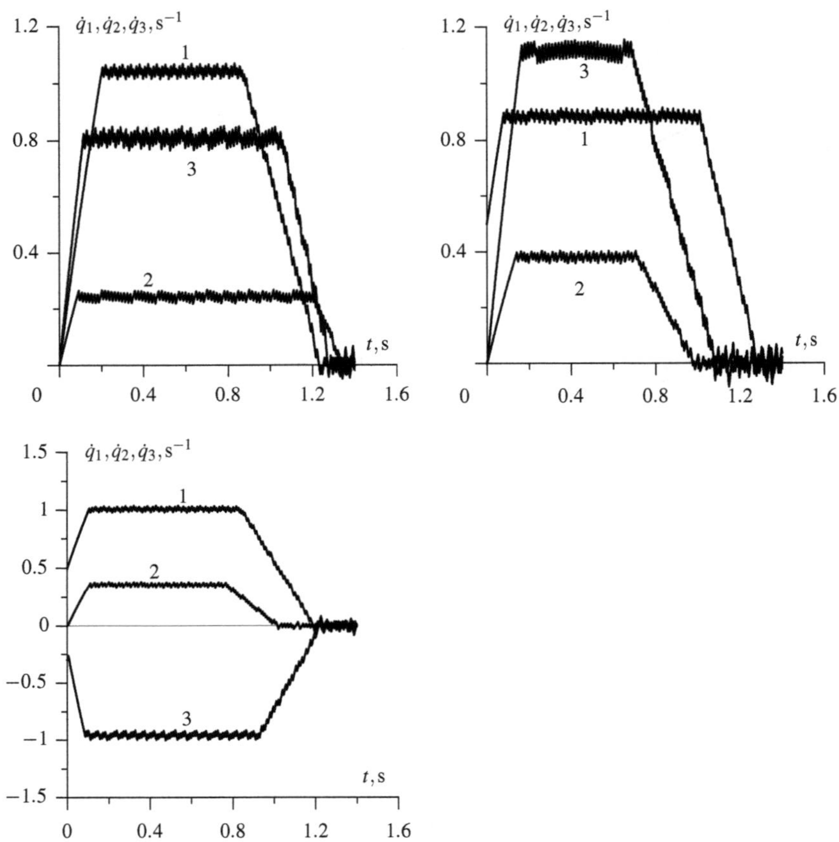

Fig. 2.21 Time histories of the generalized velocities (variants 1a–1c)

Table 2.5 Variant 2: initial (q_i^0, $\dot{q}_i^0$) and terminal (q_i^*) conditions, domain of possible motions ($[q_i^-, q_i^+]$), control parameter (X_i), estimated (τ_i^*) and real(τ_i) motion times for the ith subsystem

i	q_i^0	$\dot{q}_i^0$, s^{-1}	q_i^*	q_i^-	q_i^+	X_i, s^{-2}	τ_i^*, s	τ_i, s
1	-0.8	1	0	-0.85	0.05	1.392	1.145	1.109
2	-0.3	-0.2	0	-0.35	0.05	0.619	1.788	1.023
3	-1	0	0	-1	0	1.547	1.607	1.172

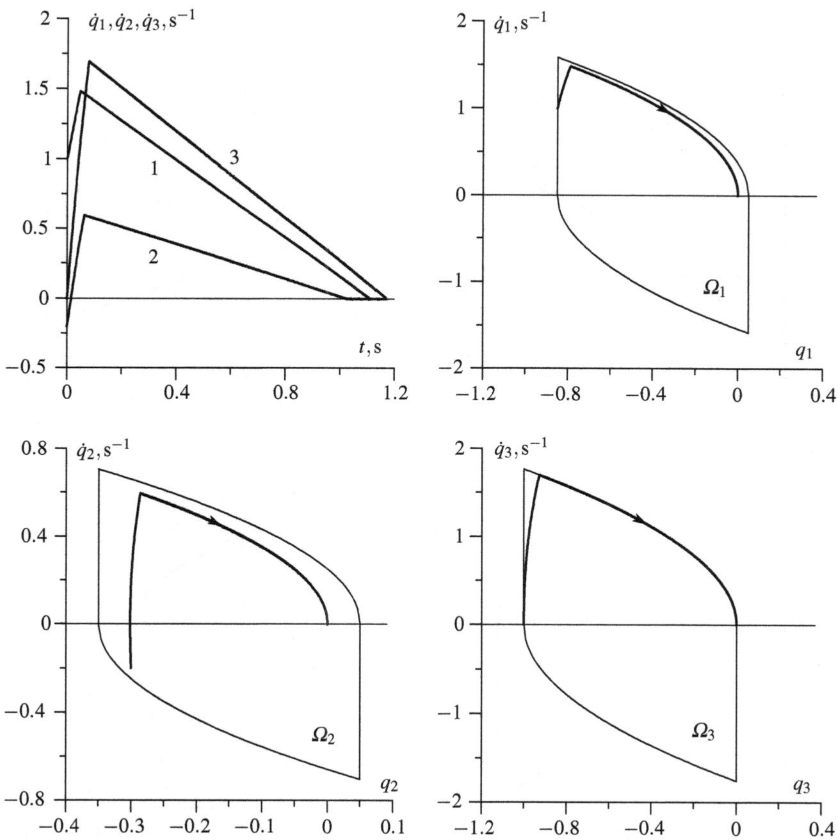

Fig. 2.22 Time histories of the generalized velocities and phase trajectories of the subsystems (variant 2)

Table 2.6 Variant 3: initial (q_i^0, $\dot{q}_i^0$) and terminal (q_i^*) conditions, domain of possible motions ($[q_i^-, q_i^+]$), control parameter (X_i), estimated (τ_i^*) and real(τ_i) motion times for the ith subsystem

i	q_i^0	$\dot{q}_i^0$, s^{-1}	q_i^*	q_i^-	q_i^+	X_i, s^{-2}	τ_i^*, s	τ_i, s
1	1	0	-0.2	-0.4	1.1	0.413	3.407	2.413
2	3	0	2	2	3	0.275	3.809	2.697
3	-0.25	0	0.3	-0.5	0.5	0.413	2.825	1.994

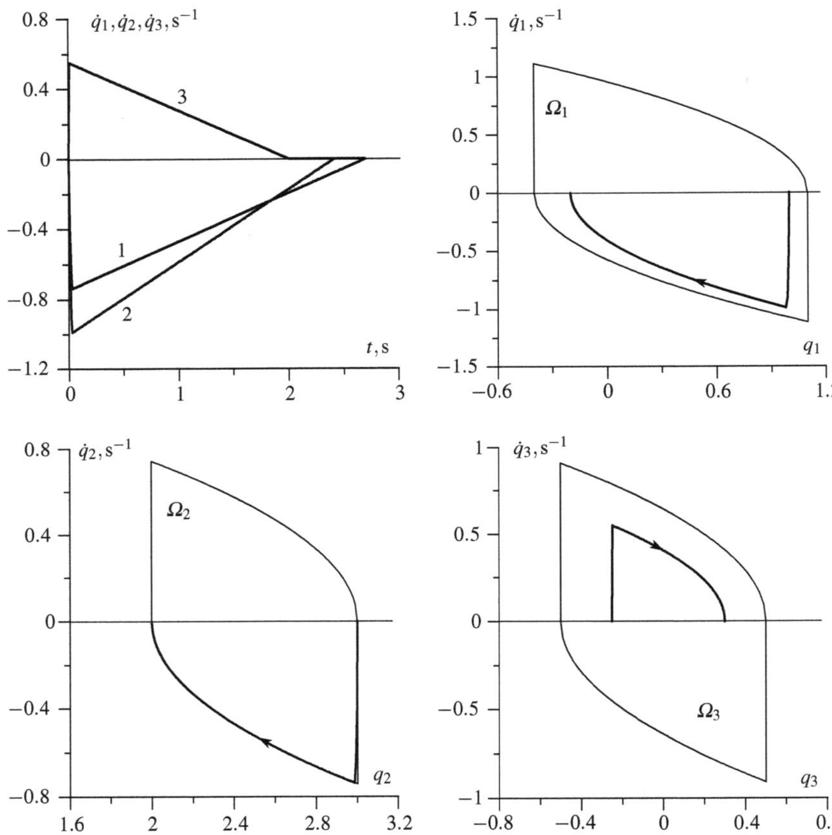

Fig. 2.23 Time histories of the generalized velocities and phase trajectories of the subsystems (variant 3)

Table 2.7 Variant 4: initial (q_i^0, $\dot{q}_i^0$) and terminal (q_i^*) conditions, domain of possible motions ($[q_i^-, q_i^+]$), control parameter (X_i), estimated (τ_i^*) and real(τ_i) motion times for the ith subsystem

i	q_i^0	$\dot{q}_i^0$, s^{-1}	q_i^*	q_i^-	q_i^+	X_i, s^{-2}	τ_i^*, s	τ_i, s
1	-1.5	0.7	-1	-1.6	-0.7	0.317	4.052	1.766
2	-0.3	-0.2	-0.2	-0.5	0	0.176	3.332	1.071
3	-0.5	0	-1	-1.3	-0.3	0.352	2.380	1.683

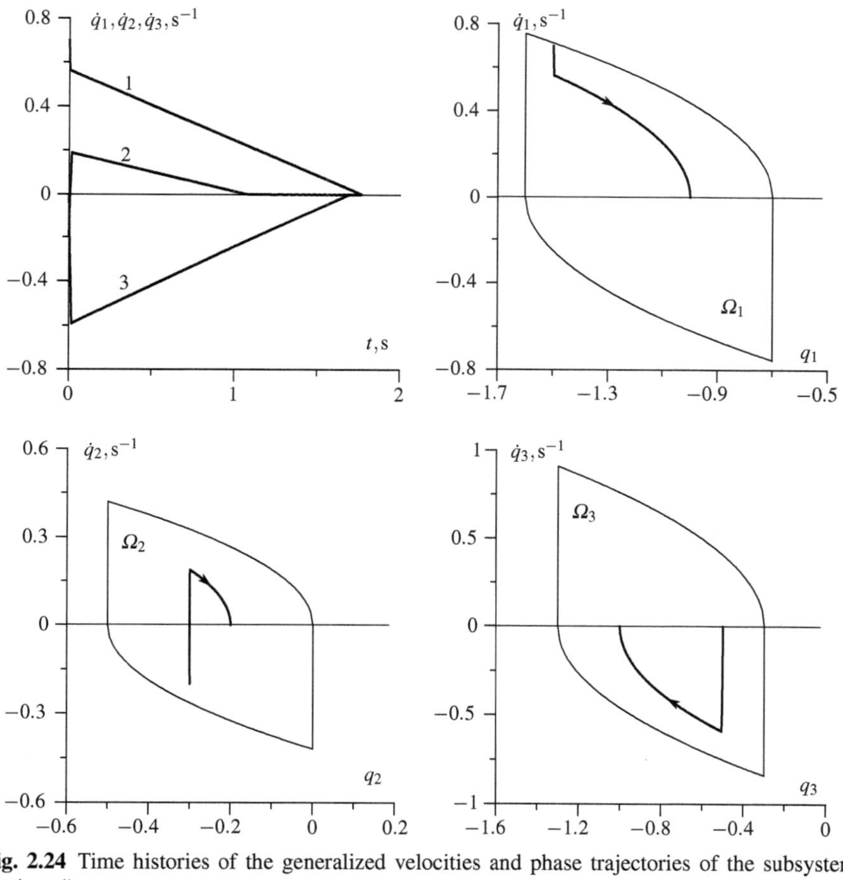

Fig. 2.24 Time histories of the generalized velocities and phase trajectories of the subsystems (variant 4)

Chapter 3
Method of decomposition (the second approach)

This chapter describes another approach to the method of decomposition for the construction of the feedback control for nonlinear Lagrangian systems. The chapter is based, mostly, on papers [31, 32, 14, 15, 54, 100, 103, 45].

3.1 Problem statement and game approach

3.1.1 Controlled mechanical system

Let us return to the system described in Sect. 2.1.1 but make a different set of assumptions. Consider a non-linear dynamical system described by Lagrange's equations

$$\frac{d}{dt}\frac{\partial T}{\partial \dot{q}_i} - \frac{\partial T}{\partial q_i} = U_i + Q_i, \quad i = 1,\ldots,n. \tag{3.1.1}$$

Here and below, the dot stands for differentiation with respect to time t, $q = (q_1,\ldots,q_n)$ is the vector of generalized coordinates, T is the kinetic energy of the system, Q_i are uncontrollable generalized forces, and U_i are control forces. We shall assume that all relevant motions of system (3.1.1) take place in a domain D in n-space R^n, so that $q \in D$ always. In particular, D may coincide with R^n.

We will now state our basic assumptions concerning the kinetic energy

$$T(q,\dot{q}) = \frac{1}{2}\langle A(q)\dot{q},\dot{q}\rangle = \frac{1}{2}\sum_{j,k=1}^{n} a_{jk}(q)\dot{q}_j\dot{q}_k \tag{3.1.2}$$

of the system and the generalized forces. Here, $A(q)$ is a symmetric positive-definite $(n\times n)$-matrix with elements a_{jk} that are continuously differentiable functions of q for $q \in D$. Throughout this chapter, $\langle\cdot,\cdot\rangle$ stands for the scalar product. It is assumed that, for any $q \in D$, all the eigenvalues of $A(q)$ lie in an interval $[m,M]$, where $M \geq m > 0$. Thus, for any n-vector z

$$m\langle z,z\rangle \le \langle A(q)z,z\rangle \le M\langle z,z\rangle, \qquad \forall q \in D. \tag{3.1.3}$$

In addition, we will assume that

$$|\partial a_{jk}/\partial q_i| \le C, \qquad \forall q \in D, \qquad C = \text{const} > 0, \tag{3.1.4}$$

and that the uncontrollable generalized forces Q_i in (3.1.1) consist of three terms, each subject to different restrictions:

$$Q_i = P_i + R_i + G_i. \tag{3.1.5}$$

The forces $P_i(q,\dot{q},t)$ are given functions of the generalized coordinates and time.

The terms $R_i(q,\dot{q},t)$ in (3.1.5) represent dissipative forces. The exact form of $R_i(q,\dot{q},t)$ may be unknown. Our only requirement is that these forces possess the property of dissipativeness, and that they be sufficiently small at low velocities. The former property means that the power of the dissipative forces is non-positive:

$$\sum_{i=1}^{n} R_i\dot{q}_i \le 0 \tag{3.1.6}$$

for all $q \in D$, all $\dot{q}$, and all $t \ge t_0$, where t_0 is the initial time instant. The second property may be stated as follows: there exists a sufficiently small number $\vartheta_0 > 0$ such that, if $|\dot{q}_i| \le \vartheta \le \vartheta_0$ for all i, then

$$|R_i| \le R_i^0(\vartheta). \tag{3.1.7}$$

Here, $R_i^0(\vartheta)$ are certain continuous monotone increasing functions defined for $\vartheta \in [0,\vartheta_0]$ and such that $R_i^0(0) = 0$.

The terms $G_i(q,\dot{q},t)$ in (3.1.5) represent uncertain external perturbations, the only restriction being that they are bounded

$$|G_i| \le G_i^0 \tag{3.1.8}$$

for all $q \in D$, all $\dot{q}$, and $t \ge t_0$. Here, $G_i^0 > 0$ are specified constants.

As for control forces U_i in (3.1.1), we will assume that they are large enough to balance the given external forces P_i, and the remaining part of the control may be chosen from a certain domain. Thus, we assume that U_i can be written as

$$U_i = -P_i(q,\dot{q},t) + w_i. \tag{3.1.9}$$

The vector $w = (w_1,\dots,w_n)$ may be chosen from some set W that will generally depend on q, $\dot{q}$, and t, i.e.,

$$w \in W(q,\dot{q},t) \subset R^n. \tag{3.1.10}$$

We will assume that, for all $q \in D$, all $\dot{q}$, and all $t \ge t_0$, the set W contains a neighbourhood W_0 of the origin:

$$W(q,\dot{q},t) \supset W_0, \quad 0 \in W_0. \tag{3.1.11}$$

We will assume that W_0 is either a sphere of radius $r > 0$:

$$W_0 = \{w : |w| \leq r\}, \tag{3.1.12}$$

or a rectangular parallelepiped corresponding to independent constraints on w_i:

$$W_0 = \{w : |w_i| \leq w_i^0\}. \tag{3.1.13}$$

In the case of constraints (3.1.13), we define

$$r = \min_i w_i^0. \tag{3.1.14}$$

3.1.2 Statement of the problem

Substituting (3.1.5) and (3.1.9) into system (3.1.1), we get

$$\frac{d}{dt}\frac{\partial T}{\partial \dot{q}_i} - \frac{\partial T}{\partial q_i} = R_i + G_i + w_i. \tag{3.1.15}$$

Suppose we are given initial conditions

$$q(t_0) = q^0, \qquad \dot{q}(t_0) = \dot{q}^0 \tag{3.1.16}$$

and terminal conditions corresponding to the state of rest

$$q(t_*) = q^*, \qquad \dot{q}(t_*) = 0, \tag{3.1.17}$$

where $q^0 \in D$, $q^* \in D$, and $t_* > t_0$. The control problem may be formulated as follows.

Problem 3.1. It is required to find a feedback control $w(q, \dot{q})$ that satisfies the condition

$$w \in W_0 \tag{3.1.18}$$

and steers system (3.1.15) from any initial state (3.1.16) to a given terminal state (3.1.17) in finite (but not fixed) time. The set W_0 is given as (3.1.12) or (3.1.13) and in either case, by (3.1.14), contains the sphere $|w| \leq r$. The kinetic energy of system (3.1.15) is defined by (3.1.2) and satisfies conditions (3.1.3) and (3.1.4), while the forces R_i and G_i in (3.1.15) satisfy constraints (3.1.6)–(3.1.8).

Note that if the control w satisfies constraint (3.1.18), it follows from (3.1.11) that it also satisfies the initial constraint (3.1.10).

We will first construct a solution of Problem 3.1 on the assumption that system (3.1.15) involves no dissipative forces or perturbations, that is, $R_i = G_i = 0$. The general case will be considered later.

3.1.3 Control in the absence of external forces

If $R_i = G_i = 0$, system (3.1.15) becomes

$$\frac{d}{dt}\frac{\partial T}{\partial \dot{q}_i} - \frac{\partial T}{\partial q_i} = w_i. \tag{3.1.19}$$

Let $\varepsilon > 0$ be some given positive number and let Ω_1 denote the set of all points of the $2n$-dimensional phase space $(q, \dot{q})$ where $q \in D$ and $|\dot{q}_i| > \varepsilon$ for at least one i. Let Ω_2 denote the set of all points $(q, \dot{q})$ where $q \in D$ and $|\dot{q}_i| \leq \varepsilon$ for all i. Thus,

$$\begin{aligned} \Omega_1 &= \{(q, \dot{q}) : \; q \in D; \exists i, \; |\dot{q}_i| > \varepsilon\} \\ \Omega_2 &= \{(q, \dot{q}) : \; q \in D; \forall i, \; |\dot{q}_i| \leq \varepsilon\} \end{aligned} \tag{3.1.20}$$

We will construct the control $w(q, \dot{q})$ separately for each of the domains Ω_1 and Ω_2, and also specify the number ε. By the theorem on the variation of kinetic energy, applied to system (3.1.19), we have

$$\frac{dT}{dt} = \sum_{i=1}^{n} w_i \dot{q}_i = \langle w, \dot{q} \rangle. \tag{3.1.21}$$

Let us choose a control w in Ω_1 so as to satisfy constraints (3.1.18) and so that derivative (3.1.21) is negative. To that end, we define

$$w = -r\dot{q}|\dot{q}|^{-1}; \qquad w_i = -w_i^0 \operatorname{sign}|\dot{q}_i|, \quad i = 1, \ldots, n, \tag{3.1.22}$$

for cases (3.1.12) and (3.1.13), respectively. Substituting (3.1.22) into (3.1.21), we obtain, respectively,

$$\frac{dT}{dt} = -r|\dot{q}|, \quad \frac{dT}{dt} = -\sum_{i=1}^{n} w_i^0 |\dot{q}_i|. \tag{3.1.23}$$

In view of notation (3.1.14), we see that, in both cases (3.1.12) and (3.1.13),

$$\frac{dT}{dt} \equiv 2T^{1/2}\frac{dT^{1/2}}{dt} \leq -r|\dot{q}|. \tag{3.1.24}$$

The upper bound (3.1.3) for the kinetic energy (3.1.2) gives

$$|\dot{q}| \geq \left(\frac{2T}{M}\right)^{1/2}. \tag{3.1.25}$$

Substituting (3.1.25) into the right-hand side of inequality (3.1.24) and noting that $T > 0$ in Ω_1 [see (3.1.20)], we obtain

$$\frac{dT^{1/2}}{dt} \leq -r(2M)^{-1/2}. \tag{3.1.26}$$

Integrating inequality (3.1.26), we have

$$T^{1/2} - T_0^{1/2} \leq -r(2M)^{-1/2}(t - t_0), \tag{3.1.27}$$

where T_0 is the kinetic energy at the initial instant t_0. It follows from (3.1.27) that in finite time the kinetic energy will become as small as desired. Consequently, at some time t_1, the system will reach the border between Ω_1 and Ω_2.

We shall need bounds for the time t_1 and generalized coordinates $q(t_1)$. By (3.1.3) and (3.1.20), if T_1 is the kinetic energy at time t_1, then

$$T_1 \geq \frac{1}{2}m\langle \dot{q}, \dot{q}\rangle \geq \frac{1}{2}m\varepsilon^2. \tag{3.1.28}$$

Inequalities (3.1.27) and (3.1.28) yield the required bound for t_1:

$$t_1 - t_0 \leq \tau_1, \quad \tau_1 = (2M)^{1/2} r^{-1}[T_0^{1/2} - (m/2)^{1/2}\varepsilon]. \tag{3.1.29}$$

To estimate $q(t_1)$, we write the obvious inequalities

$$|q_i(t_1) - q_i^0| \leq \int_{t_0}^{t_1} |\dot{q}_i| dt \leq \int_{t_0}^{t_1} |\dot{q}| dt. \tag{3.1.30}$$

We will use the following inequalities that arise from (3.1.3) and (3.1.27):

$$|\dot{q}| \leq \left(\frac{2T}{m}\right)^{1/2} \leq \left(\frac{2}{m}\right)^{1/2} [T_0^{1/2} - r(2M)^{-1/2}(t - t_0)]. \tag{3.1.31}$$

Substituting (3.1.31) into (3.1.30) and integrating, we obtain

$$\begin{gathered} |q_i(t_1) - q_i^0| \leq \phi(t_1 - t_0), \\ \phi(\tau) = \left(\frac{2T_0}{m}\right)^{1/2} \tau - \frac{1}{2} r(Mm)^{-1/2}\tau^2. \end{gathered} \tag{3.1.32}$$

A direct check will show that $\phi(\tau)$ is a strictly increasing function in the interval $[0, \tau_1]$, where τ_1 is defined in (3.1.29). Since $t_1 - t_0 \leq \tau_1$ [see (3.1.29)], it follows that $\phi(t_1 - t_0) \leq \phi(\tau_1)$, and therefore, using (3.1.29), we deduce from (3.1.32) that

$$|q_i(t_1) - q_i^0| \leq \phi(\tau_1) = \left(\frac{M}{m}\right)^{1/2} r^{-1} \left(T_0 - \frac{1}{2}m\varepsilon^2\right). \tag{3.1.33}$$

3.1.4 Decomposition

Thus, at the instant t_1, the system is on the boundary of Ω_1 and Ω_2. We construct the control in Ω_2 so that the system, having once entered Ω_2, will never leave it again but will reach the terminal state (3.1.17) in finite time.

We will write Lagrange's equations (3.1.19) in expanded form substituting T from (3.1.2):

$$\sum_{j=1}^{n} a_{ij}\ddot{q}_j + \sum_{j,k=1}^{n} \Gamma_{ijk}\dot{q}_j\dot{q}_k = w_i. \tag{3.1.34}$$

Expression for Γ_{ijk} is given by (2.1.5), and Γ_{ijk} may be regarded as the components of n-vectors

$$\Gamma_{jk} = (\Gamma_{1jk}, \ldots, \Gamma_{njk}). \tag{3.1.35}$$

We rewrite (3.1.34) in vector notation and solve it for $\ddot{q}$. This gives

$$\ddot{q} = U' + V', \tag{3.1.36}$$

where

$$U' = A^{-1}w, \quad V' = -\sum_{j,k=1}^{n} A^{-1}\Gamma_{jk}\dot{q}_j\dot{q}_k. \tag{3.1.37}$$

It follows from condition (3.1.3) that the eigenvalues of the inverse A^{-1} lie in the interval $[M^{-1}, m^{-1}]$. Consequently, for any n-vector z,

$$|Az| \le M|z|, \quad |A^{-1}z| \le m^{-1}|z|. \tag{3.1.38}$$

We subject the components U_i' of the vector U' to the constraints

$$|U_i'| \le U_0, \quad U_0 = rM^{-1}n^{-1/2}. \tag{3.1.39}$$

The truth of constraints (3.1.39) implies that of the inequality $|U'| \le rM^{-1}$ that, in turn, by (3.1.37) and (3.1.38), implies $|w| = |AU'| \le M|U'| \le r$. Consequently, w satisfies (3.1.18) whether W_0 is taken to be (3.1.12) or (3.1.13). Thus, constraint (3.1.39) implies the truth of condition (3.1.18).

To estimate the vector V' in (3.1.37), we use the second inequality of (3.1.38)

$$|V'| \le m^{-1}\sum_{j,k}^{n} |\Gamma_{jk}||\dot{q}_j||\dot{q}_k|. \tag{3.1.40}$$

Inequalities (3.1.4) imply estimates for the quantities Γ_{ijk} introduced in (2.1.5): $|\Gamma_{ijk}| \le (3/2)C$. Hence, using (3.1.35), we have

$$|\Gamma_{jk}| = \left(\sum_{i=1}^{n} \Gamma_{ijk}^2\right)^{1/2} \le \frac{3}{2}Cn^{1/2}.$$

We substitute these bounds for Γ_{jk} and also the inequalities $|\dot{q}_i| \leq \varepsilon$—that are true in Ω_2 by virtue of (3.1.20)—into (3.1.40). This gives $|V'| \leq (3/2)Cn^{5/2}m^{-1}\varepsilon^2$. Consequently, we have the following bounds for the components V_i' of vector V'

$$|V_i'| \leq V_0, \quad V_0 = \frac{3}{2}Cn^{5/2}m^{-1}\varepsilon^2. \tag{3.1.41}$$

Equations (3.1.36) and constraints (3.1.39) and (3.1.41) may be rewritten as

$$\ddot{q}_i = U_i' + V_i', \quad |U_i'| \leq U_0, \quad |V_i'| \leq V_0, \tag{3.1.42}$$

where U_0 and V_0 are defined in (3.1.39) and (3.1.41).

Assuming that

$$\rho = \frac{V_0}{U_0} < 1, \tag{3.1.43}$$

we will construct a control U_i' separately for each degree of freedom of system (3.1.42).

To do this, we will admit that V_i' may be arbitrary functions satisfying constraints (3.1.42). We will use the minimax (guaranteed) approach that is characteristic of the theory of differential games.

Considering the ith equation of (3.1.42), we define

$$q_i - q_i^* = x, \quad \dot{q}_i = \dot{x}_i = y, \quad U_i' = u, \quad V_i' = v \tag{3.1.44}$$

and rewrite (3.1.42) and (3.1.43) as

$$\dot{x} = y, \quad \dot{y} = u + v, \quad |u| \leq U_0, \quad |v| \leq \rho U_0, \quad 0 < \rho < 1. \tag{3.1.45}$$

At time t_1, by assumption, the system is at the boundary of the domains Ω_1 and Ω_2 [see (3.1.20)]. Taking (3.1.44) into account, we have the following initial conditions for system (3.1.45):

$$x(t_1) = x^1 = q_i(t_1) - q_i^*, \quad y(t_1) = y^1 = \dot{q}_i(t_1), \quad |y^1| \leq \varepsilon. \tag{3.1.46}$$

The terminal conditions (3.1.17) become

$$x(t_*) = 0, \quad y(t_*) = 0. \tag{3.1.47}$$

To ensure that the system, having reached Ω_2 at the instant t_1, will not leave the domain again, we require that

$$|y(t)| \leq \varepsilon, \quad t > t_1. \tag{3.1.48}$$

System (3.1.45) is similar to system (2.1.30) for $\lambda = 0$. The only distinction is an additional phase constraint (3.1.48).

Thus, we have the following decomposition of Problem 3.1 in Ω_2: instead of the problem for the initial system with n degrees of freedom, we obtain n analogous problems for systems with one degree of freedom each. To solve Problem 3.1 in Ω_2, therefore, we need only solve the following problem.

Problem 3.2. Find a control $u(x,y)$ for system (3.1.45) that satisfies constraints (3.1.45) and (3.1.48) and will steer the system from the initial state (3.1.46) to a terminal state (3.1.47) in finite time for any admissible v satisfying (3.1.45).

To construct the control, we can, just as in Sect. 2.2.1, use the game approach and find the solution of the corresponding differential game with phase constraints. Instead, we suggest a simpler, though not optimal, control of the form

$$u(x,y) = U_0 \operatorname{sign}[\psi(x) - y], \quad y \neq \psi(x);$$
$$u(x,y) = U_0 \operatorname{sign} x = -U_0 \operatorname{sign} y, \quad y = \psi(x), \tag{3.1.49}$$

where the function $\psi(x)$ is defined by the relations

$$\psi(x) = -[2U_0(1-\rho)|x|]^{1/2} \operatorname{sign} x, \quad |x| \leq x^*;$$
$$\psi(x) = -\delta \operatorname{sign} x, \quad |x| > x^*. \tag{3.1.50}$$

Here, $\delta > 0$ is any number from the interval $0 < \delta < \varepsilon$, and the parameter x^* is defined by the condition of continuity of the function $\psi(x)$. According to this condition, (3.1.50) yields

$$x^* = \delta^2[2U_0(1-\rho)]^{-1}. \tag{3.1.51}$$

The switching curve $\psi(x)$ for control (3.1.49) and (3.1.50) is symmetrical about the origin and is the union of two arcs of parabolas and two rays. It is depicted by the thick curve in Fig. 3.1. Note that the parabolic arcs of the switching curve are identical with those of the switching curve constructed in Sect. 2.2.1, see (2.2.10) for $\lambda = 0$. Since $\delta < \varepsilon$, this curve lies within the strip $|y| \leq \varepsilon$ and divides it into two symmetrical parts: the domain X^+, where $y < \psi(x)$ and $u = U_0$, and the domain X^-, where $y > \psi(x)$ and $u = -U_0$ [see (3.1.49)].

We shall prove that control (3.1.49) and (3.1.50) solves Problem 3.2.

The initial conditions (3.1.46) hold at the instant t_1.

According to (3.1.45) and the control law (3.1.49), we have

$$\dot{y} \geq U_0(1-\rho), \quad (x,y) \in X^+;$$
$$\dot{y} \leq -U_0(1-\rho), \quad (x,y) \in X^-. \tag{3.1.52}$$

The width of the domains X^+ and X^- in the y direction is at most $\varepsilon + \delta$ (see Fig. 3.1), while the velocity of motion in that direction is finite and directed toward the switching curve according to (3.1.52). Consequently, the phase point will never leave the strip $|y| \leq \varepsilon$ but at a certain time $t_2 > t_1$ will be incident on the switching curve $y = \psi(x)$.

Suppose that at time t_2 the phase point has hit the straight part $y = \pm\delta$ of the switching curve $y = \psi(x)$. After that, the point will move along the straight part of

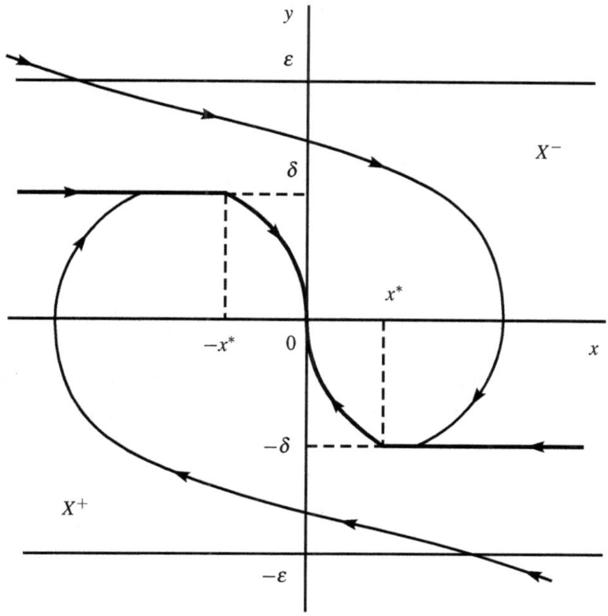

Fig. 3.1 Switching curve and phase trajectories

the curve in a sliding regime. This follows from the fact that the phase velocities on both sides of this part of the curve are finite and directed toward the switching curve. The motion will take place along these parts of the curve at an appropriate constant velocity $y = \dot{x} = \pm\delta$ in the direction of decreasing $|x|$. Consequently, at some time $t_3 > t_2$, the phase point will reach one of the points $(\pm x^*, \mp\delta)$ at the junction of the straight and curved parts of the switching curve. The curved (parabolic) parts are the phase trajectories of system (3.1.45), if u is selected in accordance with (3.1.49) and $v = -\rho u$. If $v \neq -\rho u$, the motion induced by control (3.1.49) will nevertheless take place along these parts of the parabolas, but in a sliding regime. Therefore, at some time t_*, the phase point will reach the origin.

The thin curves in Fig. 3.1 represent some possible phase trajectories. The arrows indicate the direction of increasing time t.

The entire motion, from time t_1 to time t_*, is divided into three stages: motion in the domain X^+ or X^-, motion along the straight lines $y = \pm\delta$, and motion along the parabolas. Some of these stages may be missing. For example, at the initial time t_1 the phase point may either lie on the switching curve or proceed directly from X^+ or X^- to the parabolic part of the curve. In all cases, however, the duration $t_* - t_1$ of the motion is finite.

To estimate this total time, let us assume that all three stages actually occur—this will lead us to an upper bound. The length $t_2 - t_1$ of the first stage (motion in X^+ or X^-) is estimated by dividing the maximum width $\varepsilon + \delta$ of the domains along the y axis by the velocity $\dot{y}$ of minimum absolute value as in (3.1.52). This gives

$$t_2 - t_1 \le (\varepsilon + \delta)[U_0(1-\rho)]^{-1}. \qquad (3.1.53)$$

To estimate the coordinate $x(t_2)$, we will use constraint (3.1.48) and the initial condition (3.1.46):

$$|x(t_2) - x^1| \le \int_{t_1}^{t_2} |y| dt \le \varepsilon(t_2 - t_1).$$

Hence, using (3.1.53), we get

$$|x(t_2)| \le |x^1| + \varepsilon(\varepsilon + \delta)[U_0(1-\rho)]^{-1}. \qquad (3.1.54)$$

The length $t_3 - t_2$ of the second stage (motion along straight lines $y = \pm\delta$) is obtained by dividing the distance along the x axis by the velocity that is δ in absolute value

$$t_3 - t_2 = [|x(t_2)| - x^*]\delta^{-1}.$$

Substituting (3.1.51) and (3.1.54) into this equality, we obtain

$$t_3 - t_2 \le |x^1|\delta^{-1} + \varepsilon(\varepsilon + \delta)[U_0(1-\rho)\delta]^{-1} - \delta[2U_0(1-\rho)]^{-1}. \qquad (3.1.55)$$

The length $t^* - t_3$ of the third, last stage (parabolic motion) may be estimated by dividing the velocity δ of maximum absolute value at the beginning of the stage by the acceleration of minimum absolute value defined by (3.1.52). This gives

$$t_* - t_3 = \delta[U_0(1-\rho)]^{-1}. \qquad (3.1.56)$$

Adding together (3.1.53), (3.1.55), and (3.1.56), we obtain an upper bound for the total duration of motion in Problem 3.2:

$$t_* - t_1 \le |x^1|\delta^{-1} + (2\varepsilon^2 + 4\varepsilon\delta + 3\delta^2)\delta^{-1}[2U_0(1-\rho)]^{-1}. \qquad (3.1.57)$$

The result may be summarized in the form of a theorem.

Theorem 3.1. *The control $u(x,y)$ determined by (3.1.49) and (3.1.50) in which the number x^* is defined by (3.1.51) and δ by any number in the interval $(0,\varepsilon)$ is a solution of Problem 3.2, i.e., it satisfies constraints (3.1.45) and (3.1.48) and steers system (3.1.45) from the initial state (3.1.46) to the terminal state (3.1.47) in finite time $t_* - t_1$ that is bounded as in (3.1.57).*

3.2 Feedback control design and its generalizations

3.2.1 Feedback control design

We now turn to the solution of the original Problem 3.2 in the case $R_i = G_i = 0$. The required control $w(q,\dot{q})$ in Ω_1 is defined by (3.1.22); the control in Ω_2 may be obtained from the solution $u(x,y)$ of Problem 3.2. To that end, it is sufficient to use the relations $w = AU'$ of (3.1.37) and notation (3.1.44). The result is

$$w(q,\dot{q}) = A(q)U'(q,\dot{q}), \quad U_i'(q_i,\dot{q}_i) = u(q_i - q_i^*, \dot{q}_i). \tag{3.2.1}$$

We recall that the solution $u(x,y)$ of Problem 3.1 was obtained on the assumption that $\rho < 1$ [see (3.1.43)]; with notation (3.1.39) and (3.1.41), this leads to the following restriction on ε:

$$\varepsilon < \varepsilon_0 = \left(\frac{2mr}{3MCn^3}\right)^{1/2}. \tag{3.2.2}$$

To estimate the total duration $t_* - t_0$ of the motion, we must add the times of motion in the domains Ω_1 and Ω_2. When evaluating $t_* - t_1$, we take into consideration that $|x^1|$ in (3.1.57) should be replaced by the maximum over i difference $|q_i(t_1) - q_i^*|$ [see (3.1.46)], since the system will reach the terminal state when all coordinates take their terminal values. Using estimate (3.1.33), we obtain

$$|x^1| = \max_i |q_i(t_1) - q_i^*| \le \max_i(|q_i(t_1) - q_i^0| + |q_i^0 - q_i^*|)$$

$$\le \max_i |q_i^0 - q_i^*| + \left(\frac{M}{m}\right)^{1/2} r^{-1}\left[T_0 - \frac{1}{2}m\varepsilon^2\right].$$

This expression is substituted into (3.1.57) that we then add to inequality (3.1.29):

$$\begin{aligned} t_* - t_0 \le{}& \delta^{-1}\max_i |q_i(t_1) - q_i^*| + (2M)^{1/2} r^{-1}\left[T_0^{1/2} - \left(\frac{m}{2}\right)^{1/2}\varepsilon\right] \\ &+ \left(\frac{M}{m}\right)^{1/2} r^{-1}\delta^{-1}\left(T_0 - \frac{1}{2}m\varepsilon^2\right) \\ &+ (2\varepsilon^2 + 4\varepsilon\delta + 3\delta^2)\delta^{-1}[2U_0(1-\rho)]^{-1}. \end{aligned} \tag{3.2.3}$$

The parameters U_0 and ρ are defined by (3.1.39), (3.1.41), and (3.1.43) with $\rho < 1$ by condition (3.2.2).

The result may be stated as the following theorem.

Theorem 3.2. *Problem 3.1 for system (3.1.19), i.e., when $R_i = G_i = 0$, is always solvable. For any $\varepsilon \in (0, \varepsilon_0)$ with ε_0 given by (3.2.2), the control $w(q,\dot{q})$ defined by (3.1.22) in Ω_1 [for cases (3.1.12) and (3.1.13), respectively] and by (3.2.1) in Ω_2 solves the problem, i.e., it steers system (3.1.19) from any initial state (3.1.16) to a given terminal state (3.1.17) in finite time $t_* - t_0$ that satisfies inequality (3.2.3). Under these conditions, the function $u(x,y)$ in (3.2.1) is defined by (3.1.49) and (3.1.50) in which the parameters U_0, ρ, and x^* are given by formulas (3.1.39), (3.1.41), (3.1.43), and (3.1.51) and δ is any number in the interval $(0,\varepsilon)$.*

We observe that, in order to reduce the duration of the motion, δ should be chosen as close as possible to ε. If $\delta = \varepsilon$, however, one can no longer guarantee that the system will remain in Ω_2 after reaching the boundary of Ω_1 and Ω_2. For that reason, δ should be chosen in the interval $(0, \varepsilon)$. Our solutions to Problems 3.1 and 3.2 are naturally not unique. In particular, there are other possible ways to synthesize controls in the one-dimensional system (3.1.45) obtained by the above decomposition.

3.2.2 Control in the general case

We now proceed to solve Problem 3.1 for system (3.1.15) in the general case. The approach is largely the same as in Sects. 3.1.3–3.2.1.

Letting $\varepsilon > 0$ be given, we again introduce the domains Ω_1 and Ω_2 defined by (3.1.20). By the dissipative property (3.1.6) of the forces R_i, the theorem on the variation of the kinetic energy of system (3.1.15) yields a relation similar to (3.1.21)

$$\frac{dT}{dt} \leq \sum_{i=1}^{n} (w_i + G_i)\dot{q}_i. \tag{3.2.4}$$

The control w in Ω_1 will be chosen so as to minimize the scalar product $\langle w, \dot{q} \rangle$ subject to constraint (3.1.18). Whether W_0 is defined by (3.1.12) or (3.1.13), we again obtain the appropriate expression of (3.1.22). We now substitute these expressions into inequality (3.2.4) and use constraints (3.1.8) as well as the Cauchy inequality. If W_0 is a sphere (3.1.12), we obtain

$$\frac{dT}{dt} \leq -r|\dot{q}| + \sum_{i=1}^{n} G_i^0 |\dot{q}_i| \leq -r_1 |\dot{q}|, \tag{3.2.5}$$

where

$$r_1 = r - |G^0| > 0, \qquad G^0 = (G_1^0, \ldots, G_n^0). \tag{3.2.6}$$

If W_0 is a rectangular parallelepiped (3.1.13), we obtain

$$\frac{dT}{dt} \leq -\sum_{i=1}^{n} (w_i^0 + G_i^0)|\dot{q}_i| \leq -r_2 |\dot{q}|. \tag{3.2.7}$$

Here,

$$r_2 = |w^0 - G^0|, \qquad w^0 = (w_1^0, \ldots, w_n^0), \tag{3.2.8}$$

and it is assumed that

$$w_i^0 > G_i^0, \quad i = 1, \ldots, n. \tag{3.2.9}$$

Thus, if inequalities (3.2.6) hold for the sphere (3.1.12), or inequalities (3.2.9) for the parallelepiped (3.1.13), then inequalities (3.2.5) and (3.2.7) lead directly to inequality (3.1.24) with the constant $r > 0$ replaced by $r_\alpha > 0$. Here and below, the indices $\alpha = 1,2$ correspond to cases (3.1.12) and (3.1.13), respectively. Hence, all the formulas of Sect. 3.1.3 relating to Ω_1 remain valid with the above reservation.

We will now consider Ω_2. We impose the condition

$$\varepsilon \leq \vartheta_0, \tag{3.2.10}$$

under which estimates (3.1.7) will hold in the domain Ω_2. Lagrange's equations (3.1.15) may again be converted to the form (3.1.36) by solving them for the derivatives

$$\ddot{q} = U' + V^* \tag{3.2.11}$$

with the same relation (3.1.37) holding for U' as before. The vector V^* in (3.2.11) is given by

$$V^* = V' + A^{-1}(R + G). \tag{3.2.12}$$

The vector V' is defined by (3.1.37); R and G are the vectors with components R_i and G_i, respectively.

Using inequalities (3.1.38) for A^{-1}, (3.1.7) for R_i, and (3.1.8) for G_i, we obtain the estimate

$$\begin{gathered} |A^{-1}(R+G)| \leq m^{-1}|R^0(\varepsilon) + G^0|, \\ R^0(\varepsilon) = (R_1^0(\varepsilon), \ldots, R_n^0(\varepsilon)). \end{gathered} \tag{3.2.13}$$

In accordance with the assumptions made in Sect. 3.1.1 about the functions R_i^0 [see (3.1.7)], $R_i^0(\varepsilon)$ is a continuous monotone increasing function of ε with $R_i^0(0) = 0$.

Inequalities (3.1.41) and (3.2.13) imply the following bound for the vector V^* in (3.2.12):

$$\begin{aligned} |V^*| \leq V_0^* &= V_0 + m^{-1}|R^0(\varepsilon) + G^0| \\ &= m^{-1}\left[\frac{3}{2}Cn^{5/2}\varepsilon^2 + |R^0(\varepsilon) + G^0|\right]. \end{aligned} \tag{3.2.14}$$

We impose on V_0^* the following analogue of condition (3.1.43):

$$\rho^* = \frac{V_0^*}{U_0} < 1. \tag{3.2.15}$$

The procedure for constructing the control in Ω_2 and all the subsequent estimates in that domain remain the same as in Sects. 3.1.4 and 3.2.1. The only changes are to replace ρ by ρ^* and r by r_α in estimates (3.2.3) for the time. In formula (3.1.39) for U_0, the parameter r must be retained without change: here, it is defined by (3.1.12) and (3.1.14) for cases (3.1.12) and (3.1.13), respectively. In addition, the restrictions on the choice of ε are changed: instead of (3.2.2), we now have two conditions: (3.2.10) and (3.2.15). In developed form, using (3.1.39) and (3.2.14), we obtain

$$\varepsilon \leq \vartheta_0, \quad \frac{3}{2}Cn^{5/2}\varepsilon^2 + |R^0(\varepsilon) + G^0| < mM^{-1}rn^{-1/2}. \tag{3.2.16}$$

Thus, our procedure for the control synthesis will produce a solution of Problem 3.1 provided the following conditions are satisfied: inequalities (3.2.6) or (3.2.9) in cases $\alpha = 1,2$, respectively, and both inequalities (3.2.16) for ε. A number ε satisfying (3.2.16) will always exist if there are no perturbations ($G^0 = 0$) or if the perturbations are sufficiently small

$$|G^0| < mM^{-1}rn^{-1/2}. \tag{3.2.17}$$

This follows from the continuity of $|R^0(\varepsilon)|$: $|R^0| \to 0$ as $\varepsilon \to 0$. We note that, in the case of dissipative forces proportional to the velocities, the functions R_i^0 in (3.1.7) and R^0 in (3.2.13) are linear in ε.

We summarize the results.

Theorem 3.3. *Let α be 1 or 2 depending on whether W_0 is a sphere (3.1.12) or a parallelepiped (3.1.13), respectively. Assume that conditions (3.2.6) and (3.2.9) are satisfied for $\alpha = 1,2$, respectively, and that there exists $\varepsilon > 0$ satisfying both conditions (3.2.16). Then, the control $w(q,\dot{q})$ defined by (3.1.22) in Ω_1 (for $\alpha = 1,2$, respectively) and by (3.2.1) in Ω_2 solves Problem 3.1 for system (3.1.15), i.e., it steers the system from any initial state (3.1.16) to the given terminal state (3.1.17). Under these conditions, the function $u(x,y)$ in (3.2.1) is defined by (3.1.49) and (3.1.50) in which the parameters U_0 and x^* are given by (3.1.39) and (3.1.51). The parameter ρ in formula (3.1.51) should be replaced by ρ^* as in (3.2.15) and (3.2.14); under these conditions, we have $\rho^* < 1$. The number δ may be chosen anywhere in the interval $(0,\varepsilon)$. The duration $t_* - t_0$ of the motion is finite and satisfies inequality (3.2.3) with r replaced by r_α [see (3.2.6) and (3.2.8)] and ρ by ρ^*.*

Let us note that the full duration of time depends on the value of ε. First, an increase of the number ε results in a decrease in time for reaching by the trajectory of the set Ω_2. Second, in the set Ω_2 the system goes through the straight-line trajectory segments along the rays $y = \pm\delta$ in (3.1.49) and (3.1.50) at the velocity with $|\dot{q}| = \delta < \varepsilon$. Consequently, the greater ε, the greater the number δ can be, and the higher will be the velocity of motion along these segments. According to (3.2.16), the selection of ε is determined, in particular, by the constants M and m that bound in (3.1.3) the maximal and minimal eigenvalues of the matrix $A(q)$, and the constant C from inequalities (3.1.4). Thus, as region for change of the vector of generalized coordinates q, it is reasonable to use, not the entire region D, but some subregion $D' \subset D$ in which the trajectory of motion will lie. The region D' depends on the initial and terminal conditions in the problem. A decreasing region D' in general causes the constants M and C to decrease, and the constant m to increase, which allows the parameter ε to be increased.

Remark 3.1. In this approach, the values $m, M, C, \varepsilon, \delta, \rho$, and x^* are selected earlier; they are general for the entire motion process. Nevertheless, it is clear that in real mechanical systems the eigenvalues $m(q)$, $M(q)$ and the derivatives $\partial a_{ij}(q)/\partial q_k$ of the elements of the matrix $A(q)$ by far do not always reach the boundaries of inequalities (3.1.3) and (3.1.4). For this reason, the modified control law according to which values m, M, and C are selected at each instant, depending on the current condition of the matrix $A(q)$, becomes of interest. Thus, if at time t the vector of phase coordinates for the system is q, then the control is formed according to law (3.1.22), where $m = m(q)$, $M = M(q)$, $C = \max_{i,j,k} |\partial a_{ij}(q)/\partial q_k|$, and the numbers ε, δ, ρ, and x^* are selected in accordance with (3.1.39), (3.1.51), (3.2.15), and (3.2.16). Modelling described in details in Sect. 3.3.2 has shown that the system controlled according to this modified law reaches the terminal state more quickly; nevertheless, to determine the region of applicability of such an approach, further studies are needed.

3.2.3 Extension to the case of nonzero terminal velocity

Let us generalize the suggested approach to the case of nonzero terminal velocities.

Problem 3.3. We want to construct a control $w(q,\dot{q})$ that satisfies constraints (3.1.18) and transfers system (3.1.15) from the arbitrary initial state (3.1.16) to the assigned terminal state

$$q(t_*) = q^* \in D, \quad \dot{q}(t_*) = \dot{q}^*, \quad t_* > t_0 \tag{3.2.18}$$

in finite nonfixed time.

Without loss of generality, we can say that, in the end state (3.2.18), of all phase coordinates q_i^* and $\dot{q}_i^*$ only one, $\dot{q}_1^*$, the first component of the velocity vector, is not equal to zero. To see this, we note that there exists an orthogonal matrix B, $B^T B = E$ (E is the identity matrix), such that the linear transformation $q \mapsto B(q - q^*)$ transforms conditions (3.2.18) to the form

$$q^* = 0, \quad \dot{q}_1^* = |\dot{q}^*|, \quad \dot{q}_i^* = 0, \quad i = 2, \ldots, n. \tag{3.2.19}$$

The constants that bound the components of the matrices $\partial A/\partial q_k$ and the vectors G and w as well as function R_i^0 that bounds the components of the vector R do not change more than by a factor $n^{1/2}$. For simplicity, we will assume that inequalities (3.1.4), (3.1.6)–(3.1.8), and (3.1.12) refer to the system of coordinates obtained after the transformation made.

Let us assume at the outset that the parameters of the problem satisfy (3.2.17), and the absolute value of the terminal velocity $\dot{q}^*$ is sufficiently small so that the number ε is selected on the basis of the condition

$$|\dot{q}_1^*| \le \varepsilon. \tag{3.2.20}$$

Following Sects. 3.1.3–3.2.2, the solution of Problem 3.3 will be split into two stages.

The purpose of the first stage is to lower the phase velocity to values at which the decomposition of the system is possible, namely $|\dot{q}_i| \le \varepsilon$. In order to attain the goal of the first stage, we will use the control constructed in Sects. 3.1.3 and 3.2.2 in the domain Ω_1 defined by (3.1.20).

In the region Ω_2 of small velocities in the phase space $(q,\dot{q})$, the original nonlinear system (3.1.15) of order $2n$ reduces to a set of n controllable subsystems (3.2.11) of second order in which the nonlinear terms are treated as perturbations. The purpose of the second stage is to construct for each subsystem (3.2.11) in the domain $|\dot{q}_i| \leq \varepsilon$ of the phase space $(q_i, \dot{q}_i)$ a feedback control U_i' satisfying the constraint $|U_i'| \leq U_0$ that brings the corresponding subsystem from a certain initial state $(q_i^1, \dot{q}_i^1)$ at $t = t_1$ to the assigned terminal state $(q_i^*, \dot{q}_i^*)$ in finite time. The initial state and the entire trajectory here should belong to the set $|\dot{q}_i| \leq \varepsilon$. In Sects. 3.1.4–3.2.2, this problem is solved for the case of zero terminal conditions. The control proposed there brings each subsystem (3.2.11) with indices $i = 2, \ldots, n$ to the origin of coordinates and keeps it there.

We will construct the desired control for the first equation of (3.2.11) corresponding to the component q_1 of the vector q. The terminal value for the velocity of this component is not zero. Let us designate

$$x = q_1, \quad y = \dot{x} = \dot{q}_1, \quad u = U_1', \quad v = V_1^*$$

and reduce the considered equation to the form of (3.1.45). Concerning finding the constants U_0 and ρ that figure in the constraints imposed on the functions u and v in (3.1.45), see Sects. 3.1.3–3.2.2. Let us examine there an additional problem.

Problem 3.4. Construct a control $u(x,y)$ that brings system (3.1.45) from the initial state

$$x(t_1) = q_1(t_1), \quad y(t_1) = \dot{q}_1(t_1), \quad |y(t_1)| \leq \varepsilon \tag{3.2.21}$$

to the terminal state

$$x(t_*) = 0, \quad y(t_*) = y^* \tag{3.2.22}$$

in finite time with any allowable v [the phase limits (3.1.48) are ignored for the time being].

Let us solve Problem 3.4 as a differential game using the mini-max approach. In this game, the side selecting the control u strives to reduce the time t_* for reaching the terminal state, and the second side strives to increase this time with the help of control v. The optimal control u in this game coincides [79] with the minimum time control for the system:

$$\dot{x} = y, \quad \dot{y} = (1-\rho)u, \quad |u| \leq U_0. \tag{3.2.23}$$

System (3.2.23) is obtained by substitution of the control $v = -\rho u$ optimal for the second player (and the worst for the first player) into system (3.2.11). Let us synthesize a time-optimal feedback control for system (3.2.23) with terminal condition (3.2.22). In the phase space (x,y), the time-optimal trajectories for system (3.2.23) consist of segments of two families of parabolas (see Chapter 1)

$$x = \frac{y^2}{2(1-\rho)U_0} + b, \qquad x = -\frac{y^2}{2(1-\rho)U_0} + b, \tag{3.2.24}$$

where b is an arbitrary constant. Motion along the parabolas of the first family occurs in the (x,y)-plane upward in the direction of increasing y, and along parabolas of the second family, downward in the direction of decreasing y [see Fig. 3.2]. It is not difficult to see that only two parabolas pass through the point $(0,y^*)$, the final state of the system. Consequently, the switching curve consists of segments of these parabolas and is described by

$$x = -\frac{y^2 - y^{*2}}{2(1-\rho)U_0}, \quad y \geq y^*; \qquad x = \frac{y^2 - y^{*2}}{2(1-\rho)U_0}, \quad y < y^*. \tag{3.2.25}$$

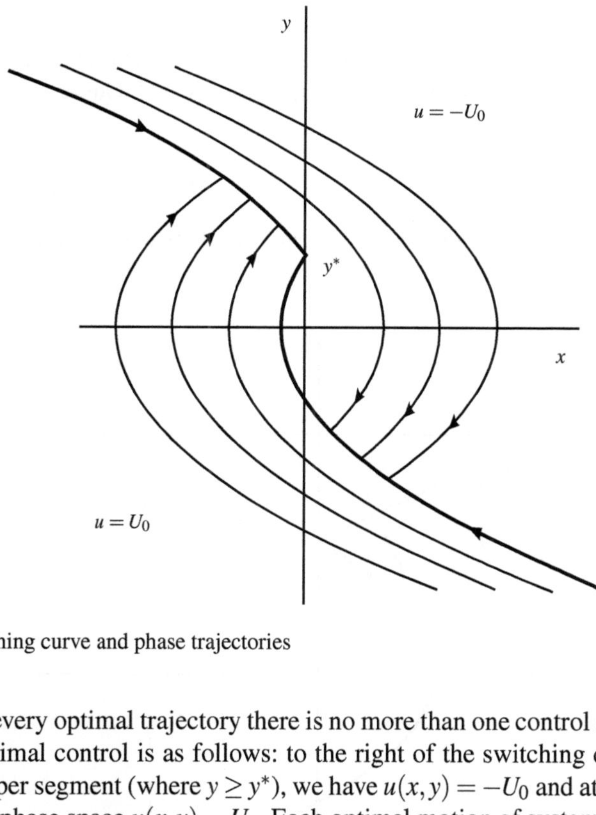

Fig. 3.2 Switching curve and phase trajectories

Since on every optimal trajectory there is no more than one control switching, the feedback optimal control is as follows: to the right of the switching curve (3.2.25) and on its upper segment (where $y \geq y^*$), we have $u(x,y) = -U_0$ and at all remaining points of the phase space $u(x,y) = U_0$. Each optimal motion of system (3.2.23) consists of two stages: at first the representative point moves along one of the parabolas (3.2.24) to the switching curve (3.2.25), then along this curve to the point $(0,y^*)$. If in the initial position the system is located on curve (3.2.25), then the first stage is missing.

Let us remember that motion of system (3.1.45) is subject to (3.2.23) in the case $v = -\rho u$, i.e., if the second player acts in a manner optimal for him. If, however, the control v is selected in some other way, then the motion along the switching curve (3.2.25) also occurs along different trajectories, while on segments of this curve where $y > y^*$ or $y < 0$, sliding regimes appear. In addition, on the segment of the switching curve where $0 \le y < y^*$, the trajectory can leave the curve returning to it again when $y > y^*$. Then, the system will come to its terminal state (3.2.22) along the upper segment of curve (3.2.25). We point out that, if the second player uses a nonoptimal method ($v = -\rho u$), the time of motion to the terminal state can only decrease.

Using the solution to Problem 3.4, we will synthesize a control (already nonoptimal) for system (3.1.45) taking into account the phase limits (3.1.48). Let the numbers δ and x^* be such that

$$y^* < \delta < \varepsilon, \quad x^* = \frac{\delta^2 - y^{*2}}{2(1-\rho)U_0}. \tag{3.2.26}$$

Let us designate by K the continuous curve located in the strip $\Omega^\varepsilon = \{(x,y) : |y| \le \varepsilon\}$ of the phase space (x,y) and passing through point $(0,y^*)$. Curve K consists of two rays

$$L_1 = \{(x,y) : x \le -x^*, y = \delta\}, \qquad L_2 = \{(x,y) : x \ge x^*, y = -\delta\},$$

and also the segment of curve (3.2.25) enclosed between the lines $y = \pm\delta$ (see Fig. 3.3).

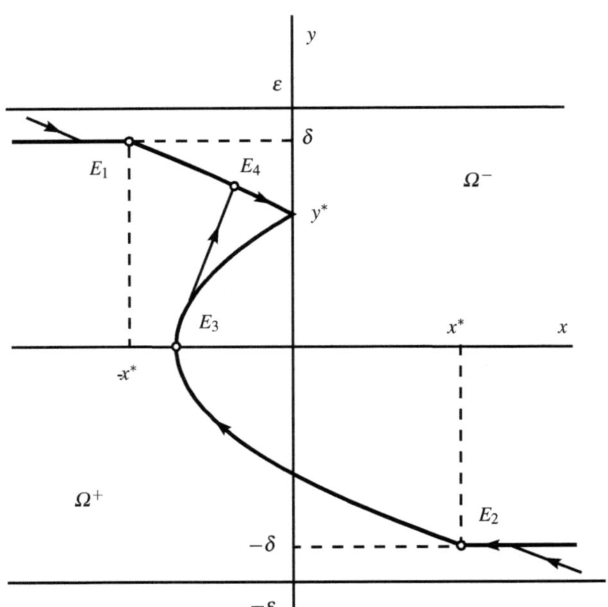

Fig. 3.3 Modified switching curve

Let us split the strip Ω^ε into two sets. The part Ω^- will represent the set of points of Ω^ε lying above and to the right of the curve K or on that segment of it, where $y > y^*$. The symbol Ω^+ will be used to designate the supplement of the set Ω^- in Ω^ε. In the strip Ω^ε, we will define the control $u(x,y)$ as follows:

$$u(x,y) = \begin{cases} -U_0, \ (x,y) \in \Omega^-, \\ U_0, \ (x,y) \in \Omega^+. \end{cases} \tag{3.2.27}$$

On the strength of (3.1.45) and (3.2.27), the derivative $\dot{y}$ on Ω^- satisfies the inequality

$$\dot{y} \leq -(1-\rho)U_0 < 0, \tag{3.2.28}$$

and on Ω^+,

$$\dot{y} \geq (1-\rho)U_0 > 0. \tag{3.2.29}$$

Inequalities (3.2.28) and (3.2.29) are analogous to inequalities (3.1.52).

Consequently, the value of $\dot{y}$ on the line $y = \varepsilon$ is negative, and, on the line $y = -\varepsilon$, it is positive. Thus, the trajectory, having fallen into the set Ω^ε, will not leave it, and condition (3.1.48) will be fulfilled.

We will show that control (3.2.27) brings system (3.1.45) to the terminal state in finite time, and we will estimate this time. At the instant $t = t_1$, let the trajectory of system (3.1.45) be located on the set Ω^- (or Ω^+), i.e., $(x^1, y^1) \in \Omega^-$ [or $(x^1, y^1) \in \Omega^+$]. Taking into consideration inequality (3.2.28) [or (3.2.29) for set Ω^+], we can conclude that the trajectory reaches the curve K in time τ_1 that does not exceed the ratio of the maximum width of the set Ω^- (or Ω^+) along the y axis and the minimal derivative $\dot{y}$ in absolute value:

$$\tau_1 \leq \frac{\varepsilon + \delta}{(1-\rho)U_0}. \tag{3.2.30}$$

In this time, the x coordinate changes no more than by $\varepsilon\tau_1$, since $|\dot{x}| = |y| \leq \varepsilon$. Thus,

$$|x(t_2)| \leq |x^1| + \frac{\varepsilon(\varepsilon + \delta)}{(1-\rho)U_0}, \tag{3.2.31}$$

where $(x(t_2), y(t_2))$ is the point at which the trajectory first reaches the switching curve. Estimates (3.2.30) and (3.2.31) are analogous to (3.1.53) and (3.1.54).

If $|x(t_2)| > x^*$, i.e., if the point $(x(t_2), y(t_2))$ lies on one of the rays L_1 or L_2, then further motion will occur along this ray in the direction of decreasing $|x|$ with the constant velocity $\dot{x} = \pm\delta$. The system will pass this segment of the trajectory in a sliding regime because on both sides the phase velocities are finite and directed toward the switching curve. The motion time τ_2 along the ray from point $(x(t_2), y(t_2))$ to point E_1 (or E_2) with the abscissa $\pm x^*$ (see Fig. 3.3) is equal to $\tau_2 = (|x(t_2)| - x^*)/\delta$. Taking into account (3.2.31), we get the estimate

$$\tau_2 \leq \left[|x^1| + \frac{\varepsilon(\varepsilon + \delta)}{(1-\rho)U_0} - x^* \right] \frac{1}{\delta} \tag{3.2.32}$$

that is analogous to (3.1.55). The described stage of motion along rays L_1 and L_2 is absent if $|y(t_2)| \leq \delta$, i.e., if the point $(x(t_2), y(t_2))$ lies on the segment of curve (3.2.25) between the lines $y = \pm\delta$.

From point E_1, the system moves, possibly in a sliding regime, along the segment $y^* < y < \delta$ of curve (3.2.25). The time τ_3 of passing this segment will be defined as follows:

$$\tau_3 = \frac{\delta - y^*}{(1-\rho)U_0}. \tag{3.2.33}$$

From point E_2, the motion occurs in time τ_4 along curve (3.2.25) to point E_3 lying on the x axis, and

$$\tau_4 = \frac{\delta}{(1-\rho)U_0}. \tag{3.2.34}$$

Finally, from point E_3, the system can move along curve (3.2.25) to the terminal state (if $v = -\rho U_0$), or it can move off it (if $v \neq -\rho U_0$). In the second case, the trajectory falls on the switching curve at some point E_4 [that lies either on the ray L_1 or, as shown in Fig. 3.3, on the segment $y > y^*$ of curve (3.2.25)], after which, along the switching curve, it will come to the terminal state. The time τ_5 of motion on this segment of the path reaches the maximum if control v is optimal from the point of view of the second player, i.e., if $v = -\rho U_0$. Then, on the strength of (3.2.29), we get

$$\tau_5 \leq \frac{y^*}{(1-\rho)U_0}. \tag{3.2.35}$$

From (3.2.33) and (3.2.34), it follows that $\tau_3 < \tau_4$ and, hence, from point E_1, the trajectory reaches the terminal state clearly more quickly than from point E_2. Thus, to obtain the estimate τ_1^* for the complete time of motion of system (3.1.45) from the initial point (x^1, y^1) to the terminal state $(0, y^*)$, it is sufficient to sum up the right-hand sides of inequalities (3.2.30), (3.2.32), (3.2.34), and (3.2.35). After some transformations taking into account (3.2.26), we obtain

$$\tau_1^* = \frac{(y^* + \delta)^2 + 2(\varepsilon + \delta)^2}{2(1-\rho)U_0\delta} + \frac{|x^1|}{\delta}, \qquad x^1 = q_1(t_1). \tag{3.2.36}$$

We note that the control constructed above can be used to bring the ith subsystem, $i \geq 2$, to the zero terminal position and keep it there. In order to do this, we need to assume that $y^* = 0$ and $u(0,0) = 0$ (we will then get the control presented in Sect. 3.1.4). The time estimate τ_i^* for motion of the ith subsystem (3.2.11) or (3.1.45) to zero can be obtained by assuming $y^* = 0$ and $x^1 = q_i(t_1)$ in (3.2.36):

$$\tau_i^* = \frac{\delta^2 + 2(\varepsilon + \delta)^2}{2(1-\rho)U_0\delta} + \frac{|q_i(t_1)|}{\delta}, \qquad i \geq 2. \tag{3.2.37}$$

Let $\tau^* = \max_i \tau_i^*$, $i = 2, \ldots, n$. Then

$$\tau^* = \frac{\delta^2 + 2(\varepsilon + \delta)^2}{2(1-\rho)U_0\delta} + \frac{\max_i |q_i(t_1)|}{\delta}, \quad i \geq 2. \tag{3.2.38}$$

In bounds (3.2.36)–(3.2.38), the values of coordinates $q_i(t_1)$ enter into the time t_1 for emergence of the system onto the boundary between sets Ω_1 and Ω_2. These coordinates can be estimated analogously to (3.1.33). We will get

$$|q_i(t_1)| \leq |q_i^0| + \frac{1}{2}\sqrt{\frac{M}{m}\frac{M\left(\dot{q}^0\right)^2 - m\varepsilon^2}{r}}. \tag{3.2.39}$$

Estimate (3.2.39) can be substituted into inequalities (3.2.36)–(3.2.38).

In contrast with other subsystems, the subsystem corresponding to the first component of the vector q cannot be maintained constantly in the terminal state $(0, y^*)$. Thus, for $i = 1$, system (3.1.36) will not stop; nevertheless, its trajectory will return to the point $(0, y^*)$ in a time not exceeding τ_1^*.

Let us return to the solution of Problem 3.3. From the above considerations, it follows that the control given in the form (3.1.22) in set Ω_1 brings system (3.1.15) from the initial state (3.1.16) to the boundary separating the sets Ω_1 and Ω_2 in a certain time t_1 for which estimate (3.1.29) is valid. Sets Ω_1 and Ω_2 are determined in (3.1.20), and the number ε is specified by (3.2.16).

In the set Ω_2, the system splits into n separate subsystems (3.2.11) or (3.1.45). Using the control of the form (3.2.27), each of these subsystems (3.1.45) having i greater than one is brought to the origin of coordinates no later than in time $t_1 + \tau^*$ and is kept there. Let us note that, for $i > 1$, during construction of sets Ω^- and Ω^+, it must be assumed that $y^* = 0$.

Subsystem (3.2.11) with $i = 1$ will be brought to the terminal state $(0, y^*)$ for the first time no later than in time $t_1 + \tau_1^*$. If $\tau^* \leq \tau_1^*$, then the entire system as a whole at this time will turn out to be in the given terminal state. If, however, $\tau^* > \tau_1^*$, then the first subsystem will come out of the terminal state, and it will again be brought to this state with the same control (3.2.27). Thus, for the time t_* necessary to bring the entire system to the given terminal state, we have the estimate

$$t_* \leq t_1 + \tau^* + \tau_1^*, \tag{3.2.40}$$

where τ_1^* and τ^* are subject to inequalities (3.2.36) and (3.2.38), respectively.

Thus, if the parameters of the original system (3.1.15) satisfy the limitations (3.1.3), (3.1.4), (3.1.6)–(3.1.8), (3.1.12), and (3.2.17), then in finite time the system can be brought from an arbitrary initial state (3.1.16) to the nonzero terminal state $(q^*, \dot{q}^*)$, if conditions (3.2.16) and (3.2.20) are satisfied. These conditions restrict the choosing of the parameter ε: on the one hand, it should be sufficiently low so that inequalities (3.2.16) are satisfied, on the other hand, the fulfillment of inequality (3.2.20) is required. Thus, the realization of the suggested control approach is possible only in the case of the sufficiently low terminal velocity $\dot{q}_1^*$. The feedback control law for solving the stated problem is determined by formulas (3.1.22) in the set Ω_1 and by (3.2.1) and (3.2.27) in the set Ω_2. This control law ensures, under conditions mentioned above, the steering of system (3.1.15) to the terminal state at the instant t_*, for which estimate (3.2.40) holds. Remarks concerning the choosing of the parameter ε, given at the end of Sect. 3.2.2, remain valid also for the case of nonzero terminal velocity, under a restriction: here, condition (3.2.20) should be satisfied.

3.2.4 Tracking control for mechanical system

Statement of the problem

Following [15], we apply the decomposition method to the problem of tracking the given trajectory of mechanical system. Consider a system whose dynamics is governed by (3.1.1) and (3.1.2), where forces Q_i consist of the given generalized forces $P_i(t)$ and the unknown generalized forces $G_i(q,\dot{q},t)$ called hereafter perturbations:

$$Q_i = P_i + G_i. \tag{3.2.41}$$

We assume that the perturbations $G_i(t,q,\dot{q})$ satisfy the conditions

$$|G_i(t,q,\dot{q})| \leq G_i^0, \tag{3.2.42}$$

while the control forces U_i are subject to the constraints

$$|U_i| \leq U_i^0, \qquad i = 1,\dots,n, \tag{3.2.43}$$

where U_i^0 and G_i^0 are given constants such that

$$U_i^0 > G_i^0 > 0. \tag{3.2.44}$$

It is assumed that, for any $q \in R^n$, the eigenvalues of the positive definite symmetric matrix $A(q)$ lie on the segment $[m,M]$, where $0 < m \leq M$, i.e., condition (3.1.3) holds. We also assume that the matrix $A(q)$ is twice differentiable and its partial derivatives of the first and second order are uniformly bounded in the norm, i.e.,

$$\left\| \frac{\partial A}{\partial q_i}(q) \right\| \leq C_1, \quad \left\| \frac{\partial^2 A}{\partial q_i \partial q_j}(q) \right\| \leq C_2, \quad C_1, C_2 > 0, \quad i,j = 1,\dots,n. \tag{3.2.45}$$

By $\|Z\|$, we denote the induced norm of a matrix, i.e., the norm of the corresponding linear operator in the Euclidean space:

$$\|Z\| = \max_{|z|=1} |Zz|,$$

where z is a vector of the appropriate dimension. If the matrix Z is symmetric then its norm is equal to the maximal absolute value of all eigenvalues of the matrix. If the matrix Z is not symmetric then its norm is equal to the square root or maximal eigenvalue of the symmetric nonnegative definite matrix $Z^\top Z$.

Since the matrices mentioned in (3.2.45) are symmetric, the constants C_1 and C_2 may be chosen as the maximal absolute value of eigenvalues of the matrices $\partial A/\partial q_i$ and $\partial^2 A/\partial q_i \partial q_j$, $i, j = 1, \ldots, n$, respectively.

Suppose that the vector functions $\tilde{q}(t)$ and $\dot{\tilde{q}}(t)$ define the motion trajectory that starts at the moment $t = t_0$ at the point

$$\tilde{q}^0 = \tilde{q}(t_0), \qquad \dot{\tilde{q}}^0 = \dot{\tilde{q}}(t_0)$$

and is realized under the action of generalized forces $P_i(t)$ on the unperturbed system for

$$U_i(t) \equiv 0, \qquad G_i(t) \equiv 0, \qquad t \geq t_0$$

Henceforth, we call such a trajectory *nominal.*

Denote by x and $\dot{x}$ the deviations of the phase coordinates and velocities of the perturbed trajectory from the nominal one, i.e.,

$$x(t) = q(t) - \tilde{q}(t), \qquad \dot{x}(t) = \dot{q}(t) - \dot{\tilde{q}}(t), \tag{3.2.46}$$

and let

$$x^0 = x(t_0) = q^0 - \tilde{q}^0, \qquad \dot{x}^0 = \dot{x}(t_0) = \dot{q}^0 - \dot{\tilde{q}}^0 \tag{3.2.47}$$

be the initial deviation of the trajectory of the system from the nominal one, where $q^0 = q(t_0)$ and $\dot{q}^0 = \dot{q}(t_0)$.

Problem 3.5. Construct a control $U = (U_1, \ldots, U_n)$ as a vector function of the phase variables q and $\dot{q}$ that satisfies condition (3.2.43) and find a domain $\Omega^x \subset R^{2n}$ of admissible initial deviations x^0 and $\dot{x}^0$ such that any trajectory of the perturbed control system (3.1.1) starting in this domain reaches the nominal trajectory in finite time and will move along this trajectory under any perturbations $G = (G_1, \ldots, G_n)$ subject to constraints (3.2.42).

Equations in deviations

Let us write the equation of motion along the nominal trajectory as

$$\sum_{j=1}^{n} a_{ij}(\tilde{q})\ddot{\tilde{q}}_j = - \sum_{j,k=1}^{n} \Gamma_{ijk}(\tilde{q})\dot{\tilde{q}}_j\dot{\tilde{q}}_k + P_i(t) \tag{3.2.48}$$

and the equation of motion along the perturbed trajectory as

$$\begin{aligned} \sum_{j=1}^{n} a_{ij}(\tilde{q}+x)(\ddot{\tilde{q}}_j + \ddot{x}_j) = - \sum_{j,k=1}^{n} \Gamma_{ijk}(\tilde{q}+x)(\dot{\tilde{q}}_j + \dot{x}_j)(\dot{\tilde{q}}_k + \dot{x}_k) \\ + P_i(t) + U_i + G_i, \quad i = 1, \ldots, n. \end{aligned} \tag{3.2.49}$$

Here, the functions Γ_{ijk} are defined by (2.1.5). Applying the Taylor formula with the remainder term in the Lagrangian form:

$$a_{ij}(\tilde{q}+x) = a_{ij}(\tilde{q}) + \sum_{m=1}^{n} \frac{\partial a_{ij}(\bar{q})}{\partial q_m} x_m,$$

where $\bar{q} = \tilde{q} + \bar{\theta}x$, $0 < \bar{\theta} < 1$, we transform the left-hand side of (3.2.49) to

$$\begin{aligned}\sum_{j=1}^{n} a_{ij}(\tilde{q}+x)(\ddot{\tilde{q}}_j + \ddot{x}_j) &= \sum_{j=1}^{n} a_{ij}(\tilde{q}+x)\ddot{x}_j \\ &+ \sum_{j=1}^{n} a_{ij}(\tilde{q})\ddot{\tilde{q}}_j + \sum_{j,m=1}^{n} \frac{\partial a_{ij}(\bar{q})}{\partial q_m} x_m \ddot{\tilde{q}}_j.\end{aligned} \tag{3.2.50}$$

Using the equality

$$\Gamma_{ijk}(\tilde{q}+x) = \Gamma_{ijk}(\tilde{q}) + \sum_{m=1}^{n} \frac{\partial \Gamma_{ijk}(\bar{\bar{q}})}{\partial q_m} x_m, \quad \bar{\bar{q}} = \tilde{q} + \bar{\bar{\theta}}x,\ 0 < \bar{\bar{\theta}} < 1,$$

we reduce the expression on the right-hand side of (3.2.49) to

$$\begin{aligned}\sum_{j,k=1}^{n} \Gamma_{ijk}(\tilde{q}+x)\left(\dot{\tilde{q}}_j\dot{\tilde{q}}_k + \dot{\tilde{q}}_j\dot{x}_k + \dot{\tilde{q}}_k\dot{x}_j + \dot{x}_j\dot{x}_k\right) &= \sum_{j,k=1}^{n} \Gamma_{ijk}(\tilde{q})\dot{\tilde{q}}_k\dot{\tilde{q}}_j \\ + \sum_{j,k=1}^{n} \Gamma_{ijk}(\tilde{q}+x)\left(\dot{\tilde{q}}_j\dot{x}_k + \dot{\tilde{q}}_k\dot{x}_j + \dot{x}_j\dot{x}_k\right) &+ \sum_{j,k,m=1}^{n} \frac{\partial \Gamma_{ijk}(\bar{\bar{q}})}{\partial q_m} x_m \dot{\tilde{q}}_j\dot{\tilde{q}}_k.\end{aligned} \tag{3.2.51}$$

Taking into account relations (3.2.48)–(3.2.51), we write the equations in deviations as

$$\begin{aligned}\sum_{j=1}^{n} a_{ij}(\tilde{q}+x)\ddot{x}_j &= -\sum_{j,m=1}^{n} \frac{\partial a_{ij}(\bar{q})}{\partial q_m} x_m \ddot{\tilde{q}}_j - \sum_{j,k,m=1}^{n} \frac{\partial \Gamma_{ijk}(\bar{\bar{q}})}{\partial q_m} x_m \dot{\tilde{q}}_j\dot{\tilde{q}}_k \\ &- \sum_{j,k=1}^{n} \Gamma_{ijk}(\tilde{q}+x)\left(\dot{\tilde{q}}_j\dot{x}_k + \dot{\tilde{q}}_k\dot{x}_j + \dot{x}_j\dot{x}_k\right) + G_i + U_i\end{aligned}$$

and, then, in the vector form

$$
\begin{aligned}
A(\tilde{q}+x)\ddot{x} = &-\left(\sum_{m=1}^{n} \frac{\partial A(\bar{q})}{\partial q_m} x_m\right)\ddot{\tilde{q}} - \left(\sum_{k,m=1}^{n} \frac{\partial^2 A(\bar{\bar{q}})}{\partial q_k \partial q_m} x_m \dot{\tilde{q}}_k\right)\dot{\tilde{q}} \\
&+\frac{1}{2}\frac{\partial}{\partial q}\left\langle\left(\sum_{m=1}^{n} \frac{\partial A(\bar{\bar{q}})}{\partial q_m} x_m\right)\dot{\tilde{q}},\dot{\tilde{q}}\right\rangle - \left(\sum_{k=1}^{n} \frac{\partial A(\tilde{q}+x)}{\partial q_k}\dot{x}_k\right)\dot{\tilde{q}} \\
&-\left(\sum_{k=1}^{n} \frac{\partial A(\tilde{q}+x)}{\partial q_k}\dot{\tilde{q}}_k\right)\dot{x} - \left(\sum_{k=1}^{n} \frac{\partial A(\tilde{q}+x)}{\partial q_k}\dot{x}_k\right)\dot{x} \\
&+\frac{\partial}{\partial q}\left(\langle A(\tilde{q}+x)\dot{\tilde{q}},\dot{x}\rangle + \frac{1}{2}\langle A(\tilde{q}+x)\dot{x},\dot{x}\rangle\right) + G + U.
\end{aligned}
\tag{3.2.52}
$$

Let us estimate the individual terms on the right-hand side of (3.2.52). We assume that the phase coordinates, velocities, and accelerations along the nominal trajectory are subject to the following constraints:

$$
|\dot{\tilde{q}}| \le Q_1, \quad |\ddot{\tilde{q}}| \le Q_2. \tag{3.2.53}
$$

In view of (3.2.45), (3.2.53), and the inequality

$$
\sum_{m=1}^{n} |z_m| \le \sqrt{n}|z|, \quad z \in R^n, \tag{3.2.54}
$$

the following estimates hold:

$$
\begin{aligned}
&\left|\left(\sum_{m=1}^{n} \frac{\partial A(\bar{q})}{\partial q_m} x_m\right)\ddot{\tilde{q}}\right| \le C_1 \sum_{m=1}^{n} |x_m|\,|\ddot{\tilde{q}}| \le \sqrt{n} C_1 Q_2 |x|, \\
&\left|\left(\sum_{k=1}^{n} \frac{\partial A(\tilde{q}+x)}{\partial q_k}\dot{x}_k\right)\dot{\tilde{q}}\right| \le C_1 \sum_{k=1}^{n} |\dot{x}_k|\,|\dot{\tilde{q}}| \le \sqrt{n} C_1 Q_1 |\dot{x}|, \\
&\left|\left(\sum_{k=1}^{n} \frac{\partial A(\tilde{q}+x)}{\partial q_k}\dot{\tilde{q}}_k\right)\dot{x}\right| \le C_1 \sum_{k=1}^{n} |\dot{\tilde{q}}_k|\,|\dot{x}| \le \sqrt{n} C_1 Q_1 |\dot{x}|, \\
&\left|\left(\sum_{k=1}^{n} \frac{\partial A(\tilde{q}+x)}{\partial q_k}\dot{x}_k\right)\dot{x}\right| \le C_1 \sum_{k=1}^{n} |\dot{x}_k||\dot{x}| \le \sqrt{n} C_1 |\dot{x}|^2.
\end{aligned}
\tag{3.2.55}
$$

Conditions (3.2.45), inequality (3.2.54), and the Cauchy inequality imply the relations

$$
\left|\frac{\partial}{\partial q_i}\left\langle\left(\sum_{m=1}^{n} \frac{\partial A(\bar{\bar{q}})}{\partial q_m} x_m\right)\dot{\tilde{q}},\dot{\tilde{q}}\right\rangle\right| = \left|\left\langle\left(\sum_{m=1}^{n} \frac{\partial^2 A(\bar{\bar{q}})}{\partial q_i \partial q_m} x_m\right)\dot{\tilde{q}},\dot{\tilde{q}}\right\rangle\right|
$$

$$\leq C_2 \sum_{m=1}^{n} |x_m| \left|\dot{\tilde{q}}\right|^2 \leq \sqrt{n} C_2 Q_1^2 |x|,$$

$$\left|\frac{\partial}{\partial q_i} \langle A(\tilde{q}+x)\dot{\tilde{q}}, \dot{x}\rangle\right| \leq C_1 \left|\dot{\tilde{q}}\right| |\dot{x}| \leq C_1 Q_1 |\dot{x}|,$$

$$\left|\frac{\partial}{\partial q_i} \langle A(\tilde{q}+x)\dot{x}, \dot{x}\rangle\right| \leq C_1 |\dot{x}|^2, \quad i = 1, \ldots, n,$$

whence we obtain

$$\left|\frac{1}{2}\frac{\partial}{\partial q}\left\langle \left(\sum_{m=1}^{n} \frac{\partial A(\bar{\tilde{q}})}{\partial q_m} x_m\right) \dot{\tilde{q}}, \dot{\tilde{q}}\right\rangle\right| \leq \frac{n}{2} C_2 Q_1^2 |x|,$$

$$\left|\frac{\partial}{\partial q}\left(\langle A(\tilde{q}+x)\dot{\tilde{q}}, \dot{x}\rangle + \frac{1}{2}\langle A(\tilde{q}+x)\dot{x}, \dot{x}\rangle\right)\right| \leq \sqrt{n}\left(C_1 Q_1 |\dot{x}| + \frac{1}{2} C_1 |\dot{x}|^2\right). \tag{3.2.56}$$

Here, we used the following assertion that is valid for an arbitrary vector $z \in R^n$: if $|z_i| \leq h$, $i = 1, \ldots, n$, then $|z| \leq \sqrt{n} h$.

Applying again relations (3.2.45) and (3.2.54), we estimate the remaining term in (3.2.52) as follows:

$$\left|\left(\sum_{k,m=1}^{n} \frac{\partial^2 A(\bar{\tilde{q}})}{\partial q_k \partial q_m} x_m \dot{\tilde{q}}_k\right) \dot{\tilde{q}}\right| \leq \left|\dot{\tilde{q}}\right| \sum_{k,m=1}^{n} C_2 \left|x_m \dot{\tilde{q}}_k\right|$$
$$\leq C_2 \left|\dot{\tilde{q}}\right| \sum_{m=1}^{n} |x_m| \sum_{k=1}^{n} \left|\dot{\tilde{q}}_k\right| \leq n C_2 |x| \left|\dot{\tilde{q}}\right|^2 \leq n C_2 |x| Q_1^2. \tag{3.2.57}$$

Let us denote by v the sum of all terms on the right-hand side of (3.2.52) except for the control forces U and perturbations G and rewrite the equations in deviations as

$$A(\tilde{q}+x)\ddot{x} = v + G + U. \tag{3.2.58}$$

Formulas (3.2.55)–(3.2.57) yield the following estimate:

$$|v| \leq v^0(x, \dot{x}) = \left(\sqrt{n} C_1 Q_2 + \frac{3n}{2} C_2 Q_1^2\right) |x|$$
$$+ 3\sqrt{n} C_1 Q_1 |\dot{x}| + \frac{3}{2}\sqrt{n} C_1 |\dot{x}|^2. \tag{3.2.59}$$

Decomposition of the system

Let us resolve (3.2.58) for $\ddot{x}$. We obtain

$$\ddot{x} = U' + V'. \tag{3.2.60}$$

Here,

$$U' = A^{-1}U, \qquad V' = A^{-1}(v+G). \tag{3.2.61}$$

We will interpret U' as a new control vector and impose the following constraints on its components:

$$|U_i'| \le U_0, \qquad U_0 \le rM^{-1}n^{-1/2},$$
$$r = \min_i U_i^0, \qquad 1 \le i \le n. \tag{3.2.62}$$

Constraints (3.2.62) guarantee that the original constraints (3.2.43) are fulfilled.

We define control U' satisfying (3.2.62) in the same way as before in Sect. 2.3.2 in the feedback form (2.3.13):

$$U_i' = -U_0 \operatorname{sign}(\dot{x}_i - \psi_i), \quad \dot{x}_i \ne \psi_i;$$
$$U_i' = -U_0 \operatorname{sign} \dot{x}_i, \qquad \dot{x}_i = \psi_i; \tag{3.2.63}$$
$$\psi_i(x_i, X_i) = -(2X_i|x_i|)^{1/2} \operatorname{sign} x_i, \qquad i = 1, \ldots, n.$$

Here, X_i are positive control parameters to be identified.

Let us define a set Ω^x in the form analogous to (2.3.16):

$$\Omega^x = \Omega_1^x \times \ldots \times \Omega_n^x, \quad \Omega_i^x = \{(x_i, \dot{x}_i) : x_i^- \le x_i \le x_i^+,$$
$$\psi_i(x_i - x_i^-, X_i) \le \dot{x}_i \le \psi_i(x_i - x_i^+, X_i)\}. \tag{3.2.64}$$

Here, the values of $x_i^- < 0$ and $x_i^+ > 0$ are unknown and also to be found. We note that the terminal state $x = \dot{x} = 0$ lies in the domain Ω^x so that $x_i = 0 \in [x_i^-, x_i^+]$.

If, at the intitial instant, conditions $(x_i^0, \dot{x}_i^0) \in \Omega_i^x$ hold for all i, then during the control process the following relations are true:

$$|x_i| \le d_i, \qquad |\dot{x}_i| \le \psi_i^d, \qquad d_i = x_i^+ - x_i^-, \qquad \psi_i^d = \psi_i(-d_i, X_i). \tag{3.2.65}$$

Using the technique of Sect. 2.3.3, we obtain the system of inequalities [analogous to (2.3.33)] for finding admissible parameters X_i and d_i, $i = 1, \ldots, n$, in the form

$$X_i + m^{-1}v^0\left(d, \psi^d\right) \le U_0 - m^{-1}|G^0|, \quad i = 1, \ldots, n,$$
$$d = (d_1, \ldots, d_n), \quad \psi^d = (\psi_1^d, \ldots, \psi_n^d), \tag{3.2.66}$$
$$X = (X_1, \ldots, X_n), \quad G^0 = (G_1^0, \ldots, G_n^0).$$

Note that for a fixed d_i, the values x_i^- and x_i^+ may be chosen arbitrarily; it is only needed that conditions $x_i^- < 0$, $x_i^+ > 0$, and $x_i^+ - x_i^- = d_i$, $i = 1, \ldots, n$ are satisfied.

The expressions on the left-hand side of the system of inequalities (3.2.66) increase monotonically with d_i and X_i and vanish when $d_i = X_i = 0$. Therefore, a solution $d_i > 0$ and $X_i > 0$ exists, if the capabilities of the correcting control are sufficiently large and if the following condition holds:

$$U_0 \geq m^{-1}|G^0|, \quad i=1,\dots,n. \tag{3.2.67}$$

Let us summarize the results obtained. Suppose that condition (3.2.67) holds and that we have found positive parameters d_i and X_i, $i=1,\dots,n$, that satisfy inequality (3.2.66). Then, the feedback control $U(q,\dot{q})$ that solves our problem is defined by the relations $U(q,\dot{q})=A(\tilde{q}+x)U'(x,\dot{x})$, (3.2.46) and (3.2.63). This control steers system (3.1.1), (3.1.2), and (3.2.41) to the nominal trajectory in finite time provided that the initial deviations $(x^0,\dot{x}^0)$ lie in domain Ω^x defined by constraints (3.2.64).

Note that, if the lower boundary m of the eigenvalues of the matrix $A(q)$ is small, then constraint (3.2.67) may prove to be too stringent. In this case, it is expedient to apply another modification of the suggested control method. Let us introduce the notation

$$y=A(\tilde{q})x \tag{3.2.68}$$

and represent system (3.2.58) as

$$\begin{aligned} \ddot{y} &= U+V, \qquad V=G+v-[A(\tilde{q}+x)-A(\tilde{q})]A^{-1}(\tilde{q}+x)\,(U+G+v) \\ &+\left(\sum_{k,m=1}^{n}\frac{\partial^2 A(\tilde{q})}{\partial q_k \partial q_m}\dot{\tilde{q}}_k\dot{\tilde{q}}_m+\sum_{m=1}^{n}\frac{\partial A(\tilde{q})}{\partial q_m}\ddot{\tilde{q}}_m\right)x+2\left(\sum_{m=1}^{n}\frac{\partial A(\tilde{q})}{\partial q_m}\dot{\tilde{q}}_m\right)\dot{x}. \end{aligned} \tag{3.2.69}$$

Taking into account that

$$\|A(\tilde{q}+x)-A(\tilde{q})\| \leq C_1\sqrt{n}|x|,$$

we obtain analogously to (3.2.59)

$$\begin{aligned} |V_i| &\leq G_i^0+v^0(x,\dot{x})+C_1\sqrt{n}m^{-1}|x|\left(v^0(x,\dot{x})+|U^0|+|G^0|\right) \\ &+\left(nC_2Q_1^2+\sqrt{n}C_1Q_2\right)|x|+2C_1Q_1\sqrt{n}|\dot{x}|. \end{aligned} \tag{3.2.70}$$

Here, $U^0=(U_1^0,\dots,U_n^0)$ and $G^0=(G_1^0,\dots,G_n^0)$ are vectors with the components introduced in (3.2.42) and (3.2.43). In view of notation (3.2.68) and constraint (3.1.3), the following relations hold:

$$\begin{aligned} |x| &= |A^{-1}(\tilde{q})y| \leq m^{-1}|y|, \\ |\dot{x}| &= \left|A^{-1}(\tilde{q})\dot{y}-A^{-1}(\tilde{q})\left(\sum_{m=1}^{n}\frac{\partial A(\tilde{q})}{\partial q_m}\dot{\tilde{q}}_m\right)A^{-1}(\tilde{q})y\right| \\ &\leq m^{-1}|\dot{y}|+\sqrt{n}m^{-2}C_1Q_1|y|. \end{aligned} \tag{3.2.71}$$

Let us replace $|x|$ and $|\dot{x}|$ in inequalities (3.2.70) by their upper bounds (3.2.71). We obtain

$$|V_i| \leq G_i^0 + \tilde{v}^0(y,\dot{y}).$$

The specific form of the function $\tilde{v}^0(y,\dot{y})$ is determined from (3.2.70) and (3.2.71).

Let us apply the control law (3.2.63) to system (3.2.69) replacing x_i and U_i' in it by y_i and U_i, respectively. To determine the admissible control parameters X_i and $d_i = y_i^+ - y_i^-$, $i = 1,\ldots,n$, we obtain the following system of inequalities, analogous to (3.2.66):

$$X_i + \tilde{v}^0(d,\psi^d) \leq U_i^0 - G_i^0, \quad i = 1,\ldots,n. \tag{3.2.72}$$

The expressions on the right-hand sides of inequalities (3.2.72) are positive, while the function $\tilde{v}^0(d,\psi^d)$ increases monotonically with in the variables d_i and X_i and vanishes when $d_i = 0$ and $X_i = 0$. Therefore, a solution $d_i > 0$ and $X_i > 0$ to the system of inequalities (3.2.72) always exists.

After choosing parameters $X_i > 0$ and $d_i > 0$ that satisfy (3.2.72), we find the corresponding values of $y_i^- < 0$ and $y_i^+ > 0$ and admissible set Ω^y of initial values $(y_i^0, \dot{y}_i^0)$. Note that, for a fixed d_i, the values y_i^- and y_i^+ may be chosen arbitrarily; it is only needed that conditions $y_i^- < 0$, $y_i^+ > 0$, and $y_i^+ - y_i^- = d_i$, $i = 1,\ldots,n$ are satisfied. Now, the set Ω^y is defined by relations (3.2.64), where x should be replaces by y. Further, using change of a variable (3.2.68) and returning to the original variables x, we obtain the set Ω^x of the admissible initial deviations $(x_i^0, \dot{x}_i^0)$ from the nominal trajectory.

3.3 Applications to robots

3.3.1 Symbolic generation of equations for multibody systems

Control methods proposed above can be applied to various controlled mechanical systems. The most interesting application of these methods is that to the robotic systems and, primarily, to the manipulation robots. The manipulator has several degrees of freedom, each of which is controlled, as a rule, by its own motor. Therefore, the number of the control functions here is equal to the number of degrees of freedom as assumed in system (3.1.1). The motion equations of the manipulation robot may be generated by different ways, in particular, using Lagrangian or Hamiltonian equations. Composing the system of equations by hand implies cumbersome calculations requiring considerable time and efforts, and this way is also not guaranteed from mistakes. That is why the symbolic generation of equations for multibody systems is widespread [76]. A series of software tools are developped allowing one automatically generate the motion equations for multibody systems [121, 120, 108, 109].

The description of the scheme for forming the motion equations of the holonomic systems [54] is given below. Note that for its realization it is not required to write programs on the special-purpose language of symbolic computation. It is sufficient to use any universal software tool, which allows one to create significantly more simple user interface, often not assuming the special experience in programming.

Kinetic energy

Let the state of the mechanical system with n degrees of freedom be described by the generalized coordinates q_i, $i = 1,\dots,n$.

Motion of the system of N rigid bodies is considered with respect to the fixed coordinate system $OXYZ$. Let us introduce successively N local coordinate systems $O_ix_iy_iz_i$, the ith of which is rigidly connected with the ith body so that axes O_ix_i, O_iy_i, and O_iz_i are the main axes of inertia of the ith body, $i = 1,\dots,N$. Position of the trihedron $O_ix_iy_iz_i$ in the cordinate system $OXYZ$ is defined by the following way.

The pair $\{r_{O_i}, NUM\}$ gives coordinates $r_{O_i} = (x_{O_i}, y_{O_i}, z_{O_i})$ of the point O_i in the system $O_{NUM}x_{NUM}y_{NUM}z_{NUM}$, where $NUM < i$, $i = 1,\dots,N$, i.e., in the preceding local system. The case $NUM = 0$ corresponds to the assignment of the coordinates of the point O_i in the fixed system $OXYZ$.

Orientation of the trihedron $O_ix_iy_iz_i$ relative to $O_{NUM}x_{NUM}y_{NUM}z_{NUM}$ is given by the sequence of the pairs $\{\gamma_1, K_1\}$, $\{\gamma_2, K_2\}$, $\{\gamma_3, K_3\}$, which determines the sequence of the rotations of the trihedron $O_{NUM}x_{NUM}y_{NUM}z_{NUM}$ bringing it to the position where its axes become parallel to the axes of $O_ix_iy_iz_i$. Parameter K_j, $j = 1, 2, 3$, takes the values 1, 2, 3, which conventionally designate the axis with respect to which the rotation occurs. Axes $O_{NUM}x_{NUM}$, $O_{NUM}y_{NUM}$, $O_{NUM}z_{NUM}$ correspond to the values $K_1 = 1$, $K_2 = 2$, $K_3 = 3$, respectively. The first turn occurs relative to the axis K_1 by angle γ_1, the second turn—relative to the new position of the axis K_2 by angle γ_2, and the third—relative to the new position of the axis K_3 by angle γ_3.

The matrix of transition from the coordinate system $O_ix_iy_iz_i$ to the coordinate system $OXYZ$ is formed for every ith body:

$$\Gamma_i = \begin{pmatrix} g_{i11} & g_{i12} & g_{i13} \\ g_{i21} & g_{i22} & g_{i23} \\ g_{i31} & g_{i32} & g_{i33} \end{pmatrix}.$$

Finding the matrix Γ_i, and also the absolute angular velocity of the trihedron $O_ix_iy_iz_i$, is performed gradually for the given triplet of the parameters $\{\gamma_j, K_j, NUM\}$, $j = 1, 2, 3$:

Step 1. $K_1 = 1$. The auxiliary variable Δ is assigned to

$$\Delta = \begin{pmatrix} 1 & 0 & 0 \\ 0 & \cos\gamma_1 & -\sin\gamma_1 \\ 0 & \sin\gamma_1 & \cos\gamma_1 \end{pmatrix}.$$

The auxiliary variable ω_r is assigned to

$$\omega_r = (\dot{\gamma}_1, 0, 0),$$

where $\dot{\gamma}_1$ means the derivative of γ_1 with respect to time. Go to Step 4.

Step 2. $K_2 = 2$. The auxiliary variable Δ is assigned to

$$\Delta = \begin{pmatrix} \cos\gamma_2 & 0 & \sin\gamma_2 \\ 0 & 1 & 0 \\ -\sin\gamma_2 & 0 & \cos\gamma_2 \end{pmatrix}.$$

The auxiliary variable ω_r is assigned to

$$\omega_r = (0, \dot{\gamma}_2, 0).$$

Go to Step 4.

Step 3. $K_3 = 3$. The auxiliary variable Δ is assigned to

$$\Delta = \begin{pmatrix} \cos\gamma_3 & -\sin\gamma_3 & 0 \\ \sin\gamma_3 & \cos\gamma_3 & 0 \\ 0 & 0 & 1 \end{pmatrix}.$$

The auxiliary variable ω_r is assigned to

$$\omega_r = (0, 0, \dot{\gamma}_3).$$

Go to Step 4.

Step 4. New values of Γ_i and ω_i are determined:

$$\Gamma_i := \Gamma_i \Delta,$$

$$\omega_i := \omega_i + \Gamma_i \omega_r.$$

Assign $\Gamma_i = \Gamma_{NUM}$ in the case of the first calling the procedure.

Find the absolute velocity of the center of mass of the ith body by the formula

$$v_{C_i} = \frac{d}{dt}\Gamma_{NUM} r_{O_i} + \frac{d}{dt}\Gamma_i r_{C_i}, \tag{3.3.1}$$

where the vector r_{C_i} defines the coordinates of the center of mass of the ith body in the coordinate system $O_i x_i y_i z_i$.

Assuming that the moments of inertia of the ith body in the coordinate system $O_i x_i y_i z_i$ are known:

$$I_{x_i} = \int\limits_{m_i} (y_i^2 + z_i^2) dm_i, \quad I_{y_i} = \int\limits_{m_i} (x_i^2 + z_i^2) dm_i, \quad I_{z_i} = \int\limits_{m_i} (x_i^2 + y_i^2) dm_i, \tag{3.3.2}$$

we obtain elements of the matrix of the inertia tensor in the coordinate system $OXYZ$. It follows from (3.3.2) that

$$\int\limits_{m_i} x_i^2 dm_i = \frac{1}{2}(I_{y_i} + I_{z_i} - I_{x_i}),$$

$$\int_{m_i} y_i^2 dm_i = \frac{1}{2}(I_{x_i} + I_{z_i} - I_{y_i}),$$

$$\int_{m_i} z_i^2 dm_i = \frac{1}{2}(I_{x_i} + I_{y_i} - I_{z_i}).$$

Taking into account that

$$X_i = g_{i11}x_i + g_{i12}y_i + g_{i13}z_i,$$

$$Y_i = g_{i21}x_i + g_{i22}y_i + g_{i23}z_i,$$

$$Z_i = g_{i31}x_i + g_{i32}y_i + g_{i33}z_i,$$

where g_{ijk}, $j,k = 1,2,3$ are elements of the transition matrix Γ_i for the ith body, we obtain

$$\begin{aligned}
I_{iX} = {} & \frac{1}{2}[(g_{21}^2 + g_{31}^2)(I_{y_i} + I_{z_i} - I_{x_i})^2 + (g_{22}^2 + g_{32}^2)(I_{x_i} + I_{z_i} - I_{y_i})^2 \\
& + (g_{23}^2 + g_{33}^2)(I_{x_i} + I_{y_i} - I_{z_i})^2], \\
I_{iY} = {} & \frac{1}{2}[(g_{11}^2 + g_{31}^2)(I_{y_i} + I_{z_i} - I_{x_i})^2 + (g_{12}^2 + g_{32}^2)(I_{x_i} + I_{z_i} - I_{y_i})^2 \\
& + (g_{13}^2 + g_{33}^2)(I_{x_i} + I_{y_i} - I_{z_i})^2], \\
I_{iZ} = {} & \frac{1}{2}[(g_{11}^2 + g_{21}^2)(I_{y_i} + I_{z_i} - I_{x_i})^2 + (g_{12}^2 + g_{22}^2)(I_{x_i} + I_{z_i} - I_{y_i})^2 \\
& + (g_{13}^2 + g_{23}^2)(I_{x_i} + I_{y_i} - I_{z_i})^2],
\end{aligned} \tag{3.3.3}$$

$$\begin{aligned}
I_{iXY} = {} & \frac{1}{2}[g_{12}^2 g_{21}^2 (I_{y_i} + I_{z_i} - I_{x_i})^2 + g_{12}^2 g_{22}^2 (I_{x_i} + I_{z_i} - I_{y_i})^2 \\
& + g_{13}^2 g_{23}^2 (I_{x_i} + I_{y_i} - I_{z_i})^2, \\
I_{iXZ} = {} & \frac{1}{2}[g_{12}^2 g_{31}^2 (I_{y_i} + I_{z_i} - I_{x_i})^2 + g_{12}^2 g_{32}^2 (I_{x_i} + I_{z_i} - I_{y_i})^2 \\
& + g_{13}^2 g_{33}^2 (I_{x_i} + I_{y_i} - I_{z_i})^2, \\
I_{iYZ} = {} & \frac{1}{2}[g_{22}^2 g_{31}^2 (I_{y_i} + I_{z_i} - I_{x_i})^2 + g_{22}^2 g_{32}^2 (I_{x_i} + I_{z_i} - I_{y_i})^2 \\
& + g_{23}^2 g_{33}^2 (I_{x_i} + I_{y_i} - I_{z_i})^2.
\end{aligned}$$

The kinetic energy of the ith body is obtained by the formula

$$T_i = \frac{1}{2} m_i v_{C_i}^2 + \frac{1}{2}(\omega_i, I_i \omega_i), \tag{3.3.4}$$

where m_i is the mass of the ith body, v_{C_i} is the absolute velocity of the center of mass of the ith body determined by (3.3.1), ω_i is the absolute angular velocity of the trihedron $O_i x_i y_i z_i$, and I_i is the matrix of the inertia tensor of the ith body in the coordinate system $OXYZ$:

$$I_i = \begin{pmatrix} I_{iX} & I_{iXY} & I_{iXZ} \\ I_{iXY} & I_{iY} & I_{iYZ} \\ I_{iXZ} & I_{iYZ} & I_{iZ} \end{pmatrix},$$

the elements of which are presented in (3.3.3).

The kinetic energy of the system of N bodies is equal to

$$T = \sum_{i=1}^{N} T_i. \tag{3.3.5}$$

Thus, the procedure of finding the total kinetic energy is subdivided into several steps.

At the first step, the matrix of transition from the local coordinate system $O_i x_i y_i z_i$ to the fixed one $OXYZ$ is found for every body. Then, the kinetic energy of the ith body is obtained by (3.3.4).

At the last step, the kinetic energy T of the system is determined in accordance with (3.3.5).

Forming the Lagrangian equations of the second kind

Motion of the multibody system can be described by the Lagrangian equations:

$$\frac{d}{dt}\frac{\partial T}{\partial \dot{q}_i} - \frac{\partial T}{\partial q_i} = Q_i, \quad i = 1, \ldots, n, \tag{3.3.6}$$

where generalized force Q_i is defined by the following expession:

$$Q_i = \sum_{j=1}^{k} F_j \frac{\partial r_j}{\partial q_i}.$$

Here, F_j, $j = 1, \ldots, k$ are forces acting upon the system; these forces are applied at the points $r_1, \ldots, r_k$.

Finding the derivatives

$$\frac{\partial T}{\partial q_i}, \qquad \frac{\partial T}{\partial \dot{q}_i}, \qquad \frac{d}{dt}\frac{\partial T}{\partial \dot{q}_i} = \sum_{j=1}^{n} \left(\frac{\partial}{\partial q_j}\frac{\partial T}{\partial \dot{q}_i}\dot{q}_j + \frac{\partial}{\partial \dot{q}_j}\frac{\partial T}{\partial \dot{q}_i}\ddot{q}_j \right),$$

we form the Lagrangian equations. Algorithm for obtaining the expression of the kinetic energy T is given above.

3.3.2 *Modelling of control for a two-link mechanism (with three degrees of freedom)*

Description of the dynamical system

Let us apply the control law proposed in Sect. 3.2.3 to the system describing the dynamics of a two-link mechanism that models transport motions of a manipulation robot. The elements of the two-link mechanism are connected by a cylindrical joint , while the two-link mechanism itself is fastened to an immovable base by a two-degree joint (see Fig. 3.4), whose movable axis is parallel to the axis of the joint connecting the links. It is assumed that the links are homogeneous thin-walled straight rods with a round cross section. The links have the following parameters: masses of the rods m_1 and m_2, lengths of the rods l_1 and l_2, and the radii of the round cross sections R_1 and R_2. The principal central moments of inertia of the links with respect to their longitudinal and transverse axes are equal to J_{x1}, J_{x2} and J_1, J_2, respectively.

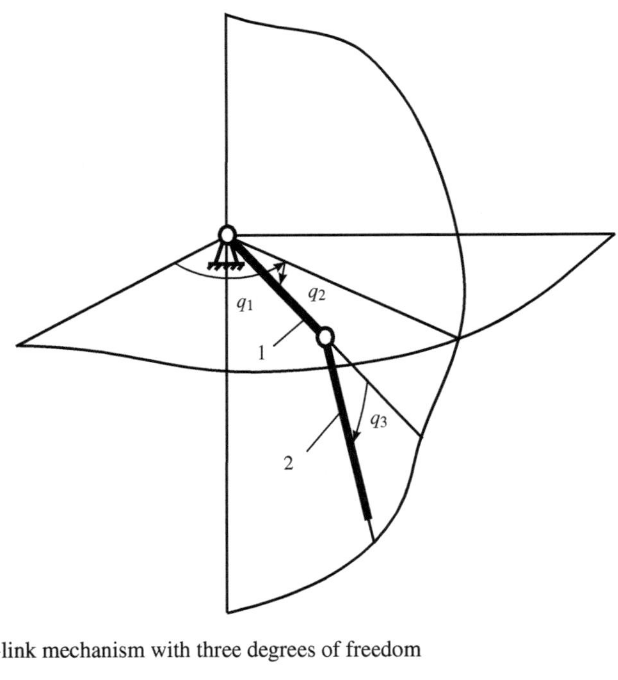

Fig. 3.4 Two-link mechanism with three degrees of freedom

Kinetic energy and equations of motion

The system has three degrees of freedom. For the first two generalized coordinates $q_1 = \psi$ and $q_2 = \theta$, the angles of rotation around the axes of the two-degree joint are selected while the third coordinate $q_3 = \beta$ is the angle between the axes of the links. The control is effected independently for each degree of freedom; the control torques are applied to the appropriate axes of the links. In addition, certain moments G_i, $i = 1,2,3$, of the force of gravity and uncertain external perturbations act on the two-link mechanism.

For establishing the matrix A of the kinetic energy, the eigenvalues λ_i of the matrix A, and Lagrange's equations for the two-link mechanism, MAPLE computer algebra system was used. The procedure to obtain these equations is presented in Sect. 3.3.1.

The matrix A has the form

$$A = \begin{pmatrix} a_{11} & 0 & 0 \\ 0 & a_{22} & a_{23} \\ 0 & a_{32} & a_{33} \end{pmatrix},$$

where

$$\begin{aligned} a_{11} = & \tfrac{1}{2}\left\{(m_2 l_1^2 + J_1 - J_{x1})\cos 2\theta + (J_2 - J_{x2})\cos 2(\theta+\beta) \right. \\ & \left. + m_2 l_1 l_2 [\cos\beta + \cos(2\theta+\beta)] + J_1 + J_2 + J_{x1} + J_{x2} + m_2 l_1^2 \right\} \\ & + \tfrac{1}{8}\left\{m_1 l_1^2 (1 + \cos 2\theta) + m_2 l_2^2 [\cos 2(\theta+\beta) + 1]\right\}, \end{aligned}$$

$$a_{22} = J_1 + J_2 + l_1^2\left(m_2 + \tfrac{1}{4}m_1\right) + m_2 l_2\left(l_1\cos\beta + \tfrac{1}{4}l_2\right),$$

$$\begin{aligned} a_{23} &= J_2 + \tfrac{1}{2}m_2 l_2\left(l_1\cos\beta + \tfrac{1}{4}l_2\right), \\ a_{32} &= a_{23}, \\ a_{33} &= J_2 + \tfrac{1}{4}m_2 l_2^2. \end{aligned}$$

The eigenvalues of the matrix A are equal to

$$\begin{aligned} \lambda_1 = & \tfrac{1}{2}\left\{\left(J_1 - J_{x1} + m_2 l_1^2 + \tfrac{1}{4}m_1 l_1^2\right)\cos 2\theta \right. \\ & + \left(J_2 - J_{x2} + \tfrac{1}{4}m_2 l_2^2\right)\cos 2(\theta+\beta) + m_2 l_1 l_2 [\cos\beta + \cos(2\theta+\beta)] \\ & \left. + J_1 + J_2 + J_{x1} + J_{x2} + m_2 l_1^2 + \tfrac{1}{4}(m_1 l_1^2 + m_2 l_2^2)\right\}, \end{aligned}$$

$$\begin{aligned}\lambda_2 = {} & \frac{1}{2}\left[J_1+2J_2+m_2l_1(l_1+l_2\cos\beta)+\frac{1}{2}m_2l_2^2+\frac{1}{4}m_1l_1^2\right]\\ & +\frac{1}{2}\left[(J_1+m_2l_1l_2\cos\beta)^2+(2J_2+m_2l_1l_2\cos\beta)^2\right.\\ & +\frac{1}{2}m_1l_1^2(J_1+m_2l_1^2+m_2l_1l_2\cos\beta)+2m_2l_2(l_2J_2+m_2l_1^3\cos\beta)\\ & \left.+2m_2l_1^2J_1+16m_1^2l_1^4+\frac{1}{2}m_2^2l_2^2+m_2^2l_1(l_1^3+l_2^3\cos\beta)\right]^{1/2},\end{aligned}$$

$$\begin{aligned}\lambda_3 = {} & \frac{1}{2}\left[J_1+2J_2+m_2l_1(l_1+l_2\cos\beta)+\frac{1}{2}m_2l_2^2+\frac{1}{4}m_1l_1^2\right]\\ & -\frac{1}{2}\left[(J_1+m_2l_1l_2\cos\beta)^2+(2J_2+m_2l_1l_2\cos\beta)^2\right.\\ & +\frac{1}{2}m_1l_1^2(J_1+m_2l_1^2+m_2l_1l_2\cos\beta)+2m_2l_2(l_2J_2+m_2l_1^3\cos\beta)\\ & \left.+2m_2l_1^2J_1+16m_1^2l_1^4+\frac{1}{2}m_2^2l_2^2+m_2^2l_1(l_1^3+l_2^3\cos\beta)\right]^{1/2}.\end{aligned}$$

Lagrange's equations for the system under consideration have the form:

$$\begin{aligned} & \frac{\ddot\psi}{2}\left\{J_1+J_2+J_{x1}+J_{x2}+m_2l_1^2+\frac{1}{4}(m_1l_1^2+m_2l_2^2)\right.\\ & +\left(J_2-J_{x2}+\frac{1}{4}m_2l_2^2\right)\cos 2(\theta+\beta)\\ & \left.+m_2l_1l_2\left[\cos\beta+\cos(2\theta+\beta)\right]+\left(J_1-J_{x1}+m_2l_1^2+\frac{1}{4}m_1l_1^2\right)\cos 2\theta\right\}\\ & -\dot\theta\dot\psi\left\{\left(j_2-J_{x2}+\frac{1}{4}m_2l_2^2\right)\sin 2(\theta+\beta)\right.\\ & \left.+m_2l_1l_2\sin(2\theta+\beta)+\left(J_1-J_{x1}+m_2l_1^2+\frac{1}{4}m_1l_1^2\right)\sin 2\theta\right\}\\ & -\dot\beta\dot\psi\left\{\left(J_2-J_{x2}+\frac{1}{4}m_2l_2^2\right)\sin 2(\theta+\beta)\right.\\ & \left.+\frac{1}{2}m_2l_1l_2\left[\sin\beta+\sin(2\theta+\beta)\right]\right\}=M_\psi+G_1,\end{aligned}$$

$$\begin{aligned} & \ddot\theta\left[\frac{1}{4}(m_1l_1^2+m_2l_2^2)+m_2l_1(l_1+l_2\cos\beta)+J_1+J_2\right]\\ & +\ddot\beta\left[J_2+\frac{1}{2}m_2l_2\left(\frac{1}{2}l_2+l_1\cos\beta\right)\right]-\left(\dot\theta\dot\beta-\frac{\dot\beta^2}{2}\right)m_2l_1l_2\sin\beta\\ & \frac{\dot\psi^2}{2}\left\{\left(J_2-J_{x2}+\frac{1}{4}m_2l_2^2\right)\sin 2(\theta+\beta)+m_2l_1l_2\sin(2\theta+\beta)\right.\\ & \left.+\left(J_1-J_{x1}+m_2l_1^2+\frac{1}{4}m_1l_1^2\right)\sin 2\theta\right]=M_\theta+G_2,\end{aligned} \tag{3.3.7}$$

$$\begin{aligned} & \ddot\beta\left(J_2+\frac{1}{4}m_2l_2^2\right)+\ddot\theta\left[J_2+\frac{1}{2}m_2l_2\left(\frac{1}{2}l_2+l_1\cos\beta\right)\right]\\ & +\dot\theta^2\frac{1}{2}m_2l_1l_2\sin\beta+\frac{\dot\psi^2}{2}\left\{\left(J_2-J_{x2}+\frac{1}{4}m_2l_2^2\right)\sin 2(\theta+\beta)\right.\\ & \left.+\frac{1}{2}m_2l_1l_2\left[\sin\beta+\sin(2\theta+\beta)\right]\right\}=M_\beta+G_3.\end{aligned}$$

Results of modelling for a nonzero terminal state

The modelling is carried out for the following parameters of the system: masses of the rods $m_1 = 8$ kg and $m_2 = 2$ kg, lengths of the rods $l_1 = 0.4$ m and $l_2 = 0.5$ m, and the radii of the round cross sections $R_1 = R_2 = 0.05$ m.

Consider one of the variants of the initial and terminal states:

$$q^0 = (45°; -30°; -100°), \quad \dot{q}^0 = (57.3°/\text{s}; -40°/\text{s}; 30°/\text{s}),$$

$$q^* = (0; 0; -80°), \quad \dot{q}^* = (5.73°/\text{s}; 0; 0).$$

The perturbations G_i are taken in the form:

$$G_1(t) = \cos(10\pi t), \quad G_2(t) = 2\cos(8\pi t), \quad G_3(t) = 3\cos(6\pi t),$$

(dimensions of the torques $G_i(t)$ are N·m).

The region D of change of the generalized coordinates in this case is:

$$0 \le q_1 \le 180°, \quad -90° \le q_2 \le 90°, \quad -180° \le q_3 \le 180°.$$

We remind that

$$q_1 = \psi, \qquad q_2 = \theta, \qquad q_3 = \beta.$$

The eigenvalues of the matrix $A(q)$ with q changing in the entire region D lie between

$$m = 2.5 \cdot 10^{-2}\ \text{kg} \cdot \text{m}^2 \qquad \text{and} \qquad M = 1.4\ \text{kg} \cdot \text{m}^2.$$

The derivatives $\partial a_{ij}(q)/\partial q_k$ of the elements of the matrix $A(q)$ in the region D satisfy inequality (3.1.4) for $C = 1.3$ kg·m^2. The number r in bounds (3.1.39) and the parameters ε and δ are selected taking into account (3.2.16), (3.2.20), and (3.2.26). We have

$$r = 6.5 \cdot 10^2\ \text{N} \cdot \text{m}, \quad \varepsilon = 0.19\ \text{s}^{-1}, \quad \delta = 0.17\ \text{s}^{-1}.$$

According to (3.1.39), (3.2.14), (3.2.15), and (3.2.26), for these values for m, M, C, r, and ε, we get

$$\rho = 0.96, \qquad x^* = 1.4 \cdot 10^{-3}.$$

Figures 3.5 and 3.6 illustrate the behavior of the phase trajectories of system (3.3.7) controlled in the set Ω_1 according to law (3.1.22) and in the set Ω_2 according to law (3.2.27). Figure 3.5 shows the behavior of all phase coordinates on the entire interval of motion $t \in [0, t_*]$, and Fig. 3.6, using a different scale, shows their behavior near the terminal state. For each generalized coordinate, there is a corresponding curve designated with the number of the coordinate. From Fig. 3.5, it is clear that, in the set Ω_1, the components of the phase velocity vector $\dot{q}$ quickly drop. The absolute value of the velocity $|\dot{q}_1| = |\dot{\psi}|$ reaches the value ε last of all velocities, after that the trajectory enters the set Ω_2, and the control law (3.1.22) changes to law (3.2.27). Further, each of the curves represents a phase trajectory appropriate to a subsystem of type (3.1.45) and behaves as described in Sect. 3.2.3 (see Fig. 3.6).

The first coordinate q_1 reaches the final state in 4.67 s; the second arrives at 3.12 s, and the third at 2.07 s. Thus, the duration of the process is 4.67 s.

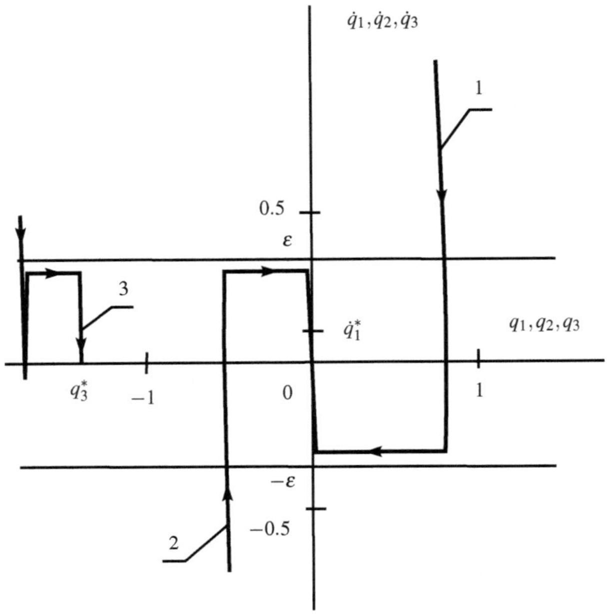

Fig. 3.5 Projections of the phase trajectory of the system onto the planes $(q_i, \dot{q}_i)$, $i = 1,2,3$

The modelling of the dynamics for the two-link mechanism controlled according to the law presented above is carried out also for the case, where the parallelepiped

$$-35^\circ \le q_2 \le 5^\circ, \quad -105^\circ \le q_3 \le -75^\circ, \quad 0 \le q_1 \le 180^\circ$$

was chosen as region D [the eigenvalues of matrix $A(q)$ do not depend on the variable $q_1 = \psi$]. In this parallelepiped, inequalities (3.1.4) and (3.1.3) are fulfilled with

$$m = 0.12\ \text{kg} \cdot \text{m}^2, \quad M = 1.08\ \text{kg} \cdot \text{m}^2, \quad C = 0.84\ \text{kg} \cdot \text{m}^2.$$

The number ε is taken equal to $0.34\ \text{s}^{-1}$, and it turned out that it is possible to reduce the constant r to $1.3 \cdot 10^2$ N·m.

For such values of parameters, the behavior of the trajectory of the system does not change drastically; nevertheless, the time of transition from the initial state to the terminal state reduces to 2.63 s.

Figure 3.7 shows the results of modelling for the motion of system (3.3.7) controlled by the modified law (see Remark 3.1 in Sect. 3.2.3). In contrast to Figs. 3.5 and 3.6, straight-line segments of motion are absent here, which is due to the dependence of the value δ on time. As to be expected, the given control method reaches

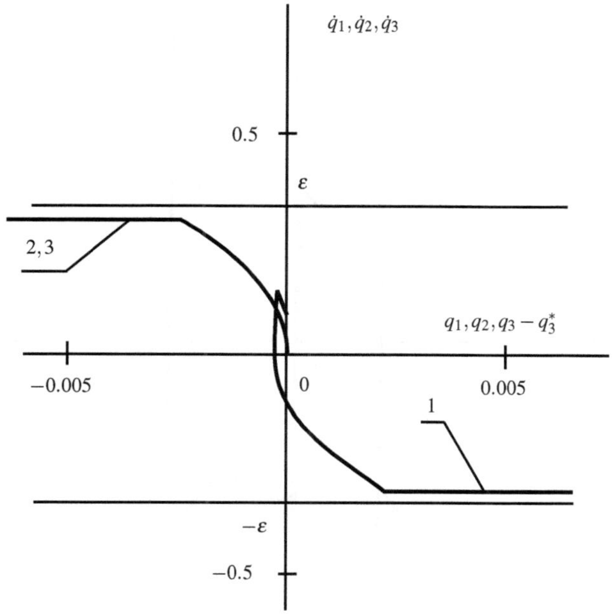

Fig. 3.6 Behavior of the phase trajectory near the terminal state

the target faster than the preceding ones: the complete time for moving the two-link mechanism from the initial to the terminal position is 1.8 s.

Results of modelling for the zero terminal state

In this case, the following parameters of the system are chosen:

$$l_1 = l_2 = 1\ \text{m}, \qquad m_1 = m_2 = 20\ \text{kg}, \qquad R_1 = R_2 = 5 \cdot 10^{-2}\ \text{m},$$

$$M_\psi^0 = M_\theta^0 = M_\beta^0 = r = 200\ \text{N} \cdot \text{m}.$$

The moments of inertia and variables entering the expressions for the control are

$$J_1 = J_2 = 1.67\ \text{kg} \cdot \text{m}^2, \qquad J_{x1} = J_{x2} = 2.5\ \text{kg} \cdot \text{m}^2,$$

$$m = 7.5 \cdot 10^{-2}\ \text{kg} \cdot \text{m}^2, \qquad M = 58.67\ \text{kg} \cdot \text{m}^2, \qquad C = 53.26\ \text{kg} \cdot \text{m}^2,$$
$$\varepsilon = 9.8 \cdot 10^{-3}\ \text{s}^{-1}, \qquad \delta = 8.82 \cdot 10^{-3}\text{s}^{-1}, \qquad U_0 = 1.97\text{s}^{-2},$$
$$\rho = 0.81, \qquad x^* = 1.04 \cdot 10^{-4}.$$

The identical values of parameters δ and x^* are chosen for all three degrees of freedom. Therefore, the switching curves for all control torques in the domain Ω_2

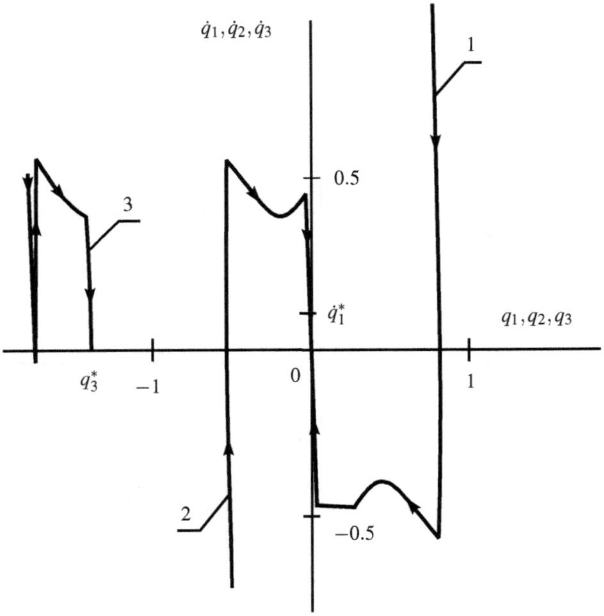

Fig. 3.7 Projections of the phase trajectory of the system under the modified control law

are also identical, and the final sections of the phase trajectories for all three degrees of freedom lie on these switching curves.

Some typical phase trajectories of the system are shown in Figs. 3.8 and 3.9, where solid, dashed, and dot-and-dash lines correspond to different degrees of freedom (angles ψ, θ, and β, respectively). Here, the final sections of the trajectories—inside the domain Ω_2 and alongside it—are shown.

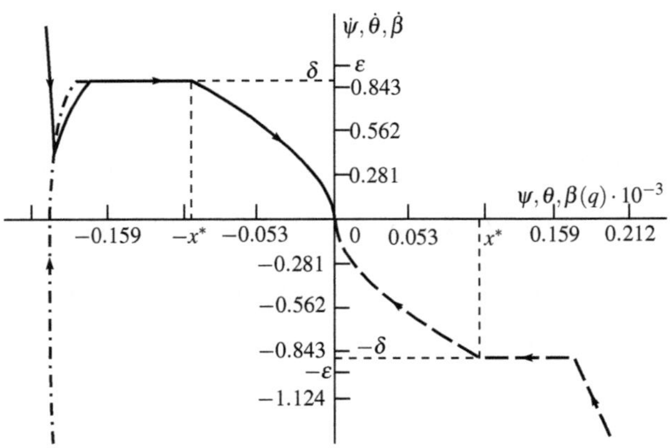

Fig. 3.8 Projections of the phase trajectory of the system in the case of zero terminal state

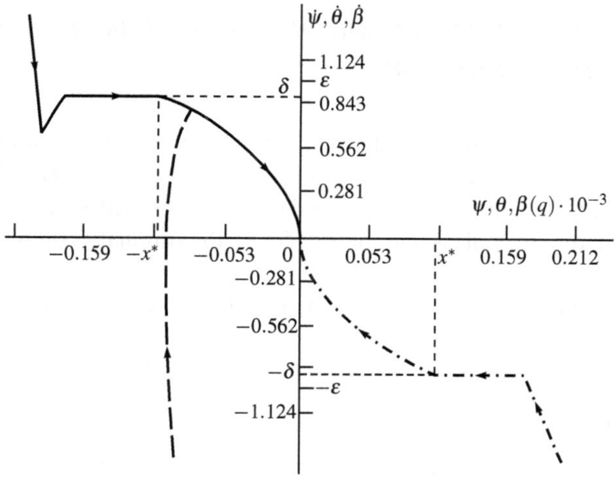

Fig. 3.9 Projections of the phase trajectory of the system in the case of zero terminal state

The time histories of the angular velocities $\dot{\psi}$, $\dot{\theta}$, and $\dot{\beta}$ are presented in Fig. 3.10 for one of the variants of simulations. At the final stage, the angular velocities vary linearly, which agrees with the motion along the parabolic sections of the switching curves in the domain Ω_2. The terminal states for different degrees of freedom are reached at different times.

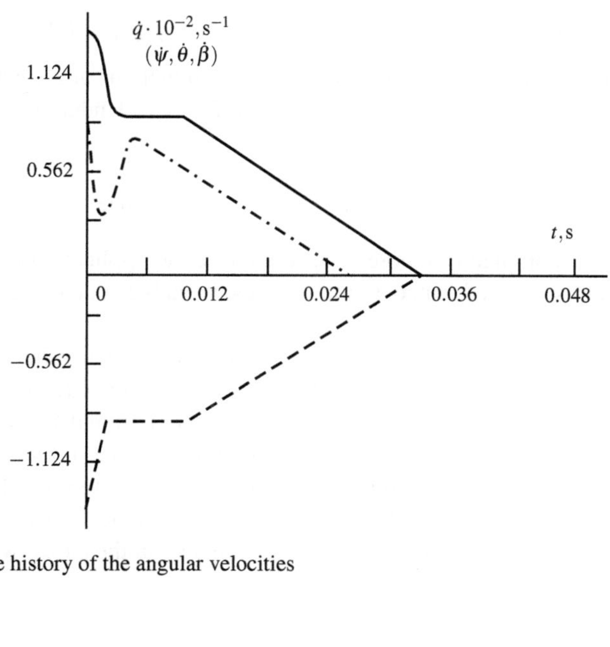

Fig. 3.10 Time history of the angular velocities

3.3.3 Modelling of tracking control for a two-link mechanism (with two degrees of freedom)

The modelling is carried out for system (2.5.16) with the characteristics used earlier and given in (2.5.29). For this example, we suppose $Q = P(t) + G$, where $P(t)$ are given forces, and G are unknown forces and disturbances (see Sect. 3.2.4). After the transition to the dimensionless variables (2.5.18), the original system (2.5.16) takes the form of (2.5.19).

The nominal trajectory $(\tilde{q}_i, \dot{\tilde{q}}_i)$ is obtained by the numerical integration of system (2.5.19) for

$$U_1 = U_2 = 0, \qquad G_1 = G_2 = 0,$$

$$P_1(t) = 1 - \tilde{q}_1(t) - \dot{\tilde{q}}_1(t), \qquad P_2(t) = 1 - \tilde{q}_2(t) - \dot{\tilde{q}}_2(t),$$

$$\tilde{q}_1(0) = \tilde{q}_2(0) = 0.8, \qquad \dot{\tilde{q}}_1(0) = 0.45, \qquad \dot{\tilde{q}}_2(0) = 0.15.$$

We determine particular values of the constants introduced in (3.1.3), (3.2.45), and (3.2.53). It turns out that

$$m = 0.13, \qquad M = 5.87, \qquad C_1 = C_2 = 2.41,$$

$$|\dot{\tilde{q}}| \le Q_1 = 0.47, \qquad |\ddot{\tilde{q}}| \le Q_2 = 0.28$$

for such a trajectory.

Then, we set the following initial values of the generalized coordinates and velocities:

$$q_1(0) = q_2(0) = 1.8, \quad \dot{q}_1(0) = 1.45, \quad \dot{q}_2(0) = 1.15.$$

System (2.5.19) is integrated with the control designed in accordance with the method suggested at the end of Sect. 3.2.4 with $X_i = U_i^0$. Under such a simplified control, the nonlinearities and the perturbations

$$G_1 = -(q_1 - \tilde{q}_1) - (\dot{q}_1 - \dot{\tilde{q}}_1), \qquad G_2 = -(q_2 - \tilde{q}_2) - (\dot{q}_2 - \dot{\tilde{q}}_2)$$

in the system are completely ignored. Nevertheless, the application of such a simplified approach is justified, since, in many cases, it enables one to steer the system to the nominal trajectory.

Figure 3.11 shows the graphs of the time history of the generalized coordinates of the system, and Fig. 3.12 represents similar graphs for the generalized velocities. The dashed curves correspond to the motion along the nominal trajectory, and the solid curves correspond to the motion of the perturbed system. One can see that, approximately 7 s after the beginning of the process, the system reaches the nominal trajectory and then moves along it in the sliding mode. Thus, the algorithm described allows one to reach the control objective also in the cases, where the sufficient conditions (3.2.72) are not fulfilled.

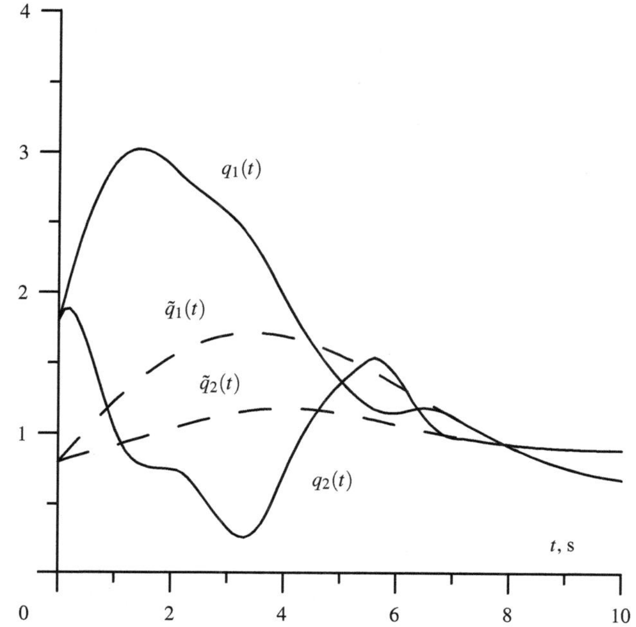

Fig. 3.11 Time history of the generalized coordinates

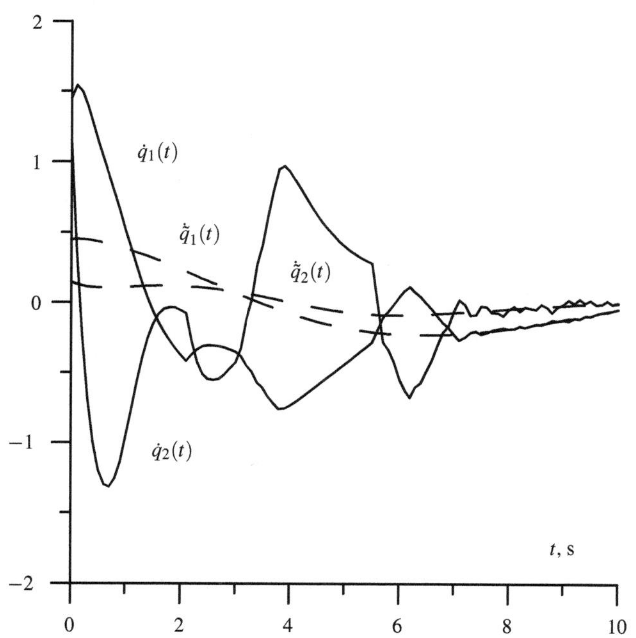

Fig. 3.12 Time history of the generalized velocities

Chapter 4
Stability based control for Lagrangian mechanical systems

Two general approaches to the synthesis of control laws for nonlinear dynamical systems are discussed in this book. One of them, based on the decomposition of a Lagrangian system, is considered Chapters 2 and 3. This approach is associated with the theory of optimal control.

Another approach is based on the methods of the theory of stability of motion. In Chapters 5 and 6, this approach will be utilized for designing control algorithms for various mechanical systems. Both scleronomic and rheonomic systems will be considered. In the present chapter, we recall the concepts of scleronomic and rheonomic mechanical systems, as well as some notions of the theory of stability of motion.

4.1 Scleronomic and rheonomic mechanical systems

In classical mechanics, a system is said to be holonomic if all constraints of the system can be expressed as functions of the coordinates and time only. They do not depend on the velocities.

A holonomic mechanical system is called *scleronomic*, if the equations of the constraints imposed on it do not contain time as an explicit variable; otherwise it is said to be *rheonomic*.

In our book, we deal with holonomic mechanical systems whose dynamics is described by Lagrange's equations of the second kind

$$\frac{d}{dt}\frac{\partial T}{\partial \dot{q}} - \frac{\partial T}{\partial q} = Q, \tag{4.1.1}$$

where $q, \dot{q}$ are the generalized coordinates and velocities, T is the kinetic energy of the system, Q is the vector of generalized forces acting upon the system.

In the scleronomic case the kinetic energy of the system is a quadratic form of the generalized velocities $\dot{q}$ with coefficients depending on the generalized coordinates q, i.e.,

$$T(q,\dot{q}) = \frac{1}{2}\langle A(q)\dot{q},\dot{q}\rangle, \tag{4.1.2}$$

where $A(q)$ is a symmetric positive definite matrix called the matrix of the kinetic energy or the matrix of inertia.

A two-link manipulator on a stationary base represents an example of a scleronomic system (see Fig.2.15).

In the rheonomic case, the kinetic energy of the system has the form of a full quadratic polynomial

$$T = \frac{1}{2}\langle A(t,q)\dot{q},\dot{q}\rangle + \langle a_1(t,q),\dot{q}\rangle + a_0(t,q), \tag{4.1.3}$$

where the matrix of the kinetic energy $A(t,q)$, the vector-valued function $a_1(t,q)$, and the function $a_0(t,q)$ depend on time explicitly.

A body with a moment of inertia depending on time provides an example of the rheonomic system.

A body with a variable moment of inertia

Let us consider a system consisting of a weightless bar and a particle of mass m_0 that can slide along the bar (see Fig. 4.1). We assume that the bar rotates in a horizontal plane about one of its ends under the action of a torque Q.

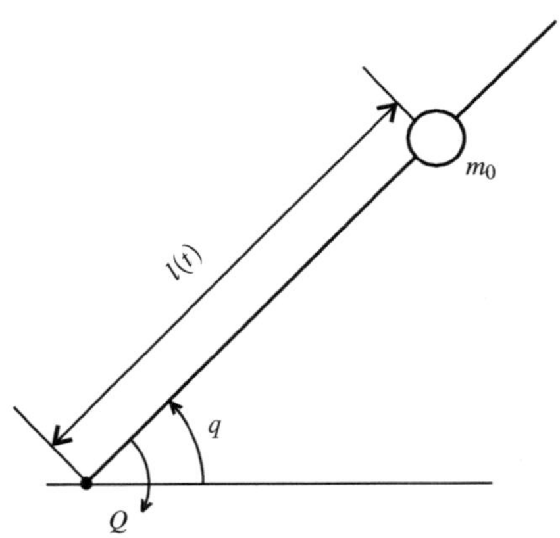

Fig. 4.1 A bar and a particle

We denote the angular coordinate and the angular velocity of the bar by q and $\dot{q}$, respectively, and the distance from the axis of rotation to the particle by $l(t)$. With this notation, the individual terms in expression (4.1.3) for the kinetic energy of the system take the form

$$A(t) = m_0 l^2(t),\ a_1 \equiv 0,\ a_0(t) = m_0 \dot{l}^2(t)/2,$$

and the equation of motion can be written as follows:

$$m_0 l^2(t)\ddot{q} + 2m_0 l(t)\dot{l}(t)\dot{q} = Q. \tag{4.1.4}$$

Here, the moment of inertia (which is the matrix of inertia of the dimension 1×1) equals $m_0 l^2(t)$ and depends on time.

Another example of a rheonomic system is a two-link manipulator on a movable base (see Fig. 4.2). In Chapters 6, the dynamics of such a manipulator will be simulated numerically.

A two-link manipulator on a movable base

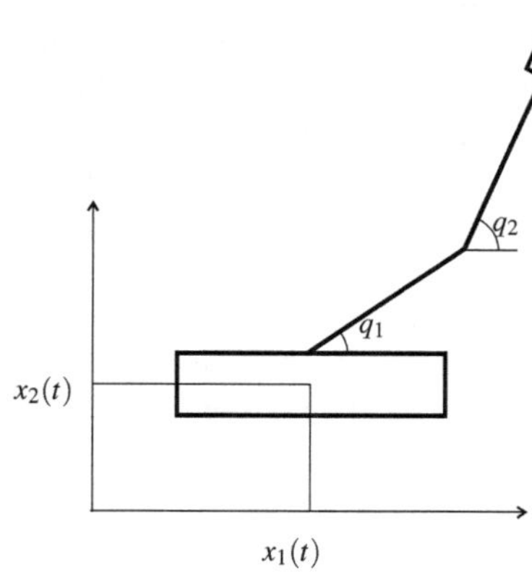

Fig. 4.2 Two-link manipulator on a movable base

We assume that the manipulator moves in a horizontal plane, and the base performs a translational motion. Denote by x_1 and x_2 the coordinates of the base, and by q_1 and q_2 the angular coordinates of the links.

At first, let us note that such a manipulator can be described as a scleronomic mechanical system with four degrees of freedom. Its kinetic energy (4.1.2) has the form

$$\begin{aligned} T = &\left(m_0 + \frac{m_1}{3} + m_2\right) l_1^2 \dot{q}_1^2 + \left(m_0 + \frac{m_2}{3}\right) l_2^2 \dot{q}_2^2 \\ &+ \left(2m_0 + \frac{m_2}{2}\right) l_1 l_2 \cos(q_1 - q_2)\dot{q}_1\dot{q}_2 \\ &+ \frac{m_0 + M_0 + m_1 + m_2}{2} \left(\dot{x}_1^2(t) + \dot{x}_2^2(t)\right) \\ &+ \left(m_0 + \frac{m_1}{2} + m_2\right) l_1 \left(\dot{x}_2(t)\cos q_1 - \dot{x}_1(t)\sin q_1\right)\dot{q}_1 \\ &+ \left(m_0 + \frac{m_2}{2}\right) l_2(\dot{x}_2(t)\cos q_2 - \dot{x}_1(t)\sin q_2)\dot{q}_2, \end{aligned} \tag{4.1.5}$$

where m_1 and m_2 are the masses of the links, l_1 and l_2 are the lengths of the links, M_0 is the mass of the base, and m_0 is the mass of a load into the gripper of the manipulator. The motion of this system is governed by the equations

$$\frac{d}{dt}\frac{\partial T}{\partial \dot{q}} - \frac{\partial T}{\partial q} = Q_1, \quad \frac{d}{dt}\frac{\partial T}{\partial \dot{x}} - \frac{\partial T}{\partial x} = Q_2, \tag{4.1.6}$$

where Q_1 and Q_2 are the projections of the vector of generalized forces Q onto the subspaces $q = (q_1, q_2)$ and $x = (x_1, x_2)$.

Since the motion of the links affects the dynamics of the base and vice versa, in order to investigate the dynamics of the links, it is necessary to consider equations (4.1.5) in the aggregate. However, in some cases, for investigating the dynamics of the links, it suffices to take into account only the first group of equations (4.1.5). For instance, if the mass of the base is much bigger than the mass of the links (the manipulator performs its operations onboard a rolling ship), then the influence of the links on the base may be neglected. Another such possibility happens if the base moves according to a certain fixed law, independent of the motion of the links. Then, the law of motion of the base can be considered as constraint imposed on the system and the functions $x_1(t)$ and $x_2(t)$ may be regarded either as known (if the dynamics of the base is given or if the position of the base in the space is measured) or as unknown.

Consequently, such a manipulator can be described as a rheonomic system with two degrees of freedom corresponding to the joint angles q_1 and q_2. In this case, the kinetic energy of the manipulator can be represented as a full quadratic polynomial of the generalized velocities $\dot{q}_1$ and $\dot{q}_2$ with coefficients, depending on the functions $x_1(t)$ and $x_2(t)$. Regarding these functions as arbitrary functions of time, we come to the following expressions for the inertia matrix $A(t,q)$, the vector-valued function $a(t,q)$ and the function $a_0(t,q)$ in (4.1.3):

$$A = \begin{bmatrix} \left(m_0 + \frac{m_1}{3} + m_2\right) l_1^2 & \left(m_0 + \frac{m_2}{2}\right) l_1 l_2 \cos(q_1 - q_2) \\ \left(m_0 + \frac{m_2}{2}\right) l_1 l_2 \cos(q_1 - q_2) & \left(m_0 + \frac{m_2}{3}\right) l_2^2 \end{bmatrix},$$

$$a_1 = \begin{bmatrix} \left(m_0 + \frac{m_1}{2} + m_2\right) l_1 \left(\dot{x}_2(t)\cos q_1 - \dot{x}_1(t)\sin q_1\right) \\ \left(m_0 + \frac{m_2}{2}\right) l_2 (\dot{x}_2(t)\cos q_2 - \dot{x}_1(t)\sin q_2) \end{bmatrix}, \tag{4.1.7}$$

$$a_0 = \frac{m_0 + M_0 + m_1 + m_2}{2} \left(\dot{x}_1^2(t) + \dot{x}_2^2(t)\right).$$

Let us note that, in this particular case, the matrix A does not depend on time but in the case of a generic rheonomic system it does, and so do the vector-valued function a_1 and the function a_0.

In Chapters 6, we will investigate the problem of control of mechanical systems with respect to some part of the variables; hence, we will consider the first group of equations (4.1.5) only. In our consideration, we will assume that the functions $x_1(t)$ and $x_2(t)$ are unknown and so are the coefficients of polynomial (4.1.3).

4.2 Lyapunov stability of equilibrium

The state $(q,\dot{q}) \in R^{2n}$ is called a state of rest (or an equilibrium point) of system (4.1.1) if $q(t) = \bar{q},\ \dot{q}(t) = 0$ is a solution of system (4.1.1).

Lyapunov stability of an equilibrium point means that, if solutions of system (4.1.1) start close enough to the equilibrium, than it remain close enough forever. Asymptotic stability means that all solutions of system (4.1.1) that start close enough to the equilibrium remain close enough to it and, furthermore, tend to the equilibrium as time goes to infinity. This properties can be formalize in the following definitions [88].

The state of rest $q = \bar{q},\ \dot{q} = 0$ of system (4.1.1) is called *stable* if for every $\varepsilon > 0$ there exists $\delta > 0$ such that $(|q(t_0) - \bar{q}|^2 + |\dot{q}(t_0)|^2)^{1/2} < \delta$ implies $(|q(t) - \bar{q}|^2 + |\dot{q}(t)|^2)^{1/2} < \varepsilon$ for all $t \geq t_0$.

The state of rest $q = \bar{q},\ \dot{q} = 0$ of system (4.1.1) is called *asymptotically stable* if it is stable and, in addition, there exists $\delta_1 > 0$ such that $(|q(t_0) - \bar{q}|^2 + |\dot{q}(t_0)|^2)^{1/2} < \delta_1$ implies

$$\lim_{t\to\infty} q(t) = \bar{q},\ \lim_{t\to\infty} \dot{q}(t) = 0. \tag{4.2.1}$$

The initial states $(q(t_0), \dot{q}(t_0))$ of the trajectories with property (4.2.1) make *a domain of attraction* $G \subset R^{2n}$ of the state of rest $q = \bar{q},\ \dot{q} = 0$.

The state of rest $q = \bar{q},\ \dot{q} = 0$ is called *globally asymptotically stable* if $G = R^{2n}$.

Lyapunov's direct method (also called the second method of Lyapunov) provides us with a powerful tool for investigating the stability in nonlinear dynamical systems without solving them [88, 77, 107]. The method is based on the use of Lyapunov functions. The idea of this method is to find a function that does not increase along solutions if the system is stable.

4.3 Lyapunov's direct method for autonomous systems

At first, we present the basic theorems of Lyapunov's direct method for autonomous systems, i.e., for the systems the equations of motion of which do not contain time as an explicit variable. An example of such a system is scleronomic mechanical system (4.1.1) with the kinetic energy (4.1.2) and the time-independent vector of the generalized force $Q(q,\dot{q})$.

Let $\Omega \subset R^{2n}$ be an open neighborhood of the state of rest $q = \bar{q},\ \dot{q} = 0$ of system (4.1.1). A continuous scalar function $V : \Omega \to R$ is called *positive [negative] definite* if $V(\bar{q},0) = 0$ and $V(q,\dot{q}) > 0$ [respectively, $V(q,\dot{q}) < 0$] in $\Omega \setminus \{(\bar{q},0)\}$.

Theorem 4.1. *Suppose that there exists a continuously differentiable positive definite function V in Ω such that its derivative calculated by virtue of system (4.1.1) satisfies the inequality $\dot{V} \leq 0$. Then the state of rest $q = \bar{q},\ \dot{q} = 0$ is stable.*

Theorem 4.2. *Suppose that there exists a continuously differentiable positive definite function V in Ω such that its derivative calculated by virtue of system (4.1.1) is negative definite. Then the state of rest $q = \bar{q}$, $\dot{q} = 0$ is asymptotically stable.*

Theorem 4.3. *Suppose that there exists a continuously differentiable positive definite function V in Ω such that its derivative calculated by virtue of system (4.1.1) satisfies the inequality $\dot{V} \leq 0$. Suppose, in addition, that the set*

$$\{(q,\dot{q}) \in \Omega \setminus \{(\bar{q},0)\} : \dot{V}(q,\dot{q}) = 0\}$$

does not contain positive semitrajectories of (4.1.1). Then the state of rest $q = \bar{q}$, $\dot{q} = 0$ is asymptotically stable.

Theorem 4.4. *Let $\Omega = R^{2n}$. Suppose that there exists a continuously differentiable positive definite function V in Ω such that $V(q,\dot{q}) \to \infty$ as $|q|^2 + |\dot{q}|^2 \to \infty$. Suppose also that the derivative of the function V calculated by virtue of system (4.1.1) satisfies the inequality $\dot{V} \leq 0$ and the set*

$$\{(q,\dot{q}) \in \Omega \setminus \{(\bar{q},0)\} : \dot{V}(q,\dot{q}) = 0\}$$

does not contain positive semitrajectories of (4.1.1). Then the state of rest $q = \bar{q}$, $\dot{q} = 0$ is globally asymptotically stable.

The functions $V(q,\dot{q})$ that appear in the theorems presented are called *Lyapunov functions* of system (4.1.1).

The classic example of a Lyapunov function is the total energy of a conservative mechanical system. The total energy of a system without any external energy source does not increase. Such a system remains in a neighborhood of an equilibrium point as it illustrates the following *Lagrange–Dirichlet Theorem.*

Theorem 4.5. *Let*

$$Q(q) = -\frac{\partial P}{\partial q}(q),$$

where $P(q)$ is a continuously differentiable, in a neighborhood of $\bar{q}$, function of the potential energy. Suppose that $q = \bar{q}$ is the point of a local strict minimum of $P(q)$. Then the state of rest $q = \bar{q}$, $\dot{q} = 0$ of the scleronomic system (4.1.1) with the kinetic energy (4.1.2) is stable.

Proof. Without loss of generality one may assume that $P(\bar{q}) = 0$. The total energy $E(q,\dot{q}) = T(q,\dot{q}) + P(q)$ may be chosen as a Lyapunov function of system (4.1.1). The function E is, obviously, positive definite in a neighborhood of the state of rest $q = \bar{q}$, $\dot{q} = 0$. Its derivative along the solutions of system (4.1.1) equals zero. Hence, by virtue of Theorem 4.1, the state of rest $q = \bar{q}$, $\dot{q} = 0$ of system (4.1.1) is stable. □

4.4 Lyapunov's direct method for nonautonomous systems

Since the kinetic energy (4.1.3) of the rheonomic mechanical system depends on time in an explicit form the dynamics of such a system is described by nonautonomous differential equations (4.1.1). In this section, we present Lyapunov function method technique conformably to nonautonomous systems. Now, Lyapunov functions depend on time explicitly.

As above, let $\Omega \subset R^{2n}$ be an open neighborhood of the state of rest $q = \bar{q}$, $\dot{q} = 0$ of system (4.1.1).

A continuous scalar function $V(t,q,\dot{q})$ is called positive [negative] definite if $V(t,\bar{q},0) = 0$, for any $t \geq t_0$, and there exists a positive [negative] definite function $W_1(q,\dot{q})$ such that $V(t,q,\dot{q}) \geq W_1(q,\dot{q})$ [respectively, $V(t,q,\dot{q}) \leq W_1(q,\dot{q})$], for any $t \geq t_0$ and $(q,\dot{q}) \in \Omega \setminus \{(\bar{q},0)\}$.

A positive definite function $V(t,q,\dot{q})$ is said to have an infinitesimal upper limit if there exists a positive definite function $W_2(q,\dot{q})$ such that $V(t,q,\dot{q}) \leq W_2(q,\dot{q})$, for any $t \geq t_0$ and $(q,\dot{q}) \in \Omega \setminus \{(\bar{q},0)\}$, i.e., the function $V(t,q,\dot{q})$, uniformly in t for $t \geq t_0$, tends to zero as $(q,\dot{q}) \to (\bar{q},0)$.

Theorem 4.6. *Suppose that there exists a continuously differentiable positive definite function $V(t,q,\dot{q})$ such that its derivative calculated by virtue of system (4.1.1) satisfies the inequality $\dot{V}(t,q,\dot{q}) \leq 0$. Then the state of rest $q = \bar{q}$, $\dot{q} = 0$ is stable.*

Theorem 4.7. *Suppose that there exists a continuously differentiable positive definite function $V(t,q,\dot{q})$ that has an infinitesimal upper limit and such that its derivative calculated by virtue of system (4.1.1) is negative definite. Then the state of rest $q = \bar{q}$, $\dot{q} = 0$ is asymptotically stable.*

Theorem 4.8. *Let $\Omega = R^{2n}$. Suppose that there exists a continuously differentiable positive definite function $V(t,q,\dot{q})$ that has an infinitesimal upper limit and tends to infinity, uniformly in t for $t \geq t_0$, as $|q|^2 + |\dot{q}|^2 \to \infty$. Suppose also that the derivative of the function V calculated by virtue of system (4.1.1) is negative definite. Then the state of rest $q = \bar{q}$, $\dot{q} = 0$ is globally asymptotically stable.*

4.5 Stabilization of mechanical systems

Very often, in control theory and in applications, the objective of control design is to stabilize a system, i.e., to design a control that makes the motion of the system (asymptotically) stable. The stability theory based approach to designing control algorithms has being developed, for instance, in the framework of the classical theory of automatic control and consists in constructing simple regimes of feedback control which ensure the asymptotic stability of the required motion (in particular, the terminal state). In so doing, the form of the feedback is given in advance, most often in the form of a linear function of phase variables.

Let us consider a scleronomic mechanical system subjected to *PD-controller*, i.e, the control vector-function chosen in the form of a linear feedback

$$U(q,\dot{q}) = -\alpha\dot{q} - \beta q, \tag{4.5.1}$$

where the feedback factors α and β are some positive real constants. It is not difficult to prove that, for such a system, in the absence of other forces, the origin of the phase space is globally asymptotically stable. Really, in this case the equations of motion can be written as follows:

$$\frac{d}{dt}\frac{\partial T}{\partial \dot{q}} - \frac{\partial T}{\partial q} = -\alpha\dot{q} - \beta q, \tag{4.5.2}$$

where the kinetic energy $T(q,\dot{q}$ is given by (4.1.2). The first term of the right-hand side in (5.1.12) is a dissipative force, and the second term plays the role of the potential force, i.e.,

$$\beta q = \frac{\partial P}{\partial q}(q), \quad P(q) = \frac{\beta}{2}q^2.$$

The total energy of system (5.1.12)

$$E(q,\dot{q}) = T(q,\dot{q}) + P(q)$$

is a positive definite Lyapunov function whose derivative along the trajectory of the system satisfies the inequality

$$\dot{E} = -\alpha\dot{q}^2 \leq 0,$$

and the set

$$\{(q,\dot{q}) \in R^{2n} : \dot{E}(q,\dot{q}) = 0\}$$

does not contain entire positive semitrajectories (with the exception for the trivial solution). By Theorem 4.4, this implies the asymptotic stability of the state of rest $q = \dot{q} = 0$.

Aside from simplicity, the PD-controller has some other important advantages. It has a closed loop form, does not depend on the parameters of the system, and can be applied for stabilizing the state of rest of system (5.1.12) with an arbitrary matrix of the kinetic energy $A(q)$. To utilize this control, it is sufficient to know the current phase state only.

However, the above linear feedback control has some disadvantages. First, the control force is not bounded and does not meet constraint which is present, as a rule, in practice. The control generated by the PD-controller is too large when the current state of the system is far away from the origin of the phase space.

Second, the asymptotic stability of the state of rest of system (5.1.12) means that it takes for the system infinite time to approach this state. The closer the system to the terminal state, the smaller the control force. The control force tends to zero as the trajectory tends to the origin of the phase space, therefore, in a small neighborhood of the origin, the PD-controller does not use the control possibilities to full extent, which implies infinite time of steering. In addition, in the presence of even small final force disturbances in the system, such steering becomes impossible because, in a small neighborhood of the origin of the phase space, the disturbances exceed the control.

4.6 Modification of Lyapunov's direct method

As we already mentioned, in our book, we are searching for the control algorithms which are bounded, capable of coping with uncertain bounded disturbances, and steer the system to the prescribed terminal state in finite time. For this purpose, we use a modification of Lyapunov's direct method.

In Chapters 5 and 6 the stability theory based control laws are presented that can be treated as linear feedback control (4.5.1) with variable feedback factors α and β. To make the control bounded and more effective, and to speed up steering, in Chapters 5, we change the feedback factors in a jump-like manner while the trajectory approaches the terminal state. In Chapters 6, we specify the feedback factors α and β as continuously differentiable functions of the phase variables q and $\dot{q}$, and time t. In both cases, the feedback factors increase and tend to infinity as the trajectory approaches the terminal state; nevertheless, the control force remains bounded and meets the imposed constraint.

Below, the following theorem will be applied for justifications of the controls to be design.

Theorem 4.9. *Let $\Omega \subset R^{2n}$ be an open neighborhood of the state $q = \bar{q}$, $\dot{q} = 0$, and $V(t,q,\dot{q})$ be a positive definite scalar function, $(q,\dot{q}) \in \Omega$, $t \geq t_0$, which is continuously differentiable, for all $(q,\dot{q}) \in \Omega \setminus \{(\bar{q},0)\}$, $t \geq t_0$, has an infinitesimal upper limit, and tends to infinity, uniformly in t for $t \geq t_0$, if $|q|^2 + |\dot{q}|^2 \to \infty$. Suppose that the derivative of the function V calculated by virtue of system (4.1.1), along the trajectory starting at the point $(t_0, q_0, \dot{q}_0)$, satisfies the inequality*

$$\dot{V} \leq -\delta V^{1/2},$$

where δ is a positive real constant. Then this trajectory approaches the state $q = \bar{q}$, $\dot{q} = 0$ in finite time.

This theorem concerns both scleronomic and rheonomic cases and is an obvious modification of the theorems of Lyapunov's direct method stated above.

Chapter 5
Piecewise linear control for mechanical systems under uncertainty

In the fifth chapter, we consider a Lagrangian mechanical system (2.1.1) under the assumption that the kinetic energy matrix $A(q)$ of the system is unknown and the system is subject to uncontrollable bounded external forces. A control law is proposed that transfers the system from an arbitrary initial state to a given terminal state in finite time by a bounded force. In the algorithm proposed, a linear feedback control is used with piecewise constant coefficients: the coefficients increase and tend to infinity as the system approaches the terminal state. Nevertheless, the control force is bounded and meets the imposed constraint. The algorithm is based on the Lyapunov's direct method.

By an example of two-mass oscillatory systems, it is shown that the proposed approach can be used for control of underactuated systems, that is, in the case where the number of degrees of freedom exceeds the dimension of the control force vector. In the final part of Chapter 5, piecewise linear feedback control is applied to rheonomic systems.

The results presented in this chapter were published previously in [4, 5, 6, 8, 9].

5.1 Piecewise linear control for scleronomic systems

5.1.1 Problem statement

In the first section of this chapter, we consider, as before, a controlled scleronomic mechanical system whose kinetic energy has the form of a quadratic polynomial of the generalized velocities $\dot{q}$ with coefficients depending on the generalized coordinates q

$$T(q,\dot{q}) = \frac{1}{2}\langle A(q)\dot{q},\dot{q}\rangle, \tag{5.1.1}$$

where $A(q)$ is a symmetric positive definite matrix of the kinetic energy.

The system dynamics is described by Lagrange's equations of the second kind

$$\frac{d}{dt}\frac{\partial T}{\partial \dot{q}} - \frac{\partial T}{\partial q} = U + Q, \tag{5.1.2}$$

where U is the vector of control forces, Q is the vector of all other forces acting on the system. The vector of generalizes forces $Q(t,q,\dot{q})$ may be an arbitrary vector-valued function, including a discontinuous one, satisfying some existence conditions for the solution of system (5.1.2) and meeting the constraint

$$|Q| \le Q_0, \quad Q_0 > 0. \tag{5.1.3}$$

The vector of control forces U is also bounded

$$|U| \le U_0, \quad U_0 > 0. \tag{5.1.4}$$

The vector of the forces Q is considered to be unknown and treated as an uncertain disturbance. Along with them, other specified forces may act on the system. However, we assume that the control possibilities are large enough to compensate these specified forces. Let U_0 be the maximum admissible control magnitude remaining after such compensation.

We assume that the kinetic energy matrix $A(q)$ is continuously differentiable and unknown, its eigenvalues belong to the interval $[m,M]$, $0 < m \le M$ for any q, and the partial derivatives of $A(q)$ are bounded uniformly in q with respect to the Euclidean norm, that is,

$$mz^2 \le \langle A(q)z, z\rangle \le Mz^2, \quad z \in R^n, \tag{5.1.5}$$

$$\|\frac{\partial A(q)}{\partial q_i}\| \le D, \quad D > 0, \quad i = 1,\dots,n. \tag{5.1.6}$$

The phase variables q and $\dot{q}$ are assumed to be available for measuring at every time instant.

Problem 5.1. For given initial state $q(0) = q_*$ and $\dot{q}(0) = \dot{q}_*$, and constants m, M, D, U_0, and Q_0, it is required to construct a control that satisfies (5.1.4) and steers system (5.1.2) to a prescribed terminal state $(\bar{q}, 0)$ in finite time.

Let us note that in case of $U = Q = 0$ the terminal state is a rest point of system (5.1.2). Without loss of generality, we assume that $\bar{q} = 0$, i.e., the terminal state coincides with the phase space origin. Otherwise, we can take $q - \bar{q}$ as a vector of generalized coordinates.

The problem of control for a system of connected rigid bodies whose precise mass-inertial characteristics are unknown gives us an illustrative example of the formulation of the above problem. In this case, not only the inertia matrix of the system, but also the forces acting on the bodies remain unknown. Apart from these forces, the system may be subject to other external perturbations.

The problem of transporting a load of unknown mass by a manipulator is a special case of this problem.

5.1.2 Description of the control algorithm

We will construct the desired control on the base of liner feedback control (4.5.1). In Chapter 4, we have already mentioned the merits of such PD-controller which are the simplicity of its implementation and the robustness with respect to the disturbances of parameters of a dynamical system over a wide range. Control (4.5.1) guarantees asymptotic stability of the phase space origin, i.e., steers the system to the terminal state in infinite time. Besides, control (4.5.1) is neither bounded and nor capable of coping with disturbances in a small neighborhood of the terminal state.

To meet the constraints imposed and to use the control possibilities to full extent we will change the feedback factors α and β in a jump-like manner during the motion. Therefore, the control law proposed below uses linear feedback (4.5.1) with the feedback factors α and β as step-functions of time.

Now, we reformulate the original problem as follows.

Problem 5.2. For given initial state $q(0) = q_*$ and $\dot{q}(0) = \dot{q}_*$, and constants m, M, D, U_0, and Q_0, it is required to specify how the feedback factors α and β in the control function (4.5.1) should be varied so that for any disturbances Q that satisfy (5.1.3) the trajectory of system (5.1.2) and(4.5.1) arrives at the state $(0,0)$ in finite time and the control U meets constraints (5.1.4) along the trajectory.

Consider the function

$$W(q,\dot{q}) = M\dot{q}^2 + \left(M^2\dot{q}^4 + \frac{U_0^2}{2}q^2\right)^{1/2}. \tag{5.1.7}$$

The quantity $W(q,\dot{q})$ has the dimension of energy and characterizes the distance between the point $(q,\dot{q})$ and the terminal state $(0,0)$. The level set $W(q,\dot{q}) = C$ of the function W in the phase space R^{2n} is an ellipsoid $4CM\dot{q}^2 + U_0^2 q^2 = 2C^2$, that collapses to the phase space origin $(0,0)$ as $C \to 0$.

We put

$$D_1 = \frac{\sqrt{n}D}{2}, \quad W_0 = \frac{MU_0}{2\sqrt{2}D_1} \quad W_k = \frac{W_0}{2^k}, \tag{5.1.8}$$

and define a set of ellipsoids

$$\{(q,\dot{q}) \in R^{2n} : \; W(q,\dot{q}) = W_k\},$$

where k runs through the set of integers (see Fig. 5.1). Suppose that the point $(q_*, \dot{q}_*)$ corresponding to the phase state of the original system at the initial instant of time $t = 0$ lies on the ellipsoid $\{(q,\dot{q}) \in R^{2n} : \; W(q,\dot{q}) = W_{k_*}\}$ or inside it, but outside the ellipsoid $\{(q,\dot{q}) \in R^{2n} : \; W(q,\dot{q}) = W_{k_*+1}\}$, i.e.,

$$W_{k_*+1} < W(q_*, \dot{q}_*) \le W_{k_*}.$$

Let t_{k_*+1} be the first instant of time when the trajectory of the system hits the ellipsoid $\{(q,\dot{q}) \in R^{2n} : \; W(q,\dot{q}) = W_{k_*+1}\}$. Below, it will be shown that, for the

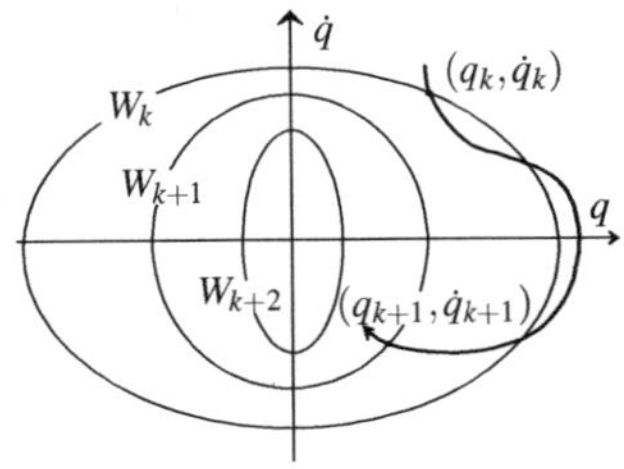

Fig. 5.1 The set of ellipsoids and the trajectory of the system

chosen control algorithm, the trajectory of the system tends to the origin, which means that such an instant of time exists.

We denote $q(t_{k_*+1}) = q_{k_*+1}$, $\dot{q}(t_{k_*+1}) = \dot{q}_{k_*+1}$. Let t_{k_*+2} be the first instant of time when the trajectory of the system hits the ellipsoid $\{(q,\dot{q}) \in R^{2n} : W(q,\dot{q}) = W_{k_*+2}\}$. We denote $q(t_{k_*+2}) = q_{k_*+2}$, $\dot{q}(t_{k_*+2}) = \dot{q}_{k_*+2}$, and so on.

The sequence $\{t_k\}, k = k_* + 1, k_* + 2, \ldots$, defines the instants of time when we change the feedback factors α and β in (4.5.1). We specify the values of these factors in the time half-interval $[t_k, t_{k+1})$ as follows:

$$\beta_k = \frac{U_0^2}{4W_k}, \quad \alpha_k^2 = m\beta_k. \tag{5.1.9}$$

The initial values of the factors are defined by formulas (5.1.9), where $k = k_*$.

In the phase space $(q,\dot{q})$, the trajectory of the mechanical system under consideration, therefore, consists of segments of trajectories of different systems of differential equations (see Fig. 5.1): the kth segment connects points $(q_k, \dot{q}_k)$ and $(q_{k+1}, \dot{q}_{k+1})$ and corresponds to a system of the form (5.1.2) and (4.5.1) in which the gains $\alpha = \alpha_k$ and $\beta = \beta_k$ are constant and given by (5.1.9). All points $(q_k, \dot{q}_k)$ lie on the corresponding ellipsoids $\{(q,\dot{q}) \in R^{2n} : W(q,\dot{q}) = W_k\}$, $k > k_*$.

Remark 5.1. The trajectory of the system tends to the space origin $(0,0)$, but the function W is not, in general, a decreasing function along the trajectory. Therefore, along with the points $(q_k, \dot{q}_k)$, the trajectory can also have other points of intersection with the ellipsoids from the above set. Suppose, for example, that once the new feedback factors are assigned at time t_k, the trajectory of the system starts "moving away" from the terminal state $(0,0)$ and intersects the ellipsoid with number $k-1$ again at some instant of time $t' > t_k$. At the instant t', the index k and the gains α_k and β_k not vary. They take new values only when the trajectory reaches the ellipsoid $W(q,\dot{q}) = W_{k+1}$. The index k increases by one and, in an accordance to formulas (5.1.8) and (5.1.9), the gain α increases by a factor of $\sqrt{2}$ and the gain β by a factor of 2.

Remark 5.2. The proposed control law is not, strictly speaking, a feedback one. To determine the control force, besides the current phase state of the system, one must know also the time history of the trajectory. For a given matrix of kinetic energy $A(q)$ and the proposed algorithm of specifying feedback factors in control law (4.5.1), the trajectory of the system is defined by the initial state $(q_*, \dot{q}_*)$ and the disturbance $Q(t, q, \dot{q})$. However, to calculate the control force at any instant of time, it is sufficient to know the current value of index k. Having known k, one can find the gains α_k and β_k through formulas (5.1.8) and (5.1.9) and, then, the control force U according to (4.5.1). Thus, the control is a function of the phase variables q and $\dot{q}$ and the integer k:

$$U(q, \dot{q}, k) = -\alpha_k \dot{q} - \beta_k q. \tag{5.1.10}$$

To implement the proposed control, it is sufficient to measure the phase state of the system and keep in store of the computer the current value of index k. At each instant of time, the index k is equal to the number of the minimum ellipsoid that has already been visited by the trajectory of the system. Every time when the gains change, the index k increases by unity.

5.1.3 Justification of the algorithm

We shall study the behavior of the trajectory of the kth system for some $k > k_*$. The segment of the trajectory that is of interest to us starts at the point $(q_k, \dot{q}_k)$ at instant of time t_k and ends, in accordance with the algorithm, at time t_{k+1} on the ellipsoid $W(q, \dot{q}) = W_{k+1}$. Since the existence of an intersection of the trajectory with the $(k+1)$st ellipsoid has not been shown so far, we shall assume that $t_{k+1} = \infty$, if there is no such intersection. Below it will be shown that $t_{k+1} < \infty$.

We introduce a family of the Lyapunov functions

$$V^k(q, \dot{q}) = T(q, \dot{q}) + \frac{\beta_k}{2} q^2 + \varepsilon_k \langle A(q) q, \dot{q} \rangle, \tag{5.1.11}$$

where k is an integer, the number $\varepsilon_k > 0$ is specified below.

The expression for the function $V^k(q, \dot{q})$ contains the inertia matrix $A(q)$, that is assumed to be unknown. We shall estimate the value of this function at the point $(q, \dot{q})$ in the phase space by means of known quantities.

Suppose that ε_k satisfies the condition

$$\varepsilon_k^2 < \frac{m\beta_k}{4M^2}. \tag{5.1.12}$$

We estimate the function $V^k(q, \dot{q})$ from below using (5.1.5) and the inequality

$$\varepsilon_k M |q| |\dot{q}| \leq \frac{\varepsilon_k^2 M^2}{m} q^2 + \frac{m}{4} \dot{q}^2$$

as follows:

$$V^k(q, \dot{q}) \geq \frac{\beta_k}{2} q^2 + \frac{m}{2} \dot{q}^2 - \varepsilon_k M |q| |\dot{q}| \geq \left(\frac{\beta_k}{2} - \frac{\varepsilon_k^2 M^2}{m} \right) q^2 + \frac{m}{4} \dot{q}^2.$$

This together with (5.1.12) yields

$$V_-^k(q,\dot q) \le V^k(q,\dot q), \quad V_-^k(q,\dot q) = \frac{1}{4}(\beta_k q^2 + m\dot q^2). \tag{5.1.13}$$

We now estimate the function $V^k(q,\dot q)$ from above using (5.1.5) and the inequality

$$\varepsilon_k |q||\dot q| \le \frac{\varepsilon_k^2}{2} q^2 + \frac{1}{2}\dot q^2$$

as follows:

$$V^k(q,\dot q) \le \frac{\beta_k}{2} q^2 + \frac{M}{2}\dot q^2 + \varepsilon_k M |q||\dot q| \le \frac{(\beta_k + \varepsilon_k^2 M)}{2} q^2 + M\dot q^2.$$

Since $m < 4M$, inequality (5.1.12) implies that $\varepsilon_k^2 M < \beta_k$, from whence we come to the inequality

$$V^k(q,\dot q) \le V_+^k(q,\dot q), \quad V_+^k(q,\dot q) = \beta_k q^2 + M\dot q^2. \tag{5.1.14}$$

We shall establish relations between the quadratic forms $V_+^k(q,\dot q)$ and the function $W(q,\dot q)$, whose level sets generate the family of ellipsoids defined above. Let us show that, for any integer k, the equality

$$2V_+^k(q_k,\dot q_k) = W_k \tag{5.1.15}$$

holds.

For the proof, we substitute the expression for β_k given by (5.1.9) into expression (5.1.14) for the function V_+^k. We obtain

$$V_+^k(q_k,\dot q_k) = \frac{U_0^2 q_k^2 + 4W_k M\dot q_k^2}{4W_k}. \tag{5.1.16}$$

By construction, the point $(q_k,\dot q_k)$ lies on the ellipsoid with the number k. Hence, by the definition (5.1.7) of the function W, it follows that

$$W_k = W(q_k,\dot q_k) = M\dot q_k^2 + \left(M^2\dot q_k^4 + \frac{U_0^2}{2} q_k^2\right)^{1/2}$$

and

$$U_0^2 q_k^2 + 4W_k M\dot q_k^2 = 2W_k^2.$$

The latter equality and (5.1.16) yield (5.1.15).

Equality (5.1.15) means that, for any k, the ellipsoid with the number k is a level set of the quadratic form $V_+^k(q,\dot q)$ corresponding to $W_k/2$.

Suppose that the system is in a state $(q,\dot q)$ at an instant of time t, $t_k < t < t_{k+1}$. We shall estimate know the value of the quadratic form $V_+^k(q,\dot q)$ at time t through its value $V_+^k(q_k,\dot q_k)$ at time t_k, i.e., when the trajectory hits the kth ellipsoid for the first time. Let us recall that, due to the function W being nonmonotone along the trajectory, the point $(q,\dot{q})$ may lie either inside or outside the kth ellipsoid (see Remark 5.1 in Sect. 5.1.2). Anyway, according to the algorithm and by virtue of condition $t < t_{k+1}$, the point $(q,\dot{q})$ lies outside the $(k+1)$st ellipsoid whereas the point $(q_{k+1},\dot{q}_{k+1})$ belongs to it. Therefore, taking into account (5.1.15), we obtain

$$V_+^{k+1}(q,\dot{q}) \geq V_+^{k+1}(q_{k+1},\dot{q}_{k+1}) = \frac{W_{k+1}}{2}.$$

By (5.1.8), (5.1.9), and definition (5.1.14) of the function V_+^k, the equalities

$$W_{k+1} = \frac{W_k}{2},\ \beta_{k+1} = 2\beta_k,\ V_+^{k+1}(q,\dot{q}) = \beta_{k+1}q^2 + M\dot{q}^2$$

hold. Hence,

$$V_+^{k+1}(q,\dot{q}) = 2\beta_k q^2 + M\dot{q}^2 \geq \frac{W_k}{4}$$

and, consequently,

$$V_+^k(q,\dot{q}) \geq \beta_k q^2 + \frac{M}{2}\dot{q}^2 = \frac{1}{2}V_+^{k+1}(q,\dot{q}) \geq \frac{W_k}{8}. \tag{5.1.17}$$

From here and using (5.1.15), we draw the following estimate for quadratic form $V_+^k(q,\dot{q})$ at time t through its value $V_+^k(q_k,\dot{q}_k)$ at time t_k:

$$V_+^k(q,\dot{q}) \geq \frac{1}{4}V_+^k(q_k,\dot{q}_k). \tag{5.1.18}$$

We now proceed to calculating the derivative $\dot{V}^k$. By differentiating V^k according to (5.1.2) and (5.1.10), we obtain

$$\begin{aligned} \dot{V}^k(q,\dot{q}) = &-\varepsilon_k\beta_k q^2 - \langle \left[\alpha_k I - \varepsilon_k A(q) - \frac{\varepsilon_k}{2}\sum_{i=1}^n q_i\frac{\partial A}{\partial q_i}\right]\dot{q},\dot{q}\rangle \\ &-\varepsilon_k\alpha_k\langle q,\dot{q}\rangle + \langle Q,\varepsilon_k q + \dot{q}\rangle, \end{aligned} \tag{5.1.19}$$

where I is the identity matrix. Let us estimate the individual terms in expression (5.1.19). By virtue of (5.1.5), (5.1.6), and (5.1.8, the inequalities

$$\|\varepsilon_k A(q)\| \leq \varepsilon_k M, \quad \|\frac{\varepsilon_k}{2}\sum_{i=1}^n q_i\frac{\partial A}{\partial q_i}\| \leq \varepsilon_k D_1|q| \tag{5.1.20}$$

hold.

Using relationships (5.1.3), (5.1.12), (5.1.14), (5.1.15), and (5.1.18), we estimate the last term in expression (5.1.19) as follows:

$$|\langle Q,\varepsilon_k q + \dot{q}\rangle| \leq Q_0|\varepsilon_k q + \dot{q}| \leq Q_0\left(5\varepsilon_k^2 q^2 + \frac{5}{4}\dot{q}^2\right)^{1/2}$$

$$\leq \frac{\sqrt{5}Q_0}{2}\left(\frac{m\beta_k}{M^2}q^2+\dot{q}^2\right)^{1/2} \leq \frac{Q_0}{2}\sqrt{\frac{5}{M}}\left(\beta_k q^2 + M\dot{q}^2\right)^{1/2} \tag{5.1.21}$$
$$= \frac{\sqrt{5}Q_0V_+^k(q,\dot{q})}{2\sqrt{MV_+^k(q,\dot{q})}} \leq \frac{\sqrt{5}Q_0V_+^k(q,\dot{q})}{\sqrt{MV_+^k(q_k,\dot{q}_k)}} = \frac{\sqrt{10}Q_0}{\sqrt{MW_k}}(\beta_k q^2 + M\dot{q}^2).$$

Substituting (5.1.20) and (5.1.21) into (5.1.19) and using the inequality

$$|\varepsilon_k\alpha_k\langle q,\dot{q}\rangle| \leq \varepsilon_k^2\alpha_k q^2 + \frac{\alpha_k}{4}\dot{q}^2,$$

we arrive at the estimate

$$\dot{V}^k(q,\dot{q}) \leq -\varepsilon_k\left(\beta_k - \varepsilon_k\alpha_k - \frac{\sqrt{10}Q_0\beta_k}{\varepsilon_k\sqrt{MW_k}}\right)q^2$$
$$-\left(\frac{3\alpha_k}{4} - \varepsilon_k M - \varepsilon_k D_1|q| - \frac{\sqrt{10M}Q_0}{\sqrt{W_k}}\right)\dot{q}^2. \tag{5.1.22}$$

We define the parameter ε_k by the formula

$$\varepsilon_k = \min\{\frac{\sqrt{m}U_0}{8M\sqrt{W_k}}, \frac{\sqrt{m}U_0^2}{16D_1W_k\sqrt{2W_k}}\} \tag{5.1.23}$$

and introduce the domain

$$G = \{(q,\dot{q}): \ |q| < \frac{2\sqrt{2}W_k}{U_0}\}.$$

Lemma 5.1. *Let the condition*

$$Q_0 \leq \min\{\frac{\sqrt{m}U_0}{16\sqrt{10M}}, \frac{\varepsilon_k\sqrt{MW_k}}{2\sqrt{10}}\} \tag{5.1.24}$$

be satisfied. Then, at those points of the trajectory that lie in the domain G, the derivative $V^k(t) = V^k(q(t),\dot{q}(t))$ calculated by virtue of system (5.1.2) and (5.1.10) satisfies the inequality

$$\dot{V}^k(q,\dot{q}) \leq -\frac{\varepsilon_k\beta_k}{4}q^2 - \frac{\alpha_k}{8}\dot{q}^2. \tag{5.1.25}$$

Proof. Definition (5.1.23) of the parameter ε_k yields (5.1.12). Therefore, relationships (5.1.13), (5.1.14), (5.1.21), and (5.1.22) are valid.

Using (5.1.9) and (5.1.23), we obtain

$$\varepsilon_k\alpha_k \leq \frac{mU_0\sqrt{\beta_k}}{8MW_k^{1/2}} = \frac{m\beta_k}{4M} \leq \frac{\beta_k}{4}, \tag{5.1.26}$$

$$\varepsilon_k M \le \frac{\sqrt{m}U_0}{8\sqrt{W_k}} = \frac{\sqrt{m\beta_k}}{4} = \frac{\alpha_k}{4}.$$

Relationship (5.1.24) and formulas (5.1.9) imply that

$$\frac{\sqrt{10}Q_0\beta_k}{\varepsilon_k\sqrt{MW_k}} \le \frac{\beta_k}{2}, \quad \frac{\sqrt{10M}Q_0}{\sqrt{W_k}} \le \frac{\sqrt{m}U_0}{16\sqrt{W_k}} = \frac{\alpha_k}{8}. \tag{5.1.27}$$

From (5.1.23) and the definition of G, it follows that

$$\varepsilon_k D_1|q| \le \frac{\sqrt{m}U_0^2}{16W_k\sqrt{2W_k}}|q| \le \frac{\sqrt{m}U_0}{8\sqrt{W_k}} = \frac{\alpha_k}{4}. \tag{5.1.28}$$

Substituting inequalities (5.1.26)–(5.1.28) into estimate (5.1.22), we obtain (5.1.25). This completes the proof of the lemma. □

Lemma 5.2. *Suppose that conditions (5.1.24) are satisfied. Then the part of the trajectory corresponding to the time interval* $[t_k, t_{k+1})$ *lies wholly in the domain* G.

Proof. We shall verify that the initial point of the trajectory $(q_k, \dot{q}_k)$ belongs to the domain G. By construction, the point $(q_k, \dot{q}_k)$ belongs to the ellipsoid with number k, i. e.,

$$M\dot{q}_k^2 + \left(M^2\dot{q}_k^4 + \frac{U_0^2}{2}q_k^2\right)^{1/2} = W_k.$$

Therefore, $q_k^2 \le 2W_k^2/U_0^2$, that implies $(q_k, \dot{q}_k) \in G$.

Suppose that the assertion of the Lemma does not hold and let t' be the first instant of time when the trajectory reaches the boundary of G, $t' > t_k$. By Lemma 5.1, the function V^k is strictly decreasing in G along the solutions of system (5.1.2) and (5.1.10). Whence, by (5.1.14) and (5.1.15), we obtain

$$V^k\left(q(t'), \dot{q}(t')\right) < V^k(q_k, \dot{q}_k) = V^k(q(t_k), \dot{q}(t_k)) \le V_+^k(q_k, \dot{q}_k) = \frac{W_k}{2}.$$

On the other hand, relationship (5.1.13) yields the inequality

$$V^k(q(t'), \dot{q}(t')) \ge V_-^k(q(t'), \dot{q}(t')) \ge \frac{\beta_k}{4}q^2(t').$$

By supposition, the point $q(t')$ lies on the boundary of G. By definition of G, we have $q^2(t') = 8W_k^2/U_0^2$. From whence, formulas (5.1.9), and the latter inequality, it follows that

$$V^k(q(t'), \dot{q}(t')) \ge \frac{W_k}{2}.$$

This contradiction completes the proof of the lemma. □

By inequality (5.1.13), for any k, Lyapunov's function (5.1.11) is positive definite, while inequality (5.1.25) implies that its derivative is negative and non-zero outside the $(k+1)$st ellipsoid. Hence, we can conclude that there exists an instant of time $t_{k+1} < \infty$, when the trajectory reaches the ellipsoid with number $k+1$.

We shall verify now that control forces (5.1.10) meet constraint (5.1.4) on the segment of the trajectory corresponding to the half-interval of time $[t_k, t_{k+1})$. Let us estimate from above the norm of the vector U using formula (5.1.9) for the gain α_k and inequality (5.1.13) as follows:

$$|U(q,\dot{q},k)|^2 = |\beta_k q + \alpha_k \dot{q}|^2 \le 2(\beta_k^2 q^2 + \alpha_k^2 \dot{q}^2)$$
$$= 2\beta_k(\beta_k q^2 + m\dot{q}^2) = 8\beta_k V_-^k(q,\dot{q}) \le 8\beta_k V^k(q,\dot{q}).$$

As it has been already shown, the function V^k does not increase along this part of the trajectory, therefore, we have $V^k(q,\dot{q}) \le V^k(q_k,\dot{q}_k)$. Taking relationships (5.1.9), (5.1.14), and (5.1.15) into account, we obtain

$$|U(q,\dot{q},k)|^2 \le 8\beta_k V^k(q_k,\dot{q}_k) \le 8\beta_k V_+^k(q_k,\dot{q}_k) = 4\beta_k W_k = U_0^2.$$

Thus, condition (5.1.4) is fulfilled.

5.1.4 Estimation of the time of motion

It follows from (5.1.9) and (5.1.23) that

$$\frac{\alpha_k}{8} = \frac{\sqrt{m}U_0}{16\sqrt{W_k}} \ge \frac{M\varepsilon_k}{2}.$$

Using this estimate, we continue inequality (5.1.25) as follows:

$$\dot{V}^k(q,\dot{q}) \le -\frac{\varepsilon_k \beta_k}{4} q^2 - \frac{M\varepsilon_k}{2}\dot{q}^2 \le -\frac{\varepsilon_k}{4} V_+^k(q,\dot{q}) \le -\frac{\varepsilon_k}{4} V^k(q,\dot{q}).$$

We integrate this inequality over the half-interval $[t_k, t_{k+1})$ to get

$$t_{k+1} - t_k \le \frac{4}{\varepsilon_k} \log \frac{V^k(q_k,\dot{q}_k)}{V^k(q_{k+1},\dot{q}_{k+1})}. \tag{5.1.29}$$

Let us estimate the expression under the logarithm sign. The numerator of this expression, obviously, satisfies the inequality

$$V^k(q_k,\dot{q}_k) \le V_+^k(q_k,\dot{q}_k) = \frac{W_k}{2}.$$

By definition, the quadratic forms (5.1.13) and (5.1.14) are connected by the relationship

$$V_-^k(q,\dot{q}) \ge \frac{m}{4M} V_+^k(q,\dot{q}),$$

whence, using (5.1.15) and the equalities $\beta_k = \beta_{k+1}/2$ and $W_{k+1} = W_k/2$, we obtain the following estimate for the denominator:

$$V^k(q_{k+1},\dot{q}_{k+1}) \geq V_-^k(q_{k+1},\dot{q}_{k+1}) \geq \frac{m}{4M}V_+^k(q_{k+1},\dot{q}_{k+1})$$

$$= \frac{m}{4M}(\frac{\beta_{k+1}}{2}q_{k+1}^2 + M\dot{q}_{k+1}^2) \geq \frac{m}{8M}V_+^{k+1}(q_{k+1},\dot{q}_{k+1}) = \frac{mW_k}{32M}.$$

Using the obtained estimates of the numerator and denominator, the inequality (5.1.29) can be transformed as follows:

$$t_{k+1} - t_k \leq \frac{4}{\varepsilon_k}\log\frac{16M}{m}, \tag{5.1.30}$$

ε_k being given by (5.1.23). It is easy to see that the expressions under the min sign in (5.1.23) are identical for $k = 0$. If the point $(q_k,\dot{q}_k)$ lies outside the ellipsoid with number 0, i.e., $k < 0$, then

$$\frac{\sqrt{m}U_0}{8M\sqrt{W_k}} > \frac{\sqrt{m}U_0^2}{16D_1W_k\sqrt{2W_k}},$$

and if $(q_k,\dot{q}_k)$ lies inside or on the null ellipsoid, i.e., $k \geq 0$, then the reverse inequality holds.

Let us suppose first that $k < 0$. We substitute the expressions for ε_k and W_k into (5.1.30) to obtain the following estimate for the time of motion along the kth segment of the trajectory:

$$t_{k+1} - t_k \leq \tau 2^{-3k/2}, \quad \tau = \frac{16\sqrt[4]{2}M\sqrt{M}}{\sqrt{m}D_1U_0}\log\frac{16M}{m}. \tag{5.1.31}$$

Therefore, the time of motion τ_1 of the system from the state $(q_k,\dot{q}_k)$ to the state $(q_0,\dot{q}_0)$, i. e. from the ellipsoid with number k to that with number zero does not exceed

$$\tau_1 = \tau\sum_{i=k}^{-1}2^{-3i/2} = \tau 2\sqrt{2}\frac{(2\sqrt{2})^{-k}-1}{2\sqrt{2}-1}. \tag{5.1.32}$$

Let us suppose now that $k \geq 0$. In this case inequality (5.1.30) takes the form

$$t_{k+1} - t_k \leq \tau 2^{-k/2}, \tag{5.1.33}$$

and the time of motion τ_2 from the ellipsoid with number 0 to the terminal position $(0,0)$ does not exceed the sum of the series

$$\tau_2 = \tau\sum_{i=0}^{\infty}2^{-i/2} = \tau\frac{\sqrt{2}}{(\sqrt{2}-1)}. \tag{5.1.34}$$

We have assumed until now that $k > k_*$, and the segment of trajectory whose end-points lie on two nearby ellipsoids from the the family of ellipsoids specified above was considered. Inequalities (5.1.33) and (5.1.34) provide an estimate for the time of motion of system (5.1.2) and (5.1.10) along such a segment. Now, let $k = k_*$. At the point $(q_*, \dot{q}_*)$ corresponding to the initial state of the system, the function W satisfies the inequality

$$W_{k_*+1} < W(q_*, \dot{q}_*) \leq W_{k_*}.$$

Therefore, the point $(q_*, \dot{q}_*)$ does not, in general, lie on the ellipsoid with number k_*. Nevertheless, at the initial instant of time $t = 0$, we determine the gains α_k and β_k in control (5.1.10) according to formulas (5.1.9) for $k = k_*$. Using the reasoning similar to the used above, it can be shown that the trajectory of system (5.1.2) and (5.1.10) reaches the ellipsoid with number $k_* + 1$, and the time of motion towards this ellipsoid either satisfies inequality (5.1.31) if $k_* < 0$ or inequality (5.1.33) if $k_* \geq 0$. The total time τ_* of motion of the system from the point $(q_*, \dot{q}_*)$ to the terminal state $(0,0)$ satisfies the inequality $\tau_* \leq \tau_1 + \tau_2$, where τ_1 and τ_2 can be computed from (5.1.32) and (5.1.34) for $k = k_*$.

5.1.5 Sufficient condition for steering the system to the prescribed state

We now consider the restrictions imposed on the external perturbations Q. One can see easily that for $k \geq 0$, i.e., inside the ellipsoid with number 0, condition (5.1.24) is equivalent to the enequality

$$Q_0 \leq \frac{\sqrt{m}U_0}{16\sqrt{10M}}, \tag{5.1.35}$$

and outside the null ellipsoid (5.1.24), i.e., for $k < 0$, it is equivalent to the enequality

$$Q_0 \leq \frac{2^k\sqrt{m}U_0}{16\sqrt{10M}}. \tag{5.1.36}$$

The least value of index k along the trajectory starting at the point $(q_*, \dot{q}_*)$ is equal to k_*. Thus, if the point $(q_*, \dot{q}_*)$ lies inside or on the null ellipsoid, then $k_* \geq 0$ and inequality (5.1.35) provides a sufficient condition for the system under consideration to be taken from this point to the phase space origin in finite time using the above control law. But if $(q_*, \dot{q}_*)$ lies outside the null ellipsoid and $k_* < 0$, then such a sufficient condition is provided by inequality (5.1.36) for $k = k_*$.

The proposed sufficient conditions for the system to be taken to the phase space origin are such that the maximum admissible magnitude Q_0 of external perturbations depends on the initial state of the system: the further away $(q_*, \dot{q}_*)$ is from $(0,0)$, the smaller should the value Q_0 be. However, these conditions may be weakened, if the control law is modified. We will show that condition (5.1.35) is sufficient for steering the system from $(q_*, \dot{q}_*)$ to the origin $(0,0)$.

It has been mentioned above that any point of the form $(\bar{q},0)$ in the phase space of the system can be chosen as the terminal state. Then the family of ellipsoids on which the gains are changed turns out to be shifted by the vector $\bar{q}$, while the parameters of the ellipsoids remain the same. Let us assume first that at the initial instant of time the velocity of the system satisfies the inequality

$$\dot{q}_*^2 \leq \frac{U_0}{4\sqrt{2}D_1}, \tag{5.1.37}$$

that is, the point $(q_*,\dot{q}_*)$ lies on or inside the ellipsoid $W(q-q_*,\dot{q}) = W_0$ [this is the null ellipsoid with the centre moved to $(q_*,0)$]. We apply the control algorithm presented and transfer the system to the state $(q_*,0)$. It follows from the above reasoning that condition (5.1.35) is sufficient for such transferring.

We choose a finite sequence of points $(\bar{q}_j,0)$ such that $\bar{q}_0 = q_*$, $\bar{q}_J = 0$, and

$$|\bar{q}_j - \bar{q}_{j-1}| \leq \frac{M}{2D_1}, \quad j = 1,\ldots,J. \tag{5.1.38}$$

We transfer the system from the state $(q_*,0)$ to the phase space origin in J steps applying the above control algorithm again each time. At the jth step, the point $(\bar{q}_{j-1},0)$ corresponds to the initial state and $(\bar{q}_j,0)$ to the final state of the system (see Fig. 5.2). Inequality (5.1.38) means that for any j the point $(\bar{q}_{j-1},0)$ lies on or inside the null ellipsoid with the centre at $(\bar{q}_j,0)$. Consequently, the value Q_0 satisfying (5.1.35) is sufficient for transferring the system from $(\bar{q}_{j-1},0)$ to $(\bar{q}_j,0)$.

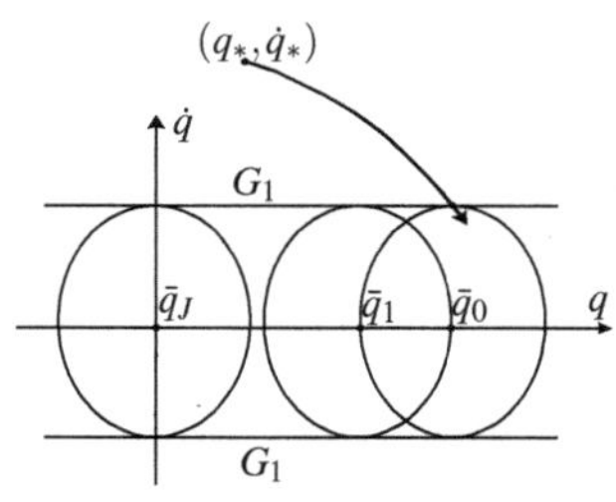

Fig. 5.2 Step-by-step transferring of the system

Let us assume now that (5.1.37) is not satisfied at the initial instant of time. We supplement the control algorithm by one more stage preceding all the others. The purpose of this preliminary stage is to reduce the velocity of motion of the system to the value satisfying inequality (5.1.37).

We introduce the domain

$$G_1 = \{(q,\dot{q}) : \dot{q}^2 > \frac{U_0}{4\sqrt{2}D_1}\}$$

and define the control in it as follows:

$$U = -\frac{U_0}{|\dot{q}|}\dot{q}.$$

By the theorem on the variation of the kinetic energy of the system, conditions (5.1.35), and the definition of the domain G_1, it follows that the estimates

$$T(q,\dot{q}) \geq \frac{m}{2}\dot{q}^2 > \frac{mU_0}{8\sqrt{2}D_1},$$

$$\dot{T}(q,\dot{q}) = \langle U + Q, \dot{q} \rangle \leq -(U_0 - Q_0)|\dot{q}|$$

$$\leq -\left(1 - \frac{\sqrt{m}}{16\sqrt{10M}}\right)\left(\frac{U_0^3}{2\sqrt{2}D_1}\right)^{1/2} < 0$$

hold in G_1. Therefore, the system leaves the domain G_1 in finite time. As soon as the trajectory reaches the boundary of G_1, the preliminary stage of control is completed, and the realization of the above algorithm of the step-by-step transferring of the system to the terminal state begins.

Thus, inequality (5.1.35) provides a sufficient condition for steering the system from an arbitrary initial state $(q_*, \dot{q}_*)$ to the phase space origin $(0,0)$.

Remark 5.3. The control law proposed does not depend on the value Q_0 and can therefore also be formally applied in cases where constraint (5.1.35) does not hold. Computer simulation of the dynamics of various mechanical systems shows that the control law is also effective far beyond the limits of the sufficient condition stated. To explain this phenomenon, let us note that the sufficient condition (5.1.35) guarantees monotone decrease of the Lyapunov functions V^k along the trajectory of system (5.1.2) controlled by law (5.1.10). However, it may happen that the functions V^k are not monotone while the trajectories tend to the terminal state. The results of the computer simulation presented below show such behavior of the system.

5.2 Applications to mechanical systems

5.2.1 Control of a two-link manipulator

The control law proposed will be used for the numerical simulation of the controlled motion of a two-link manipulator on a fixed base (see Fig. 2.15). The manipulator is assumed to move in a horizontal plane, hence, it is not subject to gravity. The hinge angles of the links in the stationary reference system are chosen as the generalized coordinates of the system. The matrix of the kinetic energy of the manipulator has the form

$$A(q) = \begin{pmatrix} A_1 & A_3\cos(q_1 - q_2) \\ A_3\cos(q_1 - q_2) & A_2 \end{pmatrix}.$$

Calculations are performed for the following parameter values:

$$A_1 = 13.9\,\text{kg}\cdot\text{m}^2,\ A_2 = 2.1\,\text{kg}\cdot\text{m}^2,\ A_3 = 4\,\text{kg}\cdot\text{m}^2.$$

The eigenvalues of the inertia matrix lie between constants $m = 1.8\,\text{kg}\cdot\text{m}^2$ and $M = 14.2\ \text{kg}\cdot\text{m}^2$. The norm of the partial derivatives of the matrix is bounded by $D = 3$. The maximum admissible magnitude of the vector of control torques is chosen to be equal to $U_0 = 500\,\text{N}\cdot\text{m}$. When modelling, we specify the disturbances by the constant vector-function $Q(t) = (0; 30)\ \text{N}\cdot\text{m}$. The manipulator moves from the initial state

$$q_{*1} = 0.5\,\text{rad},\ q_{*2} = 1\,\text{rad},\ \dot{q}_{*1} = \dot{q}_{*2} = 0\,\text{rad/s}$$

to the "stretched arm" position

$$q_1 = q_2 = \dot{q}_1 = \dot{q}_2 = 0.$$

The value of W at the initial state of the system is equal to $W(q_*,\dot{q}_*) = 395$, and the quantity determining the null ellipsoid is equal to $W_0 = 837$. Since $W_0/4 < W(q_*,\dot{q}_*) < W_0/2$, the first value of index k is equal to 1. The initial point of the trajectory lie inside the null ellipsoid, so the preliminary stage of the control (slowing down and step-by-step transferring the system into the null ellipsoid) is absent. Sufficient condition (5.1.35) in the case in question takes the form $Q_0 \le 3.47$ and does not hold for the chosen vector of the disturbances. Nevertheless, the manipulator reaches the terminal state in finite time (see Remark 5.3 in Sect. 5.1.5).

System (5.1.2) was integrated by the Runge – Kutta method. Integrating was stopped, when, in phase space $(q,\dot{q})$, the Euclidean distance

$$\rho(t) = \left(q_1^2(t) + q_2^2(t) + \dot{q}_1^2(t) + \dot{q}_2^2(t)\right)^{1/2}$$

between the current state of the system and the terminal state became less than 0.01.

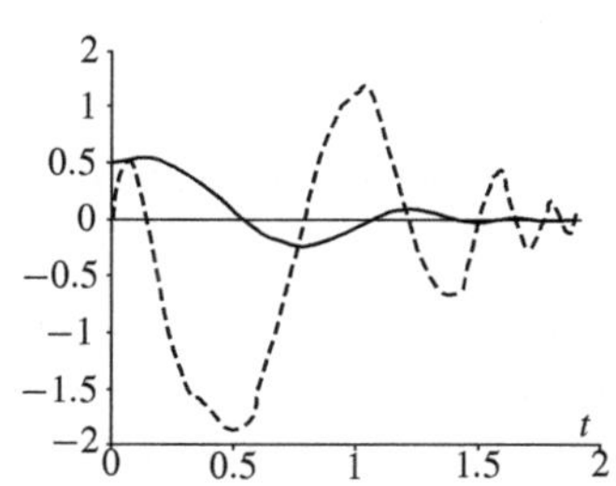

Fig. 5.3 The first link

Figures 5.3 and 5.4 depict the graphs of the time histories of the phase variables of the system. The solid lines correspond to the generalized coordinates (rad), and the dashed lines correspond to the generalized velocities (rad/s). Figure 5.3 describes the motion of the first link, and Fig. 5.4 the motion of the second link.

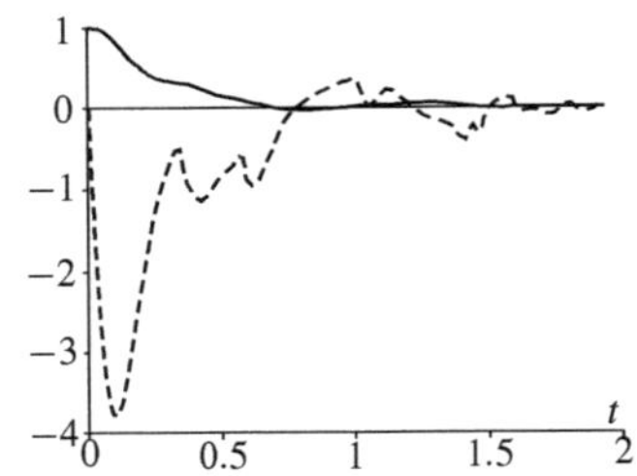

Fig. 5.4 The second link

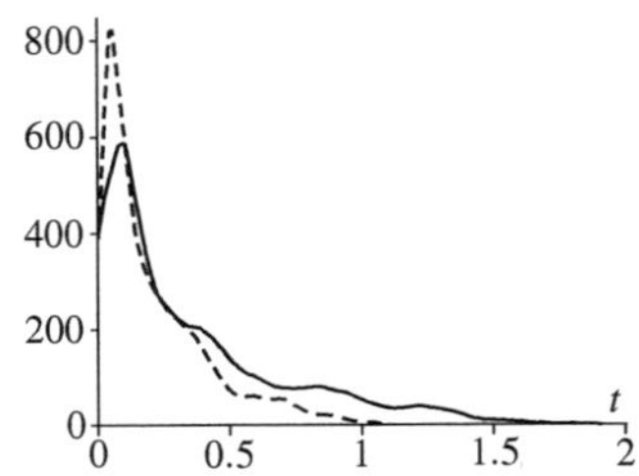

Fig. 5.5 The function W

The solid line in Fig. 5.5 shows the behavior of the function W along the trajectory. One can see that the function W does not depend monotonically on time.

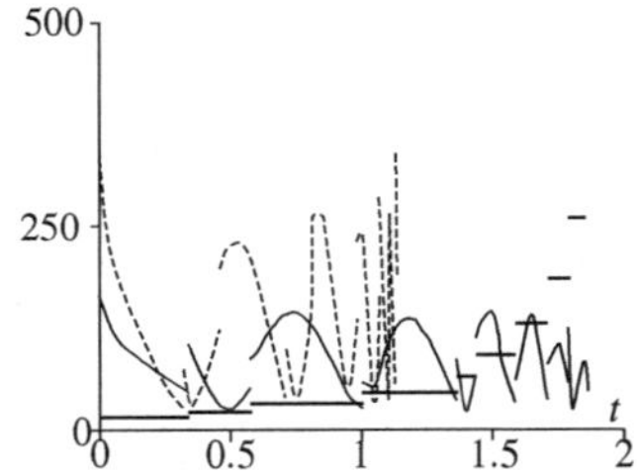

Fig. 5.6 The absolute value of the control torque and the gain α

Figure 5.6 shows the time history of the absolute value of the vector of control torques (the thin line) and the magnitude of the gain α (the step function). During the time of integration of the equations, the gains in (5.1.10) changed 12 times.

In accordance with the algorithm, the feedback factors α_k and β_k are chosen in such a way that for any admissible values of the unknown parameters, that is, the elements of the inertia matrix and the components of the vector of disturbances, the control constraint (5.1.4) holds along the resulting trajectory of motion. For a specified mechanical system, the domain of variation of these parameters contracts significantly and the choice of the gains may turn out to be unnecessarily conservative. One can see that, in the case under consideration, the control torques obtained are much smaller than the maximum allowed magnitude U_0 equal to 500. This is why we could simulate the motion of the two-link manipulator controlled by the same law, but with gains α_k and β_k twice as large as those prescribed by the algorithm. The results of such simulation are depicted by dashed lines in Fig. 5.5 and 5.6.

In Fig. 5.5, the dashed line represents the time history of W, and in Fig. 5.6 it represents the absolute magnitude of the vector of control torques for this control method. The time it takes to bring the system to the terminal position has been reduced by a factor of two, while the control satisfies (5.1.4) as before, with a substantial margin.

5.2.2 Control of a two-mass system with unknown parameters

Control of mechanical systems with so-called structural perturbations is an important area of control theory. The term 'structural perturbations' implies the appearance of additional degrees of freedom that are not controlled directly but affect the motion of the system and its controlled part. The aim of the control, as a rule, is to bring controlled coordinates to a given terminal set, whereas the values of the variables corresponding to these additional degrees of freedom are not important.

Below, we consider some simple systems of this kind. The first system consists of two concentrated masses (bodies) placed on a horizontal plane and connected by a spring (see Fig. 5.7). Both bodies are subject to dry friction forces with varying coefficients (depending on the position of the bodies). The masses of the bodies, the spring stiffness, and the coefficients of friction are assumed to be unknown but belong to certain, a priori specified, intervals. The first (controlled) mass is subject to a bounded control force. The controlled mass must be brought to a given terminal state in finite time (the state of the other mass at this moment is of no importance).

What distinguishes the second system from the first one is that one body rests on the other (a load on a cart, Fig. 5.8).

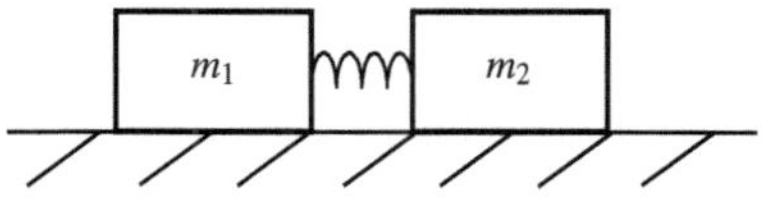

Fig. 5.7 A two-mass elastic system

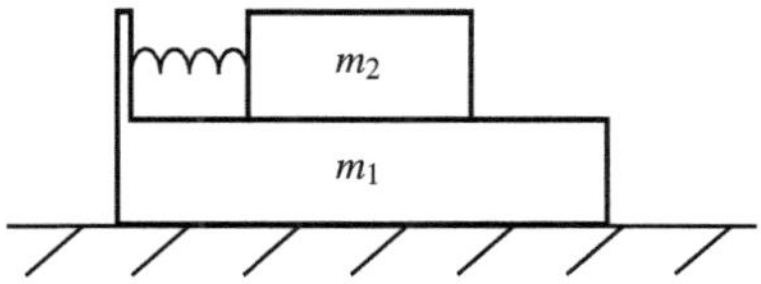

Fig. 5.8 A load on a cart

The third system under consideration also consists of two masses: a body is suspended from another one moving along a horizontal plane (a pendulum on a cart, see Fig. 5.9). The initial assumptions and the aim of the control are the same as in the previous cases. In all cases under consideration, there are "stagnation" regions because of dry friction. Hence, there exist segments of trajectories where the original system can be viewed as a system with only one degree of freedom.

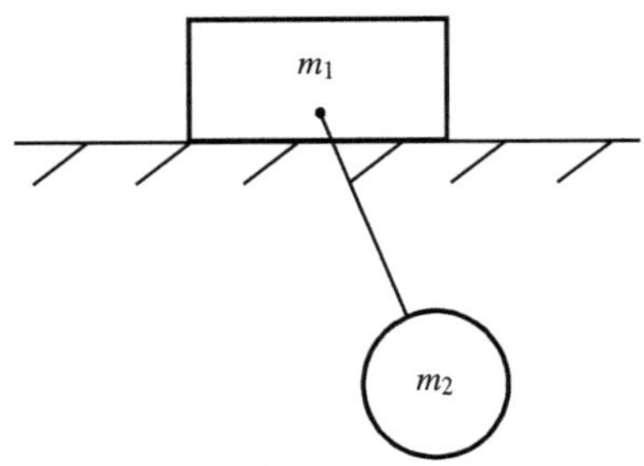

Fig. 5.9 A pendulum on a cart

We will apply the control laws based on the algorithm developed in Sect 5.1 that utilize a linear feedback with piecewise constant gains.

Consider a system consisting of two bodies placed on a horizontal plane and connected by a spring. Masses m_1 and m_2 of the bodies and the spring stiffness c_0 are assumed to be unknown but belong to the given intervals:

$$m \leq m_1, m_2 \leq M, \tag{5.2.1}$$

$$c \leq c_0 \leq C. \tag{5.2.2}$$

The first body of mass m_1 and the second body of mass m_2 will be referred to as carrying and carried bodies, respectively. A control force u is applied to the first body. Let us choose an immovable frame of reference on the horizontal line and denote a coordinate of the first mass in this frame by x. Let the variable ϕ describe the position of the second mass with respect to the first one so that $\phi = 0$ corresponds to the undeformed spring. Thus, ϕ is the extension of the spring.

Suppose that coefficients of the dry friction forces acting on both bodies are not constant and depend on the positions of the bodies (the "roughness" of the plane is different in different segments). Denote the friction forces acting on the carrying and carried bodies by f_1 and f_2, respectively.

In terms of the generalized coordinates x and y, the equations of the motion of the system are given by

$$\begin{aligned} m_1\ddot{x} &= c_0\phi + u + f_1, \\ m_2(\ddot{x} + \ddot{\phi}) &= -c_0\phi + f_2. \end{aligned} \tag{5.2.3}$$

Here,

$$f_1 = -\mathrm{sign}(\dot{x})\gamma_1(x)m_1 g,$$
$$f_2 = -\mathrm{sign}(\dot{x}+\dot{\phi})\gamma_2(x+\phi)m_2 g,$$

and g is the acceleration of gravity. The friction coefficients $\gamma_1(x)$ and $\gamma_2(x+\phi)$ are assumed to be unknown but satisfy the conditions

$$0 < \gamma \le \gamma_1(x), \gamma_2(x+\phi) \le \Gamma.$$

Hence, it follows that

$$|f_1|, |f_2| \le F, \quad F = \Gamma M g. \tag{5.2.4}$$

We assume that the control force u satisfies the inequality

$$|u| \le U_0, \tag{5.2.5}$$

and

$$U_0 > 3F. \tag{5.2.6}$$

We also assume that the phase coordinates x, ϕ and velocities $\dot{x}, \dot{\phi}$ are available for measuring.

Problem 5.3. At the initial instant of time $t = 0$, let the system be at the state

$$x(0) = x_0,\ \dot{x}(0) = \dot{x}_0,\ \phi(0) = \phi_0,\ \dot{\phi}(0) = \dot{\phi}_0.$$

It is required to bring system (5.2.3) to the prescribed terminal set

$$x = x_*,\ \dot{x} = 0$$

in finite time; i.e., the carrying body is to be brought to the state $(x_*, 0)$.

Without loss of generality, we can assume that $x_* = 0$, because the origin can be placed at the point the first mass should be brought to.

The required control will be constructed on the basis of a linear feedback with respect to the generalized coordinates and velocities (PD-regulator):

$$u = -\alpha\dot{x} - \beta x, \tag{5.2.7}$$

where the coefficients $\alpha, \beta > 0$ are considered, for now, to be constant. This control is equivalent to the incorporation of a spring of stiffness β and viscous damping with coefficient α into the system. This spring connects the carrying body with the immovable foundation, and the spring is undeformed when $x = 0$.

The total energy of the system inclusive of the elastic energy of this "fictitious" spring is given by

$$E(x,\dot{x},\phi,\dot{\phi}) = \frac{1}{2}\left(m_1\dot{x}^2 + m_2(\dot{x}+\dot{\phi})^2 + \beta x^2 + c_0\phi^2\right).$$

For brevity, we will denote the total energy at the moment t along the trajectory under consideration by $E(t)$. Differentiating $E(t)$ by virtue of (5.2.3) and (5.2.7), we obtain

$$\begin{aligned}\dot{E} &= -\alpha\dot{x}^2 + f_1\dot{x} + f_2(\dot{x}+\dot{\phi}) \\ &= -\alpha\dot{x}^2 - \gamma_1(x)m_1g|\dot{x}| - \gamma_2(x+\phi)m_2g|\dot{x}+\dot{\phi}|.\end{aligned}$$

It is easy to see that the total energy is not negative and does not increase along the trajectory. Because of the dry friction forces, the system can have stagnation regions, that is, the states at which both masses are at rest while the springs are deformed. The total energy $E(t)$ of the system at these states is positive, and its derivative $\dot{E}(t)$ is zero. Hence, it follows that

$$\lim_{t\to\infty} E(t) = E_*, \quad E_* \geq 0. \tag{5.2.8}$$

Let us prove some auxiliary propositions.

Lemma 5.3. *The following relationships are valid:*

$$\lim_{t\to\infty} \dot{x}(t) = 0, \quad \lim_{t\to\infty} \dot{\phi}(t) = 0. \tag{5.2.9}$$

Proof. Note that the second derivatives $\ddot{x}$ and $\ddot{\phi}$ are bounded along the trajectory that starts at the point $(x_0, \dot{x}_0, \phi_0, \dot{\phi}_0)$. Indeed, the total energy of the system satisfies the inequality $E(t) \leq E(0)$, $t \geq 0$. Thus, the trajectory of the system lies inside the ellipsoid $E(x, \dot{x}, \phi, \dot{\phi}) = E(0)$, and the phase coordinates and velocities are bounded. This means that the right-hand sides of equations (5.2.3), and, hence, the second derivatives $\ddot{x}$ and $\ddot{\phi}$ are bounded, that is, there exists a number $D > 0$ such that

$$|\ddot{x}|, |\ddot{\phi}| \leq D. \tag{5.2.10}$$

Let us prove that the first of equations (5.2.9) is valid. Assume the contrary. Let there exist a number $\delta > 0$ and a sequence $\{t_k\}$, $t_k \to \infty$ as $k \to \infty$, such that $|\dot{x}(t_k)| \geq \delta$. This, together with (5.2.10), implies that, on the time intervals $I_k = [t_k, t_k + \delta/(2D)]$, $k = 1, 2, \ldots$, the inequality $|\dot{x}(t)| \geq \delta/2$ holds, and the derivative of the total energy satisfies the inequality

$$\dot{E}(t) \leq -\gamma m_1 g \delta/2.$$

Hence, the total energy infinitely decreases, which contradicts (5.2.8). The first of equations (5.2.9) is proved; the second equation can be proved in the same way. □

Lemma 5.4. *The following inequality is valid:*

$$\overline{\lim_{t\to\infty}} |\phi(t)| \leq \frac{F}{c_0},$$

where F is defined in (5.2.4).

Proof. Assume the contrary. Let there exist a number $\delta > 0$ and a sequence $\{t_k\}$, $t_k \to \infty$ as $k \to \infty$, such that

$$|\phi(t_k)| \geq \frac{F}{c_0} + 2\delta.$$

Let $v > 0$ be an arbitrary number. In view of Lemma 5.3, there exists $t' > 0$ such that $|\dot{\phi}(t)| \leq v$ for $t > t'$. Then, for $t_k > t'$, the inequality

$$|\phi(t)| \geq \frac{F}{c_0} + \delta \tag{5.2.11}$$

holds on the time intervals $I_k = [t_k, t_k + \delta/v]$. From this inequality and conditions (5.2.2) and (5.2.4), it follows that the inequalities

$$|-c_0\phi + f_2| \geq c_0|\phi| - |f_2| \geq c_0\left(\frac{F}{c_0} + \delta\right) - F \geq c_0\delta \tag{5.2.12}$$

hold for sufficiently large k and for $t \in I_k$.

The force f_2 and, hence, the right-hand side of (5.2.3) are not continuous functions of time whereas the function $\phi(t)$ is. By virtue of (5.2.11), for sufficiently large k, the sign of the function $\phi(t)$ is constant on any interval I_k. It follows from (5.2.12) that the sign of the right-hand side of the second equation (5.2.3) coincides with the sign of the function $\phi(t)$ and, hence, does not change on any of these intervals. Therefore, the sign of the derivative $\ddot{x}(t) + \ddot{\phi}(t)$ also does not change (this derivative is, generally speaking, a discontinuous function of time).

Relationships (5.2.3) and (5.2.12) imply

$$|m_2\left(\ddot{x}(t) + \ddot{\phi}(t)\right)| \geq c_0\delta, \quad t \in I_k,\ t_k > t'. \tag{5.2.13}$$

Therefore, for sufficiently large k, the variation of the velocity $\dot{x}(t) + \dot{\phi}(t)$ on the interval I_k is not less than $c_0\delta^2/(m_2 v)$. Choosing v to be sufficiently small, we see that inequality $|\dot{x}(t) + \dot{\phi}(t)|$ holds for an arbitrary large interval of time, and the magnitude of the velocity $|\dot{x}(t) + \dot{\phi}(t)|$ of the second mass can be arbitrarily large. This contradicts the fact that the total energy of the system is bounded. □

Lemma 5.5. *The following inequality is valid:*

$$\overline{\lim_{t\to\infty}}|x(t)| \leq \frac{2F}{\beta}.$$

Proof. Assume the contrary. Let there exist a number $\delta > 0$ and a sequence $\{t_k\}$, $t_k \to \infty$ as $k \to \infty$, such that

$$|x(t_k)| \geq \frac{2F}{\beta} + 4\delta.$$

Let v satisfy the inequality $0 < v < \beta\delta/\alpha$. In view of Lemmas 5.3 and 5.4, there exists an instant of time $t' > 0$ such that the inequalities

$$|\dot{x}(t)| < v < \frac{\beta\delta}{\alpha}, \quad |\phi(t)| < \frac{F + \beta\delta}{c_0} \tag{5.2.14}$$

hold for $t > t'$. Then, for $t_k > t'$, the inequality

$$|x(t)| \geq \frac{2F}{\beta} + 3\delta \tag{5.2.15}$$

holds on the intervals $I_k = [t_k, t_k + \delta/v]$. Using (5.2.4), (5.2.14), and (5.2.15), the right-hand side of the first of equations (5.2.3) at the instant of time $t \in I_k$, $t_k > t'$, can be estimated as follows:

$$\begin{aligned} |c_0\phi + u + f_1| = |c_0\phi - \alpha\dot{x} - \beta x + f_1| \geq \\ \geq \beta|x| - \alpha|\dot{x}| - c_0|\phi| - |f_1| \geq \beta\delta. \end{aligned} \tag{5.2.16}$$

Hence, the inequality

$$|\ddot{x}(t)| \geq \frac{\beta\delta}{m_1} \tag{5.2.17}$$

holds for $t_k > t'$ on any intervals I_k.

The force f_1 and, hence, the both sides of the first equation (5.2.3) are not continuous functions of time whereas the function $x(t)$ is. By virtue of (5.2.15), for sufficiently large k, the sign of the function $x(t)$ is constant on any interval I_k. It follows from (5.2.16) that the sign of the right-hand side of the first equation (5.2.3) coincides with the sign of the function $x(t)$ and, hence, does not change on any of these intervals. Whence and by (5.2.17), we obtain that variation of the velocity $\dot{x}$ on any of the intervals I_k is not less than $\beta\delta^2/(m_1 v)$. Choosing v to be sufficiently small, we see that the velocity of the first mass can be arbitrarily large, which contradicts the fact that the total energy of the system is bounded. □

We will bring the system to the prescribed terminal set in two stages. First, the carrying mass will be brought to a neighborhood of the point $x = 0$ on the horizontal line; simultaneously, the total energy of the system will be reduced to a sufficiently small level. Then, considering only the first of equations (5.2.3), where the elastic force $c_0\phi$ is treated as a perturbation, we will bring the carrying mass to the point $x = 0$. At that, on each of the stage the gains α and β in control (5.2.7) are chosen as piecewise constant functions.

5.2.3 The first stage of the motion

Let us carry out some auxiliary construction. We put

$$G(\dot{x}, \phi, \dot{\phi}) = M\dot{x}^2 + M(\dot{x} + \dot{\phi})^2 + C\phi^2$$

and consider the function

$$H(x, \dot{x}, \phi, \dot{\phi}) = G + (G^2 + 2U_0^2 x^2)^{1/2}. \tag{5.2.18}$$

The function H plays the same role as the function W in the control design earlier in this section. The value of H is calculated along the trajectory of the system, and when the function H achieves certain preset values H_k, the feedback factors in control change. The difference from the described above procedure of change of these factors is that the sequence H_k does not tend to zero, as $k \to \infty$, whereas it did with sequence W_k.

Let

$$H_0 = H(x_0, \dot{x}_0, \phi_0, \dot{\phi}_0).$$

Let us introduce the notation

$$\xi = \frac{CF^2}{c^2}, \quad \eta = \frac{3F}{U_0}, \quad H_* = \frac{3\xi}{1-\eta^2}. \tag{5.2.19}$$

By virtue of (5.2.6), the number η satisfies the inequality $\eta < 1$.

We will suppose at first, that at the instant of time $t = 0$ the condition

$$H_0 > H_* \tag{5.2.20}$$

holds. We introduce the sequence H_k:

$$H_k = \xi + (\xi^2 + \eta^2 H_{k-1}^2)^{1/2}, \quad k = 1, 2, \ldots. \tag{5.2.21}$$

Lemma 5.6. *The sequence H_k satisfies the conditions*

$$H_0 > H_1 > H_2 > \ldots, \quad \lim_{k\to\infty} H_k = \frac{2\xi}{1-\eta^2}.$$

Proof. We put

$$z_* = \frac{2\xi}{1-\eta^2}$$

and consider the function

$$h(z) = \xi + (\xi^2 + \eta^2 z^2)^{1/2}$$

on the half-line $\{z \in R: \ z \geq z_*\}$. It is easy to see that $h(z_*) = z_*$. By definition (5.2.19) of the number H_* and by condition (5.2.20), the inequality $H_0 > z_*$ holds. Using definition (5.2.19) of η, condition (5.2.6), and the inequality $(\xi^2 + \eta^2 z^2)^{1/2} > \eta z$, we estimate the derivative of the function $h(z)$ as follows:

$$h'(z) = \eta^2 z(\xi^2 + \eta^2 z^2)^{-1/2} < \eta < 1.$$

Therefore, $h(z)$ is a contractive mapping, and the sequence $H_0, H_1 = h(H_0), H_2 = h(H_1), \ldots$ tends monotonically to the fixed point z_* of the mapping $h(z)$. □

Let us describe an algorithm of changing the gains in control (5.2.7) on the first stage of motion. At the initial time instant $t = 0$, we specify the gains α_0 and β_0 as follows:

$$\beta_0 = \frac{U_0^2}{H_0}, \ \alpha_0 = \sqrt{m\beta_0}. \tag{5.2.22}$$

Denote

$$\overline{\lim_{t\to\infty}} H(t) = H_0^*, \quad H(t) = H(x(t), \dot{x}(t), \phi(t), \dot{\phi}(t)).$$

Here, the limit is calculated along the trajectory of system (5.2.3) governed by control law (5.2.7) with gains (5.2.22), starting at the point $(x_0, \dot{x}_0, \phi_0, \dot{\phi}_0)$.

Lemma 5.7. *The following relationship is valid:*

$$H_0^* < H_1.$$

Proof. By definition of the function $G(t) = G(\dot{x}(t), \phi(t), \dot{\phi}(t))$, assertions of Lemmas 5.3–5.5, condition (5.2.2), and notations (5.2.19), the inequality

$$\overline{\lim_{t\to\infty}} G(t) \le \xi, \quad G(t) = G(\dot{x}(t), \phi(t), \dot{\phi}(t))$$

holds. Whence, by the definitions of β_0 and η, and also in view of assertion of Lemma 5.5, we obtain

$$H_0^* = \overline{\lim_{t\to\infty}} H(t) \le \xi + \left(\xi^2 + \frac{8F^2U_0^2}{\beta_0^2}\right)^{1/2} < \xi + (\xi^2 + \eta^2 H_0^2)^{1/2} = H_1.$$

□

Let us prove that condition (5.2.5) is fulfilled along the trajectory of system (5.2.3) and (5.2.7) with gains (5.2.22) that starts at the point $(x_0, \dot{x}_0, \phi_0, \dot{\phi}_0)$. to that end, let us introduce the notation

$$\begin{aligned} E_+^0(x, \dot{x}, \phi, \dot{\phi}) &= \frac{1}{2}\left(M\dot{x}^2 + M(\dot{x} + \dot{\phi})^2 + \beta_0 x^2 + C\phi^2\right), \\ E_-^0(x, \dot{x}, \phi, \dot{\phi}) &= \frac{1}{2}\left(m\dot{x}^2 + m(\dot{x} + \dot{\phi})^2 + \beta_0 x^2 + c\phi^2\right). \end{aligned} \tag{5.2.23}$$

In view of (5.2.1) and (5.2.2), the total energy E^0 of the system (inclusive of the elastic energy of the “fictitious” spring of stiffness β_0) satisfies the inequalities

$$E_-^0(t) \le E^0(t) \le E_+^0(t). \tag{5.2.24}$$

We put $G_0 = G(\dot{x}_0, \phi_0, \dot{\phi}_0)$. By definitions of the function H and the number H_0, we have

$$H_0 = G_0 + (G_0^2 + 2U_0^2 x_0^2)^{1/2},$$

that is, the number H_0 is a root of the equation $z^2 - 2G_0 z - 2U_0^2 {x_0}^2 = 0$ being considered with respect to z. Hence, the equality

$$H_0^2 - 2G_0 H_0 - 2U_0^2 {x_0}^2 = 0$$

is valid. We divide this equality by H_0 and transform it with the help of formula (5.2.22) for the gain β_0 to get

$$2G_0 + 2\beta_0 x_0{}^2 = H_0.$$

Whence, and by definitions of functions G and E_+^0, we obtain

$$4E_+^0(x_0, \dot{x}_0, \phi_0, \dot{\phi}_0) = H_0. \tag{5.2.25}$$

Using relationships (5.2.22), (5.2.23), and (5.2.24), we estimate the control force as follows:

$$|u|^2 = |\alpha_0 \dot{x} + \beta_0 x|^2 \leq 2(\alpha_0^2 \dot{x}^2 + \beta_0^2 x^2)$$
$$= 2\beta_0(m\dot{x}^2 + \beta_0 x^2) \leq 4\beta_0 E_-^0(x, \dot{x}, \phi, \dot{\phi}) \leq 4\beta_0 E^0(x, \dot{x}, \phi, \dot{\phi}).$$

Taking into account the fact that the total energy $E^0(t)$ of the system does not increase along the trajectory under consideration, we can further extend the sequence of inequalities by means of (5.2.24) and (5.2.25):

$$4\beta_0 E^0(t) \leq 4\beta_0 E^0(0) \leq 4\beta_0 E_+^0(0) = \beta_0 H_0 = U_0^2.$$

Thus, condition (5.2.5) is fulfilled.

The function $H(x, \dot{x}, \phi, \dot{\phi})$ defined by (5.2.18) is expressed in terms of the known parameters only, and all phase coordinates and velocities can, by assumption, be measured. Hence, $H(t)$ can be calculated at any instant of time. Denote by t_1 the first instant when the value of the function $H(t)$ on the trajectory under consideration becomes equal to H_1. It follows from the definition of number H_0^* and the assertion of Lemma 5.7 that such a moment exists.

Let

$$x(t_1) = x_1,\ \dot{x}(t_1) = \dot{x}_1,\ \phi(t_1) = \phi_1,\ \dot{\phi}(t_1) = \dot{\phi}_1.$$

At the moment t_1, we change the gains in the control law (5.2.7):

$$\beta_1 = \frac{U_0^2}{H_1}, \quad \alpha_1 = \sqrt{m\beta_1}. \tag{5.2.26}$$

Introduce the notation

$$\overline{\lim_{t \to \infty}} H(t) = H_1^*.$$

Applying similar arguments, we can show that, on the trajectory starting at the point $(x_1, \dot{x}_1, \phi_1, \dot{\phi}_1)$ of system (5.2.3) that is subject to control (5.2.7) with gains (5.2.26), condition (5.2.5) is fulfilled, and

$$H_1^* < H_2. \tag{5.2.27}$$

Remark 5.4. The numbers H_0^* and H_1^*, generally speaking, do not coincide because they are defined as limits of function $H(t)$ along the trajectories with the different initial conditions and different gains in the control law.

In view of (5.2.27), there exists a moment t_2 when the value of the function $H(t)$ becomes equal to H_2 for the first time. At the moment t_2, we redefine the gains in control law (5.2.7) as follows:

$$\beta_2 = \frac{U_0^2}{H_2}, \ \alpha_2 = \sqrt{m\beta_2},$$

and so on.

Thus, the trajectory of the motion consists of segments of the trajectories corresponding to different systems of differential equations: the kth segment connects the points $(x_k, \dot{x}_k, \phi_k, \dot{\phi}_k)$ and $(x_{k+1}, \dot{x}_{k+1}, \phi_{k+1}, \dot{\phi}_{k+1})$ and corresponds to system (5.2.3) subject to control (5.2.7) with the gains given by

$$\beta_k = \frac{U_0^2}{H_k}, \quad \alpha_k = \sqrt{m\beta_k}.$$

The numbers H_k are defined recursively by formulas (5.2.21), and

$$H_k = H(x_k, \dot{x}_k, \phi_k, \dot{\phi}_k) = H(t_k), \quad k = 0, 1, \ldots.$$

Repeating the arguments given above for the case of $k = 0$, we can show that, for any k on the kth segment of the trajectory, the control force meets constraint (5.2.5).

Theorem 5.1. *Let condition (5.2.20) be fulfilled. Then, there exists an instant of time τ_0 when the function $H(t)$ on the trajectory of system (5.2.3) controlled by means of the algorithm described above becomes equal to H_*.*

Proof. Condition (5.2.20) implies that, at the initial instant of time $t = 0$, the inequality $H_* < H_0$ holds. In view of Lemma 5.6, the relationships

$$\lim_{k\to\infty} H_k < H_*, \quad H_0 > H_1 > H_2 > \ldots$$

are valid. Therefore, there exists such integer k that $H(t_k) > H_* > H(t_{k+1})$, and the continuous function $H(t)$, on the kth segment of the trajectory, becomes equal to H_*. □

We denote by τ_0 the first instant of time when the function $H(t)$ becomes equal to H_*. The first stage of the motion is completed at τ_0. If condition (5.2.20) is not fulfilled at the initial instant of time, then there is no first stage. In this case, we have $H(\tau_0) < H_*$ and, putting $\tau_0 = 0$, we turn to the second stage.

5.2.4 The second stage of the motion

At the second stage, we apply the control law put forward in Sect. 5.1, that brings a general-type Lagrangian system with an unknown matrix of kinetic energy to a given state.

Consider the motion of the first mass only that is governed by the equation

$$m_1\ddot{x} = u + Q, \quad Q = c_0\phi + f_1, \tag{5.2.28}$$

where Q is treated as a perturbation.

As before, the first mass is subject to the control law (5.2.7) with the gains being piecewise constant functions. Let us describe an algorithm of changing these coefficients.

We introduce the function

$$W(x,\dot{x}) = M\dot{x}^2 + (M^2\dot{x}^4 + 2U_0^2x^2)^{1/2}$$

and the notation

$$W_0 = W(x(\tau_0),\dot{x}(\tau_0)), \quad W_k = \frac{W_0}{2^k}, \quad k > 1,2,\ldots.$$

Let the gains α and β, at the initial instant τ_0, be

$$\beta_0 = \frac{U_0^2}{W_0}, \quad \alpha_0 = \sqrt{m\beta_0}.$$

Let τ_1 be the first instant when the value of the function $W(t)$ calculated along the trajectory becomes equal to W_1. We set

$$\beta_1 = \frac{U_0^2}{W_1}, \quad \alpha_1 = \sqrt{m\beta_1}.$$

Denote the first instant of time when the function $W(t)$ is equal to W_2 by τ_2, and so on.

Thus, the sequence τ_k, $k = 0,1,\ldots$, specifies the moments when the coefficients are switched. At the moment τ_k, the function $W(t)$ becomes equal to W_k for the first time, and the gains are defined as

$$\beta_k = \frac{U_0^2}{W_k}, \quad \alpha_k = \sqrt{m\beta_k}, \quad k = 0,1,\ldots,$$

that is, each time, the gains β and α are increased by the factors 2 and $\sqrt{2}$, respectively.

Remark 5.5. In Sect. 5.1.2, in the description of the control law, the sequence W_k defined by relationships (5.1.8) depends only on parameters of the problem D, M, and U_0. Therefore, the set of ellipsoids on which the gains switches, also depends only on these parameters. In the present case, the modified control law is applied. The difference is that now the ellipse with index 0 coincides with the level set of the function W corresponding to value of this function at the initial instant of time. Thus, the set of ellipsoids on which the feedback factors change is defined by the initial state of the system, and the trajectory of the controlled motion on the second stage always starts on the ellipsoid with index 0.

According to Sect. 5.1.5, the condition

$$|Q| \le \frac{\sqrt{m}U_0}{16\sqrt{10M}} \tag{5.2.29}$$

is sufficient for the first mass to be brought to the origin in finite time by means of this algorithm. Note that, in this case, condition (5.2.5) holds on the trajectory.

Let us prove that inequality (5.2.29) is satisfied. To do this, let us estimate the total energy E^k of system (5.2.3) on each interval of the second-stage motion. Denote by $\bar{E}^k$ the energy of the first mass on the kth interval of the trajectory inclusive of the elastic energy of the "fictitious" spring of stiffness β_k:

$$\bar{E}^k(x,\dot{x}) = \frac{1}{2}(m_1\dot{x}^2 + \beta_k x^2).$$

Using algebraic transformations, it is not difficult to find that

$$\begin{gathered}\bar{E}^k(x,\dot{x}) \le \bar{E}^k_+(x,\dot{x}), \quad 4\bar{E}^k_+(\tau_k) = W_k,\\ \bar{E}^k_+(x,\dot{x}) = \frac{1}{2}(M\dot{x}^2 + \beta_k x^2),\ k = 0,1,\dots .\end{gathered} \tag{5.2.30}$$

In view of definition (5.2.18) of the function G and conditions (5.2.1) and (5.2.2), the energy of the system at the beginning of the second stage (except the elastic energy of the "fictitious" spring) satisfies the inequalities

$$\begin{gathered}\frac{1}{2}\left(m_1\dot{x}^2(\tau_0) + m_2(\dot{x}(\tau_0) + \dot{\phi}(\tau_0))^2 + c_0\phi^2(\tau_0)\right)\\ \le G(\dot{x}(\tau_0),\phi(\tau_0),\dot{\phi}(\tau_0))/2 \le H(\tau_0)/4 \le H_*/4.\end{gathered}$$

In view of (5.2.30), the elastic energy of the "fictitious" spring satisfies the inequalities

$$\frac{\beta_0}{2}x^2(\tau_0) \le \bar{E}^0_+(\tau_0) = \frac{W_0}{4}.$$

Hence,

$$E^0(\tau_0) \le \frac{H_* + W_0}{4}. \tag{5.2.31}$$

On each time interval $[\tau_k, \tau_{k+1})$, $k = 0,1,\dots$, the gains in the control law (5.2.7) are constant; therefore, the total energy of the system does not increase. At the moment τ_{k+1}, the stiffness β of the "fictitious" spring is doubled; hence, the elastic energy of the spring is increased by

$$\pi_{k+1} = \frac{1}{2}(\beta_{k+1} - \beta_k)x^2(\tau_{k+1}) = \frac{\beta_{k+1}}{4}x^2(\tau_{k+1}).$$

By the definition of the total energy $\bar{E}^k$ and by virtue of (5.2.30), the quantity π_{k+1} satisfies the inequality

$$\pi_{k+1} \le \frac{1}{2}\bar{E}_+^{k+1}(\tau_{k+1}) = \frac{W_{k+1}}{8} = \frac{W_0}{2^{k+4}}.$$

The total increment of the elastic energy of the "fictitious" spring due to the instantaneous variation of the spring stiffness does not exceed the sum of the series

$$\sum_{k=1}^{\infty} \frac{W_0}{2^{k+4}} = \frac{W_0}{16}.$$

Whence and by (5.2.31), for any $t \ge \tau_0$, the total energy $E(t)$ of the system is bounded:

$$E(t) \le E^0(\tau_0) + \frac{W_0}{16} \le \frac{H_*}{4} + \frac{5W_0}{16}.$$

By definitions of the functions W and H, we have

$$W_0 = W(\tau_0) \le H(\tau_0) = H_*.$$

Taking this into account and using condition (5.2.2), we arrive at the conclusion that the deformation of the spring connecting the masses m_1 and m_2 satisfies the inequality

$$|\phi(t)| \le \sqrt{\frac{2E(t)}{c_0}} \le \left(\frac{4H_* + 5W_0}{8c}\right)^{1/2} \le \frac{3}{2}\sqrt{\frac{H_*}{2c}}, \quad t \ge \tau_0.$$

Hence, the value of the perturbation Q in equation (5.2.28) is bounded during the second stage:

$$|Q| \le C|\phi| + F \le F + \frac{3C}{2\sqrt{2c}}\left(\frac{3\xi}{1-\eta^2}\right)^{1/2}.$$

Taking into account expressions (5.2.19) for ξ and η, the latter can be written as

$$|Q| \le Q_0, \quad Q_0 = F\left(1 + \left(\frac{3C}{2c}\right)^{3/2} \frac{U_0}{(U_0^2 - 9F^2)^{1/2}}\right).$$

Substituting Q_0 for $|Q|$ into (5.2.29), we obtain the following sufficient condition for bringing the first mass to the origin:

$$F\left[1 + \left(\frac{3C}{2c}\right)^{3/2} \frac{U_0}{(U_0^2 - 9F^2)^{1/2}}\right] \le \frac{\sqrt{m}U_0}{16\sqrt{10M}}. \tag{5.2.32}$$

Thus, the following theorem is valid.

Theorem 5.2. *Let the parameters m, M, c, C, F, and U_0 of the problem satisfy condition (5.2.32). Then, the suggested control algorithm brings the first mass to the origin in finite time.*

5.2.5 System "a load on a cart"

Consider the mechanical system consisting of two bodies connected by a spring. The first body moves along the horizontal line; the second body is placed on the first body and can also move in the horizontal direction (see Fig. 5.8). The equations governing the motion of this system have the form

$$m_1\ddot{x} = c_0\phi + u + f_1 - f_2, \quad m_2(\ddot{x} + \ddot{\phi}) = -c_0\phi + f_2.$$

Here, x is the coordinate of the first mass, and ϕ is the coordinate of the second mass with respect to the first body, so that $\phi = 0$ corresponds to the undeformed state of the spring. Denote the friction forces between the carrying body and the foundation and between the bodies f_1 and f_2, respectively:

$$f_1 = -\mathrm{sign}(\dot{x})\gamma_1(x)(m_1 + m_2)g,$$

$$f_2 = -\mathrm{sign}(\dot{\phi})\gamma_2(\phi)m_2 g.$$

Restrictions imposed on the values of the masses, spring stiffness, and the coefficients of friction are assumed to be the same as before. Restriction (5.2.5) on the control force remains the same; however, now we assume that

$$U_0 > 2\sqrt{5}F. \tag{5.2.33}$$

It is required to bring the carrying body to the state $x = \dot{x} = 0$ in finite time.

Applying the same reasoning as in the previous case, one can prove the following relations:

$$\lim_{t\to\infty} \dot{x}(t) = 0, \quad \lim_{t\to\infty} \dot{\phi}(t) = 0,$$

$$\overline{\lim_{t\to\infty}}|\phi(t)| \le \frac{F}{c_0}, \quad \overline{\lim_{t\to\infty}}|x(t)| \le \frac{3F}{\beta}.$$

Let us apply the control algorithm described above to this system. In view of the new restriction (5.2.33) on the value of F, we take

$$\eta = \frac{2\sqrt{5}F}{U_0}.$$

The formulas for ξ and H_* remain the same. At the first stage, the system is brought to the set

$$\{(x, \dot{x}, \phi, \dot{\phi}) \in R^4 : H(x, \dot{x}, \phi, \dot{\phi}) \le H_*\}.$$

At the second stage, the carrying body reaches its terminal state. The dynamics of the carrying body at the second stage is described, as before, by equations (5.2.28); however, the perturbation Q takes the form

$$Q = c_0\phi + f_1 - f_2.$$

The value of Q can be estimated as follows:

$$|Q| \le C|\phi| + 2F \le 2F + \frac{3C}{2\sqrt{2c}}\left(\frac{3\xi}{1-\eta^2}\right)^{1/2}.$$

The sufficient conditions for bringing the carrying body to the terminal state take the form

$$F\left[2 + \left(\frac{3C}{2c}\right)^{3/2} \frac{U_0}{(U_0^2 - 20F^2)^{1/2}}\right] \le \frac{\sqrt{m}U_0}{16\sqrt{10M}}.$$

5.2.6 System "a pendulum on a cart"

Now, we consider the control problem for the mechanical system consisting of a mass m_1 moving along a horizontal line and a mass m_2 suspended from the first mass (see Fig. 5.9). We will consider the motion of the system in the vertical plane.

Let us introduce a rectangular frame of reference with the abscissa axis directed along the horizontal line. The dynamics of the system is described by the equations

$$\begin{aligned} (m_1 + m_2)\ddot{x} + m_2 l \cos\phi\,\ddot{\phi} &= u + m_2 l \sin\phi\,\dot{\phi}^2 + f(t,x), \\ m_2 l \cos\phi\,\ddot{x} + (m_2 l^2 + J)\ddot{\phi} &= -m_2 g l \sin\phi + \mu(t,\phi). \end{aligned} \tag{5.2.34}$$

Here, x is the coordinate of the first mass on the line, ϕ is the angle between the vertical line and the line connecting the suspension point and the center of inertia of the second body, l is the distance between the suspension point and the center of inertia, J is the moment of inertia of the second body with respect to its center of inertia, $f(t,x)$ is the dry friction force acting on the first body from the immovable foundation, and $\mu(t,\phi)$ is the moment due to the dry friction at the suspension point. We assume that f and μ are unknown and satisfy the inequalities

$$\begin{aligned} 0 \le |f(t,x)| \le F, \quad f(t,x)\dot{x} \le 0, \\ 0 < \mu_1 \le |\mu(t,\phi)| \le \mu_2, \quad \mu(t,\phi)\dot{\phi} \le 0. \end{aligned} \tag{5.2.35}$$

As before, we assume that masses m_1 and m_2 of the bodies are unknown but belong to given intervals (5.2.1); the control force u is subject to restrictions (5.2.5) and (5.2.6); and all phase variables $x, \dot{x}, \phi$, and $\dot{\phi}$ can be measured. The carrying body must be brought to the state $x = \dot{x} = 0$ in finite time.

At the initial instant of time $t = 0$, let the system be at the state

$$x(0) = x_0,\ \dot{x}(0) = \dot{x}_0,\ \phi(0) = \phi_0,\ \dot{\phi}(0) = \dot{\phi}_0.$$

As in the previous case, the control is sought in the form of a linear feedback (5.2.7) with piecewise constant gains. Let us introduce a "fictitious" spring of stiffness β that connects the first body and the immovable foundation with the position $x = 0$ corresponding to its undeformed state. The total energy of the system inclusive of the elastic energy of the "fictitious" spring is

$$E = \frac{1}{2}\left((m_1+m_2)\dot{x}^2 + (m_2 l^2 + J)\dot{\phi}^2 + 2m_2 l\dot{x}\dot{\phi}\cos\phi + \beta x^2\right) + m_2 g l(1-\cos\phi),$$

and the derivative of E calculated by virtue of system (5.2.34) is given by

$$\dot{E} = -\alpha\dot{x}^2 - |f\dot{x}| - |\mu\dot{\phi}|.$$

Using arguments similar to those in Sect. 5.2.2, we can prove the following relationships:

$$\lim_{t\to\infty} E(t) = E_*, \quad E_* \geq 0,$$
$$\lim_{t\to\infty} \dot{x}(t) = 0, \quad \lim_{t\to\infty} \dot{\phi}(t) = 0. \tag{5.2.36}$$

Lemma 5.8. *The following inequality is valid:*

$$\overline{\lim_{t\to\infty}} |x(t)| \leq \frac{F}{\beta}.$$

Proof. Let us assume the contrary. Let there exist a number $\delta > 0$ and a sequence $\{t_k\}$, $t_k \to \infty$ as $k \to \infty$ such that

$$|x(t_k)| \geq \frac{F}{\beta} + 4\delta. \tag{5.2.37}$$

Denote the left-hand side of the first equation in (5.2.34) by $\Psi(t)$

$$\Psi(t) = (m_1+m_2)\ddot{x} + m_2 l\ddot{\phi}\cos\phi.$$

Let a number v satisfy the conditions

$$0 < v < \frac{\beta\delta}{\alpha}, \quad v^2 < \frac{\beta\delta}{m_2 l}.$$

In view of (5.2.36), there is $t' > 0$ such that $|\dot{\phi}(t)|, |\dot{x}(t)| \leq v$ for $t > t'$. Let us substitute expression (5.2.7) for the control u into the first of equations (5.2.34) and, taking into account (5.2.37), estimate the terms in the right-hand side of the equation thus obtained on the intervals

$$I_k = [t_k, t_k + \frac{\delta}{v}], \quad t_k > t',$$

as follows:

$$|\beta x| \geq F + 3\beta\delta, \quad |\alpha \dot{x}| \leq \alpha v < \beta\delta, \quad |m_2 l \dot{\phi}^2 \sin\phi| \leq m_2 l v^2 \leq \beta\delta.$$

It follows from these inequalities and from (5.2.35) that

$$|\Psi(t)| \geq \beta\delta > 0, \; t \in I_k, \; t_k > t'.$$

Hence,

$$\int_{I_k} |\Psi(t)|dt \geq \frac{\beta\delta^2}{v}. \tag{5.2.38}$$

The function $\Psi(t)$ is not continuous; however, on each of the intervals I_k, $t_k > t'$, the sign of the function $\Psi(t)$ does not change, because it coincides with the sign of the variable $x(t)$, that is continuous and, in view of the condition

$$|\beta x| \geq F + 3\beta\delta,$$

does not vanish. Therefore,

$$\int_{I_k} |\Psi(t)|dt = \left| \int_{I_k} \Psi(t)dt \right|$$

$$= \left| \int_{I_k} m_2 l \dot{\phi}^2 \sin\phi \, dt + \left((m_1 + m_2)\dot{x} + m_2 l \dot{\phi} \cos\phi\right) \Big|_{t_k}^{t_k + \delta/v} \right|.$$

We estimate the summands in the expression for the integral of $\Psi(t)$ as follows:

$$\left| (m_1 + m_2)\dot{x}\Big|_{t_k}^{t_k+\delta/v} \right| \leq 4Mv, \quad \left| m_2 l \dot{\phi} \cos\phi \Big|_{t_k}^{t_k+\delta/v} \right| \leq 2Mlv,$$

$$\left| \int_{I_k} m_2 l \dot{\phi}^2 \sin\phi \, dt \right| \leq Ml\delta v.$$

Hence,

$$\int_{I_k} |\Psi(t)|dt \leq 2M(2 + l + \delta l)v,$$

which contradicts inequality (5.2.38) for sufficiently small v. $\square$

We will transfer the carrying body to the prescribed terminal state in two stages. Introduce the notation

$$G(\dot{x}, \phi, \dot{\phi}) = \frac{1}{2}\left(M\dot{x}^2 + M(|\dot{x}| + l|\dot{\phi}|)^2 + J\dot{\phi}^2\right) + Mgl(1 - \cos\phi)$$

and consider the function

$$H(x, \dot{x}, \phi, \dot{\phi}) = G + \left(G^2 + \frac{U_0^2}{2}x^2\right)^{1/2}. \tag{5.2.39}$$

Let

$$H(x_0,\dot{x}_0,\phi_0,\dot{\phi}_0)=H_0, \quad G(\dot{x}_0,\phi_0,\dot{\phi}_0)=G_0.$$

We define the gains α_0 and β_0 in the control law (5.2.7) at the initial instant of time by the formulas

$$\beta_0=\frac{U_0^2}{2H_0}, \quad \alpha_0=\sqrt{m\beta_0}.$$

We introduce the notation

$$\begin{aligned}E_+^0(x,\dot{x},\phi,\dot{\phi})&=G(\dot{x},\phi,\dot{\phi})+\frac{\beta_0}{2}x^2,\\ E_-^0(x,\dot{x},\phi,\dot{\phi})&=\frac{1}{2}\left(m\dot{x}^2+J\dot{\phi}^2\right)+mgl(1-\cos\phi)+\frac{\beta_0}{2}x^2.\end{aligned} \tag{5.2.40}$$

It is not difficult to see that the total energy E^0 of the system, inclusive of the elastic energy of the "fictitious" spring of stiffness β_0, satisfies the inequalities

$$E_-^0(t)\le E^0(t)\le E_+^0(t). \tag{5.2.41}$$

In view of (5.2.39), the number H_0 is a root of the quadratic equation

$$H_0^2-2G_0H_0-\frac{U^2}{2}x_0^2=0,$$

being considered with respect to H_0. Dividing this equation by H_0 and making use of the formula for β_0 and expression (5.2.40) for E_+^0, we obtain

$$2E_+^0(x_0,\dot{x}_0,\phi_0,\dot{\phi}_0)=H_0. \tag{5.2.42}$$

Using (5.2.40)–(5.2.42) and the above reasoning, let us show that restriction (5.2.5) holds on the trajectory of system (5.2.34) and (5.2.7) with the gains α_0 and β_0 that starts at the point $(x_0,\dot{x}_0,\phi_0,\dot{\phi}_0)$:

$$\begin{aligned}|u|^2&\le 2(\alpha_0^2\dot{x}^2+\beta_0^2x^2)=2\beta_0(m\dot{x}^2+\beta_0x^2)\le 4\beta_0E_-^0(x,\dot{x},\phi,\dot{\phi})\\ &\le 4\beta_0E^0(t)\le 4\beta_0E^0(0)\le 4\beta_0E_+^0(0)=2\beta_0H_0=U_0^2.\end{aligned}$$

Let us introduce the notation:

$$\xi=2Mgl, \quad \eta=\frac{3F}{U_0}, \quad H_*=\frac{3\xi}{1-\eta^2}. \tag{5.2.43}$$

In view of (5.2.6), the number η satisfies the inequality $\eta<1$. Assume first that, at the initial instant, the condition

$$H_0>H_* \tag{5.2.44}$$

holds. It follows from expression (5.2.39) for the function $H(x,\dot{x},\phi,\dot{\phi})$, Lemma 5.8, the formula for the gain β_0, and relationships (5.2.36) that

$$\overline{\lim_{t\to\infty}} H(t) \le \xi + \left(\xi^2 + \frac{F^2 U_0^2}{2\beta_0^2}\right)^{1/2} < \xi + (\xi^2 + \eta^2 H_0^2)^{1/2}.$$

We put

$$H_1 = \xi + (\xi^2 + \eta^2 H_0^2)^{1/2}.$$

Let t_1 be the first instant of time when the value of the function $H(t)$ on the trajectory under consideration becomes equal to H_1. We change the gains in the control law (5.2.7) at the moment t_1 for

$$\beta_1 = \frac{U_0^2}{2H_1}, \quad \alpha_1 = \sqrt{m\beta_1},$$

and so on, for $t_2, \dots$

As in the case of the two-mass elastic system, the trajectory of this system consists of segments of trajectories of different systems of differential equations: the kth segment corresponds to system (5.2.34) subjected to control (5.2.7) with the gains given by the formulas

$$\beta_k = \frac{U_0^2}{2H_k}, \quad \alpha_k = \sqrt{m\beta_k}.$$

At that, on each segment constraint (5.2.5) holds. The numbers H_k are defined recursively by the formulas

$$H_k = \xi + (\xi^2 + \eta^2 H_{k-1}^2)^{1/2}, \; k = 1, 2, \dots,$$

and satisfy the inequalities

$$H_0 > H_1 > H_2 > \dots, \quad \lim_{k\to\infty} H_k = \frac{2\xi}{1-\eta^2} < H_*.$$

This implies that the following theorem is valid.

Theorem 5.3. *Let condition (5.2.44) be fulfilled. Then, there exists an instant of time τ_0 such that the value of the function $H(t)$ on the trajectory of system (5.2.34) controlled by means of the algorithm discussed above is equal to H_*.*

At the instants τ_0, the second stage of the motion begins. If condition (5.2.44) is not fulfilled at the initial moment, then we set $\tau_0 = 0$ and go directly to the second stage.

Let us extract the equation of motion of the first mass from set (5.2.34). To do this, we multiply the second of the equations by $m_2 l \cos\phi/(m_2 l^2 + J)$ and subtract the resulting equation from the first one:

$$m'\ddot{x} = u + Q, \quad m' = m_1 + m_2 - m_2 \frac{m_2 l^2 \cos^2\phi}{(m_2 l^2 + J)},$$
$$Q = f - \frac{m_2 l \mu}{(m_2 l^2 + J)} \cos\phi + m_2 l \sin\phi \left(\dot{\phi}^2 + \frac{m_2 g l}{(m_2 l^2 + J)} \cos\phi\right). \tag{5.2.45}$$

The quantity Q will be treated as an unknown perturbation.

At the second stage, we use the algorithm proposed in Sect. 5.1.2 for the control of a scleronomic mechanical system. In contrast to the system considered in Sect. 5.1.2, the coefficient m' in (5.2.45) that plays a role of the kinetic energy matrix depends on $\phi(t)$. The quantity $\phi(t)$ is a phase variable for the original system (5.2.34) but not for system (5.2.45). Therefore, the applicability of the algorithm needs additional justification. We will briefly give this substantiation.

Since, by our assumptions, the quantity ϕ can be measured at every time instant, we consider $\phi(t)$ to be known.

In view of (5.2.1), m' satisfies the inequalities

$$m \le m' \le 2M.$$

We introduce the function

$$W(x,\dot{x}) = 2M\dot{x}^2 + \left(4M^2\dot{x}^4 + \frac{U_0^2}{2}x^2\right)^{1/2}$$

and put

$$W_0 = W(x(\tau_0),\dot{x}(\tau_0)),\ W_k = \frac{W_0}{2^k},\ k=1,2,\dots\ .$$

Let τ_k be the first instant of time when the function $W(t)$ calculated along the trajectory becomes equal to W_k. We define the gains α and β at the moment τ_k by the formulas

$$\beta_k = \frac{U_0^2}{4W_k}, \quad \alpha_k = \sqrt{m\beta_k}, \quad k \ge 0.$$

In the phase space $(x,\dot{x})$, the level sets $W(x,\dot{x}) = W_k$ of the function W are ellipses contracting to the origin $(0,0)$ as $k \to \infty$.

We set $x(\tau_k) = x_k$ and $\dot{x}(\tau_k) = \dot{x}_k$. Let us show that the trajectory starting at the point $(x_k,\dot{x}_k)$ at the instant τ_k will reach the $(k+1)$st ellipse.

Consider the Lyapunov function

$$V^k(x,\dot{x}) = \frac{m'}{2}\dot{x}^2 + \frac{\beta_k}{2}x^2 + \varepsilon_k m' x\dot{x}, \quad \varepsilon_k = \frac{\sqrt{m}U_0}{8M\sqrt{W_k}}.$$

For $\tau_k \le t < \tau_{k+1}$, we have the following relations:

$$\begin{gathered} V_-^k(x,\dot{x}) \le V^k(x,\dot{x}) \le V_+^k(x,\dot{x}), \\ V_-^k(x,\dot{x}) = \frac{1}{4}\left(\beta_k x^2 + m\dot{x}^2\right), \quad V_+^k(x,\dot{x}) = \beta_k x^2 + 2M\dot{x}^2; \\ V_+^k(x,\dot{x}) \ge \frac{W_k}{8}, \quad 2V_+^k(x_k,\dot{x}_k) = W_k, \quad k=0,1,\dots \end{gathered} \tag{5.2.46}$$

To prove the latter relationship, let us note that the value

$$W_k = W(x_k, \dot{x}_k) = 2M\dot{x}_k^2 + \left(4M^2\dot{x}_k^4 + \frac{U_0^2}{2}x_k^2\right)^{1/2},$$

is obviously a root of the quadratic equation

$$W_k^2 - 4M\dot{x}_k^2 W_k - \frac{U_0^2}{2}x_k^2 = 0,$$

being considered with respect to W_k. Dividing this equation by W_k and making use of the formulas for β_k and V_+^k, we obtain the desired equality.

The derivative of the function V^k calculated by virtue of system (5.2.45) and the control law (5.2.7), is given by the formula

$$\begin{aligned}\dot{V}^k(x,\dot{x}) &= -\varepsilon_k\beta_k x^2 + (\varepsilon_k\dot{m}' - \varepsilon_k\alpha_k)x\dot{x}\\ &- \left(\alpha_k - \varepsilon_k m' - \frac{\dot{m}'}{2}\right)\dot{x}^2 + Q(\varepsilon_k x + \dot{x}).\end{aligned} \tag{5.2.47}$$

The following inequalities are valid:

$$|\varepsilon_k\alpha_k x\dot{x}| \le \varepsilon_k^2\alpha_k x^2 + \frac{\alpha_k}{4}\dot{x}^2, \quad |\varepsilon_k\dot{m}' x\dot{x}| \le |\dot{m}'|\left(4\varepsilon_k^2 x^2 + \frac{1}{16}\dot{x}^2\right). \tag{5.2.48}$$

Let Q satisfy the inequality

$$|Q| \le Q_0 = \frac{\sqrt{m}U_0}{32\sqrt{5M}}. \tag{5.2.49}$$

Let us estimate the last summand in (5.2.47) as follows:

$$\begin{aligned}|Q(\varepsilon_k x + \dot{x})| &\le Q_0|\varepsilon_k x + \dot{x}| \le Q_0[5\varepsilon_k^2 x^2 + \frac{5}{4}\dot{x}^2]^{1/2}\\ = Q_0\left(\frac{5m}{16M^2}\beta_k x^2 + \frac{5}{4}\dot{x}^2\right)^{1/2} &\le Q_0\left(\frac{5}{8M}(\beta_k x^2 + 2M\dot{x}^2)\right)^{1/2}\\ = \frac{\sqrt{5}Q_0}{2\sqrt{2M}}\frac{V_+^k(x,\dot{x})}{\sqrt{V_+^k(x,\dot{x})}} &\le \frac{\sqrt{m}U_0(\beta_k x^2 + 2M\dot{x}^2)}{32M\sqrt{W_k}}.\end{aligned}$$

Substituting the inequality obtained and inequality (5.2.48) into (5.2.47), we find that

$$\begin{aligned}\dot{V}^k(x,\dot{x}) &\le -\varepsilon_k\left(\beta_k - \varepsilon_k\alpha_k - \frac{\sqrt{m}U_0\beta_k}{32M\varepsilon_k\sqrt{W_k}} - 4\varepsilon_k|\dot{m}'|\right)x^2\\ &- \left(\frac{3\alpha_k}{4} - \varepsilon_k m' - \frac{\sqrt{m}U_0}{16\sqrt{W_k}} - \frac{9|\dot{m}'|}{16}\right)\dot{x}^2.\end{aligned}$$

Let the derivative $\dot{m}'$ be bounded:

$$|\dot{m}'| \leq \frac{\alpha_0}{8}. \tag{5.2.50}$$

Since $\alpha_0 < \alpha_k$, $k = 1,2\ldots$, then $|\dot{m}'| \leq \alpha_k/8$. Taking into account (5.2.46) and the relationships

$$\varepsilon_k \alpha_k = \frac{mU_0^2}{16MW_k} \leq \frac{\beta_k}{4}, \quad |\varepsilon_k m'| \leq 2\varepsilon_k M = \frac{\alpha_k}{2},$$

$$\frac{\sqrt{m}U_0}{32M\varepsilon_k\sqrt{W_k}} = \frac{1}{4}, \quad \frac{\sqrt{m}U_0}{16\sqrt{W_k}} = \frac{\alpha_k}{8},$$

we arrive at the estimate

$$\dot{V}^k(x,\dot{x}) \leq -\frac{3\varepsilon_k\beta_k}{8}x^2 - \frac{7\alpha_k}{128}\dot{x}^2$$

$$\leq -\frac{7\varepsilon_k}{64}\left(\beta_k x^2 + 2M\dot{x}^2\right) = -\frac{7\varepsilon_k}{64}V_+^k(t) \leq -\frac{7\varepsilon_k}{64}V^k(t).$$

Hence,

$$\tau_{k+1} - \tau_k \leq \frac{64}{7\varepsilon_k}\log\frac{V^k(\tau_k)}{V^k(\tau_{k+1})}.$$

Let us estimate the expression under the logarithm sign. By (5.2.46), the numerator under the logarithm sign, satisfies the inequality

$$V^k(\tau_k) \leq V_+^k(\tau_k) = V_+^k(x_k,\dot{x}_k) = \frac{W_k}{2}.$$

Using the equality $\beta_{k+1} = 2\beta_k$ and relationships (5.2.46), we estimate the denominator as follows:

$$V^k(\tau_{k+1}) \geq V_-^k(\tau_{k+1}) = \frac{1}{4}\left(\beta_k x_{k+1}^2 + m\dot{x}_{k+1}^2\right) = \frac{1}{8}\left(\beta_{k+1}x_{k+1}^2 + 2m\dot{x}_{k+1}^2\right)$$

$$\geq \frac{m}{8M}\left(\beta_{k+1}x_{k+1}^2 + 2M\dot{x}_{k+1}^2\right) = \frac{m}{8M}V_+^{k+1}(\tau_{k+1}) = \frac{m}{32M}W_k.$$

Therefore, the time of motion from the kth ellipse to the $(k+1)$st ellipsoid satisfies the inequality

$$\tau_{k+1} - \tau_k \leq \frac{64}{7\varepsilon_k}\log\frac{16M}{m} = \frac{M\sqrt{W_0}}{7\sqrt{m}U_0}2^{(9-k/2)}\log\frac{16M}{m},$$

and the total time of the motion of the system to the origin does not exceed the sum of the series of the right-hand sides of these inequalities. This series converges; hence, the time of the motion is finite.

Let us find the conditions that guarantee inequality (5.2.49) during the whole second stage of the motion. By the definition of the function G and condition (5.2.1), at the initial moment τ_0 of the second stage of the motion, the energy of the whole system not including the elastic energy of the "fictitious" spring satisfies the inequality

$$\frac{1}{2}\left((m_1+m_2)\dot{x}^2(\tau_0)+(m_2l^2+J)\dot{\phi}^2(\tau_0)+2m_2l\dot{x}(\tau_0)\dot{\phi}(\tau_0)\cos\phi(\tau_0)\right)$$

$$+m_2gl(1-\cos\phi(\tau_0)) \le G(\tau_0) \le H(\tau_0)/2 \le H_*/2,$$

and, in view of (5.2.46), the elastic energy of the "fictitious" spring satisfies the inequality

$$\frac{\beta_0}{2}x^2(\tau_0) \le \frac{1}{2}V_+^0(\tau_0) = \frac{W_0}{4}.$$

On each time interval $[\tau_k, \tau_{k+1})$, $k = 0, 1, \ldots$, the gains in the control law (5.2.7) are constant; therefore, the total energy of the system does not increase. At the moment τ_{k+1}, the stiffness β of the "fictitious" spring is doubled; hence, the elastic energy of the "fictitious" spring is increased by

$$\pi_{k+1} = \frac{1}{2}(\beta_{k+1}-\beta_k)x^2(\tau_{k+1}) = \frac{\beta_{k+1}}{4}x^2(\tau_{k+1}).$$

In view of (5.2.46), the quantity π_{k+1} satisfies the inequality

$$\pi_{k+1} \le \frac{1}{4}\bar{V}_+^{k+1}(\tau_{k+1}) = \frac{W_{k+1}}{8} = \frac{W_0}{2^{k+4}}.$$

The total increase of the elastic energy of the "fictitious" spring due to the instantaneous change of the spring stiffness does not exceed the sum of the series

$$\sum_{k=1}^{\infty}\frac{W_0}{2^{k+4}} = \frac{W_0}{16}.$$

Hence, for any $t \ge \tau_0$, the total energy of the system $E(t)$ is bounded

$$E(t) \le \frac{H_*}{2} + \frac{5W_0}{16}.$$

By the definition of the functions G, W, and H, we have

$$M\dot{x}^2 \le G(\dot{x},\phi,\dot{\phi}), \quad W(x,\dot{x}) \le 2G + \left(4G^2 + \frac{U_0^2}{2}x^2\right)^{1/2} \le 2H(x,\dot{x},\phi,\dot{\phi}).$$

At the initial moment τ_0 of the second stage, the inequality $H(\tau_0) \le H_*$ holds, hence,

$$W_0 \le 2H(\tau_0) \le 2H_*,$$

which implies that, for any $t \ge \tau_0$, the total energy of the system $E(t)$ satisfies the inequality $E(t) \le 9H_*/8$, and the angular velocity $\dot{\phi}$ satisfies the inequality

$$\dot{\phi}^2 \le \frac{2E(t)}{m_2l^2+J} \le \frac{9H_*}{4(m_2l^2+J)}. \tag{5.2.51}$$

Let us estimate the summands in the formula for the perturbation Q in equation (5.2.45) as follows:

$$|m_2 l\dot{\phi}^2 \sin\phi| \leq \frac{9m_2 lH_*}{4(m_2 l^2+J)} \leq \frac{9H_*}{4l},$$

$$\left|\frac{m_2^2 l^2 g\cos\phi \sin\phi}{m_2 l^2+J}\right| \leq m_2 g \leq Mg, \quad \left|\frac{m_2 l\mu\cos\phi}{m_2 l^2+J}\right| \leq \frac{\mu_2}{l}.$$

Substituting the inequalities obtained and formula (5.2.43) for H_* into the formula for Q and transforming it, we find that

$$|Q| \leq= F+\frac{\mu_2}{l}+\frac{29U_0^2-18F^2}{2(U_0^2-9F^2)}Mg.$$

Comparing this inequality with (5.2.49), we conclude that, if the following inequality

$$F+\frac{\mu_2}{l}+\frac{29U_0^2-18F^2}{2(U_0^2-9F^2)}Mg \leq \frac{\sqrt{m}U_0}{32\sqrt{5M}}. \tag{5.2.52}$$

is true, then (5.2.49) holds.

Now, let us find the conditions that guarantee the fulfillment of inequality (5.2.50). To do this, we estimate the value of the derivative $\dot{m}'$ by means of (5.2.45) and (5.2.51), and the value of $\alpha_0/8$ using the formulas for the gains α_0 and β_0 and the inequality $W_0 \leq 4H_*$:

$$|\dot{m}'(t)| = 2\frac{m_2^2 l^2|\dot{\phi}\cos\phi\sin\phi|}{m_2 l^2+J} \leq M\left(\frac{7H_*}{2(ml^2+J)}\right)^{1/2},$$

$$\frac{\alpha_0}{8} = \frac{\sqrt{m\beta_0}}{8} = \frac{\sqrt{m}U_0}{16\sqrt{W_0}} \geq \frac{\sqrt{m}U_0}{32\sqrt{H_*}}.$$

The sufficient condition for the fulfillment of (5.2.50) is

$$\frac{7M^2H_*}{2(ml^2+J)} \leq \frac{mU_0^2}{2^{10}H_*}.$$

Substituting formula (5.2.43) for H_* into this inequality and transforming it, we obtain the relation

$$\frac{U_0^2}{(U_0^2-9F^2)^2} \leq \frac{m(ml^2+J)}{2^{17}M^4g^2l^2}, \tag{5.2.53}$$

which guarantees the fulfillment of (5.2.50).

Thus, the following theorem is valid.

Theorem 5.4. *Let the parameters of the problem satisfy conditions (5.2.52) and (5.2.53). Then, the suggested control algorithm brings the carrying mass to the prescribed terminal state in finite time.*

Remark 5.6. Inequalities (5.2.52) and (5.2.53) (sufficient conditions for bringing the system to the prescribed terminal state) impose rather strong restrictions on the parameters of the original problem. This can be explained by the fact that some bounds used for the justification of the algorithm are fairly rough; besides, we always suggested the "worst" behavior of the system subject to the restrictions imposed. In addition, inequalities (5.2.52) and (5.2.53) guarantee monotone decrease of the Lyapunov function V^k along the trajectory. However, it may happen that the function V^k is not monotone, while the trajectories reach the terminal state. Let us note, however, that the control algorithm itself does not contain these sufficient conditions and can be formally applied to problems for which these restrictions on the parameters are not satisfied. The results of the computer simulation show that this control law is efficient far beyond the sufficient conditions derived.

5.2.7 Computer simulation results

Let us illustrate how the suggested control algorithm works.

At first, we present the results of the computer simulation for the dynamics of the system depicted in Fig. 5.7. The equations of its motion (5.2.3) were integrated by the Runge — Kutta method for the following values of the parameters:

$$M = m_1 = 10\,\text{kg}, \quad m = m_2 = 5\,\text{kg}, \quad C = c_0 = 10\,\text{N/m}, \quad \gamma = \gamma_1 = \gamma_2 = 0.2.$$

The quantity U_0 was chosen to be equal to 100N. The system was transferred from the initial state

$$x_0 = 1\,\text{m}, \quad \phi_0 = -0.5\,\text{m}, \quad \dot{x}_0 = \dot{\phi}_0 = 0\,\text{m/s}$$

to the terminal set $x = \dot{x} = 0$, that is, the first mass had to be brought to the origin. Integrating was stopped when the quantity $(x^2 + \dot{x}^2)^{1/2}$, that is the Euclidean distance between the projection of the current state of the phase trajectory onto subspace $(x, \dot{x})$ and the origin, became less than 0.001.

It is not difficult to see that conditions (5.2.29) (sufficient conditions for bringing the system to the terminal state) are not fulfilled for the above values of the parameters. Nevertheless, the trajectory of system (5.2.3) controlled by means of the proposed algorithm comes to the terminal set in finite time.

The main characteristics of the motion were as follows:

$$H_0 = 143.9\,\text{kg} \cdot \text{m}^2/\text{s}^2, \; H_* = 706.8\,\text{kg} \cdot \text{m}^2/\text{s}^2.$$

Since $H_0 < H_*$, there is no the first stage of the motion.

Figure 5.10 describes the behavior of the phase variables of the system. The thick and thin lines correspond to the first (carrying) and second masses, respectively. The solid lines represent the time history of the coordinates x and ϕ of the masses, the dashed lines show the time history of their velocities. One can see that the graphs of

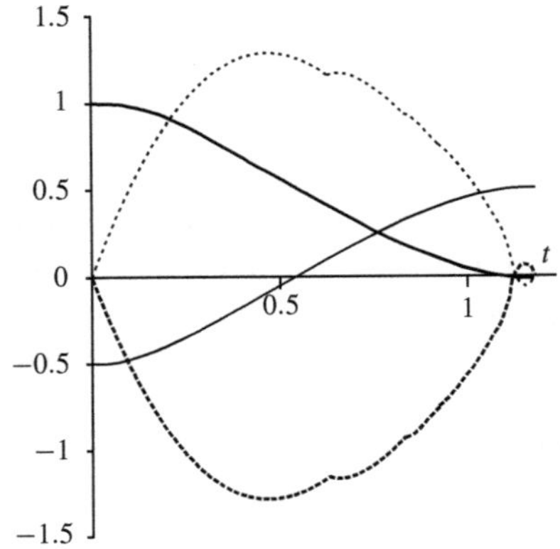

Fig. 5.10 The time dependance of the coordinates and velocities

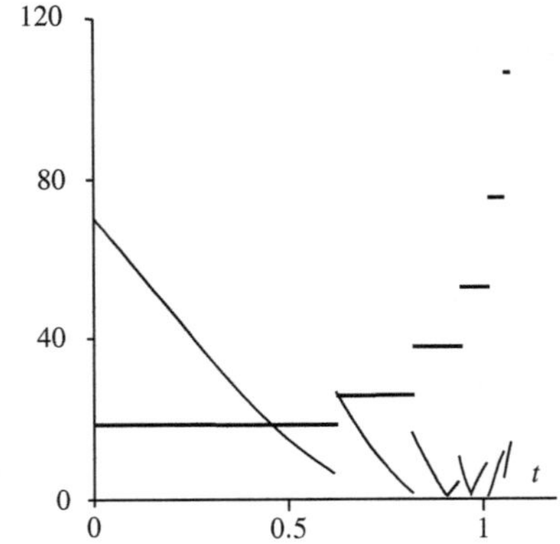

Fig. 5.11 The control force and the gain α

the velocity $\dot{x}$ of the first mass and the velocity $\dot{\phi}$ of the second mass with respect to the first one are broken lines due to the discontinuity of the control force.

The thin line in Fig. 5.11 shows the dependence of the control force on time, and the thick line depicts the behavior of the gain α (a step function). Although the gains α and β in the control law (5.2.7) increase (during integration of the equations the gains in the control have changed 16 times), the value of the control force u, as can be seen from the figure, satisfies condition (5.2.5) and has a substantial margin.

Figure 5.12 shows the results of the computer simulation for the system "a load on a cart". As before, the solid lines correspond to the coordinates x and ϕ of the masses, the dashed lines correspond to the velocities $\dot{x}$ and $\dot{\phi}$, the thick and thin lines correspond to the first (carrying) and second masses, respectively.

As we have already mentioned, the system under consideration has "stagnation" regions because of dry friction. One can see in Fig. 5.12 that, in the process of motion, the second mass sticks and remains motionless with respect to the first mass, again coming in motion on the final stage.

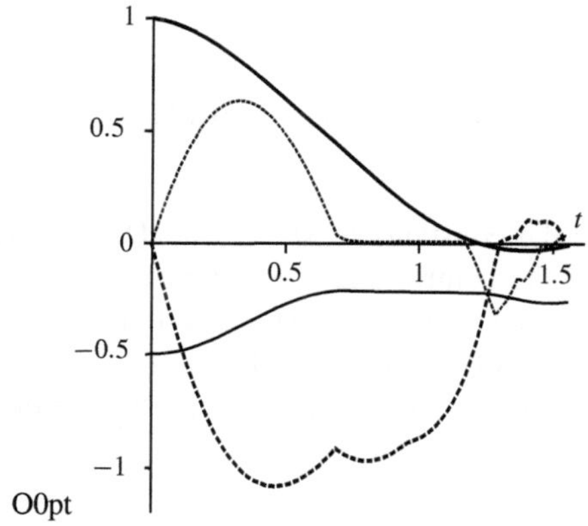

Fig. 5.12 The system "a load on a cart": coordinates and velocities

5.3 Piecewise linear control for rheonomic systems

5.3.1 Problem statement

In the following subsections a rheonomic mechanical system is considered, that is, the system whose kinetic energy is represented by expression (4.1.3).

We assume that the symmetric positive definite matrix $A(t,q) \in C^1$ is unknown, its eigenvalues for any t and q belong to the interval $[m,M]$, $0 < m \leq M$, and the partial derivatives are uniformly bounded with respect to the norm, that is,

$$
\begin{gathered}
mz^2 \leq \langle A(t,q)z,z\rangle \leq Mz^2, \quad \forall z \in R^n \\
\|\frac{\partial A}{\partial q_i}\| \leq D_1, \quad \|\frac{\partial A}{\partial t}\| \leq D_2, \quad i=1,\ldots,n, \quad D_1,D_2>0.
\end{gathered}
\tag{5.3.1}
$$

The vector function $a(t,q) \in C^1$ and the function $a_0(t,q) \in C^1$ are also assumed to be unknown and to satisfy the conditions

$$
|\left(\frac{\partial a}{\partial q}\right)^{\top} - \frac{\partial a_1}{\partial q}| \leq D_3, \quad |\frac{\partial a_0}{\partial q} - \frac{\partial a_1}{\partial t}| \leq D_4, \quad D_3,D_4>0. \tag{5.3.2}
$$

The dynamics of the system under consideration is described by equations (5.1.2).

We still assume that the system is directly controlled with respect to each degree of freedom, the constraint

$$
|U| \leq U_0, \; U_0 > 0, \tag{5.3.3}
$$

is imposed on the n-dimensional control forces vector U, and the generalized forces Q are unknown and satisfy the condition

$$|Q| \leq Q_0, \; Q_0 > 0. \tag{5.3.4}$$

The phase variables q and $\dot{q}$ are assumed to be available for measuring at every time instant.

Problem 5.4. Suppose the constants m, M, U_0, and D_j $(j = 1, \ldots, 4)$ are given. It is required to construct the control that meets constraint (5.3.3) and to specify the domain of admissible initial states from which system (5.1.2), under the action of this control, reaches the prescribed terminal rest state $(\bar{q}, 0)$ in finite time, whatever the matrix A, the vector a, the function a_0, and the perturbations Q that satisfy conditions (5.3.1), (5.3.2), and (5.3.4) be.

5.3.2 Control algorithm for rheonomic systems

Without loss of generality, we shall assume that the final state coincides with the phase space origin, that is, $\bar{q} = 0$ (this can be achieved using an appropriate variables transformation).

We will construct the control in the form of a linear feedback with respect to the generalized coordinates and velocities

$$U = -\alpha_k \dot{q} - \beta_k q, \quad \alpha_k, \beta_k > 0, \tag{5.3.5}$$

with the gains in the form of piecewise constant functions. To describe the algorithm for changing these gains, we will set the sequence of numbers α_k and β_k.

Let $q_0 = q(0)$ and $\dot{q}_0 = \dot{q}(0)$ be the initial state of the system. We introduce the function

$$W(q, \dot{q}) = M\dot{q}^2 + \left(M^2 \dot{q}^4 + U_0^2 q^2\right)^{1/2}. \tag{5.3.6}$$

The quantity $W(q, \dot{q})$ has the dimension of energy and characterizes the distance between the point $(q, \dot{q})$ and the terminal state $(0, 0)$: the level set $W(q, \dot{q}) = C$ of the function W in the phase space $(q, \dot{q}) \in R^{2n}$ is the ellipsoid $2CM\dot{q}^2 + U_0^2 q^2 = C^2$, that contracts to the phase space origin $(0, 0)$ as $C \to 0$.

We put

$$W_0 = W(q_0, \dot{q}_0), \quad W_k = \frac{W_0}{2^k}, \quad k = 1, 2, \ldots. \tag{5.3.7}$$

The level sets of the function $W(q, \dot{q})$ corresponding to the constants W_k represent a family of ellipsoids that contract to zero as k increases. We will denote by t_1 the instant of time when the trajectory hits the ellipsoid $W(q, \dot{q}) = W_1$ for the first time, and we put $q_1 = q(t_1)$ and $\dot{q}_1 = \dot{q}(t_1)$. It will be shown below that, in the case of the chosen control algorithm, the trajectory of the system tends to the phase space origin and, therefore, such an instant of time exists. We will denote by t_2 the instant of time when the trajectory of the system hits the ellipsoid $W(q, \dot{q}) = W_2$ for the first time and we put $q_2 = q(t_2)$ and $\dot{q}_2 = \dot{q}(t_2)$, and so on.

The sequence $\{t_k\}$ defines the instants of time when the coefficients α_k and β_k in control (5.3.5) are changed. We will specify the values of these coefficients in the half-interval of time $[t_k, t_{k+1})$, $k = 0, 1, \ldots$, as follows:

$$\beta_k = \frac{U_0^2}{2W_k}, \quad \alpha_k^2 = m\beta_k. \tag{5.3.8}$$

In the phase space R^{2n}, the trajectory of the mechanical system under consideration comprises segments of the trajectories of different systems of differential equations: the kth segment joins the points $(q_k, \dot{q}_k)$ and $(q_{k+1}, \dot{q}_{k+1})$ and corresponds to a system of form (5.1.2) and (5.3.5) in which the feedback factors α_k and β_k are constant and are defined by formula (5.3.8). All the points $(q_k, \dot{q}_k)$ lie on the respective ellipsoids $W(q, \dot{q}) = W_k$, $k = 0, 1, \ldots$, (see Fig. 1).

Remark 5.7. As we have already seen before, the function W may not be a monotonically decreasing along the trajectory of the system, despite the fact that the trajectory tends to the phase space origin. Hence, the trajectory may have more than one point of intersection with some ellipsoids (see Remark 5.1 in Sect. 5.1.2). Besides, unlike the approach used above for scleronomic systems, the family of ellipsoids in this case is chosen from the very beginning, so that the initial state of the system lies on the ellipsoid with an index 0.

Thus, when implementing the algorithm, it is sufficient to measure the actual values of the phase variables of the system q and $\dot{q}$ and to store in the memory the actual value of the index k, i.e., equal to the number of the smallest ellipsoid already visited by the trajectory of the system. Since, in expression (5.3.6) for the function W, only the known parameters of the problem appear, the value of the function $W(q(t), \dot{q}(t))$ can be calculated at any instant of time. As the index k increases by unity, the value of W decreases by a factor of two, the gain α increases by a factor of $\sqrt{2}$, and the gain β increases by a factor of 2.

5.3.3 Justification of the control

We will use Lyapunov's direct method to validate the algorithm. Consider the kth segment of the trajectory for a certain fixed $k \geq 0$. This segment starts at the point $(q_k, \dot{q}_k)$ at the instant of time t_k and corresponds to system (5.1.2) and (5.3.5) with constant feedback factors specified by formulas (5.3.8). We will now show that there is such an instant t_{k+1} when the trajectory of the system hits the ellipsoid $W(q, \dot{q}) = W_{k+1}$.

The Lyapunov function

We put

$$\varepsilon_k = \frac{\sqrt{m\beta_k}}{4M} \tag{5.3.9}$$

and introduce the Lyapunov function

$$V^k(t,q,\dot{q}) = \frac{1}{2}\langle A(t,q)\dot{q},\dot{q}\rangle + \frac{\beta_k}{2}q^2 + \varepsilon_k\langle A(t,q)\dot{q},q\rangle. \tag{5.3.10}$$

The expression for the function V^k contains the kinetic energy matrix $A(t,q)$ that is assumed to be unknown. We estimate the value of this function at an arbitrary point $(t,q,\dot{q})$ of the augmented phase space in terms of the known quantities. The relationships

$$|\varepsilon_k\langle A\dot{q},q\rangle| \le \frac{1}{8}\langle A\dot{q},\dot{q}\rangle + 2\varepsilon_k^2\langle Aq,q\rangle = \frac{1}{8}\left(\langle A\dot{q},\dot{q}\rangle + \frac{m\beta_k}{M^2}\langle Aq,q\rangle\right)$$

hold by virtue of the Cauchy inequality and expression (5.3.9), and it follows from condition (5.3.1) that

$$\frac{m\beta_k}{M^2}\langle Aq,q\rangle \le \frac{m\beta_k}{M}q^2 \le \beta_k q^2.$$

Substituting the inequalities obtained into relationship (5.3.10) and again using condition (5.3.1), we get the following estimates for the function V^k:

$$V_-^k(q,\dot{q}) \le V^k(t,q,\dot{q}) \le V_+^k(q,\dot{q}), \tag{5.3.11}$$

where

$$V_-^k(q,\dot{q}) = \frac{3}{8}(m\dot{q}^2 + \beta_k q^2), \quad V_+^k(q,\dot{q}) = \frac{5}{8}(M\dot{q}^2 + \beta_k q^2). \tag{5.3.12}$$

We will now establish some relations connecting the functions $V_+^k(q,\dot{q})$ and $W(q,\dot{q})$. Substituting the formula for the gain β_k from (5.3.8) into expression (5.3.12), we obtain for the function V_+^k:

$$V_+^k(q_k,\dot{q}_k) = \frac{10M\dot{q}_k^2 W_k + 5U_0^2 q_k^2}{16W_k}. \tag{5.3.13}$$

By construction, the point $(q_k,\dot{q}_k)$ lies on the ellipsoid with number k. This and definition (5.3.6) of the function W yield

$$W_k = W(q_k,\dot{q}_k) = M\dot{q}_k^2 + (M^2\dot{q}_k^4 + U_0^2 q_k^2)^{1/2}.$$

Using this equality, the numerator in expression (5.3.13) is reduced to $5W_k^2$, which yields the relationship

$$V_+^k(q_k,\dot{q}_k) = \frac{5}{16}W_k \tag{5.3.14}$$

connecting the functions $V_+^k(q,\dot q)$ and $W(q,\dot q)$. This relationship means that, for any k, the ellipsoid with number k is the level set of the quadratic form $V_+^k(q,\dot q)$ that corresponds to the value $5W_k/16$.

In accordance with the control algorithm, the point $(q(t),\dot q(t))$, for $t\in[t_k,t_{k+1})$, lies outside the ellipsoid with number $(k+1)$, that is, outside the level set

$$\{(q,\dot q):\ V_+^{k+1}(q,\dot q)=\frac{5}{16}W_{k+1}\},$$

hence

$$V_+^{k+1}\left(q(t),\dot q(t)\right)>\frac{5}{16}W_{k+1}=\frac{5}{32}W_k,\quad t_k\le t<t_{k+1}.$$

The equality $\beta_{k+1}=2\beta_k$ holds by virtue of formulas (5.3.7) and (5.3.8), and the relationship

$$M\dot q^2+\beta_k q^2\ge\frac{1}{2}\left(M\dot q^2+\beta_{k+1}q^2\right)$$

follows from this. Consequently, the estimates

$$V_+^k(q(t),\dot q(t))\ge\frac{1}{2}V_+^{k+1}(q(t),\dot q(t))\ge\frac{5}{64}W_k \tag{5.3.15}$$

hold along the kth segment of the trajectory.

The derivative of the Lyapunov function

To calculate the derivative $\dot V^k$, let us introduce the notation

$$B(t,q)=\left(\frac{\partial a}{\partial q}(t,q)\right)^{\top}-\frac{\partial a_1}{\partial q}(t,q),\quad b(t,q)=\frac{\partial a_0}{\partial q}(t,q)-\frac{\partial a_1}{\partial t}(t,q) \tag{5.3.16}$$

and differentiate the function V^k according to (5.1.2) and (5.3.5). We obtain

$$\begin{aligned}\dot V^k(t,q,\dot q)=&-\langle\left[\alpha_k I-\varepsilon_k A+\frac{1}{2}\frac{\partial A}{\partial t}-\frac{\varepsilon_k}{2}\sum_{i=1}^n q_i\frac{\partial A}{\partial q_i}\right]\dot q,\dot q\rangle\\&-\varepsilon_k\beta_k q^2-\varepsilon_k\alpha_k\langle\dot q,q\rangle+\langle Q+b,\dot q+\varepsilon_k q\rangle-\varepsilon_k\langle B\dot q,q\rangle,\end{aligned} \tag{5.3.17}$$

where I is the identity matrix.

We will now estimate the individual terms in expression (5.3.17). Using the Cauchy inequality and relationships (5.3.2) and (5.3.16), we obtain

$$|\varepsilon_k\alpha_k\langle\dot q,q\rangle|\le\frac{\alpha_k}{4}\dot q^2+\varepsilon_k^2\alpha_k q^2,\quad|\varepsilon_k\langle B\dot q,q\rangle|\le\frac{D_3}{2}\dot q^2+\frac{\varepsilon_k^2 D_3}{2}q^2. \tag{5.3.18}$$

Using the inequality

$$|2\varepsilon_k\langle \dot q, q\rangle| \le \frac{1}{16}\dot q^2 + 16\varepsilon_k^2 q^2,$$

for ε_k and relationship (5.3.15), we estimate the quantity $|\dot q + \varepsilon_k q|$ as follows:

$$(\dot q + \varepsilon_k q)^2 \le \frac{17}{16}\dot q^2 + 17\varepsilon_k^2 q^2 \le \frac{17}{16M}(M\dot q^2 + \beta_k q^2) = \frac{17}{10M}V_+^k(q,\dot q)$$
$$= \frac{17}{10MV_+^k(q,\dot q)}\left(V_+^k(q,\dot q)\right)^2 \le \frac{1088}{50MW_k}\left(V_+^k(q,\dot q)\right)^2,$$

whence, taking into account the second expression in (5.3.12), we obtain

$$|\langle Q+b, \dot q + \varepsilon_k q\rangle| \le |Q+b|\sqrt{\frac{17}{2MW_k}}\left(M\dot q^2 + \beta_k q^2\right). \tag{5.3.19}$$

The relationship

$$\left|\frac{\varepsilon_k}{2}\sum_{i=1}^{n} q_i \frac{\partial A}{\partial q_i}\right| \le \frac{\sqrt{n}D_1}{2}\varepsilon_k|q| \tag{5.3.20}$$

holds by virtue of (5.3.1) and the inequality $\sum_{i=1}^{n}|q_i| \le \sqrt{n}|q|$.

Substituting inequalities (5.3.18)–(5.3.20) into expression (5.3.17) and making use of conditions (5.3.1), (5.3.2), and (5.3.4), we arrive at the following estimate for the derivative of the function V^k along the kth segment of the trajectory:

$$\dot V^k(t,q,\dot q) \le -\left(\varepsilon_k\beta_k - \varepsilon_k^2\alpha_k - \beta_k\sqrt{\frac{17}{2MW_k}}(Q_0 + D_4) - \frac{\varepsilon_k^2 D_3}{2}\right)q^2 -$$
$$-\left(\frac{3\alpha_k}{4} - \varepsilon_k M - \frac{D_2 + D_3}{2} - \sqrt{\frac{17M}{2W_k}}(Q_0 + D_4) - \frac{\sqrt{n}D_1}{2}\varepsilon_k|q|\right)\dot q^2. \tag{5.3.21}$$

We will now show that, under certain additional assumptions, the derivative $\dot V^k$ will be negative definite. We put

$$\Omega = \min\left\{\frac{\sqrt{15}MU_0}{8\sqrt{n}D_1}, \frac{mU_0^2}{32D_2^2}, \frac{mU_0^2}{32D_3^2}\right\}$$

and introduce the sets

$$G = \{(q,\dot q) \in R^{2n} : W(q,\dot q) \le \Omega\},$$

$$G_k = \left\{(q,\dot q) : |q| < \sqrt{\frac{5}{3}\frac{W_k}{U_0}}\right\},\ k = 0,1\dots.$$

The inequality

$$3q_k^2 U_0^2 \le 5W^2(q_k, \dot q_k)$$

results from the definition (5.3.6) of the function W. From this and by virtue of relationships (5.3.7), it follows that the point $(q_k, \dot{q}_k)$ lies in the domain G_k.

Lemma 5.9. *Suppose the initial point $(q_k, \dot{q}_k)$ of the kth segment belongs to the set G, the matrix A, the vector functions Q and a, and the function a_0 satisfy conditions (5.3.1), (5.3.2), and (5.3.4), and*

$$Q_0 + D_4 \leq \sqrt{\frac{m}{17M}\frac{U_0}{8}}. \tag{5.3.22}$$

Then, on the part of the trajectory that starts at the point $(q_k, \dot{q}_k)$ and lies outside the ellipsoid $W(q, \dot{q}) = W_{k+1}$ and in the set G_k, the derivative of the function V^k calculated by virtue of system (5.1.2), (5.3.5), and (5.3.8) satisfies the inequality

$$\dot{V}^k(t, q, \dot{q}) \leq -\frac{3\alpha_k}{40M} V^k(t, q, \dot{q}). \tag{5.3.23}$$

Proof. By the condition of the lemma, $W(q_k, \dot{q}_k) \leq \Omega$ and, consequently,

$$D_2^2 \leq \frac{mU_0^2}{32W_k}, \quad D_3^2 \leq \frac{mU_0^2}{32W_k}.$$

These inequalities and definitions (5.3.8) and (5.3.9) of the numbers ε_k, α_k, and β_k yield

$$\frac{D_2 + D_3}{2} \leq \frac{\alpha_k}{4}, \quad \frac{\varepsilon_k^2 D_3}{2} \leq \frac{\alpha_k \beta_k}{64M}. \tag{5.3.24}$$

From condition (5.3.22) and formulas (5.3.8), we obtain

$$\sqrt{\frac{17M}{2W_k}}(Q_0 + D_4) \leq \frac{\alpha_k}{8}, \quad \beta_k \sqrt{\frac{17}{2MW_k}}(Q_0 + D_4) \leq \frac{\alpha_k \beta_k}{8M}. \tag{5.3.25}$$

By virtue of relationships (5.3.8) and (5.3.9), we have

$$\varepsilon_k M = \frac{\alpha_k}{4}, \quad \varepsilon_k \beta_k - \varepsilon_k^2 \alpha_k = \frac{\alpha_k \beta_k}{4M}\left(1 - \frac{m}{4M}\right) \geq \frac{3\alpha_k \beta_k}{16M}. \tag{5.3.26}$$

The inequality $W(q_k, \dot{q}_k) \leq \Omega$ yields

$$D_1 \leq \frac{\sqrt{15}MU_0}{8\sqrt{n}W_k}.$$

Since the section of the trajectory under consideration lies in the set G_k, we have

$$\varepsilon_k |q| \leq \frac{\sqrt{5}\alpha_k W_k}{4\sqrt{3}MU_0}$$

and, consequently,

$$\frac{\sqrt{n}D_1}{2}\varepsilon_k |q| \leq \frac{5\alpha_k}{64}. \tag{5.3.27}$$

Substituting inequalities (5.3.24)–(5.3.27) into (5.3.21) and using equalities (5.3.12), we arrive at the relationships

$$\dot{V}^k(t,q,\dot{q}) \le -\frac{3\alpha_k}{64M}(M\dot{q}^2 + \beta_k q^2) \le -\frac{3\alpha_k}{40M}V_+^k(q,\dot{q}),$$

whence the assertion of the lemma follows by virtue of estimates (5.3.11). □

Lemma 5.10. *Suppose the matrix A, the vector-functions Q and a, and the function a_0 satisfy conditions (5.3.1), (5.3.2), (5.3.4), and (5.3.22), and that $(q_k,\dot{q}_k) \in G$. Then, inequality (5.3.23) holds along the kth segment of the trajectory.*

Proof. It has already been established above that $(q_k,\dot{q}_k) \in G_k$. By virtue of Lemma 5.9, to prove Lemma 5.10, it is sufficient to show that the kth segment of the trajectory lies wholly in the domain G_k.

Let us assume the opposite. Suppose t' is the first instant when the trajectory leaves the domain G_k, that is,

$$q^2(t') = \frac{5W_k^2}{3U_0^2}. \tag{5.3.28}$$

On the other hand, it follows from definition (5.3.9) of the coefficient ε_k and from relationships (5.3.11) and (5.3.12) that

$$\varepsilon_k^2 q^2(t') = \frac{m}{16M^2}\beta_k q^2(t') \le \frac{m}{16M^2}(m\dot{q}^2(t') + \beta_k q^2(t'))$$
$$= \frac{m}{6M^2}V_-^k(q(t'),\dot{q}(t')) \le \frac{m}{6M^2}V^k(t',q(t'),\dot{q}(t')).$$

Since the trajectory segment under consideration lies in the domain G_k when $t_k \le t < t'$, the function V^k, by virtue of Lemma 5.9, decreases along this segment. Hence, using relationship (5.3.14), we continue the latter estimate as follows:

$$\varepsilon_k^2 q^2(t') < \frac{m}{6M^2}V^k(t_k,q(t_k),\dot{q}(t_k)) \le \frac{m}{6M^2}V_+^k(q(t_k),\dot{q}(t_k)) = \frac{5m}{96M^2}W_k.$$

Consequently,

$$q^2(t') < \frac{5mW_k}{96M^2\varepsilon_k^2} = \frac{5W_k^2}{3U_0^2}.$$

This inequality contradicts condition (5.3.28). □

It follows from the assertions of Lemmas 5.9 and 5.10 that, outside the ellipsoid $W(q,\dot{q}) = W_{k+1}$, the function V^k strictly decreases along the trajectory of system (5.1.2), (5.3.5), and (5.3.8), and, by virtue of relationships (5.3.11)–(5.3.14), there exists an instant of time t_{k+1} when the trajectory hits the ellipsoid with number $k+1$.

It is clear that, if the initial state of the system $(q_0,\dot{q}_0)$ belongs to the set G, then the null ellipsoid $W(q,\dot{q}) = W_0$ and, together with it, all the ellipsoids $W(q,\dot{q}) = W_k$, $k = 1,2,\ldots$, lie entirely in this set. Consequently, all the points $(q_k,\dot{q}_k)$ also belong to G and the assertions of Lemmas 5.9 and 5.10 are applicable to any of the segments comprising the trajectory of the system.

Estimation of the time of motion

We will now show that the system reaches the phase space origin in finite time. In order to estimate the time of motion along the kth segment of the trajectory, we integrate inequality (5.3.23) and obtain

$$t_{k+1} - t_k \le \frac{40M}{3\alpha_k} \log \frac{V^k(t_k, q_k, \dot{q}_k)}{V^k(t_{k+1}, q_{k+1}, \dot{q}_{k+1})}. \tag{5.3.29}$$

By virtue of relationships (5.3.8) and (5.3.11)–(5.3.14), we have

$$V^k(t_k, q_k, \dot{q}_k) \le \frac{5}{16} W_k,$$

$$V^k(t_{k+1}, q_{k+1}, \dot{q}_{k+1}) \ge V_-^k(q_{k+1}, \dot{q}_{k+1}) = \frac{3}{8}\left(m\dot{q}_{k+1}^2 + \beta_k q_{k+1}^2\right)$$

$$\ge \frac{3m}{16M}\left(M\dot{q}_{k+1}^2 + \beta_{k+1} q_{k+1}^2\right) = \frac{3m}{10M} V_+^{k+1}(q_{k+1}, \dot{q}_{k+1}) = \frac{3m}{64M} W_k.$$

Substituting these relationships and expression (5.3.8) for α_k into inequality (5.3.29), we obtain the following estimate for the time of motion from the point $(q_k, \dot{q}_k)$ up to the point $(q_{k+1}, \dot{q}_{k+1})$:

$$t_{k+1} - t_k \le \tau 2^{-k/2}, \quad k = 0, 1, \dots ,$$

$$\tau = \frac{40M\sqrt{2W_0}}{3\sqrt{mU_0}} \log \frac{20M}{3m}.$$

The total time T_* of motion of the system up to the terminal state does not exceed the sum of the series

$$T_* \le \tau \sum_{k=0}^{\infty} 2^{-k/2} = \frac{\tau\sqrt{2}}{\sqrt{2} - 1}. \tag{5.3.30}$$

Consequently, the proposed control algorithm brings system (5.1.2) to the phase space origin in finite time.

We will now verity that condition (5.3.3) is satisfied along the trajectory of the motion. to that end, we estimate the modulus of the control force vector along the kth segment of the trajectory using the Cauchy inequality and relationships (5.3.5),(5.3.8),(5.3.11), and (5.3.12) as follows:

$$|U|^2 \le 2(\alpha_k^2 \dot{q}^2 + \beta_k^2 q^2) = 2\beta_k(m\dot{q}^2 + \beta_k q^2) = \frac{16}{3}\beta_k V_-^k(q, \dot{q}) \le \frac{16}{3}\beta_k V^k(t, q, \dot{q}).$$

Since the function V^k decreases in the half-interval $[t_k, t_{k+1})$, we can use relationship (5.3.14) to continue the estimate as follows:

$$|U|^2 \le \frac{16}{3}\beta_k V^k(t_k, q_k, \dot{q}_k) \le \frac{16}{3}\beta_k V_+^k(q_k, \dot{q}_k) = \frac{5}{3}\beta_k W_k = \frac{5}{6} U_0^2,$$

whence inequality (5.3.3) follows.

Modification of the algorithm

It follows from the above consideration that the system reaches the point $(0,0)$ in finite time if the initial state belongs to the ellipsoid G. Let us note that any point of the form $(\bar{q},0)$ in the phase space of the system can be chosen as a terminal state. Then, the set of ellipsoids on which the feedback factors change should to be shifted by the vector $\bar{q}$ while the parameters of the ellipsoids remain as before. We will now show that, using this fact and modifying the proposed algorithm, it is possible to extend the set of permissible initial states considerably.

Suppose that

$$(q_0,\dot{q}_0) \in G_*, \quad G_* = \left\{ (q,\dot{q}) \in R^{2n} : \dot{q}^2 \leq \frac{\Omega}{2M} \right\}. \tag{5.3.31}$$

We first transfer the system to the point $q = q_0$, $\dot{q} = 0$. to that end, we make the change of variables $q' = q - q_0$. In the new variables q' and $\dot{q}'$ the set

$$G' = \{(q',\dot{q}') : W(q',\dot{q}') \leq \Omega\},$$

that is analogous to the set G considered above, is an ellipsoid with its centre at the point $q' = \dot{q}' = 0$. The initial state of the system, that is, the point $q_0' = 0, \dot{q}_0' = \dot{q}_0$, belongs to this set by virtue of inclusion (5.3.31) and definition (5.3.6) of the function W. Consequently, the control law

$$U = -\alpha_k \dot{q}' - \beta_k q'$$

with the above algorithm for changing the coefficients α_k and β_k brings the system in finite time to the centre of this ellipsoid, that is, to the point $q = q_0, \dot{q} = 0$.

In the phase space $(q,\dot{q})$ we choose a finite sequence of points $(\bar{q}^j,0)$, $j = 1,2,\ldots,J$, such that $\bar{q}^1 = q_0$, $\bar{q}^J = 0$, and

$$|\bar{q}^j - \bar{q}^{j-1}| \leq \frac{\Omega}{U_0}. \tag{5.3.32}$$

We transfer the system from the point $(\bar{q}^1,0) = (q_0,0)$ to the point $(\bar{q}^J,0) = (0,0)$, that is, to the phase space origin, in $J-1$ steps applying the control algorithm again each time. The point $(\bar{q}^j,0)$ corresponds to the initial state of the system at the jth step and the point $(\bar{q}^{j+1},0)$ corresponds to the terminal state. It follows from inequality (5.3.32) and definition (5.3.6) of the function W that, for any j, the point $(\bar{q}^j,0)$ belongs to the ellipsoid

$$G^j = \{(q,\dot{q}) : W(q - \bar{q}^{j+1},\dot{q}) \leq \Omega\}.$$

This ellipsoid is the set of admissible initial states of the system that will reach to the terminal state $(\bar{q}^{j+1}, 0)$ at the jth step.

Consequently, the control law

$$U = -\alpha_k \dot{q} - \beta_k (q - \bar{q}^{j+1})$$

with the presented above algorithm for changing the gains α_k and β_k transfers the system from the point $(\bar{q}^j, 0)$ to the center of this ellipsoid, that is, to the point $q = \bar{q}^{j+1}$, $\dot{q} = 0$, in finite time. Hence, after $J-1$ steps, system (5.1.2) reaches the final state $(0,0)$).

The following theorem sums up the above reasoning.

Theorem 5.5. *Suppose the matrix A, the vector functions Q and a, and the function a_0 satisfy conditions (5.3.1), (5.3.2), (5.3.4), and (5.3.22), and $(q_0, \dot{q}_0) \in G_*$. Then, the proposed control law transfers system (5.1.2) from the initial state $(q_0, \dot{q}_0)$ to the phase space origin in finite time and meets constraint (5.3.3).*

Remark 5.8. As has already been mentioned, the approach used here is an extension on the rheonomic systems of the approach developed above for scleronomic systems. In the scleronomic case, the set of admissible initial states coincides with all phase space, that is, the system is brought from an arbitrary initial state to the prescribed terminal state. In the rheonomic case, the set of admissible initial states (5.3.31) is a bounded set in the phase space R^{2n} because the condition

$$\dot{q}_0^2 \leq \frac{\Omega}{2M}$$

is imposed on the initial velocities.

Remark 5.9. One can note that only the known parameters of the problem appear in the definition of the set G_* and in the expressions for the function W and the feedback factors α_k and β_k. To implement the algorithm, it is sufficient to know the values of m, M, U_0, and the phase variables of the system at each current instant of time. The constants D_1, D_2, and D_3 appear only in the conditions determining the set of permissible initial states G_*. These conditions, as well as the constraints on the vector function $a(t,q)$, function $a_0(t,q)$, and disturbances Q in relationships (5.3.22), are only sufficient conditions for transferring the system to the terminal state and guarantee monotone decrease of the Lyapunov function V^k along the trajectory. However, it may happen that the function V^k is not monotone and tends to zero, as the trajectories tend to the terminal state (see Remark 5.3 in Sect. 5.1.5). The algorithm proposed can, therefore, be practically applied also in the cases where constraints (5.3.22) are not satisfied and the initial state of the system does not belong to the set G_*. The computer simulation of the dynamics of various mechanical systems shows that the algorithm is also effective beyond the limits of the sufficient conditions that have been presented.

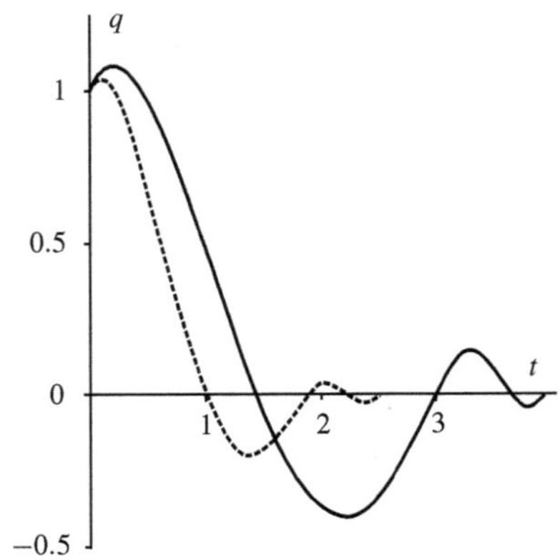

Fig. 5.13 The angular coordinate of the rod

5.3.4 Results of simulation

We will illustrate the work of the algorithm through numerical simulation of the rotation of a body with the moment of inertia depending on time. Let us consider a system consisting of a weightless bar and a particle of unknown mass m_0 that moves uncertainly along the bar (see Fig. 4.1). We assume that the bar rotates in a horizontal plane about one of its ends under the action of a control torque U and an uncertain perturbing torque Q. With the notations accepted in Sect. 4.1 the equations of motion of such a system can be written as follows:

$$m_0 l^2(t)\ddot{q} + 2m_0 l(t)\dot{l}(t)\dot{q} = Q + U. \tag{5.3.33}$$

In this case, the moment of the dry friction forces that acts on the bar serves as the unknown generalized force Q. In the simulation, the constants m, M, and U_0, the mass m_0, the perturbation Q, and the law of motion of the point mass along the bar $l(t)$ are taken as follows:

$$m = 0.25\,\text{kg},\ M = 2.25\,\text{kg},\ U_0 = 10\,\text{N}\cdot\text{m},$$

$$Q = -0.1\,\text{sign}\,(\dot{q})\,\text{H}\cdot\text{m},\ m_0 = 1\,\text{kg},\ l(t) = 1 + \frac{1}{2}\sin\omega t\ \text{m}.$$

Using the proposed control law, the bar is transferred from the initial state $q_0 = 1\,\text{rad}$ and $\dot{q}_0 = 1\,\text{rad/s}$ into the terminal state $q = \dot{q} = 0$. Integration of (5.3.33) was stopped when the Euclidean distance from the actual point of the trajectory to the terminal point in the phase space $(q, \dot{q}) \in R^2$ became less than 0.01.

The results of the simulation for the case where $\omega = 1\,\text{s}^{-1}$ are shown in Figs. 5.13 and 5.14. The solid curves correspond to the time histories of the angular coordinate in Fig. 5.13 and the absolute value of the control torque $|U|$ (the discontinuous line) in Fig. 5.14. The total time of motion is found to be equal to $T_* = 3.98\,\text{s}$.

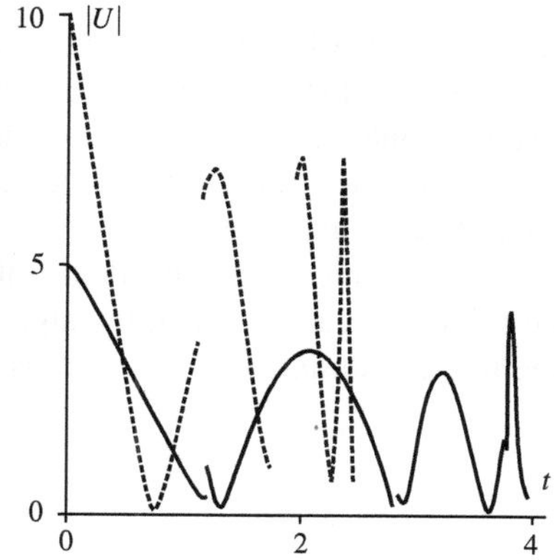

Fig. 5.14 The absolute value of the control torque

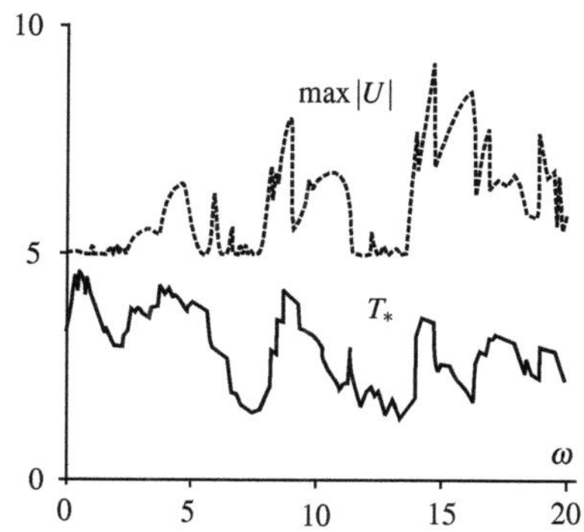

Fig. 5.15 The time of motion and the maximum magnitude of the control torque

Figure 5.14 shows that that constraint $|U| \leq 10$ is satisfied with a considerable margin. Therefore, it is reasonable to simulate the motion of the system controlled by law (5.3.5) with the feedback factors α_k and β_k twice than those prescribed by algorithm (5.3.8). The time history of the angular coordinate of the bar and the absolute value of the control torque $|U|$, for such control law, are represented by the dashed curves in Figs. 5.13 and 5.14. In this case, the time of motion was reduced to $T_* = 2.53\,\mathrm{s}$ and, as previously, the control satisfies constraint (5.3.3).

In order to estimate the efficiency of the control algorithm when condition $(q_0, \dot{q}_0) \in G_*$ of the above theorem is violated, we simulated the dynamics of system (5.3.33) for different values of ω. The solid curve in Fig. 5.15 depicts the dependence of the overall time of motion T_* up to the terminal state on the parameter ω for $\omega \in [0, 20]$. In this case,

$$\dot{A}(t) = m_0 \omega (1 + \frac{1}{2} \sin \omega t) \cos \omega t,$$

and the constant D_2 from constraints (5.3.1) satisfies the inequality $\omega \leq D_2$. Consequently, $\Omega \leq mU_0^2/32\omega^2$ and, in the case of the chosen values of the system parameters, for a large part of the interval $0 \leq \omega \leq 20\,\mathrm{s}^{-1}$, the initial state $q_0 = 1\,\mathrm{rad}$ and $\dot{q}_0 = 1\,\mathrm{rad/s}$ does not lie in the domain G_*. Nevertheless, the proposed control law does bring the system to the terminal state.

The dashed curve in Fig. 5.15 depicts the dependence of the maximum magnitude of the control torque U that is realized when the algorithm is applied, on the parameter ω. One can see that constraints (5.3.3) are satisfied for all values of ω considered.

Chapter 6
Continuous feedback control for mechanical systems under uncertainty

In the aforegoing chapters, two approaches to constructing control algorithms have been elaborated that enable a Lagrangian mechanical system to be steered to a given terminal state in finite time, on the assumption that the control forces are bounded and the system is subject to uncontrollable perturbations. The steering may be achieved through approaches based on decomposition methods, as well as on a linear feedback with piecewise constant coefficients. By using these and some other methods [95, 96], one obtains control laws that are described, in general, by discontinuous functions of time.

In this chapter, an approach to constructing continuous feedback control laws is proposed that may be used to steer a mechanical system to a terminal state in finite time. The control law proposed may be interpreted as a linear feedback with the gains being continuous functions of the phase variables. The gains increase to infinity as the trajectory approaches the terminal state, nevertheless the control forces remain bounded and satisfy the imposed constraint.

The results presented in this chapter were published previously in [10, 11, 12, 13].

6.1 Feedback control for scleronomic system with a given matrix of inertia

6.1.1 Problem statement

We return now to the consideration of a scleronomic mechanical system governed by Lagrange's equations (5.1.2). As before, the kinetic energy of the system is given by (5.1.1). We keep also the main assumptions of Sect. 5.1.1. Namely, the vector of unknown disturbances $Q(q,\dot{q})$ is an arbitrary vector-valued function satisfying some existence conditions for the solution of system (5.1.2) and meeting constraint (5.1.3). The vector of control forces U is also bounded and satisfies (5.1.4). The matrix $A(q)$ is a continuously differentiable symmetric positive definite matrix of the kinetic energy, its eigenvalues and partial derivatives are constrained by (5.1.5) and (5.1.6). The phase variables q and $\dot{q}$ are assumed to be available for measuring at every instant of time.

Let us suppose, at first, that the matrix of inertia $A(q)$ is given.

Problem 6.1. Construct a control $U(q,\dot{q})$ as a continuous vector-valued function of the phase variables $(q,\dot{q}) \in R^{2n} \setminus \{(\bar{q},0)\}$ that satisfies condition (5.1.4) and specify a domain of admissible initial states such that any trajectory of system (5.1.2) starting in that domain reaches a prescribed terminal state $(\bar{q},0)$ in finite time, whatever the perturbations $Q(q,\dot{q})$ satisfying condition (5.1.3) be.

Let us note that the terminal state is a state of rest of the unperturbed system (5.1.2), i.e., in case where $U=0$ and $Q=0$. Without loss of generality, we assume that $\bar{q}=0$, that is, the terminal state coincides with the origin of the phase space. This may be achieved by a suitable choice of the generalized coordinates.

6.1.2 Control function

We define the control as follows:

$$U(q,\dot{q}) = -\alpha(q,\dot{q})A(q)\dot{q} - \beta(q,\dot{q})q, \tag{6.1.1}$$

where

$$\alpha(q,\dot{q}) = \sqrt{\frac{\beta(q,\dot{q})}{M}}, \quad \beta(q,\dot{q}) = \frac{3U_0^2}{8V(q,\dot{q})}, \tag{6.1.2}$$

$$V(q,\dot{q}) = T + \frac{1}{2}\beta(q,\dot{q})q^2 + \frac{1}{2}\alpha(q,\dot{q})\langle A(q)\dot{q},q\rangle, \quad q^2+\dot{q}^2>0. \tag{6.1.3}$$

Relationships (6.1.2) and (6.1.3) define the functions $\alpha(q,\dot{q}), \beta(q,\dot{q})$, and $V(q,\dot{q})$ implicitly.

The function $V(q,\dot{q})$ plays a principal role in the present investigation. Given this function, one can find the feedback factors $\alpha(q,\dot{q})$ and $\beta(q,\dot{q})$ through the above relations, and, consequently, the control $U(q,\dot{q})$ according to formula (6.1.1). In addition, the function V has the dimension of energy and serves as a Lyapunov function for the system under consideration. It tends to zero as the trajectory approaches the terminal state. Since the function $V(q,\dot{q})$ appears in the denominators in relationships (6.1.2), the feedback factors tend to infinity as the trajectory approaches the origin. Nevertheless, the proposed control does not go beyond the admissible boundaries.

Theorem 6.1. *In the domain $R^{2n} \setminus \{(0,0)\}$, there exist continuously differentiable functions $\alpha(q,\dot{q}), \beta(q,\dot{q})$, and $V(q,\dot{q})$ satisfying (6.1.2) and (6.1.3) and such that $V>0$.*

Proof. Substituting the expressions for the functions $\alpha(q,\dot{q})$ and $\beta(q,\dot{q})$ into (6.1.3) and transforming equation (6.1.3), we obtain

$$16V^2(q,\dot{q}) = 16T(q,\dot{q})V(q,\dot{q}) + 3U_0^2q^2 + \frac{2\sqrt{6}}{\sqrt{M}}\langle A(q)\dot{q},q\rangle V^{1/2}(q,\dot{q}). \quad (6.1.4)$$

We introduce the notation

$$x = V^{1/2}(q,\dot{q}), \quad \xi(q,\dot{q}) = 4T^{1/2}(q,\dot{q}),$$
$$\eta(q,\dot{q}) = \frac{2\sqrt{6}U_0}{\sqrt{M}}\langle A(q)\dot{q},q\rangle, \quad \gamma(q,\dot{q}) = \sqrt{3}U_0|q| \quad (6.1.5)$$

and rewrite (6.1.4) in the form

$$F(q,\dot{q},x) = 16x^4 - \xi^2(q,\dot{q})x^2 - \eta(q,\dot{q})x - \gamma^2(q,\dot{q}) = 0. \quad (6.1.6)$$

Let us consider this equality as an equation in x. We will show that, for any $q,\dot{q} \in R^{2n} \setminus \{(0,0)\}$, there exists a unique positive root of equation (6.1.6), and its multiplicity is one.

By the Cauchy inequality and condition (5.1.5), we have

$$\eta^2(q,\dot{q}) = \frac{24U_0^2}{M}\langle A(q)\dot{q},q\rangle^2 \leq \frac{24U_0^2}{M}T(q,\dot{q})\langle A(q)q,q\rangle$$
$$\leq 24U_0^2T(q,\dot{q})q^2 = \xi^2(q,\dot{q})\gamma^2(q,\dot{q}),$$

whence it follows that

$$|\eta| \leq \gamma\xi. \quad (6.1.7)$$

We also note that, by formulas (6.1.5) and condition (5.1.2), the identity $\xi(q,\dot{q}) = \gamma(q,\dot{q}) = 0$ cannot hold in the domain $R^{2n} \setminus \{(0,0)\}$.

Considering, for a while, the coefficients ξ,η, and γ to be fixed, let us prove the following auxiliary proposition.

Lemma 6.1. *Every equation of the form*

$$f(x) = 16x^4 - \xi^2x^2 - \eta x - \gamma^2 = 0, \quad (6.1.8)$$

where the coefficients ξ,η, and γ^2 are constant and satisfy the inequalities (6.1.7) and $\xi^2 + \gamma^2 > 0$, has a unique positive real root, and its multiplicity is equal to one.

Proof. At first, we will prove that, if $\xi,\gamma > 0$, then there exist exactly two real roots, one positive and one negative, and the positive root has multiplicity one.

Since $f(0) = -\gamma^2 < 0$ and $f(x) > 0$ for large absolute values of x, equation (6.1.8) must have a positive and a negative roots. Let us verify that there are no other real roots. Suppose the contrary: there exist parameters $\gamma >, \xi 0$ and η satisfying (6.1.7) for which equation (6.1.8) has more than two real roots, i.e., three or four.

Then, let us show that there exist parameters $\gamma >, \xi 0$ and η satisfying (6.1.7) for which equation (6.1.8) has a multiple real root.

Obviously, if there are exactly three real roots, one of them is multiple.

Now, suppose that the coefficients ξ_1, η_1, and γ_1 are such that $|\eta_1| \le \xi_1 \gamma_1$ and equation (6.1.8) has four real roots. For sufficiently large γ_2, equation (6.1.8) with coefficients ξ_1, η_1, and γ_2 has exactly two real roots and $|\eta_1| \le \xi_1 \gamma_2$. Consequently, as the coefficient γ varies from γ_1 to γ_2, the number of roots changes, and there exists γ_3 at which a multiple root x_0 of equation (6.1.8) appears. Obviously, this γ_3 also satisfies condition (6.1.7).

Since x_0 is a multiple root of the equation $f(x) = 0$ with coefficients ξ_1, η_1, and γ_3, it is also a root of the equation $f'(x) = 0$ with coefficients ξ_1 and η_1 that is

$$f'(x) = 64x^3 - 2\xi_1^2 x - \eta_1 = 0. \qquad (6.1.9)$$

We multiply equation (6.1.8), where $\xi = \xi_1$, $\eta = \eta_1$, and $\gamma = \gamma_3$, by -4, and equation (6.1.9) by x, and add the resulting equations together. This gives the equation

$$2\xi_1^2 x^2 + 3\eta_1 x + 4\gamma_3^2 = 0 \qquad (6.1.10)$$

that must have x_0 as a root. By condition (6.1.7) and the inequalities $\xi_1, \gamma_3 > 0$, the discriminant $D = 9\eta_1^2 - 32\xi_1^2\gamma_3^2$ of equation (6.1.10) is negative, and therefore equation (6.1.7) has no real roots. This contradiction proves the lemma in the case where $\xi, \gamma > 0$.

Now, consider the case where $\xi = 0$ or $\gamma = 0$. If $\xi = 0$, then, by equation (6.1.7), $\eta = 0$, and equation (6.1.8) becomes $x^4 = \gamma^2/16$, $\gamma > 0$, that has a single positive root, and the multiplicity of the root is one.

In the case where $\gamma = 0$, equation (6.1.7) implies that $\eta = 0$, and equation (6.1.8) reduces to the equation $x^4 = \xi^2 x^2$, $\xi > 0$, that has three real roots. One of these roots, namely $x = 0$, has multiplicity two, and the single positive root is of multiplicity one, which completes the proof of the lemma. □

We now return to the proof of Theorem 6.1, considering the coefficients ξ, η, and γ as functions of the phase variables q and $\dot{q}$. It follows from the statement of Lemma 6.1 that, for any $(q,\dot{q}) \in R^{2n} \setminus \{(0,0)\}$, the polynomial equation (6.1.6) has a unique positive real root $x_0(q,\dot{q})$ and its multiplicity is one. Consequently,

$$\frac{\partial F}{\partial x}(q,\dot{q},x_0(q,\dot{q})) \neq 0, \quad (q,\dot{q}) \in R^{2n} \setminus \{(0,0)\},$$

and, by the Implicit Function Theorem, $V(q,\dot{q}) = x_0^2(q,\dot{q})$. Then the function $V(q,\dot{q}) = x_0^2(q,\dot{q})$, and together with it also the functions $\alpha(q,\dot{q})$ and $\beta(q,\dot{q})$ defined by (6.1.2), are continuously differentiable in the domain $R^{2n} \setminus \{(0,0)\}$, and $V > 0$. The theorem is proved. □

Bearing in mind the statement of Theorem 6.1 and formula (6.1.1), we conclude that the control function $U(q,\dot{q})$ is defined and continuously differentiable in the domain $R^{2n} \setminus \{(0,0)\}$.

The justification of the proposed control law is based on Lyapunov's second method. The peculiarity of the investigation is that we do not express the function $V(q,\dot{q})$ as well as the functions $\alpha(q,\dot{q})$ and $\beta(q,\dot{q})$ in an explicit form. We establish their properties and carry out all other necessary reasoning for the functions defined implicitly.

6.1.3 Justification of the control

We shall now find the domain of admissible initial states and show that any trajectory of system (5.1.2) and (6.1.1) beginning in that domain will reach the origin in finite time. This will be done by methods of the theory of stability, and we will show that the function V is a Lyapunov function of the system under consideration.

Let us find upper and lower bounds for $V(q,\dot{q})$. Using the Cauchy inequality, formulas (6.1.2), and conditions (5.1.5), we obtain

$$|\alpha(q,\dot{q})\langle A(q)\dot{q},q\rangle| \le |\alpha(q,\dot{q})| \left(2T(q,\dot{q})\langle A(q)q,q\rangle\right)^{1/2}$$

$$\le T(q,\dot{q}) + \frac{1}{2}\alpha^2(q,\dot{q})\langle A(q)q,q\rangle \le T(q,\dot{q}) + \frac{1}{2}\beta(q,\dot{q})q^2,$$

which implies that

$$V_-(q,\dot{q}) \le V(q,\dot{q}) \le 3V_-(q,\dot{q}), \tag{6.1.11}$$

where

$$V_-(q,\dot{q}) = \frac{1}{4}\left[2T(q,\dot{q}) + \beta(q,\dot{q})q^2\right]. \tag{6.1.12}$$

Let us substitute expressions (6.1.2) and (6.1.12) for the functions $\beta(q,\dot{q})$ and $V_-(q,\dot{q})$ into estimates (6.1.11). After some reduction, we obtain the inequalities

$$\xi(q,\dot{q}) \le 32V^2(q,\dot{q}) \le 3\xi(q,\dot{q}), \quad \xi(q,\dot{q}) = 16T(q,\dot{q})V(q,\dot{q}) + 3U_0^2q^2,$$

whence, solving for $V(q,\dot{q})$, we arrive at the following limits (for brevity, we omit the arguments q and $\dot{q}$):

$$\frac{1}{4}\left[T + \left(T^2 + \frac{3U_0^2q^2}{2}\right)^{1/2}\right] \le V \le \frac{3}{4}\left[T + \left(T^2 + \frac{U_0^2q^2}{2}\right)^{1/2}\right].$$

Using conditions (5.1.5), we finally obtain

$$\begin{aligned} &\frac{1}{8}\left[m\dot{q}^2 + \left(m^2\dot{q}^4 + 6U_0^2q^2\right)^{1/2}\right] \le V(q,\dot{q}) \\ &\le \frac{3}{8}\left[M\dot{q}^2 + \left(M^2\dot{q}^4 + 2U_0^2q^2\right)^{1/2}\right]. \end{aligned} \tag{6.1.13}$$

Let us note that the functions of the phase variables q and $\dot{q}$ on the right- and left-hand sides of (6.1.13) are expressed explicitly in terms of the known parameters of the problem. It follows from (6.1.13) that the function $V(q,\dot{q})$ can be defined as zero at $(0,0)$ while still remaining continuous, but it will not be differentiable there (see, for example, the graph of the function V for a mechanical system consisting of a point mass moving along a horizontal line presented in Fig 6.3).

We will now evaluate the derivative $\dot{V}$. Differentiating the functions $\alpha(q,\dot{q}), \beta(q,\dot{q})$, and $V(q,\dot{q})$ along trajectories of system (5.1.2) and (6.1.1), we obtain

$$\dot{\alpha} = -\frac{\alpha}{2V}\dot{V}, \quad \dot{\beta} = -\frac{\beta}{V}\dot{V},$$
$$\dot{V} = \dot{T} + \beta\langle q,\dot{q}\rangle + \alpha T + \frac{\alpha}{2}\langle\frac{d}{dt}A\dot{q},q\rangle - \frac{\dot{V}}{2V}\left(\beta q^2 + \frac{\alpha}{2}\langle A(q)\dot{q},q\rangle\right). \tag{6.1.14}$$

By the theorem on the variation of the kinetic energy of a scleronomic Lagrangian system, and by the definition of the vector-valued function $U(q,\dot{q})$, we have

$$\dot{T} = \langle u+Q,\dot{q}\rangle = -2\alpha T - \beta\langle\dot{q},q\rangle + \langle Q,\dot{q}\rangle. \tag{6.1.15}$$

It follows from expression (5.1.1) for the kinetic energy and equation (5.1.2) that

$$\frac{d}{dt}A\dot{q} = \frac{d}{dt}\frac{\partial T}{\partial \dot{q}} = \frac{\partial T}{\partial q} + U + Q. \tag{6.1.16}$$

Substituting (6.1.15), (6.1.16), and (6.1.1) into the last equation of (6.1.14), we obtain the following relation for the derivative of the function $V(q,\dot{q})$:

$$\dot{V} = -\alpha\left[T + \frac{\alpha}{2}\langle A(q)\dot{q},q\rangle + \frac{\beta}{2}q^2\right] + \frac{\alpha}{4}\langle\left(\sum_{i=1}^{n}\frac{\partial A}{\partial q_i}q_i\right)\dot{q},\dot{q}\rangle$$
$$-\frac{\dot{V}}{2V}\left[\beta q^2 + \frac{\alpha}{2}\langle A(q)\dot{q},q\rangle\right] + \langle Q,\dot{q} + \frac{\alpha}{2}q\rangle. \tag{6.1.17}$$

Let us transform and estimate the separate terms in the right-hand side of (6.1.17). It follows from definitions (6.1.2) and (6.1.3) of the functions $\alpha(q,\dot{q})$ and $V(q,\dot{q})$ that

$$\alpha\left[T + \frac{a}{2}\langle A(q)\dot{q},q\rangle + \frac{\beta}{2}q^2\right] = \alpha V = \frac{\sqrt{3}U_0}{2\sqrt{2M}}V^{1/2}. \tag{6.1.18}$$

By the Cauchy inequality, condition (5.1.5) and relations (6.1.2), (6.1.11), and (6.1.12), we have

$$\left|\dot{q} + \frac{\alpha}{2}q\right|^2 = \dot{q}^2 + \frac{\alpha^2}{4}q^2 + \alpha\langle\dot{q},q\rangle \le \frac{5}{4}(\dot{q}^2 + \alpha^2 q^2) \tag{6.1.19}$$

$$\leq \frac{5}{4}\left(\frac{2}{m}T + \frac{\beta}{M}q^2\right) \leq \frac{5}{m}V_- \leq \frac{5}{m}V.$$

Hence, by condition (5.1.3), we have the inequality

$$|\langle Q, \dot{q} + \frac{\alpha}{2}q\rangle| \leq Q_0\sqrt{\frac{5}{m}}V^{1/2}. \qquad (6.1.20)$$

Using condition (5.1.6) and the estimate

$$\sum_{i=1}^{n} |q_i| \leq \sqrt{n}|q|, \qquad (6.1.21)$$

we can conclude that

$$\left\| \sum_{i=1}^{n} \frac{\partial A}{\partial q_i} q_i \right\| \leq \sqrt{n}D|q|. \qquad (6.1.22)$$

Relations (5.1.5), (6.1.2), (6.1.11), and (6.1.12) imply the following inequalities:

$$q^2 \leq \frac{4}{\beta}V_- \leq \frac{4}{M\alpha^2}V, \quad \dot{q}^2 \leq \frac{2}{m}T \leq \frac{4}{m}V_- \leq \frac{4}{m}V. \qquad (6.1.23)$$

Hence, by inequality (6.1.22), we obtain

$$|\frac{\alpha}{4}\langle\left(\sum_{i=1}^{n} \frac{\partial A}{\partial q_i} q_i\right)\dot{q}, \dot{q}\rangle| \leq \frac{\alpha}{4}\sqrt{n}D|q|\dot{q}^2 \leq \frac{2\sqrt{n}D}{m\sqrt{M}}V^{3/2}. \qquad (6.1.24)$$

Substituting relations (6.1.18), (6.1.20), and (6.1.24) into expression (6.1.17) for the derivative $\dot{V}(q,\dot{q})$ and transposing the last term in $\dot{V}(q,\dot{q})$ to the left-hand side, we arrive at the inequality

$$B\dot{V} \leq -\delta(q,\dot{q})V^{1/2}, \qquad (6.1.25)$$

where

$$\delta(q,\dot{q}) = \frac{\sqrt{3}U_0}{2\sqrt{2M}} - Q_0\sqrt{\frac{5}{m}} - \frac{2\sqrt{n}C}{m\sqrt{M}}V(q,\dot{q}), \qquad (6.1.26)$$

$$B(q,\dot{q}) = 1 + \frac{\beta(q,\dot{q})}{2V(q,\dot{q})}q^2 + \frac{\alpha(q,\dot{q})}{4V(q,\dot{q})}\langle A(q)\dot{q}, q\rangle = \frac{1}{V}\left[T + \beta q^2 + \frac{3a}{4}\langle A(q)\dot{q}, q\rangle\right]$$

[the last equality in the chain is obtained using (6.1.3)].

By the Cauchy inequality, formula (6.1.2), and condition (5.1.5), we have

$$|\frac{3}{4}\alpha\langle A(q)\dot{q}, q\rangle| \leq \frac{1}{2}T(q,\dot{q}) + \frac{9\alpha^2}{16}\langle Aq, q\rangle \leq \frac{1}{2}T + \frac{3\beta}{4}q^2,$$

$$|\frac{3}{4}\alpha\langle A(q)\dot{q}, q\rangle| \leq \frac{3}{2}T(q,\dot{q}) + \frac{3\alpha^2}{16}\langle Aq, q\rangle \leq 2T + \frac{\beta}{2}q^2.$$

Using the first inequality to estimate the function $B(q,\dot{q})$ from below and the second to estimate it from above, we obtain

$$0 < \frac{1}{4V}\left(2T + \beta q^2\right) \le B(q,\dot{q}) \le \frac{3}{V}\left[T + \frac{\beta}{2}q^2 + \frac{\alpha}{2}\langle A(q)\dot{q}, q\rangle\right] = 3. \qquad (6.1.27)$$

Since $B(q,\dot{q}) > 0$, it follows that the sufficient condition for the derivative $\dot{V}(q,\dot{q})$ to be negative is that the expression in parentheses on the right-hand side of inequality (6.1.25) should be negative. Let us put

$$V(t) = V(q(t),\dot{q}(t)),\ B(t) = B(q(t),\dot{q}(t)),\ \delta(t) = \delta(q(t),\dot{q}(t)) \qquad (6.1.28)$$

and rewrite inequality (6.1.25) in the form

$$\dot{V}(t) \le -\frac{\delta(t)}{B(t)} V^{1/2}(t). \qquad (6.1.29)$$

Theorem 6.2. *Suppose the condition*

$$\delta(t_0) > 0 \qquad (6.1.30)$$

holds at the initial instant of time t_0. Then, the derivative of the function V along the trajectories of system (5.1.2) and (6.1.1) satisfies the inequality

$$\dot{V}(t) \le -\frac{\delta(t_0)}{3} V^{1/2}(t), \quad t \ge t_0. \qquad (6.1.31)$$

Proof. It follows from relations (6.1.27) and (6.1.29) and from condition (6.1.30) that

$$\dot{V}(t_0) \le -\frac{\delta(t_0)}{B(t_0)} V^{1/2}(t_0) \le -\frac{\delta(t_0)}{3} V^{1/2}(t_0) < 0.$$

Consequently, for $t > t_0$ in a sufficiently small neighborhood of the point t_0, the inequality $V(t) < V(t_0)$ holds. This inequality turns out to be true for all $t > t_0$.

Suppose the contrary. Let $t' > t_0$ be the first instant of time when the function V again takes the value $V(t_0)$. Then $V(t) < V(t_0)$ for $t \in (t_0, t')$. Hence, we conclude from definitions (6.1.26) and (6.1.28) of the function $\delta(t)$ and from condition (6.1.30) that

$$\delta(t) > \delta(t_0) > 0. \qquad (6.1.32)$$

Thus, using (6.1.29), we obtain the inequality $\dot{V}(t) < 0$ for $t \in (t_0, t')$.

On the other hand, since $V(t_0) = V(t')$, it follows by Lagrange's Theorem that there exists a point $t'' \in (t_0, t')$ such that $\dot{V}(t'') = 0$. This contradiction shows that $V(t) < V(t_0)$ for $t > t_0$.

The inequality just proved implies the validity of the estimate $\delta(t) > \delta(t_0) > 0$ for all $t > t_0$, whence, by relations (6.1.27) and (6.1.29), we obtain inequality (6.1.31). □

By Theorem 4.9, it follows from relations (6.1.13) and (6.1.31) that the value of the function V along the trajectory of system (5.1.2) and (6.1.1) tends to zero, while the trajectory itself approaches the origin.

To estimate the time of motion, let us integrate inequality (6.1.31) over the interval $[t_0, t]$. We obtain

$$t - t_0 \leq \frac{6}{\delta(t_0)} \left[V^{1/2}(t_0) - V^{1/2}(t) \right].$$

Taking into account that $V(t) \to 0$ as t increases, we obtain the following estimate for the time taken by system (5.1.2) and (6.1.1) to move from the initial state $q_0 = q(t_0)$, $\dot{q}_0 = \dot{q}(t_0)$ to the terminal state $q = \dot{q} = 0$:

$$\tau \leq \frac{6}{\delta(q_0, \dot{q}_0)} V^{1/2}(q_0, \dot{q}_0). \tag{6.1.33}$$

We will now verify that the control function $U(q, \dot{q})$ satisfies condition (5.1.4). By the Cauchy inequality, we have

$$\begin{aligned} u^2 &= \alpha^2 |A\dot{q}|^2 + \beta^2 q^2 + 2\alpha\beta \langle A(q)\dot{q}, q \rangle \\ &\leq \frac{4}{3} \left[\alpha^2 |A\dot{q}|^2 + \beta^2 q^2 + \alpha\beta \langle A(q)\dot{q}, q \rangle \right]. \end{aligned} \tag{6.1.34}$$

Since $A(q)$ is a symmetric positive definite matrix satisfying conditions (5.1.5), the matrix $A^{-1}(q)$ is also symmetric and positive definite, and its eigenvalues belong to the interval $[1/M, 1/m]$. Consequently,

$$z^2 M^{-1} \leq \langle A^{-1}(q) z, z \rangle, \quad \forall q, z \in R^n.$$

Substituting $z = A(q)\dot{q}$ into this inequality, we obtain the relations

$$|A\dot{q}|^2 = z^2 \leq M \langle A^{-1} z, z \rangle = 2MT$$

using which we can continue estimate (6.1.34) as follows:

$$u^2 \leq \frac{4}{3} \left[2M\alpha^2 T + \beta^2 q^2 + \alpha\beta \langle A(q)\dot{q}, q \rangle \right].$$

Making use of expressions (6.1.2) and (6.1.3) for the functions α and V, we arrive at the inequality

$$u^2 \leq \frac{8\beta}{3} V = U_0^2$$

from which it follows that constraint (5.1.4) holds along the trajectory of system (5.1.2) and (6.1.1).

We will discuss now some properties of the vector-valued control function $U(q, \dot{q})$. Let us calculate its values in the subspaces $q = 0$ and $\dot{q} = 0$ of the phase space $(q, \dot{q}) \in R^{2n}$. If $q = 0$, then

$$V(0,\dot{q}) = T(0,\dot{q}), \quad \alpha(0,\dot{q}) = \frac{\sqrt{3}U_0}{2\left(2MT(0,\dot{q})\right)^{1/2}}.$$

Consequently,

$$U(0,\dot{q}) = -\frac{\sqrt{3}U_0}{2\left(2MT(0,\dot{q})\right)^{1/2}}A(0)\dot{q} = -\frac{\sqrt{3}U_0}{2\left(2MT(0,\dot{e})\right)^{1/2}}A(0)\dot{e}, \qquad (6.1.35)$$

where e is a unit vector collinear with $\dot{q}$.

If $\dot{q} = 0$, then

$$V(q,0) = \frac{\beta(q,0)}{2}q^2, \quad \beta(q,0) = \frac{3U_0^2}{4\beta(q,0)q^2},$$

whence we get

$$\beta(q,0) = \frac{\sqrt{3}U_0}{2|q|}, \quad U(q,0) = -\frac{\sqrt{3}U_0}{2|q|}q = -\frac{\sqrt{3}U_0}{2}f, \qquad (6.1.36)$$

where f is a unit vector collinear with q.

Thus, the control force vector $U(q,\dot{q})$ is constant in the subspaces $q = 0$ and $\dot{q} = 0$ along any straight line passing through the origin of the phase space, and it points toward the origin.

Remark 6.1. The proposed control law may be formulated without using the functions $\alpha(q,\dot{q})$ and $\beta(q,\dot{q})$. To that end, we transform expression (6.1.1) for $U(q,\dot{q})$ by substituting into it formulas (6.1.2) for α and β. This gives a new definition of the vector-valued control function

$$U(q,\dot{q}) = -\frac{\sqrt{3}U_0}{2\left(2MV(q,\dot{q})\right)^{1/2}}\dot{q} - \frac{3U_0^2}{8V(q,\dot{q})}q,$$

where the function $V(q,\dot{q})$ is implicitly defined by equation (6.1.4).

6.1.4 Sufficient condition for controllability

Formulas (6.1.28) and (6.1.30) imply the following sufficient conditions for the system to reach the prescribed terminal state:

$$U_0 > Q_* + \frac{4\sqrt{2}nD}{\sqrt{3}m}V(q_0,\dot{q}_0), \quad Q_* = 2\sqrt{\frac{10M}{3m}}Q_0. \qquad (6.1.37)$$

This condition relates the maximum admissible value of the control U_0 and perturbations Q_0 with the domain of admissible initial stales of the system. In particular, in a neighbourhood of the terminal state, where the function $V(q,\dot{q})$ is small, condition (6.1.37) may be written in the form

$$U_0 > Q_*.$$

This condition characterizes the excess of the control forces over the perturbations that is sufficient for the control objective to be achieved.

If there are no perturbations, i.e., $Q_0 = 0$, the proposed control law steers system (5.1.2) to the terminal state in finite time from any point of the domain of admissible initial state sthat is given by the inequality

$$V(q,\dot{q}) \leq \sqrt{\frac{3}{2n}\frac{mU_0}{4C}}.$$

Taking into account relationship (6.1.13), we can state that this domain will certainly contain the ellipsoid

$$T(q,\dot{q}) + \left[T^2(q,\dot{q}) + \frac{U_0^2}{2}q^2\right]^{1/2} \leq \frac{mU_0}{\sqrt{6nD}}.$$

Remark 6.2. The control law defined by relations (6.1.1)–(6.1.3) does not depend on the constants Q_0 and D and on the initial state $(q_0, \dot{q}_0)$. It may, therefore, be formally applied, even if inequality (6.1.37) does not hold. Computer simulation of the dynamics of various systems shows that the control law is effective far beyond the limits of the sufficient conditions (6.1.37). This is due to the fact that condition (6.1.37) guarantees a monotone decrease of the function V along the trajectory of system (5.1.2) subjected to control (6.1.1)–(6.1.3). However, the function V may tend to zero in a non-monotone manner, while the trajectories of the system approach the terminal state as before. The simulation results presented below illustrate such behavior of the system.

6.1.5 Computer simulation results

A two-link manipulator

To verify the effectiveness of the proposed control law and to illustrate its operation, numerical simulation are carried out for controlled motions of a two-link manipulator on a fixed base (see Fig. 2.15). It is assumed that the manipulator moves in a horizontal plane, that is, the gravity force is not taken into account. The parameters of the manipulator and the initial and terminal states are taken as in Sect. 5.2.1.

With the parameters thus chosen, the sufficient condition (6.1.37) for steering the mechanical system to the terminal state using the proposed control law may be rewritten as

$$U_0 > 11.8Q_0 + 9.9V(q_0, \dot{q}_0).$$

In the simulation, the perturbation torques were defined as a constant vector-valued function $Q(t) = (0, 250)$. Consequently, the magnitude of the perturbation vector in constraint (5.1.3) does not exceed the quantity $Q_0 = 250$ in norm, and the sufficient condition for the system to be steered to the terminal state becomes

$$U_0 > 2950 + 9.9V(q_0, \dot{q}_0),$$

that is, for the value of U_0 selected in Sect. 5.2.1 ($U_0 = 500$), the condition is not satisfied for any initial values of the phase variables. Nevertheless, the proposed control law overcomes the perturbations and steers the system to the terminal state.

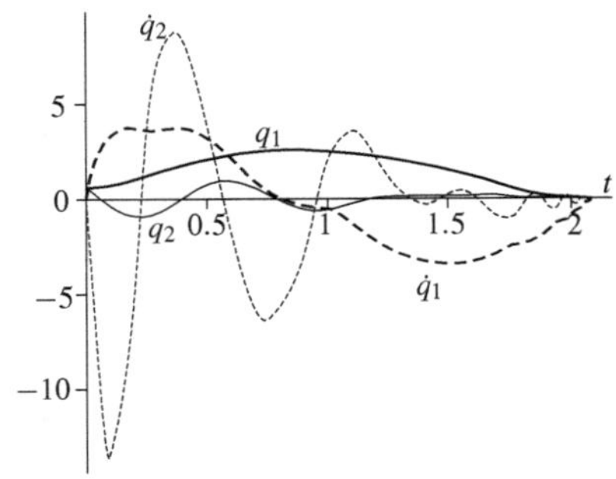

Fig. 6.1 Angular coordinates and velocities

The graphs of the phase variables of the system are presented in Fig. 6.1. The solid curves correspond to the angular coordinates (rad), the dashed curves correspond to the angular velocities (rad/sec), the thick curves describe the motion of the first link, and the thin curves describe that of the second.

The magnitude of the control vector $|U|$ as a function of time is plotted in Fig. 6.2 (the solid curve). The dashed curve in Fig. 6.2 depicts the time history of the function V along the trajectory under consideration. Clearly, V tends to zero and the control satisfies the restriction $|U| \leq 500$. As already remarked, V tends to zero non-monotonically, because the sufficient condition (6.1.37) for the derivative $\dot{V}$ to be negative is not satisfied.

Remark 6.3. In accordance with the algorithm, the control U is described in terms of the function V that is defined implicitly by equation (6.1.4). The quantity $x = \sqrt{V}$ as a root of the fourth-order polynomial equation (6.1.6), and hence also the function V, may be expressed analytically using Cardano's formulas. However, there is no need for an explicit representation of the function V when running the algorithm. From a computational point of view, it is more convenient to find the current value of the function by solving equation (6.1.6) numerically, say by Newton's method. At each step of the integration, it is convenient to take the value of V from the previous step as the initial approximation. Since the function V decreases monotonically along the trajectory and is continuous, the old value is slightly larger than the new (unknown) one.

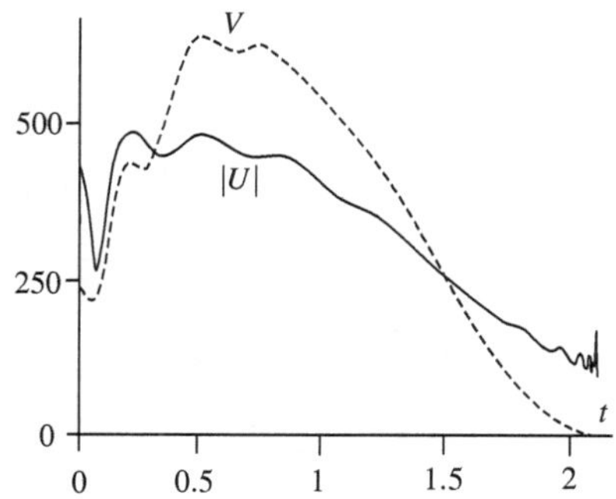

Fig. 6.2 Magnitude of the control U and the function V

A system with one degree of freedom

Let us investigate the limiting possibilities of the above control law by numerical simulation and compare the proposed control with the time-optimal one. To that end, we consider the mechanical system that consists of a particle of unit mass moving along a horizontal straight line. For such a system, in the agreed notation, we have $A(q) = m = M = 1$.

The equations of motion of the point are

$$\ddot{q} = U + Q, \tag{6.1.38}$$

where q is the coordinate of the point on the line. We assume that the control force U and the perturbations Q are subject to the restrictions

$$|U| \leq U_0 = 1, \quad |Q| \leq Q_0 < 1. \tag{6.1.39}$$

Due to the simplicity and low dimensionality of the system, one can express the functions $V(q,\dot{q})$ and $U(q,\dot{q})$, and some other characteristics of the motion by graphical means. The graph of the function $V(q,\dot{q})$, in terms of which the control law (6.1.1)–(6.1.3) is expressed and which is a Lyapunov function for system (6.1.38), is depicted in Fig. 6.3.

Figure 6.4 shows a phase portrait of system (6.1.38) for the case where there are no perturbations, that is, $Q \equiv 0$. The solid curves are the phase trajectories of the motion of the particle and the dashed curves are the level sets of the function V.

By its definition, the function V is symmetrical about the origin, that is, $V(q,\dot{q}) = V(-q,-\dot{q})$. The functions α, β, and U have the same property. Hence, the phase portrait of the unperturbed system is also symmetrical about the point $(0,0)$.

A graph of the control function $U(q,\dot{q})$ is shown in Fig. 6.5. It follows from formulas (6.1.35) and (6.1.36) that, for the values of parameters chosen, the function U satisfies the following relations on the straight lines $q = 0$ and $\dot{q} = 0$:

$$U(0,\dot{q}) = U(q,0) = \frac{\sqrt{3}}{2}.$$

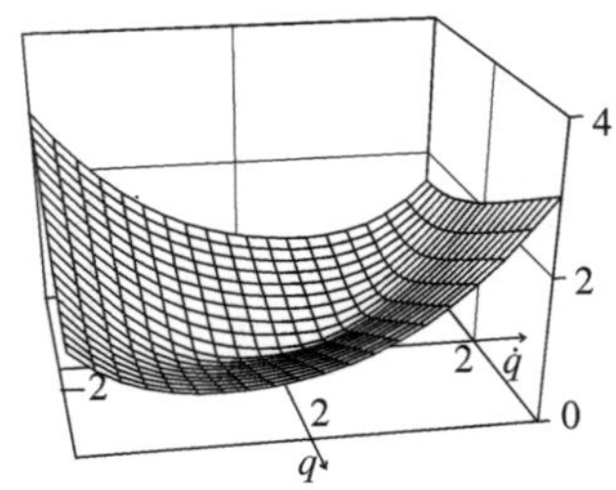

Fig. 6.3 Function $V(q,\dot{q})$

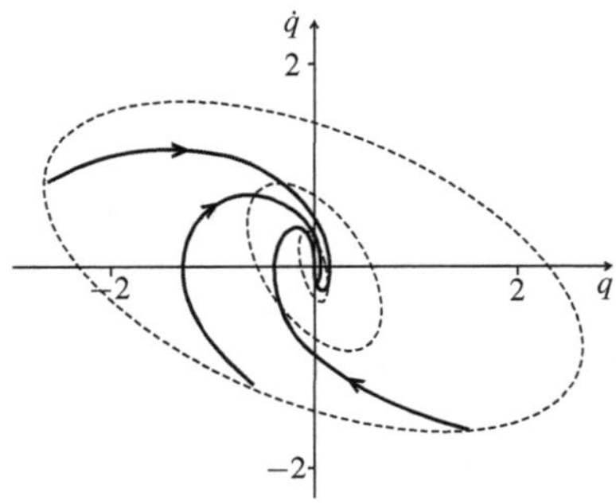

Fig. 6.4 Phase portrait of the system

One can see that, in the domains $q,\dot{q} > 0$ and $q,\dot{q} < 0$, and at values of the velocity $\dot{q}$ of large magnitude, the surface shown in Fig. 6.5 has almost horizontal parts corresponding to values of U close to $\pm\sqrt{3}/2$.

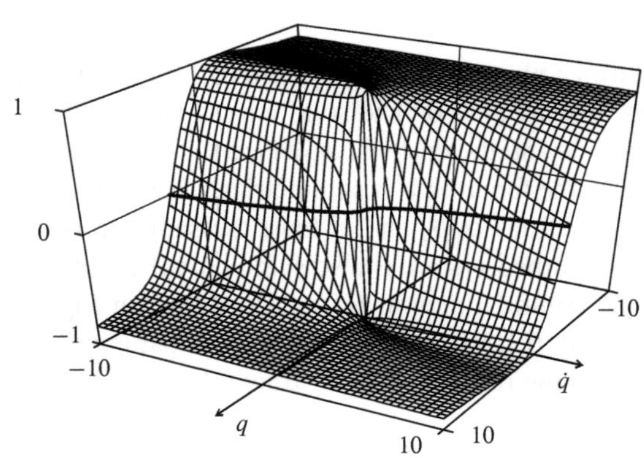

Fig. 6.5 Control function $U(q,\dot{q})$

As might have been expected, in the neighbourhood of the origin the function $U(q,\dot{q})$ has partial derivatives of arbitrarily large magnitude. This is because the control force must cope with any perturbations (including discontinuous ones) satisfying condition (6.1.39) and guarantee the monotone decrease of the function V along the trajectory. Consequently, the closer to the origin, the higher the possible rate of variation of the control force U [at the point $(0,0)$, the function U is undefined].

Let us find the curve on which the control changes its sign. This curve is defined by the equation $U(q,\dot{q}) = 0$. Using (6.1.1) and taking into account that $A(q) = m = M = 1$ for system (6.1.38), we have

$$\alpha(q,\dot{q}) = -\frac{\dot{q}}{q} \tag{6.1.40}$$

on this curve.

Using relations (6.1.2), let us express, in equation (6.1.3), the functions $V(q,\dot{q})$ and $\beta(q,\dot{q})$ in terms of $\alpha(q,\dot{q})$ and substitute expression (6.1.40) into the resulting equality. Taking into account that $U_0 = 1$, we obtain the equation $3q^2 = 4\dot{q}^4$. The function $\alpha(q,\dot{q})$ is positive by definition, hence, by (6.1.40), the coordinate q and velocity $\dot{q}$ have different signs at each point of the desired curve. Therefore, the curve itself is defined by the relation

$$q = \begin{cases} -\dfrac{2\sqrt{3}}{3}\dot{q}^2, \text{ if } \dot{q} > 0; \\ \dfrac{2\sqrt{3}}{3}\dot{q}^2, \text{ if } \dot{q} < 0. \end{cases} \tag{6.1.41}$$

It consists of two branches of parabolas that are symmetrical about the origin (the thick curve in Fig. 6.5).

Comparison with the time-optimal control

Let us compare the control law proposed with the control that minimizes the time of the motion. For system (6.1.38) with $Q \equiv 0$, the time-optimal control has the form

$$U_{\text{opt}} = \begin{cases} -1, \text{ if } \dot{q} > 0 \text{ and } q \geq -\dfrac{1}{2}\dot{q}^2; \\ -1, \text{ if } \dot{q} < 0 \text{ and } q > \dfrac{1}{2}\dot{q}^2; \\ 1, \text{ otherwise} \end{cases} \tag{6.1.42}$$

[see formulas (1.4.12) and (1.4.13) in Sect. 1.4].

The function $U(q,\dot{q})$, whose graph is shown in Fig. 6.5, and the time-optimal control function U_{opt} given by (6.1.42) are readily seen to be qualitatively similar. The switching curve for the time-optimal control U_{opt} and its analogue (6.1.41) for the proposed control U are each the union of two branches of parabolas in which the coefficients at $\dot{q}^2$ are $1/2$ and $2\sqrt{3}/3$, respectively.

If perturbations appear in system (6.1.38), that is, the assumpton $Q \equiv 0$ is not fulfilled, then the time-optimal control law becomes (see Sect. 2.2.1)

$$U'_{\text{opt}} = \begin{cases} -1, \text{ if } \dot{q} > 0 \text{ and } q \geq -\dfrac{\dot{q}^2}{2(1-Q_0)}; \\ -1, \text{ if } \dot{q} < 0 \text{ and } q > \dfrac{\dot{q}^2}{2(1-Q_0)}; \\ 1, \text{ otherwise.} \end{cases} \tag{6.1.43}$$

In that case, the switching curve is the union of two branches of parabolas $q = \pm\dot{q}^2/[2(1-Q_0)]$. If $Q_0 = 1-\sqrt{3}/4$, this switching curve coincides with the curve $U(q,\dot{q}) = 0$ given by formulas (6.1.41).

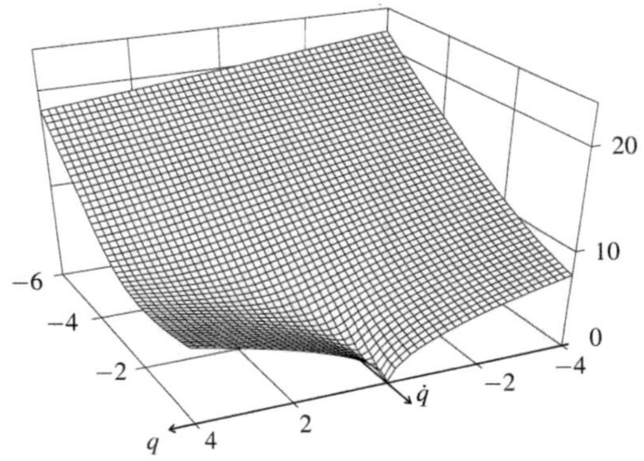

Fig. 6.6 Time of motion $\tau(q,\dot{q})$

Figure 6.6 shows the graph of the function $\tau(q,\dot{q})$ for system (6.1.38) subjected to the control prescribed by law (6.1.1)–(6.1.3) when there are no perturbations, that is, when $Q \equiv 0$. By definition, the value of this function at each point of the phase space equals the time it takes for the system to move from that point to the terminal state. For comparison, Fig. 6.7 shows the graph of the function equal at each point to the minimum possible time of motion from that point to the terminal state. It can be seen that the time of motion of the system controlled using the proposed algorithm is approximately 1.5 times greater than the minimum time.

In order to determine the limiting possibilities of the control algorithm proposed, the motion of system (6.1.38) is simulated numerically for the case of perturbations specified by

$$Q(q,\dot{q}) = -Q_0 U_{\text{opt}},$$

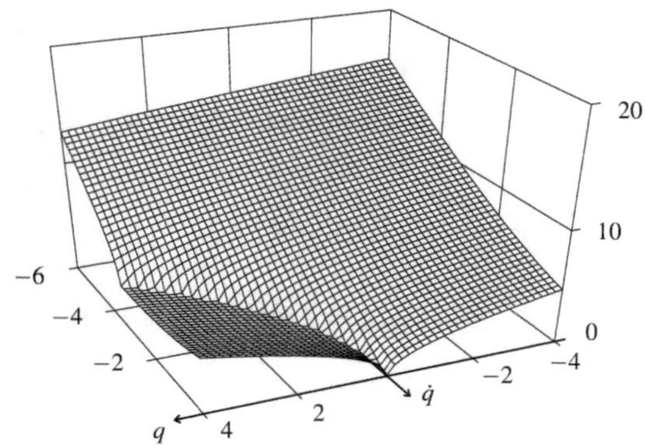

Fig. 6.7 Optimal time of motion

where U_{opt} is defined by formula (6.1.42); this is done for different values of Q_0. It turns out that the limiting value of the perturbations Q_0 for which the system is brought to the terminal state by algorithm (6.1.1)–(6.1.3) is approximately $\sqrt{3}/2$, i.e., it is equal to the value of the function U on the straight lines $q = 0$ and $\dot{q} = 0$. We recall (see Sect. 1.4) that, for the optimal control law, the system may be brought to the terminal time if and only if $Q_0 < 1$.

Thus, the approach proposed above enables one to construct algorithms that yield bounded controls as smooth functions of the phase variables. These algorithms can be used to control any scleronomic mechanical system and enable it to be steered to the given terminal state in finite time. The results of the numerical simulation of controlled motions of a point mass along a horizontal straight line demonstrate that the proposed control law is qualitatively similar to the time-optimal control law.

6.2 Control of a scleronomic system with an unknown matrix of inertia

6.2.1 Problem statement

Consider now a scleronomic mechanical system (5.1.2) under the assumption that the matrix of the kinetic energy of the system is represented in the form

$$A(q) = A_0(q) + A_1(q), \tag{6.2.1}$$

where $A_0(q)$ and $A_1(q)$ are symmetric continuously differentiable matrices, and $A_0(q)$ is known and positive definite, whereas $A_1(q)$ is unknown.

We put

$$T_0(q,\dot{q}) = \frac{1}{2}\langle A_0(q)\dot{q},\dot{q}\rangle, \quad T_1(q,\dot{q}) = \frac{1}{2}\langle A_1(q)\dot{q},\dot{q}\rangle.$$

Then,

$$T(q,\dot{q}) = T_0(q,\dot{q}) + T_1(q,\dot{q}). \tag{6.2.2}$$

As above, we assume that the eigenvalues of matrices $A(q)$ and $A_0(q)$ belong to the interval $[m,M]$, $0 < m \leq M$, for any q, i.e.,

$$\begin{aligned} &mz^2 \leq \langle A(q)z,z\rangle \leq Mz^2, \\ &mz^2 \leq \langle A_0(q)z,z\rangle \leq Mz^2, \quad \forall q,z \in R^n, \end{aligned} \tag{6.2.3}$$

and the matrix A_1 and the partial derivatives of the matrices A_1 and A are uniformly bounded in norm, i.e.,

$$\begin{aligned} &\|A_1(q)\| \leq M_1, \quad \|\frac{\partial A_1}{\partial q_i}(q)\| \leq C_1, \\ &\|\frac{\partial A}{\partial q_i}(q)\| \leq C, \quad M_1, C_1, C > 0, \quad i = 1,\dots,n. \end{aligned} \tag{6.2.4}$$

Problem 6.2. Construct a control $U(q,\dot{q})$ as a continuous vector-valued function of the phase variables $q,\dot{q} \in R^{2n}$, $q^2+\dot{q}^2 > 0$, that satisfies condition (5.1.4), and specify a domain of admissible initial states such that any trajectory of system (5.1.2) starting in that domain will reach the prescribed terminal state $(0,0)$ in finite time, whatever the matrix A_1 and the perturbations Q satisfying conditions (6.2.4) and (5.1.3) be.

We apply the approach proposed above for the control of a scleronomic mechanical system with a given matrix of the kinetic energy. We define the control as follows:

$$U(q,\dot{q}) = -\alpha(q,\dot{q})A_0(q)\dot{q} - \beta(q,\dot{q})q, \quad q^2+\dot{q}^2 \neq 0, \tag{6.2.5}$$

$$\alpha(q,\dot{q}) = \sqrt{\frac{\beta(q,\dot{q})}{M}}, \quad \beta(q,\dot{q}) = \frac{3U_0^2}{8V(q,\dot{q})}, \tag{6.2.6}$$

$$V(q,\dot{q}) = T_0(q,\dot{q}) + \frac{1}{2}\beta(q,\dot{q})q^2 + \frac{1}{2}\alpha(q,\dot{q})\langle A_0(q)\dot{q},q\rangle. \tag{6.2.7}$$

By contrast to control (6.1.1)–(6.1.3), the matrix A_0 that occurs in (6.2.5) and (6.2.7) does not coincide with the matrix of inertia A of the system; it represents "the known part" of A only.

Relationships (6.2.5)–(6.2.7) define the functions $\alpha(q,\dot{q}), \beta(q,\dot{q})$, and $V(q,\dot{q})$ implicitly. By Theorem 6.1, in the domain $q^2+\dot{q}^2 > 0$, there exist continuously differentiable functions $\alpha(q,\dot{q}), \beta(q,\dot{q})$, and $V(q,\dot{q})$ satisfying these relationships. It has already been proved also that the function V satisfies inequalities (6.1.13).

Let us show that $V(q,\dot{q})$ is the Lyapunov function for system (5.1.2). The derivative $\dot{V}(q,\dot{q})$ along the trajectories of this system has the form

$$\dot{V} = \dot{T}_0 + \beta\langle q,\dot{q}\rangle + \alpha T_0 + \frac{\alpha}{2}\langle\frac{d}{dt}(A_0\dot{q}),q\rangle - \frac{\beta}{2V}q^2\dot{V} - \frac{\alpha}{4V}\langle A_0\dot{q},q\rangle\dot{V}. \tag{6.2.8}$$

We evaluate the terms in expression (6.2.8) separately. We set

$$L(T_1) = \frac{d}{dt}\frac{\partial T_1}{\partial \dot{q}} - \frac{\partial T_1}{\partial q}. \tag{6.2.9}$$

Taking (6.2.2) into account, we rewrite (5.1.2) in the form

$$\frac{d}{dt}\frac{\partial T_0}{\partial \dot{q}} - \frac{\partial T_0}{\partial q} = u + Q - L(T_1). \tag{6.2.10}$$

By the theorem on the variation of the kinetic energy of a scleronomic Lagrangian system, and by the definition of the vector-valued function $U(q,\dot{q})$, we have

$$\dot{T}_0 = \langle u + Q - L(T_1),\dot{q}\rangle = -2\alpha T_0 - \beta\langle q,\dot{q}\rangle + \langle Q - L(T_1),\dot{q}\rangle. \tag{6.2.11}$$

By virtue of equations (6.2.10), we obtain

$$\frac{d}{dt}(A_0\dot{q}) = \frac{d}{dt}\frac{\partial T_0}{\partial \dot{q}} = \frac{\partial T_0}{\partial q} + u + Q - L(T_1). \tag{6.2.12}$$

We substitute relations (6.2.11) and (6.2.12) into expression (6.2.8) for the derivative $\dot{V}$ and use the definition (6.2.5) of the control vector-valued function U. After some reducing, we obtain the following expression for the derivative of the function V along the trajectories of system (5.1.2) and (6.2.5):

$$\begin{aligned}\dot{V} = -\alpha\left(T_0 + \frac{\alpha}{2}\langle A_0\dot{q},q\rangle + \frac{\beta}{2}q^2\right) - \left(\beta q^2 + \frac{\alpha}{2}\langle A_0\dot{q},q\rangle\right)\frac{\dot{V}}{2V}\\ + \frac{\alpha}{2}\langle\frac{\partial T_0}{\partial q},q\rangle + \langle Q - L(T_1),\frac{\alpha}{2}q + \dot{q}\rangle.\end{aligned} \tag{6.2.13}$$

Definitions (6.2.6) and (6.2.7) of functions α and V imply

$$\alpha\left(T_0 + \frac{\alpha}{2}\langle A_0\dot{q},q\rangle + \frac{\beta}{2}q^2\right) = \alpha V = \frac{\sqrt{3}U_0}{2\sqrt{2M}}V^{1/2}. \tag{6.2.14}$$

By virtue (5.1.2), we have the inequalities

$$\dot{A}\dot{q} + A\ddot{q} = \frac{d}{dt}(A\dot{q}) = \frac{d}{dt}\frac{\partial T}{\partial \dot{q}} = \frac{\partial T}{\partial q} + u + Q$$

that imply

$$\ddot{q} = A^{-1}\left(\frac{\partial T}{\partial q} + U + Q - \dot{A}\dot{q}\right). \tag{6.2.15}$$

We write down the expression for $L(T_1)$, using (6.2.10) and equality (6.2.15). Then,

$$\begin{aligned} L(T_1) &= \frac{d}{dt}A_1\dot{q} - \frac{\partial T_1}{\partial q} = \dot{A}_1\dot{q} + A_1\ddot{q} - \frac{\partial T_1}{\partial q} \\ &= \dot{A}_1\dot{q} + A_1 A^{-1}\left(\frac{\partial T}{\partial q} + U + Q - \dot{A}\dot{q}\right) - \frac{\partial T_1}{\partial q}. \end{aligned} \tag{6.2.16}$$

We introduce the function

$$B(q,\dot{q}) = 1 + \frac{\beta(q,\dot{q})}{2V(q,\dot{q})}q^2 + \frac{\alpha(q,\dot{q})}{4V(q,\dot{q})}\langle A_0(q)\dot{q}, q\rangle. \tag{6.2.17}$$

Substituting relations (6.2.14), (6.2.16), and (6.2.17) into (6.2.13), we write down the final expression for the derivative of the function $V(q,\dot{q})$ along the trajectories of system (5.1.2) and (6.2.5):

$$\begin{aligned} B\dot{V} &= -\frac{\sqrt{3}U_0}{2\sqrt{2M}}V^{1/2} + \frac{\alpha}{2}\langle\frac{\partial T_0}{\partial q}, q\rangle\langle Q - \dot{A}_1\dot{q} \\ &- A_1 A^{-1}\left(\frac{\partial T}{\partial q} + U + Q - \dot{A}\dot{q}\right) + \frac{\partial T_1}{\partial q}, \frac{\alpha}{2}q + \dot{q}\rangle. \end{aligned} \tag{6.2.18}$$

Let us estimate the terms in the right-hand side of equality (6.2.18). From relations (6.2.2), (6.2.4), (6.2.6), (6.1.21), and (6.1.23), it follows that

$$\begin{aligned} &|\frac{\alpha}{2}\langle\frac{\partial T_0}{\partial q} + \frac{\partial T_1}{\partial q}, q\rangle| = |\frac{\alpha}{2}\langle\frac{\partial T}{\partial q}, q\rangle| = |\frac{\alpha}{4}\sum_{i=1}^{n}\langle\frac{\partial A}{\partial q_i}\dot{q}, \dot{q}\rangle q_i| \\ &\le \frac{\alpha}{4}C\dot{q}^2\sum_{i=1}^{n}|q_i| \le \frac{\alpha}{4}\sqrt{n}C\dot{q}^2|q| \le \frac{2\sqrt{n}C}{m\sqrt{M}}V^{3/2}. \end{aligned} \tag{6.2.19}$$

Inequalities (6.1.21), (6.1.23), and (6.2.4) imply

$$\begin{aligned} &|\langle\frac{\partial T_1}{\partial q}, \dot{q}\rangle| = |\frac{1}{2}\sum_{i=1}^{n}\langle\frac{\partial A_1}{\partial q_i}\dot{q}, \dot{q}\rangle\dot{q}_i| \\ &\le \frac{1}{2}\sqrt{n}C_1\dot{q}^2|\dot{q}| \le \frac{4\sqrt{n}C_1}{m\sqrt{m}}V^{3/2} \le \frac{2\sqrt{5}nC_1}{m\sqrt{m}}V^{3/2}. \end{aligned} \tag{6.2.20}$$

By virtue of (6.2.3), the inequality

$$\|A^{-1}\| \le \frac{1}{m}$$

holds. This inequality, conditions (5.1.3), (6.2.4), and estimate (6.1.19) imply

$$|\langle Q - A_1 A^{-1}Q, \dot{q} + \frac{\alpha}{2}q\rangle| \le Q_0\left(1 + \frac{M_1}{m}\right)\sqrt{\frac{5}{m}}V^{1/2}. \tag{6.2.21}$$

Since the control function $U(q,\dot{q})$ satisfies constraint (5.1.4), we have

$$|\langle A_1 A^{-1} u, \dot{q} + \frac{\alpha}{2} q\rangle| \le \frac{\sqrt{5} M_1 U_0}{m\sqrt{m}} V^{1/2}. \tag{6.2.22}$$

Similarly, we obtain

$$\begin{aligned} |\langle A_1 A^{-1} \frac{\partial T}{\partial q}, \dot{q} + \frac{\alpha}{2} q\rangle| &\le |\frac{M_1}{2m} \sum_{i=1}^{n} \langle \frac{\partial A}{\partial q_i} \dot{q}, \dot{q}\rangle \left(\dot{q}_i + \frac{\alpha}{2} q_i\right)| \\ &\le \frac{M_1}{2m} \sqrt{n} C \dot{q}^2 |\dot{q} + \frac{\alpha}{2} q| \le \frac{2\sqrt{5n} C M_1}{m^2 \sqrt{m}} V^{3/2}, \end{aligned} \tag{6.2.23}$$

$$\begin{aligned} |\langle \dot{A}_1 \dot{q}, \dot{q} + \frac{\alpha}{2} q\rangle| &\le |\langle \left(\sum_{i=1}^{n} \frac{\partial A_1}{\partial q_i} \dot{q}_i\right) \dot{q}, \dot{q} + \frac{\alpha}{2} q\rangle| \\ &\le \sqrt{n} C_1 \dot{q}^2 |\dot{q} + \frac{\alpha}{2} q| \le \frac{4\sqrt{5n} C_1}{m\sqrt{m}} V^{3/2}, \end{aligned} \tag{6.2.24}$$

$$|\langle A_1 A^{-1} \dot{A} \dot{q}, \dot{q} + \frac{\alpha}{2} q\rangle| \le \frac{4\sqrt{5n} C M_1}{m^2 \sqrt{m}} V^{3/2}. \tag{6.2.25}$$

By substituting (6.2.19)–(6.2.25) into expression (6.2.18) for the derivative $\dot{V}$, we obtain that

$$\dot{V}(q,\dot{q}) \le -\frac{\delta_1 U_0 - \delta_2 Q_0 - \delta_3 V(q,\dot{q})}{B(q,\dot{q})} V^{1/2}(q,\dot{q}), \tag{6.2.26}$$

where

$$\begin{aligned} \delta_1 = \frac{\sqrt{3}}{2\sqrt{2M}} - \frac{\sqrt{5} M_1}{m\sqrt{m}}, \quad \delta_2 = \left(1 + \frac{M_1}{m}\right) \sqrt{\frac{5}{m}}, \\ \delta_3 = \frac{2\sqrt{nC}}{m\sqrt{M}} + \frac{6\sqrt{5n}}{m\sqrt{m}} \left(C_1 + \frac{C M_1}{m}\right), \end{aligned} \tag{6.2.27}$$

Hence, the function $B(q,\dot{q})$ from (6.2.17) satisfies inequality (6.1.27).

Using reasoning similar to the above, we proved the following theorem.

Theorem 6.3. *Suppose that at the initial instant of time t_0, system (5.1.2), (6.2.5) is in the state $(q_0, \dot{q}_0)$ and the condition*

$$\delta_1 U_0 - \delta_2 Q_0 - \delta_3 V(q_0, \dot{q}_0) > 0 \tag{6.2.28}$$

holds [the constants δ_1, δ_2, and δ_3 are defined in (6.2.27)]. Then, the derivative of the function V along the trajectory, starting at $(q_0, \dot{q}_0)$, satisfies the inequality

$$\dot{V} \le -\frac{\delta_1 U_0 - \delta_2 Q_0 - \delta_3 V(q_0, \dot{q}_0)}{3} V^{1/2}. \tag{6.2.29}$$

By virtue of (6.2.17) and the assertion of Theorem 6.3, we can conclude that the value of the function V along the trajectory under consideration approaches zero, and the trajectory itself approaches the origin. Integrating (6.2.29) on the interval $[t_0, t]$, we obtain the following estimate from above for the time of motion $\tau(q_0, \dot{q}_0)$ of the system from the initial state $(q_0, \dot{q}_0)$ to the terminal state $q_0 = \dot{q}_0 = 0$:

$$\tau(q_0, \dot{q}_0) \leq \frac{6V^{1/2}(t_0, q_0, \dot{q}_0)}{\delta_1 U_0 - \delta_2 Q_0 - \delta_3 V(q_0, \dot{q}_0)}.$$

Relations (6.2.27) and the condition of Theorem 6.3 imply the following sufficient controllability condition for steering the system to the terminal state by the proposed control law

$$U_0 > \frac{\delta_2}{\delta_1} Q_0 + \frac{\delta_3}{\delta_1} V(q_0, \dot{q}_0). \tag{6.2.30}$$

Inequality (6.2.30) connects the maximum admissible values of control U_0 and perturbations Q_0, the size of the domain of admissible initial states, and also the estimate of the error, with which the inertia matrix of the system is known (the constraints on the matrix A_1 and its partial derivatives are involved in the expressions for δ_i). This inequality characterizes the superiority of the control forces over the perturbations that is sufficient for reaching the target of control, if the value of the function V at the initial state and the error of knowing the inertia matrix of the system are not large enough.

Remark 6.4. As in the case of a precisely known matrix of the kinetic energy, the control law, defined by relations (6.2.5)–(6.2.7), depends neither on the initial state $(q_0, \dot{q}_0)$ nor on the constants Q_0 and δ_i, $i = 1, 2, 3$, that are used in the formulation of the sufficient controllability conditions (6.2.30). Therefore, this law may also be applied formally in the case where inequality (6.2.30) does not hold. Computer simulation of the dynamics of various systems has shown that the proposed control law is effective far beyond the sufficient condition stated (see Remark 6.3).

6.2.2 Computer simulation of the motion of a two-link manipulator

We present the results of the numerical simulation of controlled motions of a two-link manipulator on a fixed base (see Fig. 2.15). We assume now that the gripper of the manipulator holds a load of unknown mass. Such a mechanical system is governed by Lagrange's equations (5.1.2) with the kinetic energy given by (5.1.1), where the matrix $A(q)$ has form (4.1.5).

The computations were performed for the following values of parameters: $m_1 = 20$ kg and $m_2 = 10$ kg are the masses of the links; $l_1 = 0.8$ m and $l_2 = 0.5$ m are the lengths of the links, respectively. In what follows, the mass m_0 of the load of the manipulator is taken equal to 3 kg or 5 kg. The maximum admissible magnitude of the vector of control torques is chosen equal to $U_0 = 500$ N $\cdot$ m. When modelling, we specify the disturbances by the constant vector-function $Q = (0; 30)\ \mathrm{N \cdot m}$. The manipulator moves from the initial state

$$q_{01} = 0.5\,\mathrm{rad},\ q_{02} = 1\,\mathrm{rad},\ \dot{q}_{01} = \dot{q}_{02} = 0\,\mathrm{rad/s}$$

to the "stretched arm" position

$$q_1 = q_2 = \dot{q}_1 = \dot{q}_2 = 0.$$

As the computations have shown, for the given set of parameters, sufficient conditions (6.2.30) do not hold for all the initial states. Nevertheless, the control proposed steers the system to the terminal state.

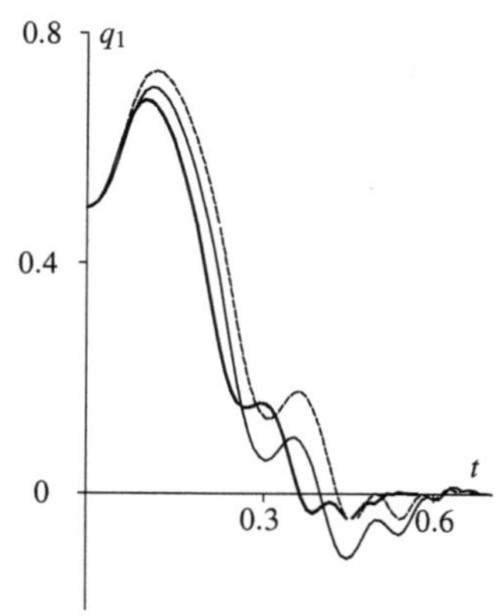

Fig. 6.8 Angular coordinate of the first link

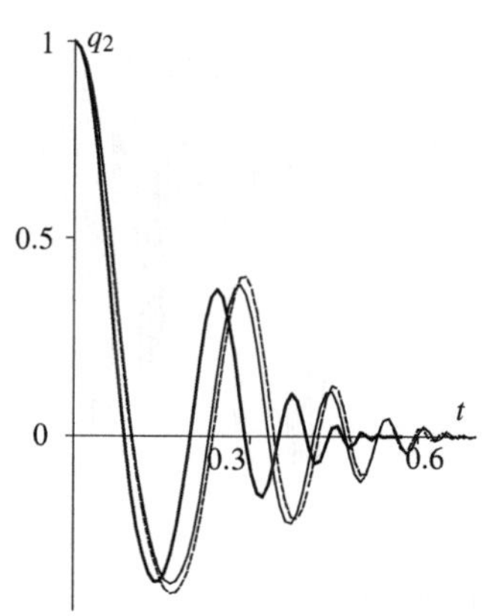

Fig. 6.9 Angular coordinate of the second link

In Figs. 6.8 and 6.9, the graphs of the time histories of the angular coordinates of the first and second links are shown (in radians). Figure 6.10 illustrates the behavior of the function V along the trajectories, and the magnitude of the control torque vector U as a function of time is shown in Fig. 6.11. In all figures, the thin solid lines correspond to the motion of the manipulator with a load of mass $m_0 = 5$ kg in its gripper subjected to the control law constructed under the assumption that the mass of the load is $m_0 = 3$ kg. The results of the modelling of the dynamics of the manipulator for the case where the mass of the load is known and equal to $m_0 = 3$ kg are presented for comparison (bold solid lines). The dashed lines in Figs. 6.8–6.11 are related to the contents of Sect. 6.3.2.

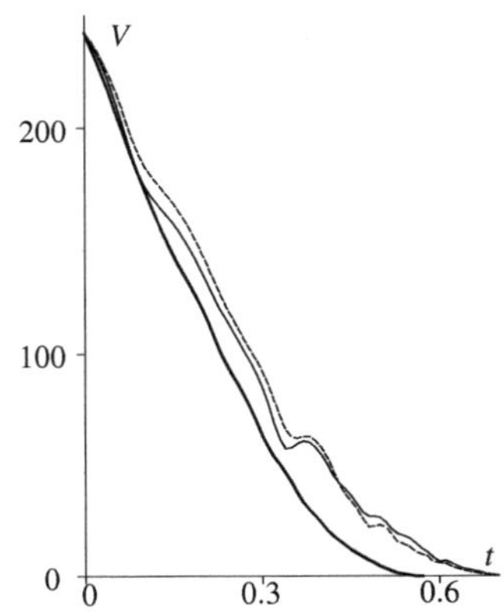

Fig. 6.10 Function V

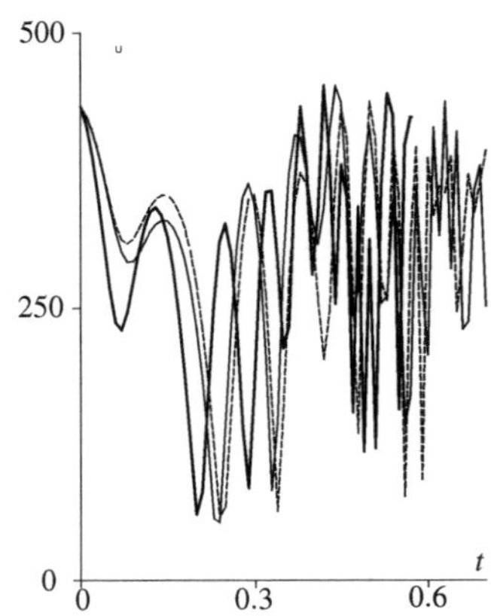

Fig. 6.11 Magnitude of the control torque vector U

From Fig. 6.10, we see that the function V approaches zero along the trajectories in a non-monotone way, which is explained by the fact that sufficient condition

(6.2.30) is not satisfied. Despite the fact that the derivative of the function V takes not only negative values, and that the function itself has intervals of increase, it vanishes after passing a finite time interval, and the system comes to the terminal state in finite time. In this case, as Fig. 6.11 shows, the control U satisfies constraint (5.1.4) everywhere.

6.3 Control of rheonomic systems under uncertainty

6.3.1 Problem statement

We have shown that the approach proposed can be applied to a scleronomic system under the assumption that the inertia matrix of the system is unknown. Below, we extend this approach to rheonomic systems in which the elements of the inertia matrix are known only with some error, whereas the other coefficients of the kinetic energy polynomial are not known at all.

Consider a generic rheonomic mechanical system with the kinetic energy given by expression (4.1.3). Suppose that the symmetric positive definite continuously differentiable matrix $A(t,q)$ can be represented in the form

$$A(t,q) = A_0(t,q) + A_1(t,q), \tag{6.3.1}$$

where $A_0(t,q)$ and $A_1(t,q)$ are also continuously differentiable matrices, the symmetric matrix $A_0(t,q)$ is positive definite and known, and $A_1(t,q)$ is an unknown symmetric matrix. Denote

$$T_0(t,q,\dot{q}) = \frac{1}{2}\langle A_0(t,q)\dot{q},\dot{q}\rangle, \quad T_1(t,q,\dot{q}) = \frac{1}{2}\langle A_1(t,q)\dot{q},\dot{q}\rangle.$$

Then, in accordance with (4.1.3), we have

$$T(t,q,\dot{q}) = T_0(t,q,\dot{q}) + T_1(t,q,\dot{q}) + \langle a_1(t,q),\dot{q}\rangle + a_0(t,q). \tag{6.3.2}$$

As in the case of a scleronomic system (see Sect. 6.2.1), we suppose that the eigenvalues of the matrices $A(t,q)$ and $A_0(t,q)$ belong to the interval $[m,M]$, $0 < m \le M$, for all t,q; i.e., the inequalities

$$\begin{aligned} &mz^2 \le \langle A(t,q)z,z\rangle \le Mz^2, \\ &mz^2 \le \langle A_0(t,q)z,z\rangle \le Mz^2, \quad \forall t,\ \forall q,z \in R^n, \end{aligned} \tag{6.3.3}$$

hold. The matrix A_1 and the partial derivatives of the matrices A_0, A_1, and A are assumed to be bounded by the norm uniformly for all t and q:

$$\|A_1\| \le M_1, \quad \|\frac{\partial A_1}{\partial q_i}\| \le C_1, \quad \|\frac{\partial A}{\partial q_i}\| \le C, \quad i = 1,\dots,n,$$
$$\|\frac{\partial A_0}{\partial t}\| \le D_0, \quad \|\frac{\partial A_1}{\partial t}\| \le D_1, \quad M_1, C, C_1, D_0, D_1 > 0. \tag{6.3.4}$$

Moreover, it is assumed that the continuously differentiable function $a_0(t,q)$ and the vector-valued function $a_1(t,q)$ are known and satisfy the conditions

$$|\frac{\partial a_0}{\partial q} - \frac{\partial a_1}{\partial t}| \le D_2, \quad \|\left(\frac{\partial a_1}{\partial q}\right)^{\top} - \frac{\partial a_1}{\partial q}\| \le D_3, \quad D_2, D_3 > 0. \tag{6.3.5}$$

The dynamics of the system is governed by Lagrange's equations (5.1.2), in which the kinetic energy $T(t,q,\dot{q})$ is given by the expression (4.1.3). In this case, as above, constraints (5.1.3) and (5.1.4) are imposed on unknown n-dimensional force vectors $Q(t,q,\dot{q})$ and the control forces $U(t,q,\dot{q})$, respectively.

Consider the problem of the synthesis of the control that ensures steering the system to a given terminal state $q = \bar{q}$ and $\dot{q} = 0$ in finite time. Without loss of generality, we will assume that the terminal state coincides with the origin of phase space $(q,\dot{q})$.

Problem 6.3. Construct the control U as a continuous vector-valued function of time t and the phase variables q and $\dot{q}$ satisfying condition (5.1.4), and specify the domain of admissible initial states such that any trajectory of system (5.1.2) starting from this domain will come to the origin of the phase space $q_0 = \dot{q}_0 = 0$ in finite time, whatever the matrix A_1, the vector-valued function a_1, the function a_0, and the perturbation Q satisfying the conditions stated above be.

Consider the domain G of the extended phase space

$$G = \{(t,q,\dot{q}) \in R^{2n+1} : q^2 + \dot{q}^2 \neq 0\}$$

and, in G, define the control in the form

$$U(t,q,\dot{q}) = -\alpha(t,q,\dot{q})A_0(t,q)\dot{q} - \beta(t,q,\dot{q})q, \tag{6.3.6}$$

where

$$\alpha(t,q,\dot{q}) = \sqrt{\frac{\beta(t,q,\dot{q})}{M}}, \quad \beta(t,q,\dot{q}) = \frac{3U_0^2}{8V(t,q,\dot{q})}, \tag{6.3.7}$$

$$V(t,q,\dot{q}) = T_0(t,q,\dot{q}) + \frac{1}{2}\beta(t,q,\dot{q})q^2 + \frac{1}{2}\alpha(t,q,\dot{q})\langle A_0(t,q)\dot{q}, q\rangle. \tag{6.3.8}$$

Relations (6.3.7) and (6.3.8) define the functions $\alpha(t,q,\dot{q}), \beta(t,q,\dot{q})$, and $V(t,q,\dot{q})$ in an implicit form.

By the reasoning, similarly to that used in Theorem 6.1, we can prove that, in the domain G, there exist continuously differentiable positive functions $\alpha(t,q,\dot{q})$, $\beta(t,q,\dot{q})$, and $V(t,q,\dot{q})$ satisfying relations (6.3.7) and (6.3.8). The fact that the functions α, β, and V depend on time now virtually does not influence the proof.

It is easy to show also that the function $U(t,q,\dot q)$ meets constraint (5.1.4), and the function $V(t,q,\dot q)$ satisfies the inequalities

$$V_-(t,q,\dot q) \le V(t,q,\dot q) \le 3V_-(t,q,\dot q), \tag{6.3.9}$$

where

$$V_-(t,q,\dot q) = \frac{1}{4}\left(2T_0(t,q,\dot q) + \beta(t,q,\dot q)q^2\right). \tag{6.3.10}$$

As in the case of the scleronomic system considered above, these statements imply the following two-sided estimates for the function $V(t,q,\dot q)$:

$$\begin{aligned} &\frac{1}{8}\left[m\dot q^2 + \left(m^2\dot q^4 + 6U_0^2 q^2\right)^{1/2}\right] \le V(t,q,\dot q) \\ &\le \frac{3}{8}\left[M\dot q^2 + \left(M^2\dot q^4 + 2U_0^2 q^2\right)^{1/2}\right]. \end{aligned} \tag{6.3.11}$$

We introduce the notation

$$\begin{aligned} \Omega(t,q) &= \left(\frac{\partial a_1}{\partial q}\right)^{\mathsf T}(t,q) - \frac{\partial a_1}{\partial q}(t,q), \\ \omega(t,q,\dot q) &= \frac{\partial a_0}{\partial q}(t,q) - \frac{\partial a_1}{\partial t}(t,q) + Q(t,q,\dot q), \\ B(t,q,\dot q) &= 1 + \frac{\beta(t,q,\dot q)}{2V(t,q,\dot q)}q^2 + \frac{\alpha(t,q,\dot q)}{4V(t,q,\dot q)}\langle A_0(t,q)\dot q, q\rangle. \end{aligned} \tag{6.3.12}$$

Taking (6.3.12) into account, the expression for the derivative of the function $V(t,q,\dot q)$ along the trajectories of system (5.1.2) with kinetic energy (4.1.3) may be written in the form

$$\begin{aligned} B\dot V = &-\alpha V + \frac{\alpha}{2}\langle\frac{\partial T_0}{\partial q} - \Omega\dot q, q\rangle + \langle\omega, \frac{\alpha}{2}q + \dot q\rangle - \frac{\partial T_0}{\partial t} \\ &-\langle\dot A_1\dot q - \frac{\partial T_1}{\partial q} + A_1A^{-1}\left(\frac{\partial T}{\partial q} + U + \omega - \Omega\dot q - \dot A\dot q\right), \frac{\alpha}{2}q + \dot q\rangle. \end{aligned} \tag{6.3.13}$$

Let us estimate the terms in (6.3.13) separately. For this purpose, note that relations (6.2.14), (6.1.19)–(6.1.23), (6.2.19), (6.2.20), and (6.2.22)–(6.2.25) remain valid, though some functions in them now depend on time explicitly.

By virtue of inequalities (6.1.23), the second condition of (6.3.5), and notation (6.3.12), we have

$$\left|\frac{\alpha}{2}\langle\Omega\dot q, q\rangle\right| \le \frac{\alpha}{2}D_3|\dot q||q| \le \frac{2D_3}{\sqrt{mM}}V. \tag{6.3.14}$$

Inequality (6.1.19), constraint (5.1.3), the first condition of (6.3.5), and notation (6.3.12) yield

$$\left|\langle\omega, \dot q + \frac{\alpha}{2}q\rangle\right| \le (D_2 + S_0)\sqrt{\frac{5}{m}}V^{1/2}. \tag{6.3.15}$$

Conditions (6.3.4) and (6.1.23) imply

$$\left|\frac{\partial T_0}{\partial t}\right| \leq \frac{D_0}{2}\dot{q}^2 \leq \frac{2D_0}{m}V. \tag{6.3.16}$$

From relations (6.1.19), (6.1.23), (6.3.12), and (6.3.3)–(6.3.5), we obtain

$$|\langle A_1A^{-1}\Omega\dot{q}, \dot{q}+\frac{\alpha}{2}q\rangle| \leq \frac{M_1D_3}{m}|\dot{q}||\dot{q}+\frac{\alpha}{2}q| \leq \frac{2\sqrt{5}M_1D_3}{m^2}V. \tag{6.3.17}$$

Similarly, equations (6.1.19), (6.3.12), and conditions (6.3.3) and (6.3.4) imply

$$|\langle A_1A^{-1}\omega, \dot{q}+\frac{\alpha}{2}q\rangle| \leq \frac{\sqrt{5}M_1D_2}{m\sqrt{m}}V^{1/2}. \tag{6.3.18}$$

After substituting relations (6.2.14), (6.2.19), (6.2.20), (6.2.22)–(6.2.25), and (6.3.14)–(6.3.18) into expression (6.3.13) for the derivative $\dot{V}$, we come to the inequality

$$\dot{V} \leq -\frac{\delta_1U_0 - \delta_2(D_2+S_0) - \delta_3V - \delta_4V^{1/2}}{B}V^{1/2},$$

where

$$\delta_4 = \frac{2D_3}{\sqrt{mM}} + \frac{2D_0}{m} + \frac{2\sqrt{5}M_1D_3}{m^2} \tag{6.3.19}$$

and the quantities δ_1, δ_2, and δ_3 are given by the expression (6.2.27).

Note that although the function $B(t,q,\dot{q})$ depends on time explicitly now, it still satisfies inequality (6.1.27). Taking this inequality into account and arguing as in the proof of Theorem 6.2 in the scleronomic case, it is easy to show the validity of an analogous theorem for the rheonomic case too.

Theorem 6.4. *Assume that at the initial instant of time t_0, system (5.1.2) controlled by law (6.3.6)–(6.3.8) is in the state $(q_0,\dot{q}_0)$, and the condition*

$$\delta_1U_0 - \delta_2(D_2+S_0) - \delta_3V_0 - \delta_4V_0^{1/2} \geq 0 \tag{6.3.20}$$

holds, where $V_0 = V(t_0,q_0,\dot{q}_0)$. Then the derivative of the function V along the trajectory starting at the point $(t_0,q_0,\dot{q}_0)$ satisfies the inequality

$$\dot{V} \leq -\frac{1}{3}\left[\delta_1U_0 - \delta_2(D_2+Q_0) - \delta_3V_0 - \delta_4V_0^{1/2}\right]V^{1/2}. \tag{6.3.21}$$

From the assertion of Theorem 6.4, we conclude that, if condition (6.3.20) is satisfied, the value of the function V on the trajectory under consideration approaches zero, and the trajectory itself approaches the origin.

In order to estimate the time $\tau(t_0,q_0,\dot{q}_0)$ of motion of the system from the initial state $(t_0,q_0,\dot{q}_0)$ to the origin of the phase space $q_0 = \dot{q}_0 = 0$, we integrate inequality (6.3.21) on the interval $[t_0,t]$. We obtain

$$\tau(t_0, q_0, \dot{q}_0) \leq \frac{6V_0^{1/2}}{\delta_1 U_0 - \delta_2 (D_2 + Q_0) - \delta_3 V_0 - \delta_4 V_0^{1/2}}.$$

Relations (6.3.20) and the conditions of Theorem 6.4 lead to the following sufficient condition of bringing the system to the terminal state by using the proposed control law:

$$U_0 \geq \frac{\delta_2}{\delta_1}(D_2 + Q_0) + \frac{\delta_3}{\delta_1} V_0 + \frac{\delta_4}{\delta_1} V_0^{1/2}. \tag{6.3.22}$$

Inequality (6.3.22) connects the maximum admissible magnitudes of the control U_0 and perturbation Q_0, the size of the domain of admissible initial states, the estimate of the error with which the inertia matrix of the system is known, and the estimate of the rate of change for the coefficients of the kinetic energy polynomial on time. The constraints on the coefficients of the kinetic energy polynomial and the rate of their change are involved in the expressions for δ_i. One can easily see that, if the quantity U_0 is big enough compared with the parameters of the problem listed here, then the sufficient controllability condition (6.3.22) is satisfied.

Remark 6.5. The proposed control law, as in the scleronomic case (see Remark 6.4), does not depend on the parameters listed above; therefore, it may be applied formally, even if inequality (6.3.22) is not satisfied. The results of computer modelling presented below have shown that the proposed control law also remains effective beyond the range of sufficient condition (6.3.22).

6.3.2 Computer simulation results

A body with a variable moment of inertia

To illustrate the work of the algorithm let us return to the rotation of a body with the moment of inertia depending on time (Fig. 4.1). The equations of motion of the system has form (5.3.33). In the simulation, the constants m, M, and U_0, the mass m_0, the law of motion of the particle along the rod $l(t)$, and the initial and the terminal states were taken as in Sect. 5.3.4.

Figures 6.12 and 6.13 illustrate the behavior of the system subjected to the control described above. Figure 6.12 shows the phase portrait of the system. The thin solid line depicts the trajectory in case $\omega = 0$, that is, in the case where the moment of inertia is constant and known. The thick line corresponds to the trajectory in case $\omega = 2$, that is, for the system with the variable moment of inertia. The dashed lines depict the level sets of the Lyapunov function V.

The graphs in Fig. 6.13 show the time history of the function V (the solid line) and the absolute value of the control torque U (the dashed line) along the trajectory under consideration.

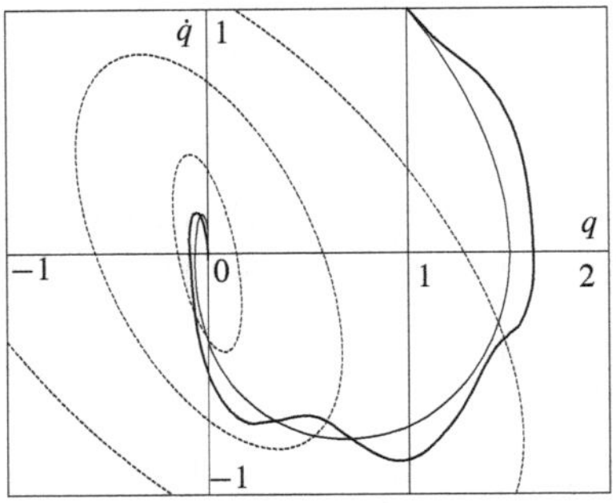

Fig. 6.12 Phase portrait of the system

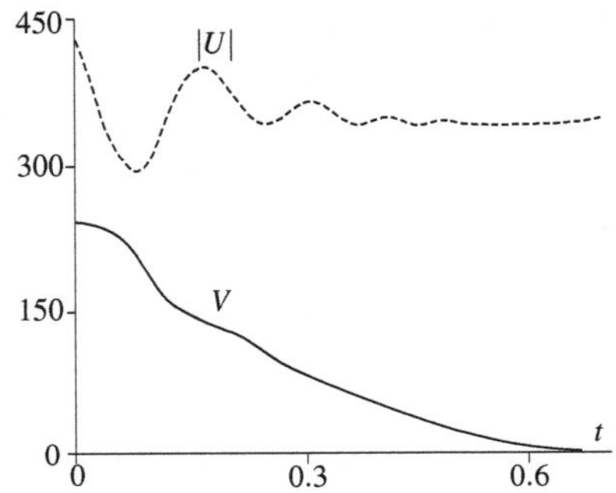

Fig. 6.13 Function V and the absolute value of the control torque U

A two-link manipulator on a movable base

In Figs. 6.8–6.11, the results of computer simulation of the dynamics of a two-link manipulator on a movable base are presented. We assume that the base of the manipulator makes a translational motion (see Fig. 4.2) according to the law

$$x_1(t) = H\cos\omega t, \quad x_2(t) = H\sin\omega t.$$

In Sect. 4.1, such manipulator was described as a rheonomic system with two degrees of freedom. The corresponding inertia matrix A, the vector-valued function a and the function a_0 are given by expressions (4.1.7). The parameters of the manipulator are chosen as in Sect. 6.2.2 for simulating the motion of a manipulator on a fixed base. The dashed lines in Figs. 6.8–6.11 correspond to the motion of the manipulator for the case of $H = 1$ m and $w = 2$ rad/s. In this case, as in Sect. 6.2.2, it is assumed that a load of mass $m_0 = 5$ kg is held in the gripper of the manipulator and that the control law is constructed for a load of mass $m_0 = 3$ kg. In Figs. 6.8 and 6.9, the graphs describe the time histories of the angular coordinates of the first and second links. In Figs. 6.10 and 6.11, the graphs show the time history of the function V and the magnitude of the control torque vector U, respectively.

One can easily see that, for the values of the parameters chosen, the sufficient controllability condition, i.e., inequality (6.3.22) is not satisfied. Nevertheless, as the results of simulation have shown, the manipulator is steered to the terminal state by the control law proposed in finite time. In this case, the motion of the base increases the total time of the process, as is quite natural.

Chapter 7
Control in distributed-parameter systems

Many publications have been devoted to systems with distributed parameters, for example [25, 86, 113, 122, 87, 117]. The control method proposed below differs from the earlier ones. It enables one to construct a constrained control in closed form and ensures that the system is brought to a given state in a finite time. This method published before in [30, 39, 36] uses a decomposition of the original system into simple subsystems and in this sense is close in spirit to the approaches described above in the present book, where systems with a finite number of degrees of freedom were considered.

In this chapter, we deal with linear elastic systems controlled by distributed bounded forces. Before constructing control for such distributed-parameters systems with infinite number of degrees of freedom, we consider first the control for an oscillatory system with a finite number of degrees of freedom.

7.1 System of linear oscillators

7.1.1 Equations of motion

We consider a dynamical system with n degrees of freedom described by equations

$$A\ddot{x} + Cx = Bv + f(x,\dot{x},t); \qquad v(t) \in V, \qquad f(x,\dot{x},t) \in F. \tag{7.1.1}$$

Here, $x \in R^n$ is the vector of generalized coordinates, A and C are constant symmetric positive definite $n \times n$ matrices of the kinetic and potential energy, respectively, B is a constant $n \times m$ matrix $(m \geq n)$, f is a given n-vector of nonlinear terms, and $v \in R^m$ is the vector of controls. The values of v and f are bounded by the given sets $V \subset R^m$ and $F \subset R^n$, respectively. We look for a feedback control law $v(x,\dot{x})$ which satisfies the imposed constraints and drives the system (7.1.1) from any given initial state

$$x(0) = x^0, \qquad \dot{x}(0) = \dot{x}^0 \tag{7.1.2}$$

to the zero terminal state $x(T) = \dot{x}(T) = 0$ in finite time T (not fixed a priori). Let us introduce normal coordinates $q = (q_1, \dots, q_n)$ defined by the transformation [64]

$$x = Hq. \tag{7.1.3}$$

Here, the $n \times n$ invertible matrix H consists of columns $h_1, \dots, h_n$ which are the eigenvectors of the eigenvalue problem:

$$(C - \lambda A)h = 0. \tag{7.1.4}$$

It is well known [64] that problem (7.1.4) has n positive eigenvalues $\lambda_1, \dots, \lambda_n$ satisfying the characteristic equation

$$\det(C - \lambda A) = 0. \tag{7.1.5}$$

In the case of multiple roots of (7.1.5), there are coincident values among $\lambda_1, \dots, \lambda_n$. The number of coincident λ_i is equal to the multiplicity of the corresponding root. The eigenvalues correspond to linearly independent eigenvectors $h_1, \dots, h_n$, and each multiple root correspondes to as many vectors as its multiplicity. The transformation (7.1.3) reduces (7.1.1) to the system of linear oscillators

$$\ddot{q}_k + \omega_k^2 q_k = w_k + \zeta_k, \qquad \omega_k = \lambda_k^{1/2}, \qquad k = 1, \dots, n. \tag{7.1.6}$$

Here, ω_k is the eigenfrequency of the kth oscillator, whereas w_k and ζ_k are the components of the n-vectors w and ζ defined by

$$w = H^{-1}A^{-1}Bv, \qquad \zeta = H^{-1}A^{-1}f. \tag{7.1.7}$$

The oscillators in (7.1.6) are coupled only through the control and nonlinear terms. On the strength of (7.1.1), the vectors w and ζ belong to the following sets in R^n:

$$w \in W = H^{-1}A^{-1}BV, \qquad \zeta \in Z = H^{-1}A^{-1}F. \tag{7.1.8}$$

From (7.1.2) and (7.1.3), we have the following initial conditions for system (7.1.6)

$$q_k(0) = q_k^0 = (H^{-1}x^0)_k, \qquad \dot{q}_k(0) = \dot{q}_k^0 = (H^{-1}\dot{x}^0)_k. \tag{7.1.9}$$

7.1.2 Decomposition

Let us consider w_k and ζ_k in each equation (7.1.6) as controls of two independent players. The first player which chooses w_k tends to bring the kth equation (7.1.6) to the zero terminal state $q_k = \dot{q}_k = 0$ in finite time, whereas the second player choosing ζ_k counteracts. We assume that the master player, i.e., the first player, is well-informed about the current values of the function f in system (7.1.1). Consequently, the first player knows the values of vector ζ at every instant. Thus, we come to the following conditions.

Let the following inclusion

$$Z + S_\varepsilon \subset W \tag{7.1.10}$$

hold for some ε (see Fig. 7.1).

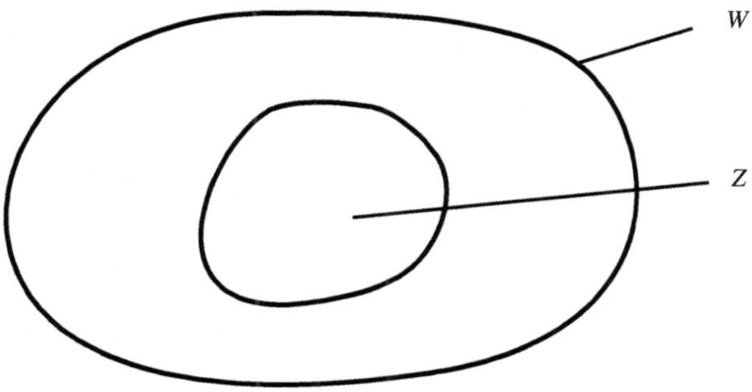

Fig. 7.1 Inclusion (7.1.10)

Here, the sets W and Z are defined by (7.1.8), and S_ε is an n-dimensional ball of the radius ε and with the center at the origin. Under the condition (7.1.10), we take

$$w = -\zeta + u, \tag{7.1.11}$$

where u is a new n-dimensional vector of control. Substituting (7.1.11) into (7.1.6), we obtain

$$\ddot{q}_k + \omega_k^2 q_k = u_k, \qquad k = 1, \ldots, n. \tag{7.1.12}$$

The inclusion (7.1.10) ensures that there exists an n-dimensional rectangular parallelepiped

$$U: \quad |u_k| \le U_k, \qquad k = 1, \ldots, n, \tag{7.1.13}$$

such that any values $u \in U$ are admissible. It means that, for any $u \in U$ and any $\zeta \in Z$, the vector w from (7.1.11) satisfies the constraints (7.1.8). In other words, for any $u \in U$ and any $\zeta \in Z$, there exists $v \in V$ such that the corresponding w given by (7.1.7) satisfies (7.1.8) and is presented in the form (7.1.11).

Thus, the inclusion (7.1.10) can be regarded as a sufficient controllability condition for the system (7.1.1). Under this condition, the control design for the system (7.1.1) is reduced to the control of simple linear subsystems (7.1.12) with one degree of freedom each by means of independent control forces u_k bounded by constraints (7.1.13).

7.1.3 Time-optimal control problem

We will now consider the optimal control problem for (7.1.12) under constraint (7.1.13) and initial conditions (7.1.9). We have

$$\ddot{q}_k + \omega_k^2 q_k = u_k, \qquad |u_k| \le U_k, \qquad \omega_k > 0,$$
$$q_k(0) = q_k^0, \qquad \dot{q}_k(0) = \dot{q}_k^0, \qquad q_k(T_k) = \dot{q}_k(T_k) = 0, \qquad T_k \to \min. \tag{7.1.14}$$

We introduce non-dimensional variables and parameters

$$t = \omega_k^{-1}\tau, \qquad q_k = U_k\omega_k^{-2}y, \qquad \dot{q}_k = U_k\omega_k^{-1}z,$$
$$u_k = U_k u, \qquad T_k = \omega_k^{-1}T_*, \qquad T_* \to \min. \tag{7.1.15}$$

After transformations (7.1.15), relations (7.1.14) acquire the normalized form

$$\frac{dy}{d\tau} = z, \qquad \frac{dz}{d\tau} = -y + u, \qquad |u| \le 1,$$
$$y(0) = y^0, \qquad z(0) = z^0, \qquad y(T_*) = z(T_*) = 0, \qquad T_* \to \min. \tag{7.1.16}$$

The solution of the time-optimal problem (7.1.16) is known [93] and presented as Example 2 in Sect. 1.4. The optimal control synthesis for this problem is given by (1.4.13) and (1.4.19). In the notation of this chapter, we have

$$u(y,z) = \operatorname{sign}\psi(y,z) \qquad \text{if} \qquad \psi(y,z) \ne 0;$$
$$u(y,z) = \operatorname{sign} y = -\operatorname{sign} z \qquad \text{if} \qquad \psi(y,z) = 0. \tag{7.1.17}$$

Here, the function $\psi(y,z)$ is given by

$$\psi(y,z) = (-y^2 - 2y)^{1/2} - z \qquad \text{if} \qquad -2 \le y \le 0;$$
$$\psi(y,z) = \psi(y+2,z) \qquad \text{if} \qquad y < -2; \tag{7.1.18}$$
$$\psi(y,z) = -\psi(-y,-z) \qquad \text{if} \qquad y > 0.$$

The switching curve $\psi(y,z) = 0$ given by (7.1.17) and (7.1.18) possesses central symmetry and consists of semicircles of unit radii with centers at the points

$$z = 0, \qquad y = \pm(2i+1), \qquad i = 0,1,\ldots. \tag{7.1.19}$$

The plus (minus) sign in (7.1.19) corresponds to semicircles in the fourth (second) quadrant of the y,z phase plane.

The optimal phase trajectory corresponding to the feedback control (7.1.17) consists of circular arcs with centers at the points $y = \pm 1$, $z = 0$. In the domain

$\psi(y,z) < 0$, where $u = -1$, the center of these circles lies at the point $y = -1, z = 0$, while in the domain $\psi(y,z) > 0$, where $u = 1$, it is at the point $y = 1, z = 0$. The semicircles of the switching curve with centers at the points $y = \pm 1, z = 0$ are themselves segments of the phase trajectories.

In Fig. 7.2, similar to Fig. 1.4, the solid lines represent the switching curve, and the thin line shows one of the optimal trajectories. The arrows indicate the time growth.

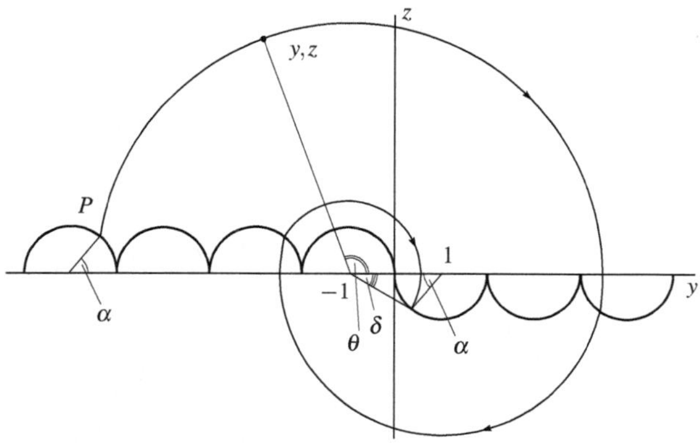

Fig. 7.2 Switching curve and optimal phase trajectory

7.1.4 Upper bound for the optimal time

We will estimate the time of motion $T_*(y,z)$ along the optimal phase trajectory that starts at some point y,z. This estimate will be used hereafter. Suppose that this point lies in the domain $z > \psi(y)$. We will first make some auxiliary constructions.

We denote by r, θ the polar coordinates of the initial point y,z, the pole being the point $y = -1, z = 0$. We have

$$y = r\cos\theta - 1, \qquad z = r\sin\theta. \tag{7.1.20}$$

The initial segment of the phase trajectory is a circular $r = \text{const}$. We continue this arc in an anticlockwise direction until it intersects the switching curve $z = \psi(y)$. Suppose the point of intersection P lies on the ith (counting from the origin of coordinates) semicircle of the switching curve (see Fig. 7.2, where $i = 4$). This means that the coordinates of P can be taken in the form

$$y_P = -2i + 1 + \cos\alpha, \qquad z_P = \sin\alpha,$$
$$i = 2, 3, \ldots, \qquad \alpha \in [0, \pi). \tag{7.1.21}$$

The angle α corresponds to the arc cut out by the point P from the semicircle of the switching curve on which it lies. We note that such arcs α are cut out by the optimal trajectory from all the semicircles of the switching curve which it intersects. The final arc of the phase trajectory also has angular dimensions α, see Fig. 7.2.

Since the point P with coordinates (7.1.21) lies on a circle $r = \text{const}$, we have, according to the elementary geometry, see Fig. 7.2

$$r^2 = (y_P + 1)^2 + z_P^2 = 4(i-1)^2 + 1 - 4(i-1)\cos\alpha. \tag{7.1.22}$$

We denote by R the length of the radius-vector of the phase point y, z. Using relation (7.1.20), we obtain

$$R^2 = y^2 + z^2 = (r-1)^2 + 2r(1-\cos\theta). \tag{7.1.23}$$

The inequalities

$$R \geq r - 1 \geq [4(i-1)^2 - 4(i-1) + 1]^{1/2} - 1 = 2i - 4. \tag{7.1.24}$$

follow from (7.1.23) and (7.1.22).

The time of motion along any arc of the optimal trajectory is, according to Sect. 1.4, equal to the angular length of this arc. Each arc between neighbouring switches of the control is either equal to π, or (for the first and second sections) does not exceed π, and the total number of sections is equal to the integer i introduced above. Hence we have $T_* \leq \pi i$. Using inequality (7.1.24) we obtain the estimate

$$T_* \leq \pi\left(\frac{R}{2} + 2\right) \equiv T^0(R). \tag{7.1.25}$$

Estimate (7.1.25) holds for all $R \geq 0$, but it does not imply that $T_* \to 0$ as $R \to 0$. Hence, we will obtain yet another estimate for sufficiently small R.

Suppose $i = 2$, i.e., there is only one switch of the control, see Fig. 7.2. In this case the optimal trajectory consists of an arc of radius r and angular dimension $\theta + \delta$ and an arc of radius 1 and angular dimension α, coinciding with a segment of the switching curve. Here, we denote by δ the angle between the y axis and the ray coming from the point $y = -1$, $z = 0$ to the point of the trajectory where the switch occurs. Thus

$$T_* = \theta + \delta + \alpha, \tag{7.1.26}$$

where, as can be determined with the help of Fig. 7.2, we have

$$\sin\delta = r^{-1}\sin\alpha, \qquad \delta \in \left[0, \frac{\pi}{2}\right]. \tag{7.1.27}$$

Let us obtain some auxiliary relations, which we will require in order to estimate the time (7.1.26). Putting $i = 2$ in (7.1.22), we find

$$r = \left[1 + 8\sin^2\left(\frac{\alpha}{2}\right)\right]^{1/2}. \tag{7.1.28}$$

Equations (7.1.27) and (7.1.28) determine the relation between angles δ and α. The analysis of this relation shows that as the angle α varies within the interval $[0,\pi]$ in (7.1.21), the angle δ varies within the interval $[0,\pi/6]$, and always $\delta \leq \alpha$. Thus, we have

$$0 \leq \delta \leq \frac{\pi}{6}, \qquad \delta \leq \alpha, \qquad 0 \leq \alpha < \pi. \tag{7.1.29}$$

The following inequality holds

$$\sin\left(\frac{\gamma}{2}\right) \geq \frac{\gamma}{\pi}, \qquad \gamma \in [0,\pi). \tag{7.1.30}$$

Putting $\gamma = \alpha$ in inequality (7.1.30), we obtain from (7.1.28) the relation

$$r \geq (1 + 8\pi^{-2}\alpha^2)^{1/2}, \qquad \alpha \in [0,\pi),$$

which we rewrite in the form

$$r \geq g(\xi) = (1+\xi)^{1/2}, \qquad \xi = 8\pi^{-2}\alpha^2, \qquad \xi \in [0,8). \tag{7.1.31}$$

Since $g(\xi)$ is a concave function, the inequality

$$[g(\xi) - g(0)]\xi^{-1} \geq \frac{1}{8}[g(8) - g(0)], \qquad \xi \in [0,8].$$

is satisfied. Substituting the values $g(0) = 1$ and $g(8) = 3$ from (7.1.31) into the last equality, we obtain

$$r \geq (1+\xi)^{1/2} \geq 1 + \frac{\xi}{4}, \qquad \xi \in [0,8].$$

This inequality makes it possible to simplify (7.1.31) as follows

$$r \geq 1 + 2\pi^{-2}\alpha^2, \qquad \alpha \in [0,\pi). \tag{7.1.32}$$

We now transform relation (7.1.23) using inequality (7.1.30) for $\gamma = \theta$. We have

$$R^2 = (r-1)^2 + 4r\sin^2\left(\frac{\theta}{2}\right) \geq (r-1)^2 + 4\pi^{-2}r\theta^2.$$

Let us substitute (7.1.32) into the latter inequality. We obtain

$$R^2 \geq 4\pi^{-4}\alpha^4 + 4\pi^{-2}\theta^2.$$

The latter relation implies the following two inequalities

$$R \geq 2\pi^{-2}\alpha^2, \qquad R \geq 2\pi^{-1}|\theta|. \tag{7.1.33}$$

We now transform (7.1.26) for T_*, using inequalities (7.1.29) and (7.1.33)

$$T_* = \theta + \delta + \alpha \le 2\alpha + \theta \le 2|\alpha| + |\theta| \le \pi \left[(2R)^{1/2} + \frac{R}{2} \right] \equiv T^1(R). \quad (7.1.34)$$

Let us compare estimates (7.1.25) and (7.1.34). We recall that estimate (7.1.34) was obtained for $i = 2$, and estimate (7.1.25) for all $i \ge 2$. But, according to (7.1.24), we have $R \ge 2$ for $i \ge 3$. From (7.1.25) and (7.1.34) it follows that $T^0(R) \le T^1(R)$ for $R \ge 2$. Consequently, for all $i \ge 3$ we have $T^0(R) \le T^1(R)$.

It has thus been established that the upper estimate (7.1.34) for the optimal time

$$T_* \le T^1(R) = \pi \left[\frac{R}{2} + (2R)^{1/2} \right], \qquad R = (y^2 + z^2)^{1/2} \quad (7.1.35)$$

holds for all y, z.

Returning to the original dimensional variables (7.1.15), we obtain the following upper estimate for our optimal time for problem (7.1.14) in the form

$$T_k(q_k^0, \dot{q}_k^0) \le \pi U_k^{-1} \left[\frac{\rho_k}{2} + (2U_k^{-1} \omega_k^{-1} \rho_k)^{1/2} \right],$$
$$\rho_k = \left[\omega_k^2 (q_k^0)^2 + (\dot{q}_k^0)^2 \right]^{1/2}; \qquad k = 1, 2, \ldots; \qquad \omega_k > 0. \quad (7.1.36)$$

The time T of steering the total system (7.1.12) with n degrees of freedom to the origin is equal to the largest T_k, i.e.,

$$T = \max_k T_k(q_k^0, \dot{q}_k^0), \qquad k = 1, \ldots, n. \quad (7.1.37)$$

The obtained inequality (7.1.36) allows one to estimate the time T from above.

7.2 Distributed-parameter systems

7.2.1 Statement of the control problem for a distributed-parameter system

Let us turn to the consideration of control systems with distributed parameters described by linear partial differential equations. We shall consider in tandem the equation

$$w_t = Aw + v, \quad (7.2.1)$$

solved with respect to the first time derivative, and the equation

$$w_{tt} = Aw + v, \quad (7.2.2)$$

solved with respect to the second derivative. In (7.2.1) and (7.2.2), $w(x,t)$ is the scalar function of the n-dimensional spatial coordinate vector $x = (x_1, \ldots, x_n)$ and time t; w describes the state of the system, v is the required control, and A is a linear differential operator containing partial derivatives with respect to the coordinates x_i, $i = 1,\dots,n$. The coefficients of the operator A do not depend on t, and its order $\operatorname{ord} A$ is assumed to be even and equal to $2m$.

The most important and frequently encountered examples of (7.2.1) and (7.2.2), which we shall have in mind in the following, are: 1) the heat-conduction equation, which is obtained from (7.2.1) if $m = 1$ and $A = \Delta$ is the Laplace operator; 2) the wave equation obtained from (7.2.2) with $m = 1$ and $A = \Delta$; 3) the equation for the vibrations of an elastic beam or plate, obtained from (7.2.2) with $m = 2$, $A = -\Delta^2$ and $n = 1,2$, respectively. Equations (7.2.1) and (7.2.2) describe also heat-conduction processes and vibrations in an inhomogeneous medium, if

$$Aw = \sum_{i=1}^{n} \frac{\partial}{\partial x_i}\left[a(x)\frac{\partial w}{\partial x_i}\right], \qquad m = 1,$$

where $a(x)$ is a specified function describing the inhomogeneity of the medium.

Equations (7.2.1) and (7.2.2) are considered in some bounded domain of variation for the spatial variables $x \in \Omega$ and for $t \geq 0$. At the boundary Γ of the domain Ω, a homogeneous boundary condition of the following form should be satisfied

$$Mw = 0, \qquad M = (M_1,\dots,M_m), \qquad x \in \Gamma. \tag{7.2.3}$$

Here, M_j is a linear differential operator of order $\operatorname{ord} M_j < 2m$ $(j = 1,\dots,m)$ with coefficients independent of t. In particular, for $m = 1$ the operator M is scalar and has the form

$$Mw = b_0(x)w + b_1(x)\frac{\partial w}{\partial x},$$

where $b_0(x)$ and $b_1(x)$ are functions given on Γ. Condition (7.2.3) can, in particular, become the Dirichlet condition (for $b_0 = 1$, $b_1 = 0$) or the Neumann condition (for $b_0 = 0$, $b_1 = 1$).

The initial conditions have the form

$$w(x,0) = w_0(x), \qquad x \in \Omega \tag{7.2.4}$$

for (7.2.1) and

$$w(x,0) = w_0(x), \qquad w_t(x,0) = w_{t0}(x), \qquad x \in \Omega \tag{7.2.5}$$

for (7.2.2).

The constraint

$$|v(x,t)| \leq v^0, \qquad x \in \Omega, \qquad t \geq 0, \tag{7.2.6}$$

is imposed on the control function v in (7.2.1) and (7.2.2), where $v^0 > 0$ is a given constant.

We will now formulate the control problem.

Problem 7.1. It is required to construct a control $v(x,t)$ satisfying condition (7.2.6) and such that the corresponding solution of (7.2.1) or (7.2.2) with boundary condition (7.2.3) and the corresponding initial condition (7.2.4) or (7.2.5) vanishes at some finite (unspecified) time instant $T > 0$. More precisely, everywhere in Ω the condition $w(x,T) = 0$ should be satisfied for (7.2.1) and $w(x,T) = w_t(x,T) = 0$ should be satisfied for (7.2.2).

Obviously, if one puts $v \equiv 0$ for $t \geq T$, the solution remains identically equal to zero for $t > T$.

The boundary of the domain Ω is assumed to be piecewise-smooth. Requirements on the initial functions and the function classes to which the solutions of the problems belong in various cases, are considered in Sect. 7.3.4.

7.2.2 Decomposition

The solution of Problem 7.1 is based on the Fourier method. To apply it, we will first consider the following eigenvalue problem corresponding to the initial-boundary-value problems (7.2.1)—(7.2.5) for $v = 0$.

The problem is to find the functions $\varphi(x), x \in \Omega$, and the corresponding constants λ that satisfy the following linear homogeneous equation with boundary conditions

$$A\varphi = -\lambda\varphi, \qquad x \in \Omega; \qquad M\varphi = 0, \qquad x \in \Gamma. \tag{7.2.7}$$

It is known that, under specified conditions (for self-conjugate elliptic equations and, in particular, for the Laplace equation, i.e., when $A = \Delta$), the eigenvalue problem (7.2.7) has the following properties.

There is a discrete denumerable spectrum of positive eigenvalues λ_k, which can be numbered in non-decreasing order: $\lambda_1 \leq \lambda_2 \leq \ldots$, with $\lambda_k \to \infty$ as $k \to \infty$. In certain cases, for example, for the Laplace operator $A = \Delta$ with Neumann conditions, there is also a zero eigenvalue $\lambda = 0$. That case will also be considered. To these eigenvalues, there corresponds an orthogonal system of eigenfunctions $\varphi_k(x)$ complete in the domain Ω. Normalizing these functions, we obtain an orthonormal system of functions $\varphi_k(x)$ possessing the following properties

$$\begin{gathered} A\varphi_k = -\lambda_k\varphi_k, \qquad x \in \Omega; \qquad M\varphi_k = 0, \qquad x \in \Gamma, \\ (\varphi_k, \varphi_i) = \int_\Omega \varphi_k(x)\varphi_i(x)dx = \delta_{ki}, \qquad k,i = 1,\ldots,n. \end{gathered} \tag{7.2.8}$$

Here, δ_{ki} is the Kronecker delta. The index k in (7.2.8) and below, unless otherwise stated, runs over values from 0 to ∞ when there is a zero eigenvalue $\lambda_0 = 0$ and from 1 to ∞ when there is none. Summation over k will also be performed over the ranges given above.

We now use the Fourier method to separate the time (t) and space (x) dependence. Solutions of (7.2.1) and (7.2.2) will be sought in the form of eigenfunction expansions

$$w(x,t) = \sum q_k(t)\varphi_k(x). \tag{7.2.9}$$

where $q_k(t)$ are certain functions of time. The control v in (7.2.1) and (7.2.2) is also represented in the form of an expansion

$$v(x,t) = \sum u_k(t)\varphi_k(x). \tag{7.2.10}$$

where $u_k(t)$ are currently unknown functions of time.

Substituting expansions (7.2.9) and (7.2.10) into (7.2.1), we obtain

$$\sum \dot{q}_k(t)\varphi_k(x) = \sum (q_k A\varphi_k + u_k\varphi_k).$$

Here and below, the dots denote time derivatives.

We use the equations for φ_k from (7.2.8) together with the orthogonality of φ_k. As a result, we obtain the system of equations

$$\dot{q}_k + \lambda_k q_k = u_k. \tag{7.2.11}$$

Similarly, substituting expansions (7.2.9) and (7.2.10) into (7.2.2), we obtain

$$\ddot{q}_k + \omega_k^2 q_k = u_k. \tag{7.2.12}$$

Here and below, ω_k are the frequencies of the natural modes given by

$$\omega_k = \lambda_k^{1/2}, \qquad 0 = \omega_0 \leq \omega_1 \leq \omega_2 \leq \dots. \tag{7.2.13}$$

We note that a solution of the form (7.2.9) satisfies, by construction, the boundary condition (7.2.3) because, according to (7.2.8), all the eigenfunctions satisfy this condition.

We substitute solution (7.2.9) into the initial conditions (7.2.4) and (7.2.5) and use the orthonormality of the eigenfunctions (7.2.8). We obtain initial conditions for (7.2.11) in the form

$$q_k(0) = q_k^0 = \int_\Omega w_0(x)\varphi_k(x)dx \tag{7.2.14}$$

and for (7.2.12) in the form

$$\begin{aligned} q_k(0) &= q_k^0 = \int_\Omega w_0(x)\varphi_k(x)dx, \\ \dot{q}_k(0) &= (\dot{q}_k)^0 = \int_\Omega w_{t0}(x)\varphi_k(x)dx. \end{aligned} \tag{7.2.15}$$

The original control problem for the partial differential equations (7.2.1) and (7.2.2) has thus been reduced to a control problem for linear control systems of infinite order (7.2.11) and (7.2.12). We impose the constraint on the control functions u_k of these systems

$$|u_k(t)| \leq U_k, \qquad t \geq 0. \tag{7.2.16}$$

The values of the constants U_k should be chosen so that the imposed constraint (7.2.6) is satisfied.

From (7.2.10) and (7.2.16), we obtain the following estimate

$$|v(x,t)| \leq \sum U_k |\varphi_k(x)|. \tag{7.2.17}$$

Consequently, to satisfy the original constraint (7.2.6), it is sufficient to require that for all $x \in \Omega$ the inequality

$$\sum U_k |\varphi_k(x)| \leq v^0, \qquad x \in \Omega \tag{7.2.18}$$

is satisfied.

We introduce the notation

$$\Phi_k = \max_{\Omega} |\varphi_k(x)|. \tag{7.2.19}$$

Inequality (7.2.18) is clearly satisfied under the condition

$$\sum U_k |\Phi_k| \leq v^0. \tag{7.2.20}$$

Thus, to solve the control problems for (7.2.1) and (7.2.2) (Problem 7.1), it is sufficient to solve the following control problems for systems (7.2.11) and (7.2.12).

Problem 7.2. It is required to construct the feedback controls $u_k(q_k)$ in system (7.2.11) and $u_k(q_k, \dot{q}_k)$ in system (7.2.12) for $k = 0, 1, \ldots$, satisfying constraints (7.2.16) and bringing these systems to the zero terminal state [$q_k = 0$ for (7.2.11) and $q_k = \dot{q}_k = 0$ for (7.2.12)] in a finite time for any initial conditions of the form (7.2.14) or (7.2.15), respectively. Here, the constants U_k in (7.2.16) should satisfy inequality (7.2.18) for all x, or, which is sufficient, the stronger inequality (7.2.20).

We note that, as a result of applying the Fourier method, we have achieved a decomposition of the system: each mode of motion is described by its own (7.2.11) or (7.2.12), with the corresponding control u_k. Thus, Problem 7.1 is reduced to Problem 7.2 which was considered for the second order systems in Sect. 7.1.3 [see system (7.1.12) with constraints (7.1.13)]. However, the constants U_k in the constraints (7.2.16) are associated with inequalities (7.2.18) or (7.2.20), which is a substantial difficulty in solving the problem.

For each of (7.2.11) and (7.2.12), we shall construct the time-optimal feedback control u_k under constraint (7.2.16) for arbitrary fixed U_k. These controls are well known, they are given for the systems of the second order in Sect. 7.1.3.

Below, these time-optimal controls are analyzed in connection to inequalities (7.2.18) and (7.2.20), and estimates obtained in Sect. 7.1.3 are used.

7.2.3 First-order equation in time

Firstly, let us consider (7.2.1) which contains the first time derivative of the sought-for function $w(x,t)$. Due to the decomposition of this equation, the system of the first order equations (7.2.11) was obtained.

Consider the problem of time-optimal control for one of (7.2.11) under constraint (7.2.16) and initial condition (7.2.14). We have

$$\begin{gathered} \dot{q}_k + \lambda_k q_k = u_k, \qquad |u_k(t)| \le U_k, \qquad \lambda_k \ge 0, \\ q_k(0) = q_k^0, \qquad q_k(T_k) = 0, \qquad T_k \to \min. \end{gathered} \tag{7.2.21}$$

The solution of problem (7.2.21) is very simple. Integrating (7.2.21) and satisfying the initial condition, we find that

$$q_k(t) = \left[q_k^0 + \int_0^t u_k(\tau)\exp(\lambda_k\tau)d\tau\right]\exp(-\lambda_k t). \tag{7.2.22}$$

Hence, it follows that, for the fastest vanishing of the solution $q_k(t)$, the control u_k should be maximal in modulus and opposite to the sign of the initial value q_k^0, or, equivalently, of the solution $q_k(t)$.

The synthesis of the time-optimal control thus has the form

$$u_k(q_k) = \begin{cases} -U_k \operatorname{sign} q_k, & q_k \ne 0 \\ 0, & q_k = 0. \end{cases} \tag{7.2.23}$$

The control (7.2.23) is constant along any phase trajectory. Substituting it into (7.2.22) and integrating, we obtain

$$q_k(t) = \left\{|q_k^0| - U_k\lambda_k^{-1}[\exp(\lambda_k t) - 1]\right\}\exp(-\lambda_k t)\operatorname{sign} q_k^0. \tag{7.2.24}$$

At the final instant, according to (7.2.21), we have $q_k(T_k) = 0$. From (7.2.24), we find the instant when the process ends

$$\begin{gathered} T_k = \lambda_k^{-1}\log\left(1 + \lambda_k|q_k^0|U_k^{-1}\right), \qquad \lambda_k > 0, \qquad k \ge 1; \\ T_0 = |q_0^0|U_0^{-1}, \qquad \lambda_0 = 0. \end{gathered} \tag{7.2.25}$$

The solution of the time-optimal control problem (7.2.21) for all $k \ge 0$ is presented in the feedback form (7.2.23). The phase trajectory and optimal time are given by formulas (7.2.24) and (7.2.25), respectively. Thus, the solution of Problem 7.2 for system (7.2.11) is obtained. Method for choosing the constants U_k will be presented in Sect. 7.2.5.

7.2.4 Second-order equation in time

We will now turn to (7.2.2) which contains the second time derivative of the function $w(x,t)$. In this case, the decomposition results in the system of the second order equations (7.2.12).

Consider the optimal control problem for one of (7.2.12) under constraint (7.2.16) and initial conditions (7.2.15).

In the case of $\omega_k > 0$, $k \geq 1$, the results obtained in Sect. 7.1 for a finite system of oscillators will be used. The optimal control synthesis $u_k(q_k, \dot{q}_k)$ for $k \geq 1$ is given by relations (7.1.17) and (7.1.18) in non-dimensional variables. To express the control through original dimensional variables, it is sufficient to use the transformation formulas (7.1.15).

We consider separately the case with the zero eigenvalue $k = 0$, $\omega_0 = 0$. In this case, the feedback optimal control for problem (7.1.14), is presented in Sect. 1.4, see Example 1, (1.4.12) and (1.4.13). We have

$$
\begin{aligned}
u_0(q_0, \dot{q}_0) &= U_0 \operatorname{sign} \psi_0(q_0, \dot{q}_0) \qquad \text{if} \qquad \psi_0 \neq 0; \\
u_0(q_0, \dot{q}_0) &= U_0 \operatorname{sign} q_0 = -U_0 \operatorname{sign} \dot{q}_0 \qquad \text{if} \qquad \psi_0 = 0; \\
\psi_0(q_0, \dot{q}_0) &= -U_0 q_0 - \frac{1}{2} \dot{q}_0 |\dot{q}_0|.
\end{aligned}
\tag{7.2.26}
$$

The optimal time for $k = 0$, $\omega_0 = 0$ is given by the formula

$$
T_0(q_0, \dot{q}_0) = U_0^{-1} \left\{ 2 \left[\frac{1}{2} (\dot{q}_0)^2 - U_0 q_0 \sigma \right]^{1/2} - \dot{q}_0 \sigma \right\},
$$

$$
\sigma = \operatorname{sign} \psi_0(q_0, \dot{q}_0) \quad \text{for} \quad \psi_0 \neq 0, \qquad \sigma = \pm 1 \quad \text{for} \quad \psi_0 = 0
$$

which can be obtained for Example 1 from Sect. 1.4, see (2.3.26). Here and below, the superscript ^{0} at q_k, $\dot{q}_k$, $k = 0, 1, \ldots$, is omitted.

Applying the inequality $(a+b)^{1/2} \leq |a|^{1/2} + |b|^{1/2}$ to the above relation, we obtain the estimate

$$
T_0(q_0, \dot{q}_0) \leq (2^{1/2} + 1) U_0^{-1} |\dot{q}_0| + 2 U_0^{-1/2} |q_0|^{1/2}. \tag{7.2.27}
$$

We have thus obtained relations for the time-optimal control for system (7.2.12) for all $k \geq 0$. Optimal phase trajectories for this system are also well known, see Sect. 1.4, Figs. 1.2 and 1.4. For the optimal time, estimates (7.1.36) and (7.2.27) has been obtained for $k \geq 1$ and $k = 0$, respectively. Thereby, the solution of Problem 7.2 for system (7.2.12) is found. The procedure for choosing the constants U_k is discussed below.

7.2.5 Analysis of the constraints and construction of the control

The relations obtained in Sects. 7.2.3 and 7.2.4 contain constants U_k that are constraints on the control for the kth mode of motion. We choose these constants so as to reduce the total time of the motion equal to

$$T = \max_k T_k, \qquad k \geq 0 \quad \text{or} \quad k \geq 1 \tag{7.2.28}$$

while satisfying constraints (7.2.18) or (7.2.20). The index k in (7.2.18), (7.2.20), and (7.2.28) takes the values $0, 1, \ldots$ when there is a zero eigenvalue $\lambda_0 = 0$ in problem (7.2.8) and values $1, 2, \ldots$ when there is none.

Since T_k increases monotonically as U_k increases, and all the U_k occur linearly with positive coefficients in constraints (7.2.18) and (7.2.20), it is natural to choose U_k by requiring that all T_k are equal: $T_0 = T_1 = \ldots$. This gives the least possible value for T in (7.2.28) under given constraints (7.2.18) or (7.2.20).

Following the idea expressed above, for the first-order equation we put, in accordance with (7.2.25)

$$T_k = \lambda_k^{-1} \log\left(1 + \lambda_k |q_k| U_k^{-1}\right) = T.$$

Here, T is a constant to be determined, and the superscript ^{0} at q_k is omitted.

From this we find

$$U_k = \lambda_k |q_k| [\exp(\lambda_k T) - 1]^{-1}, \qquad k \geq 0. \tag{7.2.29}$$

Formula (7.2.29) holds for all $\lambda_k \geq 0$. Substituting (7.2.29) into inequality (7.2.20), we obtain

$$\sum \lambda_k [\exp(\lambda_k T) - 1]^{-1} |q_k| \Phi_k \leq v^0. \tag{7.2.30}$$

It is known that, under very general assumptions, the eigenvalues λ_k and the maxima of the eigenfunctions Φ_k increase no faster than some power of k as $k \to \infty$. The moduli of the Fourier coefficients $|q_k|$ increase less rapidly than k as $k \to \infty$ for any bounded initial function $w_0(x)$. Hence, because of the presence of the exponential factor, the series on the left-hand side of inequality (7.2.30) converges for all $T > 0$. As T takes values from 0 to ∞, the sum of the series decreases monotonically from ∞ to 0. Hence, there always exists such $T > 0$ for which inequality (7.2.30) is satisfied. Thus, the stated control problem (Problem 7.1) for (7.2.1) is always solvable by the proposed method. The time T of the process can be chosen from the condition for satisfying inequality (7.2.30).

We obtain an upper estimate for the time T using the inequality

$$\lambda_k [\exp(\lambda_k T) - 1]^{-1} \leq T^{-1}. \tag{7.2.31}$$

It follows from (7.2.30) and (7.2.31) that if T is chosen from the condition

$$T = \frac{Q_1}{\nu^0}, \qquad Q_1 = \sum |q_k| \Phi_k < \infty, \tag{7.2.32}$$

then inequality (7.2.30) is clearly satisfied. Consequently, if the series for Q_1 converges, the time T can be chosen according to the simple formula (7.2.32).

We now consider (7.2.2) that is of the second order in time. In this case, instead of explicit formulas for times T_k, one only has the upper estimates (7.1.36) and (7.2.27), hence the equality condition on all the T_k cannot be satisfied exactly. Bearing this in mind, and also to simplify the subsequent formulas, we propose to choose U_k in the form

$$\begin{gathered} U_k = c\rho_k, \qquad c > 0, \qquad k = 1,2,\ldots, \\ U_0 = \max(c_1|\dot{q}_0|, c_2|q_0|), \qquad c_1 > 0, \qquad c_2 > 0. \end{gathered} \tag{7.2.33}$$

Here, c, c_1, and c_2 are constants. Substituting U_k from (7.2.33) into (7.1.36), we obtain

$$T_k \leq \pi \left[(2c)^{-1} + 2^{1/2} (\omega_k c)^{-1/2} \right], \qquad k = 1,2,\ldots.$$

The last inequality is not violated if all ω_k are replaced by $\omega_1 \leq \omega_k$. We obtain the estimate

$$T_k \leq \pi \left[(2c)^{-1} + 2^{1/2} (\omega_1 c)^{-1/2} \right]. \tag{7.2.34}$$

When substituting expression (7.2.33) for U_0 into inequality (7.2.27), we shall distinguish between two cases. In the first case, when $c_1|\dot{q}_0| \geq c_2|q_0|$, we obtain from (7.2.27) and (7.2.33)

$$T_0 \leq (2^{1/2}+1)c_1^{-1} + 2|c_1\dot{q}_0|^{-1/2} q_0^{1/2} \leq (2^{1/2}+1)c_1^{-1} + 2c_2^{-1/2}. \tag{7.2.35}$$

In the second case, when $c_1|\dot{q}_0| > c_2|q_0|$, similar estimates reduce to exactly the same result (7.2.35). We choose the constants c_1 and c_2 so that both terms on the right-hand sides of inequalities (7.2.34) and (7.2.35) are identical term by term, i.e.,

$$\pi(2c)^{-1} = (2^{1/2}+1)c_1^{-1}, \qquad \pi 2^{1/2}(\omega_1 c)^{-1/2} = 2c_2^{-1/2}.$$

From this we find the required constants

$$\begin{gathered} c_1 = \vartheta_1 c, \qquad c_2 = \vartheta_2 c, \\ \vartheta_1 = 2(2^{1/2}+1)\pi^{-1} \approx 1.53, \qquad \vartheta_2 = 2\omega_1 \pi^{-2}. \end{gathered} \tag{7.2.36}$$

Using (7.2.36), formulas (7.2.33), can be written in the form

$$U_k = c\rho_k, \qquad k \geq 1, \qquad U_0 = c\max(\vartheta_1|\dot{q}_0|, \vartheta_2|q_0|). \tag{7.2.37}$$

The quantities ϑ_1 and ϑ_2 are defined in (7.2.36) and do not depend on c. Because the right-hand sides of inequalities (7.2.34) and (7.2.35) are identical by virtue of the choice of constants c_1 and c_2, estimate (7.2.34) holds for all $k \geq 0$. Thus, in all cases we have the estimate

$$T \le \pi \left[(2c)^{-1} + 2^{1/2}(\omega_1 c)^{-1/2} \right] \tag{7.2.38}$$

for the time of the control process (7.2.28).

It remains to choose the constant c so that the constraint (7.2.18) is satisfied. Substituting (7.2.37) into (7.2.18), we obtain

$$c \le v^0 (Q^*)^{-1}. \tag{7.2.39}$$

Here, we have introduced the notation

$$Q^* = \sup_{x \in \Omega} Q_2(x), \qquad Q_2(x) = \Sigma \rho_k |\varphi_k(x)| + \max(\vartheta_1 |\dot{q}_0|, \vartheta_2 |q_0|) |\varphi_0(x)|,$$
$$\rho_k = [\omega_k^2 q_k^2 + (\dot{q}_k)^2]^{1/2}, \qquad k \ge 1, \tag{7.2.40}$$

and used formulas (7.1.36) for the ρ_k. Inequality (7.2.40) is written for the case when the zero eigenvalue is present. When it is not present one simply omits the last term (max) in formula (7.2.40) for Q_2.

Thus, a sufficient condition for the control problem (Problem 7.1) to be solvable for (7.2.2) using the proposed approach is uniform boundedness of the series for $Q_2(x)$ from (7.2.40) in the domain Ω. For this, it is sufficient to require the uniform boundedness in Ω of the following two series

$$Q_3(x) = \sum \omega_k |q_k| |\varphi_k(x)|, \qquad Q_4(x) = \sum |\dot{q}_k| |\varphi_k(x)|. \tag{7.2.41}$$

Using the notation (7.2.19), the boundedness condition on Q^* from (7.2.40) can be replaced by the stronger condition of the convergence of the numerical series

$$Q_5(x) = \sum \rho_k \Phi_k < \infty, \qquad \rho_k = [\omega_k^2 q_k^2 + (\dot{q}_k)^2]^{1/2} \tag{7.2.42}$$

or the condition of the convergence of the two series

$$Q_6(x) = \sum \omega_k |q_k| \Phi_k < \infty, \qquad Q_7(x) = \sum |\dot{q}_k| \Phi_k < \infty. \tag{7.2.43}$$

We will sum up the results obtained. For both equations (7.2.1) and (7.2.2), the solvability conditions of Problem 7.1 have been stated and upper limits have been given on the control process time T.

Problem (7.2.1) is always solvable, and its time T can be chosen from condition (7.2.30) or, when the series Q_1 converges, from the simpler condition (7.2.32).

Problem (7.2.2) is clearly solvable if one of the series convergence conditions (7.2.40)—(7.2.43) is satisfied: Q_2, or Q_3 and Q_4, or Q_5, or Q_6 and Q_7. We have the estimate (7.2.38) for the time T, in which the constant c should be chosen by condition (7.2.39). Here, the number Q^* is determined from relations (7.2.40) or one of the following relations

$$Q^* = \sup_{x \in \Omega} Q_3(x) + \sup_{x \in \Omega} Q_4(x), \qquad Q^* = Q_5, \qquad Q^* = Q_6 + Q_7$$

according to which of the series convergence conditions (7.2.41)—(7.2.43) is satisfied.

We remark that when the initial functions w_0 and w_{t0} tend uniformly to zero, all their Fourier coefficients tend to zero, and here all the series in (7.2.30), (7.2.32), (7.2.40)—(7.2.43) also tend to zero. From estimates (7.2.32), (7.2.38), (7.2.39) it follows that the process time $T \to 0$ for both (7.2.1) and (7.2.2).

In the general case, we will firstly determine time T under aforesaid solubility conditions of Problem 7.1. For (7.2.1), we take formula (7.2.32). For (7.2.2), we find c from condition (7.2.39) and after that determine T from condition (7.2.38).

After determining the time T and the constant c, we find U_k from relations (7.2.29) and (7.2.37) for (7.2.1) and (7.2.2), respectively. The coefficients u_k of the required control law (7.2.10) are found in the form of a synthesis, i.e., depending on the current values q_k and $\dot{q}_k$, in Sects. 7.2.3 and 7.2.4 for (7.2.1) and (7.2.2), respectively. Because the optimal trajectories are known for the systems (7.2.1) and (7.2.2), the controls obtained in the form of a synthesis can also be represented in the form of a program $u_k(t)$, i.e., in the form of bang-bang functions with switching points depending on the initial conditions.

Thus, the control (7.2.10) can be represented either in the form of a program control for given initial conditions, or in the feedback form, if controls u_k depending on q_k and $\dot{q}_k$ are used. In the second case, the control is organized in the form $v = v(x; w(\cdot,t))$ for system (7.2.1) and in the form $v = v(x; w(\cdot,t), w_t(\cdot,t))$ for system (7.2.1). The notation introduced shows that the control v at a point $x \in \Omega$ at time t is a functional of the functions $w(y,t)$ and $w_t(y,t)$ with $y \in \Omega$. However, here the dependence on the initial functions w_0 and w_{t0} is also preserved by means of the constants U_k which depend on the initial data, see (7.2.29) and (7.2.37). In these formulas the constants T and c also depend on the initial conditions.

The control (7.2.10) obtained is by construction such that all boundary and initial conditions together with the constraint (7.2.6) are satisfied automatically. This control is near to being time-optimal because, firstly, the controls for each subsystem are optimal, and secondly, the bounds U_k are chosen so that the control times for the subsystems are equal or nearly equal to one another.

Below we consider some specific examples in which the convergence conditions for series (7.2.32), (7.2.42), and (7.2.43) are analyzed. The conditions for Problem 7.1 to be solvable are obtained in the form of requirements on the initial functions. In Sect. 7.3.4, some general conditions for the control Problem 7.1 to be solvable for (7.2.2) are given.

7.3 Solvability conditions

7.3.1 The one-dimensional problems ($n = 1$, $A = \Delta$)

We first consider the heat-conduction and oscillation equations for the case of one spatial variable x. Equations (7.2.1) and (7.2.2) have the form

$$w_t = w_{xx} + v, \qquad w_{tt} = w_{xx} + v. \tag{7.3.1}$$

The domain Ω is the interval $[0,a]$ of the x axis, and its boundary consists of the two points $x = 0$, $x = a$. We shall consider in tandem conditions (7.2.3) of Dirichlet and Neumann type

$$w(0) = w(a) = 0, \qquad w_x(0) = w_x(a) = 0. \tag{7.3.2}$$

The eigenfunctions $\varphi_k(x)$ corresponding to problems (7.3.1) and (7.3.2) satisfy the equations

$$\varphi_k'' = -\lambda_k \varphi_k, \qquad 0 < x < a, \tag{7.3.3}$$

where the primes denote differentiation with respect to x, together with Dirichlet or Neumann conditions

$$\varphi_k(0) = \varphi_k(a) = 0, \qquad \varphi_k'(0) = \varphi_k'(a) = 0. \tag{7.3.4}$$

The eigenvalues of problems (7.3.3) and (7.3.4) are as follows:

$$\lambda_k = \omega_k^2, \qquad \omega_k = \frac{\pi k}{a}, \tag{7.3.5}$$

where $k \geq 1$ for the Dirichlet problem and $k \geq 0$ for the Neumann problem. The orthonormalized eigenfunctions for the Dirichlet and Neumann problems are, respectively, equal to

$$\begin{gathered} \varphi_k(x) = \left(\frac{2}{a}\right)^{1/2} \sin(\omega_k x), \qquad k = 1, 2 \ldots, \\ \varphi_0(x) = a^{-1/2}, \qquad \varphi_k(x) = \left(\frac{2}{a}\right)^{1/2} \cos(\omega_k x). \end{gathered} \tag{7.3.6}$$

The quantities Φ_k from (7.2.19) are bounded in this case

$$\Phi_k = \left(\frac{2}{a}\right)^{1/2}, \qquad k \geq 1, \qquad \Phi_0 = a^{-1/2}. \tag{7.3.7}$$

We shall compute the Fourier coefficients (7.2.14) and (7.2.15), assuming that the initial functions $w_0(x)$ and $w_{t0}(x)$ are differentiable with respect to x a sufficient number of times and using integration by parts. With the help of (7.3.6) we obtain

$$q_k(0) = \int_0^a w_0 \varphi_k dx$$

$$= \left(\frac{2}{a}\right)^{1/2} \omega_k^{-1} \left\{ [(-w_0)\cos(\omega_k x)]\Big|_0^a + \int_0^a w_0' \cos(\omega_k x) dx \right\}$$

$$= \left(\frac{2}{a}\right)^{1/2} \omega_k^{-1} \left\{ [(-w_0)\cos(\omega_k x)]\Big|_0^a - \omega_k^{-1} \int_0^a w_0'' \sin(\omega_k x) dx \right\}$$

$$= \left(\frac{2}{a}\right)^{1/2} \omega_k^{-1} \left\{ [(-w_0 + \omega_k^{-2} w_0'')\cos(\omega_k x)]\Big|_0^a + \omega_k^{-3} \int_0^a w_0^{\mathrm{IV}} \sin(\omega_k x) dx \right\} \tag{7.3.8}$$

for the Dirichlet problem and

$$q_k(0) = \left(\frac{2}{a}\right)^{1/2} \omega_k^{-2} \left\{ [w_0' \cos(\omega_k x)]\Big|_0^a + \omega_k^{-1} \int_0^a w_0''' \sin(\omega_k x) dx \right\}$$

$$= \left(\frac{2}{a}\right)^{1/2} \omega_k^{-2} \left\{ [(w_0' - \omega_k^{-2} w_0''')\cos(\omega_k x)]\Big|_0^a - \omega_k^{-3} \int_0^a w_0^{\mathrm{V}} \sin(\omega_k x) dx \right\}, \tag{7.3.9}$$

$$k \geq 1,$$

for the Neumann problem. From relations (7.3.8) and (7.3.9) one can derive estimates for the Fourier coefficients depending, firstly, on the degree of smoothness of the initial function w_0 and secondly, on additional conditions at the boundary points $x = 0$ and $x = a$, i.e., on Γ. We drop the argument 0 of the function q_k. Henceforth, B_j are some positive constants and C^i are classes of functions having continuous derivatives in the interval $[0, a]$ up to order i inclusive. For the Dirichlet problem we obtain, using (7.3.8),

$$\begin{aligned}
|q_k| &\leq B_1 \omega_k^{-1} \quad \text{for} \quad w_0 \in C^1; \\
|q_k| &\leq B_2 \omega_k^{-2} \quad \text{for} \quad w_0 \in C^2, \quad w_0 = 0 \ \text{on} \ \Gamma; \\
|q_k| &\leq B_3 \omega_k^{-3} \quad \text{for} \quad w_0 \in C^3, \quad w_0 = 0 \ \text{on} \ \Gamma; \\
|q_k| &\leq B_4 \omega_k^{-4} \quad \text{for} \quad w_0 \in C^4, \quad w_0 = w_0'' = 0 \ \text{on} \ \Gamma.
\end{aligned} \tag{7.3.10}$$

For the Neumann problem we similarly have from (7.3.9)

$$
\begin{aligned}
&|q_k| \le B_5 \omega_k^{-1} \quad \text{for} \quad w_0 \in C^1; \\
&|q_k| \le B_6 \omega_k^{-2} \quad \text{for} \quad w_0 \in C^2; \\
&|q_k| \le B_7 \omega_k^{-3} \quad \text{for} \quad w_0 \in C^3, \quad w_0' = 0 \text{ on } \Gamma; \\
&|q_k| \le B_8 \omega_k^{-4} \quad \text{for} \quad w_0 \in C^4, \quad w_0' = 0 \text{ on } \Gamma.
\end{aligned}
\tag{7.3.11}
$$

Obviously, estimates of the form (7.3.10) and (7.3.11) can be continued without limit. For the Fourier coefficients $\dot{q}_k(0)$ from (7.2.15) we have estimates similar to (7.3.10) and (7.3.11), with w_0 replaced by w_{t0}.

Turning to the investigation of the convergence of the series in (7.2.32) and (7.2.43), we note that according to (7.3.7) the quantities Φ_k are independent of k. Using also relation (7.3.5), we obtain the following convergence conditions for the series.

Series (7.2.32) for the Dirichlet problem converges under the conditions

$$
w_0 \in C^2, \qquad w_0 = 0 \text{ on } \Gamma, \tag{7.3.12}
$$

and for the Neumann problem under the condition

$$
w_0 \in C^2. \tag{7.3.13}
$$

The series (7.2.43) for the Dirichlet problem converges under the conditions

$$
w_0 \in C^3, \qquad w_{t0} \in C^2, \qquad w_0 = w_{t0} = 0 \text{ on } \Gamma, \tag{7.3.14}
$$

and for the Neumann problem under the conditions

$$
w_0 \in C^3, \qquad w_{t0} \in C^2, \qquad \frac{\partial w_0}{\partial n} = 0 \text{ on } \Gamma. \tag{7.3.15}
$$

We note that convergence conditions (7.3.12) and (7.3.14) for series (7.2.32) and (7.2.43) for the Dirichlet problem include, as well as smoothness requirements, Dirichlet conditions on the initial functions w_0 and w_{t0}. Generally speaking, such conditions are not necessary in the statement of initial-boundary-value problems, and they are an additional imposition. In the case of the Neumann problem, however, conditions (7.3.13) and (7.3.15) are less restrictive: for series (7.2.32) no conditions other than smoothness are imposed, while for series (7.2.43) the Neumann condition is only imposed on the initial function w_0 (and not on the function w_{t0}).

We recall that the control problem for the first equation of (7.3.1) (the heat conduction equation) is always solvable, and conditions (7.3.12) and (7.3.13) ensuring the convergence of series (7.2.32) are there only to apply the simple estimate of the control process time in (7.2.32). For the second equation of (7.3.1) (the vibrating string equation), conditions (7.3.14) and (7.3.15) are sufficient conditions for the control problem to be solvable by the proposed methods.

Remark 7.1. If conditions of the form (7.3.14) or (7.3.15) on Γ are not fulfilled at the initial instant $t = 0$, then, however, the proposed control method all the same can be applied. We notice that these conditions will be necessarily satisfied for an arbitrary small $t = \Delta t > 0$ due to the imposed boundary conditions (for an arbitrary control $v(x,t)$ on the interval $t \in [0, \Delta t]$). Therefore, the proposed control method can be applied for $t \geq \Delta t$. Hence, these conditions on Γ for $t = 0$ are not essential. Thus, sufficient solvability conditions of the control problem for the vibrating string equation with Dirichlet and Neumann conditions have the form

$$w_0 \in C^3, \qquad w_{t0} \in C^2. \tag{7.3.16}$$

Hereafter, we will also use stated above consideration, omitting not essential conditions on the boundary Γ at $t = 0$.

7.3.2 Control of beam oscillations ($n = 1$, $A = -\Delta^2$)

As an example of a fourth-order equation we consider the control of transverse oscillations of an elastic beam. Equation (7.2.2) in this case has the form

$$w_{tt} = -w_{xxxx} + v. \tag{7.3.17}$$

We will restrict ourselves to hinged support boundary conditions at both ends of a beam of length a, i.e.,

$$w = w_{xx} = 0 \ \text{ on } \Gamma, \qquad \Gamma = \{x = 0, x = a\}. \tag{7.3.18}$$

The eigenvalue problem (7.2.7) for system (7.3.17), (7.3.18) has the form

$$\varphi^{\mathrm{IV}} = \lambda \varphi, \qquad x \in \Omega = [0, a], \qquad \varphi = \varphi'' = 0 \ \text{ on } \Gamma. \tag{7.3.19}$$

It is well known that the eigenvalues of problem (7.3.19) are positive and are

$$\lambda_k = \omega_k^2, \qquad \omega_k = \left(\frac{k\pi}{a}\right)^2, \qquad k = 1, 2, \ldots, \tag{7.3.20}$$

where ω_k are interpreted as the frequencies of the natural oscillations of the beam. The corresponding eigenfunctions of problem (7.3.19) can be represented in the form of (7.3.6). Hence, estimates (7.3.7), (7.3.8), and (7.3.10) remain valid for the problem under consideration, but throughout (7.3.6), (7.3.8), and (7.3.10) the frequencies ω_k are now defined by formulas (7.3.20) [instead of (7.3.5)]. Using the given estimates, we obtain, like (7.3.14), the following sufficient conditions for series (7.2.43) to converge in the problem under consideration:

$$w_0 \in C^4, \qquad w_{t0} \in C^2. \tag{7.3.21}$$

We omit not essential conditions on Γ at $t = 0$, in accordance with Remark 7.1 stated above. It can be shown [36] that the sufficient conditions of the series convergence in the control problem for the elastic beam oscillations have the same form (7.3.21) also for other boundary conditions, namely:

$$w = w_x = 0 \quad \text{at} \quad x = 0, \qquad x = a;$$

$$w = w_x = 0 \quad \text{at} \quad x = 0, \qquad w = w_{xx} = 0 \quad \text{at} \quad x = a;$$

$$w = w_x = 0 \quad \text{at} \quad x = 0, \qquad w_x = w_{xxx} = 0 \quad \text{at} \quad x = a;$$

$$w = w_x = 0 \quad \text{at} \quad x = 0, \qquad w_{xx} = w_{xxx} = 0 \quad \text{at} \quad x = a;$$

$$w = w_{xx} = 0 \quad \text{at} \quad x = 0, \qquad w_x = w_{xxx} = 0 \quad \text{at} \quad x = a.$$

7.3.3 The two-dimensional and three-dimensional problems ($n = 2, 3$; $A = \Delta$)

We now consider the equations

$$w_t = \Delta w + v, \qquad w_{tt} = \Delta w + v; \qquad n = 2, 3 \tag{7.3.22}$$

in the two-dimensional and three-dimensional cases. Suppose the domain Ω is a rectangle when $n = 2$ and a rectangular parallelepiped when $n = 3$, i.e., specified by

$$\Omega: \qquad 0 \le x_l \le a_l; \qquad l = 1, 2 \qquad \text{or} \qquad l = 1, 2, 3. \tag{7.3.23}$$

The solutions of the eigenvalue problem (7.2.8) for (7.3.22) in domains (7.3.23) under Neumann and Dirichlet conditions are known and are obtained by separation of variables. In the two-dimensional ($n = 2$) Dirichlet case we obtain, like (7.3.5) and (7.3.6)

$$\begin{gathered} \lambda_{ik} = \omega_{ik}^2 = \pi \left[\left(\frac{i}{a_1} \right)^2 + \left(\frac{k}{a_2} \right)^2 \right]; \qquad i, k = 1, 2, \ldots, \\ \varphi_{ik}(x_1, x_2) = 2(a_1 a_2)^{-1/2} \sin\left(\frac{\pi i x_1}{a_1} \right) \sin\left(\frac{\pi k x_2}{a_2} \right). \end{gathered} \tag{7.3.24}$$

For the Neumann problem the eigenvalues are given by relations (7.3.24) for $i, k = 0, 1, \ldots$, while the eigenfunctions have a form similar to (7.3.6)

$$\begin{gathered}
\varphi_{ik}(x_1,x_2) = 2(a_1a_2)^{-1/2}\cos\left(\frac{\pi i x_1}{a_1}\right)\cos\left(\frac{\pi k x_2}{a_2}\right),\\
\varphi_{00}(x_1,x_2) = (a_1a_2)^{-1/2},\\
\varphi_{0k}(x_1,x_2) = (2a_1a_2)^{-1/2}\cos\left(\frac{\pi k x_2}{a_2}\right),\\
\varphi_{i0}(x_1,x_2) = (2a_1a_2)^{-1/2}\cos\left(\frac{\pi i x_1}{a_1}\right); \qquad i,k = 1,2\ldots.
\end{gathered} \tag{7.3.25}$$

By the virtue of (7.3.24) and (7.3.25), the quantities (7.2.19) are bounded

$$\Phi_{ik} = (2a_1a_2)^{-1/2}; \qquad i,k = 1,2,\ldots. \tag{7.3.26}$$

We will now estimate the Fourier coefficients (7.2.14) and (7.2.15), assuming that the initial functions w_0 and w_{t0} are sufficiently smooth. Replacing the multiple integrals over the domain Ω by repeated integration over x_1, x_2, and then using integration by parts, we obtain, like (7.3.8)–(7.3.11), the following estimates

$$\begin{aligned}
&|q_{ik}| \le B_1(ik)^{-1} \quad \text{for} \quad w_0 \in C^{(1)};\\
&|q_{ik}| \le B_2(ik)^{-2} \quad \text{for} \quad w_0 \in C^{(2)}, \quad w_0 = 0 \text{ on } \Gamma;\\
&|q_{ik}| \le B_3(ik)^{-3} \quad \text{for} \quad w_0 \in C^{(3)}, \quad w_0 = 0 \text{ on } \Gamma
\end{aligned} \tag{7.3.27}$$

for the Dirichlet problem and

$$\begin{aligned}
&|q_{ik}| \le B_4(ik)^{-1}, \qquad |q_{i0}| \le B_5k^{-1};\\
&|q_{i0}| \le B_6 i^{-1} \quad \text{for} \quad w_0 \in C^{(1)};\\
&|q_{ik}| \le B_7(ik)^{-1}, \qquad |q_{0k}| \le B_8k^{-2};\\
&|q_{i0}| \le B_9 i^{-2} \quad \text{for} \quad w_0 \in C^{(2)};\\
&|q_{ik}| \le B_{10}(ik)^{-3}, \qquad |q_{0k}| \le B_{11}k^{-3};\\
&|q_{i0}| \le B_{12} i^{-3} \quad \text{for} \quad w_0 \in C^{(3)}, \quad \frac{\partial w_0}{\partial n} = 0 \text{ on } \Gamma
\end{aligned} \tag{7.3.28}$$

for the Neumann problem. In (7.3.27) and (7.3.28) $i,k = 1,2,\ldots$, everywhere, while $C^{(r)}$ is the class of functions w having continuous partial derivatives of the form

$$\frac{\partial^{p+q}}{\partial x_1^p \partial x_2^q}, \qquad 0 \le p \le r, \qquad 0 \le q \le r. \tag{7.3.29}$$

in the closed domain Ω.

For the Fourier coefficients $\dot{q}_{ik}$ from (7.2.15) there are estimates similar to (7.3.27) and (7.3.28), with w_0 replaced by w_{t0}.

Using relations (7.3.24), (7.3.26)–(7.3.28) we obtain the required sufficient conditions for series (7.2.32) and (7.2.43) to converge. In the cases considered here, summation in these series is performed over two indices i and k, from 1 to ∞ for the Dirichlet problem and from 0 to ∞ for the Neumann problem.

It turns out that series (7.2.32) converges for the Dirichlet problem under the conditions

$$w_0 \in C^{(2)}, \qquad w_0 = 0 \;\text{ on }\; \Gamma, \tag{7.3.30}$$

and for the Neumann problem under the condition

$$w_0 \in C^{(2)}. \tag{7.3.31}$$

Series (7.2.43) converge for the Dirichlet problem under the conditions

$$w_0 \in C^{(3)}, \qquad w_{t0} \in C^{(2)}, \qquad w_0 = w_{t0} = 0 \;\text{ on }\; \Gamma, \tag{7.3.32}$$

and for the Neumann problem under the conditions

$$w_0 \in C^{(3)}, \qquad w_{t0} \in C^{(2)}, \qquad \frac{\partial w_0}{\partial n} = 0 \;\text{ on }\; \Gamma. \tag{7.3.33}$$

The convergence conditions (7.3.30)—(7.3.33) are completely analogous to the corresponding conditions (7.3.12)—(7.3.15) for the one-dimensional problem.

Omitting not essential conditions on Γ at $t = 0$ in accordance with Remark 7.1, we find that series (7.2.32) for the Dirichlet and Neumann problems converges under the condition (7.3.31), and series (7.2.43) for the same problems converge under the following conditions

$$w_0 \in C^{(3)}, \qquad w_{t0} \in C^{(2)}. \tag{7.3.34}$$

In the three-dimensional case ($n = 3$), which is completely analogous to the two-dimensional one, the eigenvalues are given by equalities similar to (7.3.24)

$$\lambda_{ijk} = \pi^2 \left[\left(\frac{i}{a_1}\right)^2 + \left(\frac{j}{a_2}\right)^2 + \left(\frac{k}{a_3}\right)^2 \right].$$

Here, $i, j, k \geq 1$ for the Dirichlet problem and $i, j, k \geq 0$ for the Neumann problem.

Formulas and estimates similar to (7.3.24)—(7.3.26) hold for the eigenfunctions and Fourier coefficients. Finally, we arrive at exactly the same convergence conditions (7.3.30)—(7.3.34) as in the two-dimensional case. Here, as in (7.3.29), $C^{(r)}$ is the class of functions w having continuous partial derivatives of the form

$$\frac{\partial^{p+q+s}}{\partial x_1^p \partial x_2^q \partial x_3^s}, \qquad 0 \leq p \leq r, \qquad 0 \leq q \leq r, \qquad 0 \leq s \leq r,$$

in the closed domain Ω.

7.3.4 Solvability conditions in the general case

As was pointed out in Sect. 7.2.5, no additional conditions are required for the control problem to be solvable for (7.2.1), while for the control of (7.2.2) it is sufficient, for example, that the functions $Q_3(x)$ and $Q_4(x)$ from (7.2.41) be uniformly bounded in Ω. We shall analyze these conditions.

Below we shall always assume sufficient smoothness of the coefficients of the operators of A from (7.2.2) and M from (7.2.3), and also of the boundaries Γ and initial functions w_0 and w_{t0} from (7.2.5).

We note that the series (7.2.41) contain, firstly, eigenfunctions $\varphi_k(x)$ of problem (7.2.8), and secondly, Fourier coefficients q_k and $\dot{q}_k$ of the initial functions w_0 and w_{t0}. It is therefore desirable to use the following estimates for the series (7.2.41), which follow from the Cauchy inequality and enable us to separate the contributions of the eigenfunctions and Fourier coefficients

$$
\begin{aligned}
Q_3(x) &\leq \left[\sum \lambda_k^{-\beta} \varphi_k^2(x) \cdot \sum \lambda_k^{1+\beta} q_k^2\right]^{1/2}, \\
Q_4(x) &\leq \left[\sum \lambda_k^{-\gamma} \varphi_k^2(x) \cdot \sum \lambda_k^{\gamma} (\dot{q}_k)^2\right]^{1/2}.
\end{aligned}
\tag{7.3.35}
$$

Here, β and γ are currently arbitrary numbers, which will be chosen later so that all the series in (7.3.35) are bounded.

We shall consider fractional (positive and negative) powers of the differential operator A. An operator A of order $2m$ defines a transformation $Aw = f$. Its domain of definition D_A is the class of functions w defined in the domain Ω, having square-integrable partial derivatives up to order $2m$ inclusive (this fact can be expressed in the form $D_A \subset H_{2m}(\Omega)$, where H_{2m} is the corresponding Sobolev space), and also satisfying boundary conditions (7.2.3).

According to Agmon's kernel theorem [1], for $2ms > n$ the operator A^{-s} is an integral operator with a continuous kernel equal to

$$
K(x,y) = \sum \lambda_k^{-s} \varphi_k(x) \varphi_k(y).
$$

Putting $x = y$, i.e., considering the kernel on the diagonal, we obtain the uniform boundedness of the series

$$
\sum \lambda_k^{-s} \varphi_k^2(x) \leq \text{const} < \infty, \qquad 2ms > n.
$$

It follows from this that for uniform boundedness of the first factors on the right-hand sides of (7.3.35), i.e., the series depending on x, it is sufficient that

$$
\beta > n(2m)^{-1}, \qquad \gamma > n(2m)^{-1}. \tag{7.3.36}
$$

Conditions (7.3.36) for $m = 1$ were first given in [67]. The second factors in the right-hand sides of (7.3.35) (series depending on the Fourier coefficients) can, by Parceval's equality, be represented in the form

$$
\begin{aligned}
&\sum \lambda_k^{1+\beta} q_k^2 = \int\limits_{\Omega} \left(A^{(1+\beta)/2} w_0\right)^2 dx, \\
&\sum \lambda_k^{\gamma} \dot{q}_k^2 = \int\limits_{\Omega} \left(A^{\gamma/2} w_{t0}\right)^2 dx.
\end{aligned}
\tag{7.3.37}
$$

Series (7.3.37) converge if the functions $A^{(1+\beta)/2}w_0$ and $A^{\gamma/2}w_{t0}$ are square integrable in the domain Ω, i.e., belong to the class $L_2(\Omega)$. In other words, the functions w_0 and w_{t0} should belong to the domains of definition of the corresponding operator:

$$
w_0 \in D_{A^{(1+\beta)/2}}, \qquad w_{t0} \in D_{A^{\gamma/2}}. \tag{7.3.38}
$$

It follows from results of [110] that the domain of definition D_{A^s}, for $s \in (0,1)$ lies in $H_{2ms}(\Omega)$ and is distinguished by those boundary conditions (7.2.3) whose order $\operatorname{ord} M_j = r_j < r = 2ms - 1/2$. In the case when for some j we have $r_j = r$, the corresponding boundary condition is to be understood in some integral sense.

From (7.3.38) we have, in the case under consideration

$$
\begin{aligned}
&s = \frac{1}{2}(1+\beta), \qquad r = m(1+\beta) - \frac{1}{2} \qquad \text{for} \qquad w_0, \\
&s = \frac{1}{2}\gamma, \qquad r = m\gamma - \frac{1}{2} \qquad \text{for} \qquad w_{t0},
\end{aligned}
\tag{7.3.39}
$$

where s can also be greater than unity.

Suppose, for example, $s = 1 + \sigma$, where $\sigma \in (0,1)$. Then, representing the result of the action of the operator A^s in the form $A^s w = A^{\sigma}(Aw)$ and applying Seeley's theorem [110], we arrive at the following assertion. The domain of definition D_{A^s} lies in $H_{2ms}(\Omega)$ and is distinguished by boundary conditions (7.2.3) and also those boundary conditions $M_j Aw = 0$ for which $\operatorname{ord} M_j < 2m\sigma - 1/2$. In other words, for $s \in (1,2)$, as well as the boundary conditions (7.2.3), conditions of the form $M_j Aw = 0$ for which $\operatorname{ord}(M_j A) < r = 2ms - 1/2$ are also imposed on the function w.

Similar results also follow from lemmas derived in Appendix 2 of [68].

Thus, for the convergence of series (7.3.37), the functions w_0 and w_{t0} should satisfy conditions depending on parameters s and r, the stringency of these conditions increasing with s and r. We note that for restrictions $r_j < r$ on operator orders, the fractional part of r is not significant because r_j are integers.

We determine two numbers for each of the functions w_0 and w_{t0} with the help of relations (7.3.36) and (7.3.39): the lower bound s^* on the possible values of s and the integer part r^* of the lower bound on possible values of r. The values of $\vartheta^* = 2ms^*$ and r^* for various pairs n, m for $n \le 3$, $m \le 2$ are shown in Table 7.1.

Using the values of ϑ^* and r^* obtained, one can answer the question of the convergence of series (7.3.35) and thereby obtain sufficient conditions for the control problems under consideration to be solvable. For this, it is sufficient to require that the following conditions be satisfied.

Table 7.1 The values of ϑ^* and r^* for various pairs n, m

n,m	$\vartheta^*(w_0)$	$\vartheta^*(w_{t0})$	$r^*(w_0)$	$r^*(w_{t0})$
1,1	3/2	1/2	1	0
1,2	5/2	1/2	2	0
2,1	2	1	1	0
2,2	3	1	2	0
3,1	5/2	3/2	2	1
3,2	7/2	3/2	3	1

Firstly, the functions w_0 and w_{t0} should belong to classes $H_\vartheta(\Omega)$, where ϑ is any number greater than the corresponding ϑ^*. In particular, ϑ can be chosen to be an integer, and this requirement will then indicate the existence for the functions w_0 and w_{t0} of square-integrable partial derivatives up to order ϑ inclusive.

Secondly, the functions w_0 and w_{t0} should satisfy those boundary conditions (7.3.36) on Γ for which $\operatorname{ord} M_j \le r^*$, and those of the boundary conditions $M_jAw = 0$ for which $\operatorname{ord}(M_jA) \le r^*$. Because $\operatorname{ord} M_j < \operatorname{ord} A = 2m$, the imposition of the conditions $M_jAw = 0$ is only required when $r^* \ge 2m$.

It is clear from Table 7.1 that the inequality $r^* \ge 2m$ only holds when $n = 3$, $m = 1$ for the function w_0. In this case for the Dirichlet problem ($\operatorname{ord} M = 0$) we have $\operatorname{ord} MA = 2 = r^*(w_0)$, and it is necessary to impose on w_0 the additional condition $Aw = 0$ on Γ.

In the case of the Neumann problem ($\operatorname{ord} M = 1$) for $n = 3$, $m = 1$, and also for all problems with other values of n, m, additional conditions do not appear.

The appearance of an additional boundary condition can be explained as follows. The proposed control law (7.2.10) vanishes on Γ in the case of the Dirichlet problem because here $\varphi_k = 0$ on Γ. This reduces the possibility of control on the boundary of the domain, and can require additional conditions on the initial functions on Γ.

At the same time, some of the boundary conditions (7.2.3) for the problem to be solvable need not be applied. For example, for $n = 2$, $m = 1$ we have $r^*(w_0) = 1$, $r^*(w_{t0}) = 0$. Consequently, for a second-order operator A in the case of the Dirichlet problem ($\operatorname{ord} M = 0$) the functions w_0 and w_{t0} should satisfy the Dirichlet condition, while in the case of the Neumann problem ($\operatorname{ord} M = 1$) the function w_0 should satisfy the Neumann condition, while the function w_{t0} need not satisfy it.

Comparing the data in the Table 7.1 with the results of the examples in Sects. 7.3.1–7.3.3, we see that in the examples the convergence conditions turned out to be less restrictive for $n = 1$, $m = 2$ and $n = 3$, $m = 1$. For $n = 1$, $m = 2$ in the example it is not required to impose the condition $w_0'' = 0$ on Γ, which appears in the Table 7.1: $r^*(w_0) = 2$. For $n = 3$, $m = 1$, it follows from Table 7.1 that in the Neumann problem example the condition $\partial w_{t0}/\partial n = 0$ is not required, as well as the condition is $\Delta w_0 = 0$ on Γ for the Dirichlet problem.

It should be taken into account that, according to Remark 7.1 from Sect. 7.3.1, conditions on Γ at $t = 0$ are not essential for the solution of the control problem, if they are automatically fulfilled for $t = \Delta t > 0$ due to the imposed boundary conditions.

Chapter 8
Control system under complex constraints

In this chapter a method is elaborated for constructing a control in a linear system under mixed constraints imposed at every instant of time on phase coordinates, controls, and certain integrals depending on these variables. The control method proposed is a generalization of Kalman's approach to linear systems subjected to constraints. A bounded scalar control is constructed in an explicit form for a system of oscillators, as well as for some other oscillatory systems. For certain second-order systems, the control law elaborated is compared with the time-optimal control. A control law is also constructed for some higher order systems, specifically, for fourth-order systems with mixed constraints that are models of mechanical and electromechanical systems containing oscillatory links and electric motors.

8.1 Control design in linear systems under complex constraints

8.1.1 Problem statement

We consider the linear control system

$$\dot{x} = A(t)x + B(t)u + f(t). \tag{8.1.1}$$

Here, $x = (x_1, \ldots, x_n)$ is the n-dimensional vector of phase coordinates and $u = (u_1, \ldots, u_m)$ is the m-dimensional vector of control. The $(n \times n)$-matrix $A(t)$, the $(n \times m)$-matrix $B(t)$, and the n-dimensional vector $f(t)$ are given piecewise continuous functions of time t.

Let the phase and control variables of system (8.1.1) be constrained by the inequalities that express the boundedness of the absolute values or the components of certain linear combinations of the variables x and u, and also certain integrals. To be precise, we consider constraints of the following two types:

$$|C^i(t)x(t) + D^i(t)u(t)$$

$$+\int_{t_0}^{T}[G^i(t,\tau)x(\tau)+H^i(t,\tau)u(\tau)]\,d\tau+\mu^i(t)|\leq 1,\quad i=1,\dots,r, \tag{8.1.2}$$

$$|\langle p^j(t),x(t)\rangle+\langle q^j(t),u(t)\rangle$$
$$+\int_{t_0}^{T}[\langle g^j(t,\tau),x(\tau)\rangle+\langle h^j(t,\tau),u(\tau)\rangle]\,d\tau\leq 1,\quad j=1,\dots,r. \tag{8.1.3}$$

Here, as before, brackets $\langle .,. \rangle$ denote the scalar product of vectors.

Constraints (8.1.2) and (8.1.3) must hold for all $t\in[t_0,T]$, where t_0 and T are the initial and terminal instants of time, respectively. We consider the instant t_0 to be fixed, whereas T not fixed, for a while. In relations (8.1.2) and (8.1.3), C^i and G^i are $(l\times n)$-matrices, D^i and H^i are $(l\times m)$-matrices, μ^i are l-vectors, for some integer l, p^j and g^j are n-vectors, and q^j and h^j are m-vectors. The matrices and vectors C^i,D^i,μ^i,p^j, and q^j are given piecewise continuous functions of t on the segment $[t_0,T]$, and the matrices and vectors G^i,H^i,g^j, and h^j are given piecewise continuous functions of t and τ for $t,\tau\in[t_0,T]$.

Constraints (8.1.2) and (8.1.3), in particular, include the most commonly encountered restrictions imposed on control, state, and their combinations. Specifically, if D^i is the identity $(m\times m)$-matrix and all other matrices and vectors C^i,G^i,H^i, and μ^i in (8.1.2) are equal to zero, then we obtain, from (8.1.2), the restriction $|u(t)|\leq 1$ on the absolute value of control. If the vector q^j has a single non-zero component and all other vectors p^j,g^j, and h^j in (8.1.3) are equal to zero, then we obtain, from (8.1.3), a restriction on one component of the control vector. Setting all the matrix and vector coefficients in (8.1.2) and (8.1.3) except C^i and p^j equal to zero, we obtain the phase constraints. Similarly, if we set all the matrices and vectors except H^i and h^j equal to zero, we obtain integral constraints on the control, etc.

Let us state the problem of constructing a control $u(t)$ that satisfies constraints (8.1.2) and (8.1.3) for $t\in[t_0,T]$ and brings system (8.1.1) from a given initial state

$$x(t_0)=x^0 \tag{8.1.4}$$

to a given terminal state

$$x(T)=x^1. \tag{8.1.5}$$

Here, x^0 and ^{1} are given n-dimensional vectors.

We denote by $\Phi(t)$ the fundamental matrix of the homogeneous system (8.1.1). We have

$$\dot{\Phi}=A(t)\Phi,\quad \Phi(t_0)=E_n, \tag{8.1.6}$$

where E_n is the $(n\times n)$ identity matrix. Let us write the solution of system (8.1.1) satisfying the initial condition (8.1.4) in the form

$$x(t)=\Phi(t)\{x^0+\int_{t_0}^{t}\Phi^{-1}(\tau)[B(\tau)u(\tau)+f(\tau)]\,d\tau\}. \tag{8.1.7}$$

Substituting solution (8.1.7) into the boundary condition (8.1.5), we obtain

$$\int_{t_0}^{T} \Phi^{-1}(t)B(t)u(t)\,dt = x^*. \tag{8.1.8}$$

Here, we have introduced the notation

$$x^* = \Phi^{-1}(T)x^1 - x^0 - \int_{t_0}^{T} \Phi^{-1}(t)f(t)\,dt. \tag{8.1.9}$$

Thus, the desired control must satisfy constraints (8.1.2) and (8.1.3), and also condition (8.1.8).

8.1.2 Kalman's approach

We use the method of control proposed in [72, 73] for the case where there are no constraints. We shall seek the control in the form

$$u = Q^\top c, \tag{8.1.10}$$

where c denotes an n-dimensional constant vector, $Q(t)$ denotes the $(n \times m)$-matrix

$$Q(t) = \Phi^{-1}(t)B(t), \tag{8.1.11}$$

and superscript $^\top$ denotes the transpose. Substituting (8.1.10) into (8.1.8), we obtain the equation for the vector c:

$$R(T)c = x^*. \tag{8.1.12}$$

Here,

$$R(t) = \int_{t_0}^{t} Q(\tau)Q^\top(\tau)\,d\tau. \tag{8.1.13}$$

It follows from (8.1.13) that $R(t)$ is a symmetric nonnegative definite $(n \times n)$-matrix for $t \geq t_0$. We assume that the matrix $R(t)$ is positive definite for $t \geq t_0$. As is known [78], this property implies complete controllability of the linear system (8.1.1). In this case, system (8.1.12) has the unique solution

$$c = R^{-1}(T)x^*. \tag{8.1.14}$$

Let us return to conditions (8.1.2) and (8.1.3). At first, we substitute control (8.1.10) into solution (8.1.7). Using the notation (8.1.11) and (8.1.13), we get

$$x(t) = \Phi(t)[x^0 + R(t)c + \int_{t_0}^{t} \Phi^{-1}(\tau)f(\tau)\,d\tau]. \tag{8.1.15}$$

We transform (8.1.15) with the aid of (8.1.9) and (8.1.14) as follows:

$$x(t) = \Phi(t)[\Phi^{-1}(T)x^1 + R_1(t,T)x^* - \int_t^T \Phi^{-1}(\tau)f(\tau)\,d\tau],$$
$$R_1(t,T) = R(t)R^{-1}(T) - E_n. \tag{8.1.16}$$

We now substitute expression (8.1.16) for x and the expression

$$u(t) = Q^\top(t)R^{-1}(T)x^* \tag{8.1.17}$$

for u that follows from (8.1.10) and (8.1.14), into constraints (8.1.2) and (8.1.3). Constraints (8.1.2) then take the form

$$|F^i(t,T)x^* + \phi^i(t,T)| \leq 1, \quad i = 1,\dots,r. \tag{8.1.18}$$

Here, the $(l \times n)$-matrix F^i and the l-vector ϕ^i are equal, respectively, to

$$F^i(t,T) = C^i(t)\Phi(t)R_1(t,T) + D^i(t)Q^\top(t)R^{-1}(T)$$
$$+ \int_{t_0}^T [G^i(t,\tau)\Phi(\tau)R_1(\tau,T) + H^i(t,\tau)Q^\top(\tau)R^{-1}(T)]\,d\tau,$$
$$\phi^i(t,T) = C^i(t)\Phi(t)[\Phi^{-1}(T)x^1 - \int_t^T \Phi^{-1}(\tau)f(\tau)\,d\tau] \tag{8.1.19}$$
$$+ \int_{t_0}^T G^i(t,\tau)\Phi(\tau)[\Phi^{-1}(T)x^1 - \int_\tau^T \Phi^{-1}(\tau_1)f(\tau_1)\,d\tau_1]\,d\tau + \mu^i(t), \quad i = 1,\dots,r.$$

Similarly, constraints (8.1.3) take the form

$$\langle \psi^j(t,T), x^* \rangle + \chi^j(t,T) \leq 1, \quad j = 1,\dots,s, \tag{8.1.20}$$

where the n-vector ψ^j and the scalar χ^j are given by

$$\psi^j(t,T) = R_1^\top(t,T)\Phi^\top(t)p^j(t) + R^{-1}(T)Q(t)q^j(t)$$
$$+ \int_{t_0}^T [R_1^\top(\tau,T)\Phi^\top(\tau)g^j(t,\tau) + R^{-1}(T)Q(\tau)h^j(t,\tau)]\,d\tau,$$
$$\chi^j(t,T) = \langle p^j(t), \Phi(t)[\Phi^{-1}(T)x^1 - \int_t^T \Phi^{-1}(\tau)f(\tau)\,d\tau]\rangle \tag{8.1.21}$$
$$+ \int_{t_0}^T \langle g^i(t,\tau), \Phi(\tau)[\Phi^{-1}(T)x^1 - \int_\tau^T \Phi^{-1}(\tau_1)f(\tau_1)\,d\tau_1]\rangle\,d\tau, \quad j = 1,\dots,r.$$

Let us note that the functions F^i, ϕ^i, ψ^j, and χ^j defined by (8.1.19) and (8.1.21) are expressed in terms of given functions and hence can be considered as known. For constraints (8.1.2) and (8.1.3) to be satisfied, it is necessary and sufficient that inequalities (8.1.18) and (8.1.20) hold for the given x^* and all $t \in [t_0, T]$. This imposes conditions on the time T of the process and on the vector x^*. By virtue of (8.1.9), these conditions lead (for a given terminal state x^1) to conditions on the time T and the initial state x^0. By majorizing and simplifying the left sides of inequalities

(8.1.18) and (8.1.20), we can obtain conditions on T and x^0 that guarantee constraints ((8.1.2) and (8.1.3). Thus, we will come to sufficient controllability conditions for system (8.1.1) under constraints (8.1.2) and (8.1.3).

Let us present one of the possible versions of such conditions. Suppose that, for all $T \geq t_0$ and all $t \in [t_0, T]$, the inequalities

$$|\phi^i(t,T)| \leq \phi_0^i < 1, \quad |\chi^j(t,T)| \leq \chi_0^j < 1, \quad i = 1,\dots,r, \quad j = 1,\dots,s, \tag{8.1.22}$$

are valid, where ϕ_0^i and χ_0^j are positive constants. Inequalities (8.1.18) certainly hold, if

$$|F^i(t,T)x^*| \leq 1 - \phi_0^i, \quad i = 1,\dots,r. \tag{8.1.23}$$

Using the Cauchy inequality, let us estimate the left-hand side of (8.1.23) from above as follows:

$$\begin{aligned} |F^i(t,T)x^*| &= [\sum_{j=1}^{l}(\sum_{k=1}^{n} F^i_{jk}x^*_k)^2]^{1/2} \\ &\leq [\sum_{j=1}^{l}\sum_{k=1}^{n}(F^i_{jk})^2]^{1/2}|x^*|, \quad i = 1,\dots,r. \end{aligned} \tag{8.1.24}$$

Substituting (8.1.24) into (8.1.23), we obtain a sufficient condition for inequalities (8.1.18) to hold:

$$\begin{gathered} |x^*| \leq \min_i \left\{ (1-\phi_0^i)[\max_t \sum_{j=1}^{l}\sum_{k=1}^{n}(F^i_{jk}(t,T))^2]^{-1/2} \right\}, \\ i = 1,\dots,r, \quad t \in [t_0, T]. \end{gathered} \tag{8.1.25}$$

Similarly, inequalities (8.1.20) are certainly valid, if

$$|x^*| \leq \min_j\{(1-\chi_0^j)[\max_t |\psi^j(t,T)|]^{-1}\}, \quad j = 1,\dots,s, \quad t \in [t_0, T]. \tag{8.1.26}$$

Thus, if conditions (8.1.25) and (8.1.26) are fulfilled, control (8.1.17) satisfies constraints (8.1.2) and (8.1.3) for $t \in [t_0, T]$ and brings system (8.1.1) from the initial state (8.1.4) to the terminal state (8.1.5). Therefore, conditions (8.1.25) and (8.1.26) can be regarded as sufficient conditions for controllability in finite time T. We note that, by virtue of equality (8.1.9), these conditions relate the initial and terminal states x^0 and x^1 and the time of the process T. Conditions (8.1.25) and (8.1.26) are imposed on the absolute value of the vector x^*.

Below in this chapter, following [35, 26, 53, 99, 7], we present methods of constructing a control for various linear systems under complex constraints.

The following theorem [26] provides simple sufficient conditions for the constraint (a is a positive constant)

$$|u(t)| \leq a \tag{8.1.27}$$

to be satisfied for the control law (8.1.17) in case of $f(t) \equiv 0$.

Theorem 8.1. *For some $T > t_0$, let the matrix $R(T)$ be non-singular, i.e., the condition of complete controllability holds, and let the inequalities*

$$|Q^\top(t)K(T)v| \leq \lambda_1(T)|v|, \quad t \in [t_0, T], \tag{8.1.28}$$

$$|R(T)K(T)v| \geq \lambda_2(T)|v|. \tag{8.1.29}$$

be valid for any n-dimensional vector v. Here, $K(T)$ is a non-singular $(n \times n)$-matrix, $\lambda_1(T) > 0$ and $\lambda_2(T) > 0$ are positive scalars, and v is a constant n-vector. Inequality (8.1.28) holds for all $t \in [t_0, T]$. Then, if the condition

$$|x^*| \leq a\lambda_2(T)\lambda_1^{-1}(T), \tag{8.1.30}$$

holds, then the control $u(t)$ given by (8.1.17) brings system (8.1.1) from state (8.1.4) to state (8.1.5) at the time instant T and satisfies constraint (8.1.27) for all $t \in [t_0, T]$.

Proof. Control (8.1.17) is constructed in such a way that conditions (8.1.4) and (8.1.5) hold. By (8.1.17), we have

$$|u(t)| = |Q^\top(t)R^{-1}(T)x^*| = |Q^\top(t)K(T)K^{-1}(T)R^{-1}(T)x^*|.$$

Using inequality (8.1.28), we have

$$|u(t)| \leq \lambda_1(T)|K^{-1}(T)R^{-1}(T)x^*|.$$

Now, we put $x^* = R(T)K(T)v$ and first apply (8.1.29) and then (8.1.30) to obtain

$$|u(t)| \leq \lambda_1(T)|v| \leq \lambda_1(T)\lambda_2^{-1}(T)|R(T)K(T)v| = \lambda_1(T)\lambda_2^{-1}(T)|x^*| \leq a.$$

We have thus shown that constraint (8.1.27) holds. This proves the theorem. □

Remark 8.1. The non-singular matrix $K(T)$ in (8.1.28) and (8.1.29) can be chosen arbitrarily; in particular, we can take the identity matrix $K = E_n$. The arbitrary choice of $K(T)$ can be useful, since it extends the range where our sufficient conditions are applicable. In the case of the identity matrix $K = E_n$ the number $\lambda_2(T)$ is, by (8.1.29), a lower bound for the minimal eigenvalue of the matrix $R(T)$.

Remark 8.2. To calculate the control (8.1.10), we have to solve the linear system of equations (8.1.12), whereas, in the time-optimal case, we have to solve a system of transcendental equations. Besides, control (8.1.10) is a continuous function of time, whereas the time-optimal control is, in general, discontinuous.

In further considerations, for various examples of mechanical systems, we obtain refined conditions of the controllability that take into account the contribution of different components of vector x^*.

8.2 Application to oscillating systems

8.2.1 Control for the system of oscillators

Following [26], we consider a system of harmonic oscillators subject to scalar control:

$$\ddot{\xi}_i + \omega_i^2 \xi_i = u. \tag{8.2.1}$$

Here, ξ_i are generalized coordinates, the constants $\omega_i > 0$, $i = 1, \ldots, n$, are the natural frequencies of the oscillators, u is the scalar control constrained by (8.1.27), i.e., $|u| \leq a$ where a is a constant.

As a mechanical model of system (8.2.1), one can take a system of mathematical pendulums suspended from a body G that moves horizontally with the acceleration u (see Fig. 8.1). Then, ξ_i, equal to $l_i \phi_i$, are small linear deviations of the pendulums from their points of suspension, where l_i is the length and ϕ_i is the angle of deviation of the pendulum from the vertical direction.

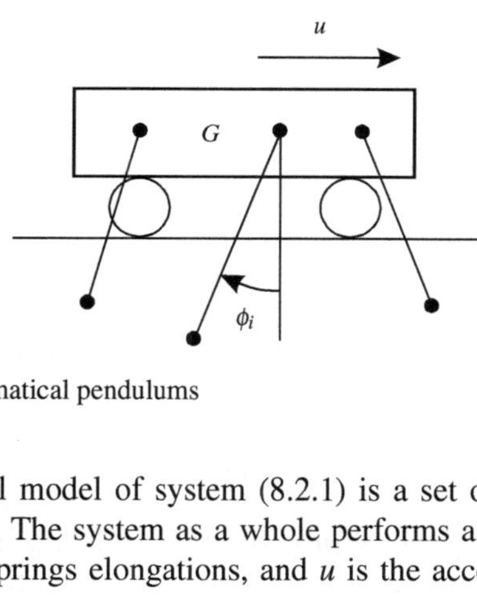

Fig. 8.1 System of mathematical pendulums

Another mechanical model of system (8.2.1) is a set of masses connected by springs to the body G. The system as a whole performs a translational horizontal motion, ξ_i being the springs elongations, and u is the acceleration of the body G (see Fig. 8.2).

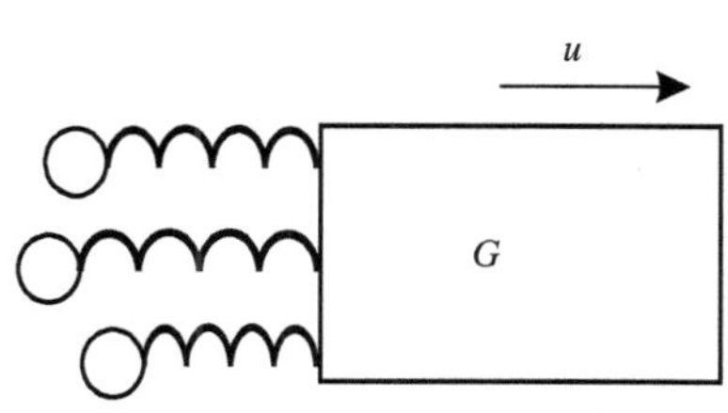

Fig. 8.2 Set of oscillators

Let us find a control $u(t)$ that satisfies constraint (8.1.27) and brings system (8.2.1) from the initial state at $t_0 = 0$:

$$\xi_i(0) = \xi_i^0, \quad \dot{\xi}_i(0) = \eta_i^0 \tag{8.2.2}$$

to the given terminal state

$$\xi_i(T) = \xi_i^1, \quad \dot{\xi}_i(T) = \eta_i^1. \tag{8.2.3}$$

We assume that the frequencies ω_i are positive and distinct. There is no loss of generality, if we number them in increasing order, put $\omega_0 = 0$, and introduce the notation

$$\Omega = \min_{0 \le k \le n-1} (\omega_{k+1} - \omega_k) > 0, \quad 0 = \omega_0 < \omega_1 < \ldots < \omega_n. \tag{8.2.4}$$

Note that, if $\Omega > 0$, system (8.2.1) is completely controllable [40]. If some frequencies are the same, the system becomes uncontrollable. Indeed, if the initial states of two oscillators with equal frequencies are different, no control can damp the oscillations of these two oscillators simultaneously: the phase difference between them remains constant.

Using the change of variables

$$\dot{\xi}_i = y_i, \quad \xi_i = \omega_i^{-1} z_i, \tag{8.2.5}$$

we reduce system (8.2.1) to the form

$$\dot{y}_i = -\omega_i z_i + u, \quad \dot{z}_i = \omega_i y_i. \tag{8.2.6}$$

The phase vector of system (8.2.6) is a $2n$-dimensional column vector composed of the components of vectors y and z. One can easily ascertain that the fundamental matrix (8.1.6) of the homogeneous system (8.2.6) is orthogonal and has the form

$$\Phi(t) = \begin{bmatrix} \mathrm{diag}(\cos\omega_i t) & \mathrm{diag}(-\sin\omega_i t) \\ \mathrm{diag}(\sin\omega_i t) & \mathrm{diag}(\cos\omega_i t) \end{bmatrix}, \quad \Phi^{-1}(t) = \Phi^\top(t). \tag{8.2.7}$$

Here, $\mathrm{diag}(a_i)$ denotes a diagonal $(n \times n)$-matrix with diagonal elements a_i.

For system (8.2.6), matrices $B(t)$ and $Q(t)$ are $2n$-dimensional column vectors. By (8.1.11), (8.2.6), and (8.2.7), their elements are

$$B_i = 1, \quad B_{n+i} = 0, \quad Q_i(t) = \cos\omega_i t, \quad Q_{n+i}(t) = -\sin\omega_i t. \tag{8.2.8}$$

From (8.1.13) and (8.2.8), we have

$$QQ^\top = \begin{bmatrix} Q^1 & Q^0 \\ Q^0 & Q^2 \end{bmatrix}, \quad R(T) = \begin{bmatrix} R^1 & R^0 \\ R^0 & R^2 \end{bmatrix},$$
$$R^k = \int_0^T Q^k\,dt, \quad k = 0,1,2. \tag{8.2.9}$$

Here, Q^k and R^k are $(n \times n)$-matrices. Their elements are calculated by means of (8.2.8) and (8.2.9) (throughout, $i, j = 1, \ldots, n$):

$$Q_{ij}^1 = \cos\omega_i t \cos\omega_j t, \quad Q_{ij}^2 = \sin\omega_i t \sin\omega_j t, \quad Q_{ij}^0 = -\cos\omega_i t \sin\omega_j t,$$
$$R_{ii}^{1,2} = \frac{T}{2} \pm \frac{\sin 2\omega_i T}{4\omega_i}, \quad R_{ij}^{1,2} = \frac{\sin(\omega_i - \omega_j)T}{2(\omega_i - \omega_j)} \pm \frac{\sin(\omega_i + \omega_j)T}{2(\omega_i + \omega_j)}, \tag{8.2.10}$$
$$R_{ii}^0 = \frac{\cos 2\omega_i T - 1}{4\omega_i}, \quad R_{ij}^0 = \frac{\cos(\omega_i - \omega_j)T - 1}{2(\omega_i - \omega_j)} + \frac{\cos(\omega_i + \omega_j)T - 1}{2(\omega_i + \omega_j)}, \quad i \neq j.$$

Note that, by condition (8.2.4), we have

$$\omega_i \geq \Omega, \quad |\omega_i - \omega_j| \geq \Omega, \quad \omega_i + \omega_j \geq 3\Omega, \quad i \neq j. \tag{8.2.11}$$

Takin into account (8.2.11), we obtain the following estimates for elements (8.2.10) of the matrix $R(T)$:

$$|R_{ii}^k - \frac{T}{2}| \leq \frac{1}{4\Omega}, \quad |R_{ii}^0| \leq \frac{1}{2\Omega}, \quad |R_{ij}^0| \leq \frac{4}{3\Omega},$$
$$|R_{ij}^k| \leq \frac{1}{2|\omega_i - \omega_j|} + \frac{1}{2|\omega_i + \omega_j|} \leq \frac{2}{3\Omega}, \quad k = 1, 2, \quad i \neq j. \tag{8.2.12}$$

In inequalities (8.1.28) and (8.1.29), we put $K(T) = E_{2n}$ and find $\lambda_1(T)$ and $\lambda_2(T)$. Let us estimate the left-hand side of inequality (8.1.28), by using the Cauchy inequality and expressions (8.2.8) for the components of the vector $Q(T)$

$$|Q^\top(t)v| \leq |Q^\top(t)||v| = n^{1/2}v.$$

Therefore, in (8.1.28), we can set

$$\lambda_1(T) = n^{1/2}. \tag{8.2.13}$$

Let us estimate the left-hand side of inequality (8.1.29). For any vector v, we have

$$|R(T)v| = \left|\frac{Tv}{2} + \left[R(T) - \frac{T}{2}E_{2n}\right]v\right| \geq \frac{T}{2}|v| - |Mv|,$$
$$M = R(T) - \frac{T}{2}E_{2n}. \tag{8.2.14}$$

Here, we introduce the symmetric $(2n \times 2n)$-matrix M. For its elements, using relations (8.2.9) and (8.2.12) for matrix $R(T)$, we obtain the estimates

$$|M_{ii}| \leq \frac{1}{4\Omega}, \quad |M_{n+i,n+i}| \leq \frac{1}{4\Omega}, \quad |M_{ij}| \leq \frac{2}{3\Omega},$$
$$|M_{n+i,n+j}| \leq \frac{2}{3\Omega}, \quad |M_{i,n+i}| \leq \frac{1}{2\Omega}, \quad |M_{i,n+j}| \leq \frac{4}{3\Omega}, \quad i \neq j. \tag{8.2.15}$$

By the Cauchy inequality, we have [the summation here and in (8.2.17) is from 1 to $2n$]

$$|Mv|^2 = \sum_i (\sum_j M_{ij} v_j)^2 \le \sum_i [(\sum_j M_{ij}^2)(\sum_j v_j^2)] = |v|^2 \sum_{i,j} M_{ij}^2. \tag{8.2.16}$$

Recalling estimates (8.2.15) and that the matrix M is symmetric, we obtain

$$\sum_{i,j} M_{ij}^2 \le \frac{2n}{16\Omega^2} + \frac{8(n^2-n)}{9\Omega^2} + \frac{2n}{4\Omega^2} + \frac{32(n^2-n)}{9\Omega^2} = \frac{5n(64n-55)}{72\Omega^2}. \tag{8.2.17}$$

Inequalities (8.2.16) and (8.2.17) yield

$$|Mv| \le \frac{k_n |v|}{\Omega}, \quad k_n = \left[\frac{5n(64n-55)}{72}\right]^{1/2}, \quad n \ge 1. \tag{8.2.18}$$

Using (8.2.14) and (8.2.18), we obtain

$$|R(T)v| \ge \left(\frac{T}{2} - \frac{k_n}{\Omega}\right)|v|. \tag{8.2.19}$$

Consequently, condition (8.1.29) holds, if $T \ge 2k_n/\Omega$. Then, comparing (8.1.29) and (8.2.19), we obtain

$$\lambda_2(T) = \left(\frac{T}{2} - \frac{k_n}{\Omega}\right) > 0. \tag{8.2.20}$$

Substituting relations (8.2.13) and (8.2.20) into (8.1.30) and solving for T, we get

$$T \ge \frac{2n^{1/2}}{a}|x^*| + \frac{2k_n}{\Omega}. \tag{8.2.21}$$

Vector x^* is given by relation (8.1.9), where the last term in the right-hand side equals zero since $f(t) \equiv 0$. By (8.2.5), (8.2.2), and(8.2.3), vectors x^0 and x^1 are:

$$x^0 = \{y_i(0), z_i(0)\}^\top = \{\eta_i^0, \omega_i \xi_i^0\}^\top,$$
$$x^1 = \{y_i(T), z_i(T)\}^\top = \{\eta_i^1, \omega_i \xi_i^1\}^\top. \tag{8.2.22}$$

We substitute the elements of the vector $Q(t)$ from (8.2.8) into control law (8.1.10):

$$u(t) = \sum_{i=1}^{n} (c_i \cos \omega_i t - c_{n+i} \sin \omega_i t). \tag{8.2.23}$$

By Theorem 8.1, we come to the following assertion [26].

Theorem 8.2. *Under condition (8.2.21), control (8.2.23), where vector c is given by (8.1.14) and matrix $R(T)$ is given by (8.2.9) and (8.2.10), satisfies constraint (8.1.27) and brings system (8.2.6) [or (8.2.1)] from the initial state (8.2.2) to the terminal state (8.2.3) in time T.*

Note that time T increases as $|x^*|$ increases, the magnitude a of the control decreases, and the natural frequencies come closer together, i.e., as Ω decreases.

As a special case, let us consider the problem of damping the initial oscillations, i.e., the problem of bringing the system to its equilibrium state. In this case, we have $x^1 = 0$, and, taking into account that $f(t) \equiv 0$, from equalities (8.1.9) and (8.2.22), we obtain [$E(t)$ is the energy of oscillations]:

$$|x^*|^2 = \sum_{i=1}^{n} [(\eta_i^0)^2 + \omega_i^2 (\xi_i^0)^2] = 2E_0, \quad E_0 = E(0), \tag{8.2.24}$$

$$E(t) = \frac{1}{2} \sum_{i=1}^{n} \left([\dot{\xi}_i(t)]^2 + \omega_i^2 [\xi_i(t)]^2 \right). \tag{8.2.25}$$

Using (8.2.24), we rewrite condition (8.2.21) as follows:

$$T \geq \frac{2(2nE_0)^{1/2}}{a} + \frac{2k_n}{\Omega}. \tag{8.2.26}$$

Under condition (8.2.26), control (8.2.23) brings system (8.2.1) from the initial state (8.2.2) to the equilibrium state $\xi_i = \dot{\xi} = 0$.

In the special case where $n = 1$, the minimal time that satisfies condition (8.2.26) is equal to [we use the second relation of (8.2.18)]

$$T^* = \frac{2(2E_0)^{1/2}}{a} + \frac{(5/2)^{1/2}}{\omega_1}. \tag{8.2.27}$$

We compare time (8.2.27) with the optimal control time under the condition

$$\varepsilon = \frac{a}{E_0^{1/2} \omega_1} \ll 1, \tag{8.2.28}$$

that means that the control is relatively small. Then, under constraint $|u| \leq 1$, the approximate optimal control for system (8.2.1) with $n = 1$ that is constructed in [40] by the method of small parameter [22] has the form

$$u = -a \operatorname{sign} \dot{\xi}_1, \tag{8.2.29}$$

while the phase coordinates are

$$\xi_1 = \frac{(2E)^{1/2}}{\omega_1} \cos(\omega_1 t + \alpha), \quad \dot{\xi}_1 = -(2E)^{1/2} \sin(\omega_1 t + \alpha). \tag{8.2.30}$$

Here, the energy E and phase α are slow variables.

We differentiate the energy E given by (8.2.25) with respect to t and use equations (8.2.1), (8.2.29), and(8.2.30):

$$\dot{E} = \dot{\xi}_1 (\ddot{\xi}_1 + \omega_1^2 \xi_1) = -\dot{\xi}_1 u = -a|\dot{\xi}_1| = -a(2E)^{1/2} |\sin(\omega_1 t + \alpha)|.$$

In accordance with the method of averaging [22], we average the right-hand side of the equation obtained with respect to t regarding E and α as constants. We come to the equation of the first approximation, which we integrate:

$$[2E(t)]^{1/2} = (2E_0)^{1/2} - \frac{2a}{\pi}t.$$

Hence, the time T^0 needed for damping oscillations (i.e., for implementation of condition $E(T^0) = 0$), is equal to

$$T^0 = \frac{\pi}{2a}(2E_0)^{1/2}. \tag{8.2.31}$$

Expressions (8.2.27) and (8.2.31) should be compared under condition (8.2.28), under which the approximate expression (8.2.31) is obtained. The second term in (8.2.27) is much smaller than the first one, while the principal parts of (8.2.27) and (8.2.31) differ by multipliers. We have

$$\frac{T^*}{T_0} \approx \frac{4}{\pi} \approx 1.273 \quad (\varepsilon \ll 1).$$

This relation permits to estimate how close are the results obtained by the control method presented above and the time-optimal control.

8.2.2 Pendulum with a suspension point controlled by acceleration

We consider now the systems shown in Figs. 8.1 and 8.2 in the case of a single oscillator ($n = 1$) but taking into account displacement ξ_0 of the body G. The equations of motion and constraint (8.1.27) take the form

$$\ddot{\xi}_1 + \omega_1^2 \xi_1 = u, \quad \ddot{\xi}_0 = u, \quad |u| \le a. \tag{8.2.32}$$

All the notation here is the same as in Sect. 8.2.1. Note that displacements ξ_0 and ξ_1 are measured in opposite directions, so that the absolute displacement of the oscillator is $\xi_0 - \xi_1$.

We also consider a modified statement of the problem, in which the systems of Figs. 8.1 and 8.2 are controlled not by the acceleration of the body G, but by force F applied to the body G and bounded in magnitude by the constant F_0. Then, instead of relations (8.2.32), we have the following equations and constraint:

$$\ddot{\xi}_1 + \omega_1^2 \xi_1 = \ddot{\xi}_0, \quad (m_0 + m_1)\ddot{\xi}_0 - m_1 \ddot{\xi}_1 = F, \quad |F| \le F_0, \tag{8.2.33}$$

where m_0 is the mass of the body G, and m_1 is the mass of the oscillator.

We introduce the coordinate of the center of mass of the system

$$\xi = \frac{(m_0 + m_1)\xi_0 - m_1 \xi_1}{m_0 + m_1}$$

and transform relations (8.2.33) to

$$\ddot{\xi}_1 + \frac{m_0+m_1}{m_0}\omega_1^2\xi_1 = \frac{F}{m_0}, \quad \ddot{\xi} = \frac{F}{m_0+m_1}, \quad |F| \le F_0. \tag{8.2.34}$$

The change of variables and constants

$$\xi' = \frac{m_0+m_1}{m_0}\xi, \quad \omega'^2 = \frac{m_0+m_1}{m_0}\omega_1^2, \quad u = \frac{F}{m_0}$$

transforms relations (8.2.34) to the form identical to (8.2.32). Thus, relations (8.2.32) also describe systems (8.2.33) controlled by a bounded force.

In order to simplify (8.2.32), we make the change of variables

$$\xi_1 = \frac{ay}{\omega_1^2}, \quad \xi_0 = \frac{az}{\omega_1^2}, \quad t = \frac{t'}{\omega_1}, \quad u = au'. \tag{8.2.35}$$

Substitution (8.2.35) transforms relations (8.2.32) to the form

$$\ddot{y} + y = u, \quad \ddot{z} = u, \quad |u| \le 1. \tag{8.2.36}$$

From now on, we consider the system in the form (8.2.36) and denote by points derivatives with respect to the new time t', the primes of t' and u' in (8.2.36) are omitted.

Let us construct the control $u(t)$ that satisfies the condition $|u| \le 1$ and brings system (8.2.36) from the given initial state

$$y(0) = y^0, \quad \dot{y}(0) = v^0, \quad z(0) = z^0, \quad \dot{z}(0) = w^0 \tag{8.2.37}$$

to the given terminal state

$$y(T) = y^1, \quad \dot{y}(T) = v^1, \quad z(T) = z^1, \quad \dot{z}(T) = w^1. \tag{8.2.38}$$

The quantities in the right-hand sides of (8.2.37) and (8.2.38) are constants, and $T > 0$ is the as yet unknown time when the process terminates.

The solution of this problem is obtained in [26, 53].

The phase vector of system (8.2.36) consists of the variables $y, \dot{y}, z$, and $\dot{z}$. Following the general idea of Sect. 8.1.2 for constructing the control, we find the fundamental matrix $\Phi(t)$ defined in (8.1.6), and then the inverse matrix $\Phi^{-1}(t)$:

$$\Phi(t) = \begin{bmatrix} \cos t & \sin t & 0 & 0 \\ -\sin t & \cos t & 0 & 0 \\ 0 & 0 & 1 & t \\ 0 & 0 & 0 & 1 \end{bmatrix}, \quad \Phi^{-1}(t) = \begin{bmatrix} \cos t & -\sin t & 0 & 0 \\ \sin t & \cos t & 0 & 0 \\ 0 & 0 & 1 & -t \\ 0 & 0 & 0 & 1 \end{bmatrix}. \tag{8.2.39}$$

The matrix $Q(t)$ defined by (8.1.11) is here the four-dimensional column vector

$$Q^\top(t) = (-\sin t, \cos t, -t, 1), \tag{8.2.40}$$

and control (8.1.10) has the form

$$u(t) = -c_1 \sin t + c_2 \cos t - c_3 t + c_4. \tag{8.2.41}$$

The expression for the matrix $R(t)$ is given by (8.1.13) and (8.2.41). The solution of system (8.1.12) is considerably simplified, if we put $T = 2\pi k,\ k = 1, 2, \ldots$. The matrix $R(T)$ then becomes

$$R(T) = \begin{bmatrix} T/2 & 0 & -T & 0 \\ 0 & T/2 & 0 & 0 \\ -T & 0 & T^3/3 & -T^2/2 \\ 0 & 0 & -T^2/2 & T \end{bmatrix}. \tag{8.2.42}$$

The case of arbitrary $T \neq 2\pi k$ will be considered in Sect. 8.2.2.

We express the components of the vector x^* through the boundary conditions (8.2.37) and (8.2.38) with the help of relations (8.1.9) and (8.2.39) and using the equality $f(t) \equiv 0$:

$$x_1^* = y^1 - y^0, \quad x_2^* = v^1 - v^0,$$
$$x_3^* = z^1 - Tw^1 - z^0, \quad x_4^* = w^1 - w^0 \quad (T = 2\pi k).$$

Using the obtained expressions for the matrix $R(T)$ and vector x^*, we solve equations (8.1.3):

$$\begin{aligned} c_1 &= \frac{2}{T(T^2-24)}\left[T^2(y^1-y^0) + 12(z^1-z^0) - 6T(w^0+w^1)\right], \\ c_2 &= \frac{2}{T}(v^1-v^0), \\ c_3 &= \frac{6}{T(T^2-24)}\left[4(y^1-y^0) + 2(z^1-z^0) - T(w^0+w^1)\right], \\ c_4 &= \frac{2}{T(T^2-24)}\left[6T(y^1-y^0) + 3T(z^1-z^0) - (T^2+12)w^1 \right. \\ &\qquad \left. -2(T^2-6)w^0\right]. \end{aligned} \tag{8.2.43}$$

It is necessary now to choose integer k in the relation $T = 2\pi k$ in such a way that the control defined by (8.2.41) and (8.2.43) meets the constraint $|u| \leq 1$ for $t \in [0, T]$. By (8.2.41) and (8.2.43), we have

$$\begin{aligned} |u(t)| \leq |c_1| + |c_2| + |c_4 - c_3 t| &\leq \frac{2}{T(T^2-24)}\left[T^2|y^1-y^0| + 12|z^1-z^0| \right. \\ &+ 6T|w^0+w^1| + (T^2-24)|v^1-v^0| + 6|y^1-y^0||T-2t| \\ &\left. + 3|z^1-z^0||T-2t| + \psi(t)\right], \\ \psi(t) = |(T^2+12)w^1 &+ 2(T^2-6)w^0 - 3Tt(w^1+w^0)|. \end{aligned} \tag{8.2.44}$$

Here, $T = 2\pi k,\ k \geq 1$, so that $T^2 > 24$.

The function $\psi(t)$ reaches its maximum at one end of the interval $[0, T]$, consequently,

$$\psi(t) \le \max\{\psi(0), \psi(T)\} = \frac{1}{2}\max\{|3T^2(w^0+w^1) - (T^2-24)(w^1-w^0)|,$$

$$|3T^2(w^0+w^1) + (T^2-24)(w^1-w^0)|\} = \frac{3}{2}T^2|w^0+w^1| + \frac{1}{2}(T^2-24)|w^1-w^0|.$$

Note also that $|T-2t| \le T$ for $t \in [0,T]$.
Using these estimates, we obtain from (8.2.44):

$$\begin{aligned} |u(t)| \le \frac{1}{T}\left(f_1(T)|y^1-y^0| + 2|v^1-v^0| + f_2(T)|w^0+w^1| + |w^1-w^0|\right) \\ + \frac{2}{T^2}f_2(T)|z^1-z^0|, \quad f_1(T) = \frac{2T^2+12T}{T^2-24}, \quad f_2(T) = \frac{3T^2+12T}{T^2-24}. \end{aligned} \tag{8.2.45}$$

On the right-hand side of (8.2.45), we replace the functions $f_1(T)$ and $f_2(T)$, that are strictly decreasing for $T \ge T_1 = 2\pi$, by their maximum values at $T \ge T_1 = 2\pi$, and, in the resulting inequality, we put $T = 2\pi k$, $T_1 = 2\pi$. We come to the inequality

$$\begin{aligned} |u(t)| \le Ak^{-1} + Bk^{-2}, \\ A = \frac{\pi+3}{\pi^2-6}|y^1-y^0| + \frac{1}{\pi}|v^1-v^0| + \frac{3(\pi+2)}{2(\pi^2-6)}|w^0+w^1| \\ + \frac{1}{2\pi}|w^1-w^0|, \quad B = \frac{3(\pi+2)}{2(\pi^2-6)}|z^1-z^0|. \end{aligned} \tag{8.2.46}$$

It follows from (8.2.46) that constraint $|u| \le 1$ holds if

$$k^2 - Ak - B \ge 1,$$

i.e., if

$$T = 2\pi k, \quad k \ge k^* = \frac{1}{2}\left[A + (A^2+4B)^{1/2}\right]. \tag{8.2.47}$$

Formulas (8.2.41) and (8.2.43), together with relations (8.2.47) for T and (8.2.46) for A and B, completely define the required control $u(t)$ in an explicit form in terms of the initial and terminal states.

We consider a special case of boundary conditions (8.2.37) and (8.2.38):

$$y^0 = v^0 = z^0 = w^0 = y^1 = v^1 = w^1 = 0 \tag{8.2.48}$$

that corresponds to the displacement of the entire system shown in Figs. 8.1 and 8.2 from one equilibrium state to another equilibrium state at a distance z^1. In the case of (8.2.48), the time-optimal control is of the bang-bang type $u = \pm 1$ and has three switching points [40]. The optimal time T^0 is the unique positive root of the equation

$$\frac{T^{0^2}}{4} - 2\left[\arccos\left(\cos^2\frac{T^0}{4}\right)\right] = |z^1|,$$

and the relations

$$T^0 \geq 2|z^1|^{1/2}, \quad T^0 \sim 2|z^1|^{1/2} \text{ as } |z^1| \to \infty \tag{8.2.49}$$

are valid.

Let us compare this result with the time of the displacement for the control law (8.2.41). By (8.2.46)–(8.2.48), we have

$$T = 2\pi(\text{ent}\, k^* + 1), \quad k^* = B^{1/2} = 0.7965|z^1|^{1/2}.$$

Hence, for large $|z^1|$, we obtain

$$T \sim 5.005|z^1|^{1/2} \text{ as } |z^1| \to \infty. \tag{8.2.50}$$

If we use estimate (8.2.45) directly in the case of (8.2.48) for $|z^1| \to \infty$, we obtain

$$T \sim \left[2f_2(\infty)|z^1|\right]^{1/2} = \sqrt{6}|z^1|^{1/2} = 2.449|z^1|^{1/2} \text{ as } |z^1| \to \infty. \tag{8.2.51}$$

Comparing formulas (8.2.49)–(8.2.51) for T^0 and T, we see that for $|z^1| \to \infty$ they differ by their coefficients. This fact is due to both the difference of control (8.2.41) from the optimal one and estimates used to obtain (8.2.46). Note that estimate (8.2.51) is much closer to (8.2.49) than estimate (8.2.50) because of the reduced "loss" in direct estimation (8.2.51).

We also note that, for arbitrary initial conditions, the time-optimal controls for the problems considered in Sects. 8.2.1 and 8.2.2 are not known.

8.2.3 Pendulum with a suspension point controlled by acceleration (continuation)

In Sect. 8.2.2, to simplify calculations, we assumed that the dimensionless time of the process is multiple of 2π, i.e., $T = 2\pi k$. Now, we give up this assumption and consider the problem for an arbitrary T.

Let us state the problem of constructing a control $u(t)$ that satisfies the constraint $|u| \leq 1$ and brings system (8.2.36) from a given initial state

$$y(0) = x_1^0, \quad \dot{y}(T) = x_2^0, \quad z(T) = x_3^0, \quad \dot{z}(T) = x_4^0 \tag{8.2.52}$$

to the equilibrium state

$$y(T) = 0, \quad \dot{y}(T) = 0, \quad z(T) = 0, \quad \dot{z}(T) = 0. \tag{8.2.53}$$

The matrix $R(T)$, for an arbitrary T, takes the form

$$R(T) = \begin{bmatrix} (T-sc)/2 & -s^2/2 & s-Tc & c-1 \\ -s^2/2 & (T+sc)/2 & 1-c-Ts & s \\ s-Tc & 1-c-Ts & T^3/3 & -T^2/2 \\ c-1 & s & -T^2/2 & T \end{bmatrix}. \tag{8.2.54}$$

Here, denotation $s = \sin T$ and $c = \cos T$ are introduced.

Let ϕ_{ij}, $i, j = 1,...,4$, be the elements of the inverse matrix $R^{-1}(T)$ of (8.2.54). Then, by virtue of (8.1.14), the expression for control (8.2.41) becomes

$$u(t) = \sum_{i=1}^{4} (\phi_{1i}x_i^0 \sin t - \phi_{2i}x_i^0 \cos t + \phi_{3i}x_i^0 t - \phi_{4i}x_i^0). \tag{8.2.55}$$

Thus, control (8.2.55) brings, for any $T > 0$, system (8.2.36) from the initial state (8.2.52) to the terminal equilibrium state (8.2.53) in time T. However, this control does not, generally speaking, satisfy the constraint $|u| \leq 1$. To take this constraint into account, we apply the Cauchy inequality to relation (8.2.55):

$$|u| \leq \left(\sum_{i=1}^{4} x_i^{0\,2} \right)^{1/2} \left[\sum_{i=1}^{4} (-\phi_{1i} \sin t + \phi_{2i} \cos t - \phi_{3i} t + \phi_{4i})^2 \right]^{1/2}. \tag{8.2.56}$$

We introduce the auxiliary functions

$$p(t,T) = \sum_{i=1}^{4} (-\phi_{1i} \sin t + \phi_{2i} \cos t - \phi_{3i} t + \phi_{4i})^2 \tag{8.2.57}$$

and

$$r(T) = \left[\max_{0 \leq t \leq T} p(t,T) \right]^{-1/2}. \tag{8.2.58}$$

Then, inequality (8.2.56) can be rewritten in the form

$$|u| \leq |x^0| \, [p(t,T)]^{1/2} \leq |x^0| / r(T). \tag{8.2.59}$$

We choose the termination time T from the condition

$$|x^0| = r(T). \tag{8.2.60}$$

It follows from (8.2.59) that, if T is chosen according to (8.2.60), constraint (8.2.36) imposed on the control is satisfied for all $t \in [0, T]$.

Thus, we arrive at the following procedure for constructing the control $u(t)$. First, we determine the elements $\phi_{ij}(T)$ of the inverse matrix $R^{-1}(T)$ and calculate the functions $p(t,T)$ and $r(T)$ with the help of equalities (8.2.54), (8.2.57), and (8.2.58). These calculations should be performed once for the given system (see below).

When they have been executed, we can construct, for any initial vector x^0, the desired bounded control that brings the system to the coordinate origin. To do this, we first determine time T from condition (8.2.60) and then find the control u from (8.2.55).

To determine the function $r(T)$, we use the REDUCE symbolic calculations language and find, using a computer, the analytical representations of the elements ϕ_{ij}, $i,j=1,...,4$, of the matrix $R^{-1}(T)$ which is the inverse of (8.2.54). The expressions for ϕ_{ij} turn out to be rather cumbersome. To illustrate, we present one element of the matrix $R^{-1}(T)$:

$$\begin{aligned}\phi_{11} = 2[T^5 + T^4 \sin T \cos T - 8T^3 \sin^2 T + 24T^2 \sin T(1-\cos T)\\ -24T(1-\cos T)^2]/[T^6 - T^4(8\cos T + \sin^2 T + 16)\\ +8T^3 \sin T(5-2\cos T) + 48T^2(1-\cos T)(1+2\cos T)\\ 240T \sin T(1-\cos T) + 192(1-\cos T)^2].\end{aligned} \tag{8.2.61}$$

Using formula (8.2.57) and the obtained expressions (8.2.61) for ϕ_{ij}, we calculate the maximum values of $p(t,T)$ from $t \in [0,T]$. Thus, the function $r(T)$ in (8.2.58) is determined. Its graph is shown in Fig. 8.3.

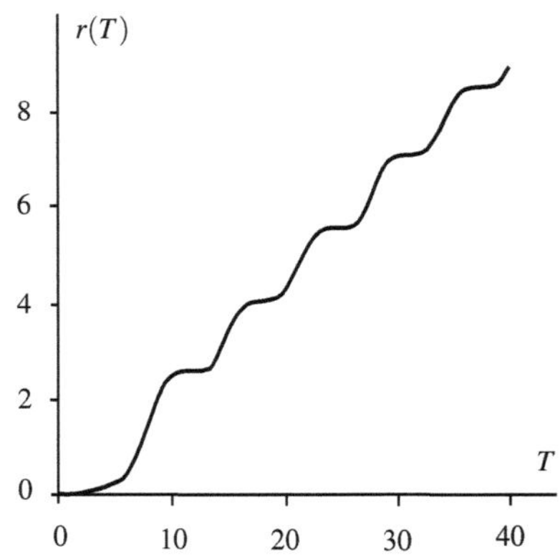

Fig. 8.3 Graph of the function $r(T)$

We analyze the behavior of the function $r(T)$ in the limiting cases.

Let the process duration T tends to 0. We expand the function $p(t,T)$ given by (8.2.57) in the Maclaurin series (we use REDUCE) in the current time t:

$$\begin{aligned}p(t,T) = (\phi_{12}+\phi_{14})^2 + (\phi_{22}+\phi_{24})^2 + (\phi_{23}+\phi_{34})^2 + (\phi_{24}+\phi_{44})^2\\ -2t[\phi_{12}(\phi_{11}+\phi_{13}+\phi_{22}+\phi_{24}) + \phi_{13}(\phi_{14}+\phi_{23}+\phi_{34})\\ +\phi_{14}(\phi_{11}+\phi_{24}+\phi_{44}) + \phi_{23}(\phi_{22}+\phi_{24}+\phi_{33}) + \phi_{34}(\phi_{24}+\phi_{33}+\phi_{44})] + \dots\end{aligned} \tag{8.2.62}$$

We then expand the numerators and denominators of the elements ϕ_{ij}, $i,j = 1,\dots,4$, of the symmetric matrix $R^{-1}(T)$ in series in T. We obtain

$$\phi_{11} = \left(\frac{T^9}{180} - \frac{T^{11}}{630} + \dots\right)\left(\frac{T^{16}}{18144000} - \dots\right)^{-1},$$
$$\phi_{12} = \left(\frac{T^{10}}{360} - \frac{11T^{12}}{18900} + \dots\right)\left(\frac{T^{16}}{18144000} - \dots\right)^{-1}. \tag{8.2.63}$$

Other elements have similar representations.

Estimates of the orders of the expansions in T in the numerators and denominators of the functions ϕ_{ij} in (8.2.63) show that, to obtain the principal term of the expansion of the function $p(t,T)$ in accordance with (8.2.62), it suffices to retain only the principal term (of the order T^{16}) in the denominators of formulas (8.2.63). In the numerators of expressions (8.2.63), one has to take into account terms of various orders.

By collecting terms of like powers, one obtains the representation

$$p(t,T) = 1411200T^{-8}f(\tau). \tag{8.2.64}$$

The notation

$$f(\tau) = 1 - 24\tau + 204\tau^2 - 760\tau^3 + 1380\tau^4 - 1200\tau^5 + 400\tau^6, \quad \tau = \frac{t}{T} \in [0,1], \tag{8.2.65}$$

is used here.

The graph of the polynomial $f(\tau)$ is shown in Fig. 8.4.

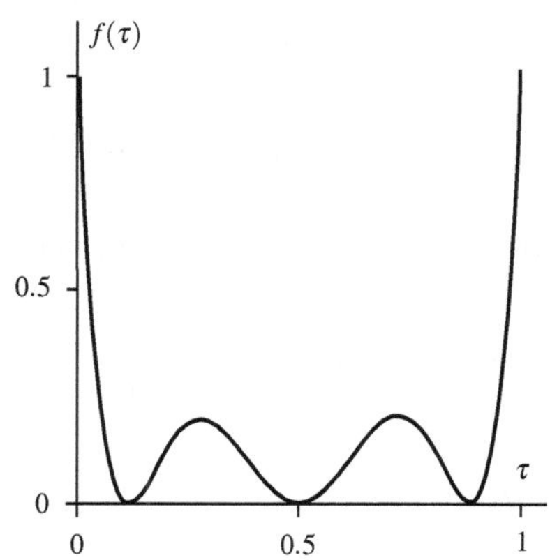

Fig. 8.4 Graph of the polynomial $f(\tau)$

One can easily see that $f(\tau)$ attains its largest value at the ends of the interval $\tau \in [0,1]$, and $f(0) = f(1) = 1$. Then, it follows from (8.2.64) and (8.2.65) that

$$\max p(t,T) = 1411200T^{-8}.$$

We substitute this result into (8.2.58) to obtain

$$r(T) = 8.4 \cdot 10^{-4} T^4, \quad T \to 0. \tag{8.2.66}$$

Equality (8.2.66) defining the function $r(T)$ for small T is confirmed by the results of the numerical calculation of this function.

Now, let the control process duration tends to infinity, i.e., $T \to \infty$. We substitute into (8.2.57) the expressions for ϕ_{ij}, $i, j = 1, \ldots, 4$, calculated by formulas (8.2.61) and expand the function $p(t,T)$ in a series of negative powers of T. Transformations made by using the REDUCE language yield the expansion

$$\begin{aligned} p(t,T) = \frac{4}{T^2}\left(\frac{9t^2}{T^2} - \frac{12t}{T} + 5 + \frac{1}{T}\left[\frac{12(T-2t)}{T}\sin(T-t)\right.\right. \\ \left.\left. - \sin 2(T-t) - \sin 2t - \frac{12(2T-3t)}{T}\sin t\right]\right) + 0\left(\frac{1}{T^4}\right), \quad T \to \infty. \end{aligned} \tag{8.2.67}$$

We rewrite expansion (8.2.67) in the form

$$\begin{aligned} p(t,T) &= \frac{4}{T^2}[p_0(\tau) + T^{-1}p_1(\tau,T)], \quad p_0(\tau) = 9\tau^2 - 12\tau + 5, \\ p_1(\tau,T) &= 12(1-2\tau)\sin T(1-\tau) - \sin 2T(1-\tau) \\ &- \sin 2T\tau - 12(2-3\tau)\sin T\tau, \quad \tau = \frac{t}{T} \in [0,1]. \end{aligned} \tag{8.2.68}$$

Let us find the maximum in (8.2.58) as $T \to \infty$ using representation (8.2.68). One can easily see that the quadratic trinomial $p_0(\tau)$ attains its maximum on the interval $[0,1]$ at $\tau = 0$. Since the contribution of the second term in (8.2.68) is small for $T \to \infty$, we have, up to higher-order infinitesimals,

$$\max_{0 \le t \le T} p(t,T) = p(0,T) = \frac{20}{T^2} + \frac{4}{T^3}(12\sin T - \sin 2T), \quad T \to \infty. \tag{8.2.69}$$

We use here expansion (8.2.67). Substituting (8.2.69) into (8.2.58) and expanding the result in a series of powers of T^{-1}, we obtain

$$r(T) = \frac{10T - 12\sin T + \sin 2T}{20\sqrt{5}} + O(\frac{1}{T}), \quad T \to \infty. \tag{8.2.70}$$

We differentiate (8.2.70) in T:

$$r'(T) = \frac{(2-\cos T)(1-\cos T)}{5\sqrt{5}} \ge 0.$$

Hence, $r(T)$ is a monotonically increasing function as $T \to \infty$.

It follows from the presented calculations and analytical expansions that the function $r(T)$ increases monotonically from 0 to ∞ as T varies from 0 to ∞. Consequently, equation (8.2.60) has, for any $|x^0|$, a unique solution.

Let us present the results of the numerical simulation. The procedure for designing the control has been described above. We dwell first on a practical numerical solution of equation (8.2.60). We divide the entire semi-infinite interval of variation of T into three parts: $[0,T_0]$, $[T_0,T_1]$, and $[T_1,\infty)$, to which three intervals of variation of $r(T)$ correspond: $[0,r_0]$, $[r_0,r_1]$, and $[r_1,\infty)$. Here, $r_i = r(T_i)$, $i = 0,1$. On the interval $[0,T_0]$, we use the asymptotic representation (8.2.66) for small T; on the interval $[T_0,T_1]$, we use the numerical values of $r(T)$, and on the interval $[T_1,\infty)$, we use the asymptotic representation (8.2.70) for large T. First, we determine, for the specified x^0, by comparing $|x^0|$ with r_0 and r_1, which of the three intervals contains the desired T. We then determine T as follows. If $T \in [0,T_0]$, then, by (8.2.66), we have $T = \left(|x^0|/0.00084\right)^{1/4}$. If $T \in [T_0,T_1]$, we find T by linear interpolation from the table of values of $r(T)$ that is stored in the computer memory. If $T \in [T_1,\infty)$, we use representation (8.2.70). In this case, it is convenient to search for T in the form

$$T = 2\sqrt{5}|x^0| + \theta. \tag{8.2.71}$$

Substituting (8.2.71) into (8.2.70), we obtain the equation for θ:

$$F(\theta) = 10\theta - 12\sin(2\sqrt{5}|x^0| + \theta) + \sin[2(2\sqrt{5}|x^0| + \theta)] = 0.$$

This equation is solved by some numerical method, for example, the method of successive interval halving.

When the process duration T has been determined for the specified initial vector x^0, the control $u(t)$ at any instant t can be calculated from (8.2.55). Here, one can use the analytical expressions for the functions ϕ_{ij}, $i,j = 1,\ldots,4$, see (8.2.61), that were obtained by analytical transformations. The control calculated in this way can be substituted into the right-hand side of system (8.2.36), that is integrated numerically or analytically for the initial conditions (8.2.52).

Some results of simulation with $x^0 = (-1,2,0.5,1)$ are presented in Fig. 8.5. The thick curve represents the projection of the phase trajectory $x(t)$ on the $(y,\dot{y})$ hyperplane, and the thin curve represents the projection of this trajectory on the $(z,\dot{z})$ hyperplane. For this example, the time of process is $T = 13.116$.

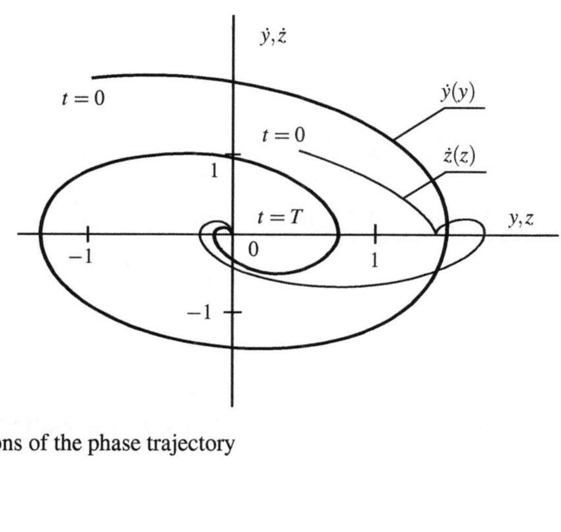

Fig. 8.5 Projections of the phase trajectory

8.2.4 Pendulum with a suspension point controlled by velocity

Consider a system that contains an oscillatory link and is subjected to a scalar control by the velocity of body G:

$$\ddot{\xi}_1 + \omega^2(\xi_1 - \xi_0) = 0, \quad \dot{\xi}_0 = u. \tag{8.2.72}$$

Here, ξ_0, ξ_1 are generalized coordinates, $\omega > 0$ is the natural frequency of the oscillator, and u is a scalar control constrained by (8.1.27).

Equations (8.2.72) describe the motion of the systems shown in Figs. 8.1 and 8.2 in the case of a single oscillator ($n = 1$) and the body G being controlled by its velocity.

In particular, this may be a two-mass system that consists of a body and a mass connected by a spring. The system as a whole performs a translational horizontal motion, with ξ_0 being the displacement of the body, u the velocity of the body, and ξ_1 the absolute displacement of the mass.

Another mechanical model of system (8.2.72) is a mathematical pendulum suspended from the body G that moves horizontally with the velocity u. Then, ξ_0 the displacement of the body, $\xi_1 = -l_1\phi_1 + \xi_0$ is the absolute linear displacement of the pendulum (here, l_1 is the length, and ϕ_i is the angle of deviation of the pendulum from the vertical direction; the deviation is assumed to be small). Relations (8.2.72) and (8.1.27) describe also other mechanical systems controlled by a bounded velocity [40].

Let us state the problem of constructing a control $u(t)$ that satisfies constraint (8.1.27) and brings system (8.2.72) from an arbitrary initial state at $t_0 = 0$:

$$\xi_0(0) = \xi_0^0, \quad \xi_1^0(0) = \xi_1^0, \quad \dot{\xi}_1(0) = \dot{\xi}_1^0 \tag{8.2.73}$$

to the given terminal state

$$\xi_0(T) = 0, \quad \xi_1(T) = 0, \quad \dot{\xi}_1(T) = 0. \tag{8.2.74}$$

The time of the process T is not fixed.

Using the change of variables

$$\xi_1 = \frac{ay}{\omega}, \quad \xi_0 = \frac{az}{\omega}, \quad t = \frac{t'}{\omega}, \quad u = au', \tag{8.2.75}$$

we reduce system (8.2.72) and (8.1.27) to the form

$$\ddot{y} + y = z, \quad \dot{z} = u, \tag{8.2.76}$$

$$|u| \leq 1. \tag{8.2.77}$$

We consider the system in the form (8.2.76) and (8.2.77) and denote by dots derivatives with respect to time t', the primes of t' and u' are omitted.

After the change of variables (8.2.75), conditions (8.2.73) and (8.2.74) take the form

$$y(0) = x_1^0, \quad \dot{y}(0) = x_2^0, \quad z(0) = x_3^0, \tag{8.2.78}$$

$$y(T) = 0, \quad \dot{y}(T) = 0, \quad z(T) = 0. \tag{8.2.79}$$

The quantities x_1^0, x_2^0, x_3^0 are given constants, and $T > 0$ is the as yet unknown time when the process terminates.

Thus, the problem stated is reduced to constructing the control $u(t)$ that satisfies condition (8.2.77) and brings system (8.2.76) from the given initial state (8.2.78) to the terminal state (8.2.79).

The solution presented below is obtained in [99].

We use the approach described in Sect. 8.1.2. We denote the phase vector by $x = (y, \dot{y}, z)$ and reduce the system to form (8.1.1), where

$$A = \begin{bmatrix} 0 & 1 & 0 \\ -1 & 0 & 1 \\ 0 & 0 & 0 \end{bmatrix}, \quad B = \begin{bmatrix} 0 \\ 0 \\ 1 \end{bmatrix}, \quad f = \begin{bmatrix} 0 \\ 0 \\ 0 \end{bmatrix}. \tag{8.2.80}$$

Initial conditions (8.2.78) and terminal ones (8.2.79) take the form

$$x(0) = x^0 = (x_1^0, x_2^0, x_3^0), \quad x(T) = 0. \tag{8.2.81}$$

The inverse of the fundamental matrix of the homogeneous system is

$$\Phi^{-1}(t) = \begin{bmatrix} \cos t & -\sin t & -\cos t \\ \sin t & \cos t & -\sin t \\ 0 & 0 & 1 \end{bmatrix}, \tag{8.2.82}$$

and the matrix Q of (8.1.11) is a three-dimensional column vector

$$Q^\top(t) = (1 - \cos t, -\sin t, 1). \tag{8.2.83}$$

Substituting (8.2.83) into expression (8.1.10) for the control u, we obtain

$$u(t) = c_1(1 - \cos t) - c_2 \sin t + c_3, \tag{8.2.84}$$

where c_1, c_2, and c_3 are the components of the vector c defined by equation (8.1.12).

We find the matrix $R(t)$ of (8.1.13) using (8.2.83):

$$R(T) = \begin{bmatrix} \frac{3T}{2} - 2\sin T + \frac{1}{2}\sin T \cos T & \frac{1}{2}\sin^2 T + \cos T - 1 & T - \sin T \\ \frac{1}{2}\sin^2 T + \cos T - 1 & \frac{T}{2} - \frac{1}{2}\sin T \cos T & \cos T - 1 \\ T - \sin T & \cos T - 1 & T \end{bmatrix}$$

and denote the elements of the inverse matrix $R^{-1}(T)$ by ψ_{ij}, $i, j = 1,2,3$. Then, taking into account (8.1.14), we rewrite expression (8.2.84) for the control $u(t)$ in the form

$$u(t) = \sum_{i=1}^{3}(-\psi_{1i}x_i^0(1-\cos t) + \psi_{2i}x_i^0 \sin t - \psi_{3i}x_i^0). \tag{8.2.85}$$

For any given $T > 0$, control (8.2.85) brings system (8.2.76) from an arbitrary initial states (8.2.78) to the terminal equilibrium state (8.2.79) in time T, but does not, generally speaking, meet constraint (8.2.77). To take this constraint into account, we apply the Cauchy inequality to relation (8.2.85):

$$|u| \leq \left(\sum_{i=1}^{3} x_i^{0\,2}\right)^{1/2} \left[\sum_{i=1}^{3}(\psi_{1i}(1-\cos t) - \psi_{2i}\sin t + \psi_{3i})^2\right]^{1/2}. \tag{8.2.86}$$

We introduce the auxiliary functions

$$p(t,T) = \sum_{i=1}^{3}\left[\psi_{1i}(1-\cos t) - \psi_{2i}\sin t + \psi_{3i}\right]^2, \tag{8.2.87}$$

$$r(T) = \left[\max_{0\leq t\leq T} p(t,T)\right]^{-1/2} \tag{8.2.88}$$

and rewrite inequality (8.2.86) in the form

$$|u| \leq |x^0|\,[p(t,T)]^{1/2} \leq \frac{|x^0|}{r(T)}. \tag{8.2.89}$$

If the time of the process is chosen from the condition

$$|x^0| = r(T), \tag{8.2.90}$$

then constraint (8.2.77) holds for all $t \in [0,T]$.

The values of the function $p(t,T)$ can be calculated numerically by formula (8.2.87). Here, elements ψ_{ij}, $i,j = 1,2,3$, of the symmetric matrix $R^{-1}(T)$ are calculated according to the formula

$$\psi_{ij} = \frac{R_{ij}}{\det R(T)}, \tag{8.2.91}$$

where R_{ij} is the cofactor of the element r_{ij} of the matrix $R(T)$. Using this procedure, one can calculate the maximum values of $p(t,T)$ for $t \in [0,T]$ and find the function $r(T)$ of (8.2.88). Since the function $p(t,T)$ is periodic in t with the period 2π, it will suffice to search for its maximum on the interval $[0,T]$, if $T < 2\pi$, and on the interval $[0,2\pi]$, if $T \geq 2\pi$.

Figure 8.6 shows the graph of the function $r(T)$ obtained by calculations with a small step $\triangle T$ with respect to T. For every fixed T, we find the maximum of the function $p(t,T)$ in t by the exhaustive search and comparison of the values.

Let us analyze the behavior of the function $r(T)$ in the limiting cases. Let time T of the control be small. We expand the elements ψ_{ij}, $i,j = 1,2,3$, of the symmetric

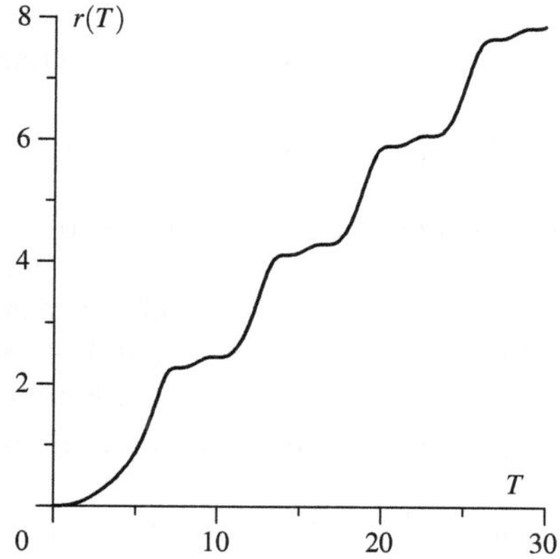

Fig. 8.6 Function $r(T)$

matrix $R^{-1}(T)$ in series of powers of T^{-1} using formula (8.2.91). The principal terms of this expansion are:

$$R^{-1}(T) = \begin{bmatrix} 720T^{-5} & 360T^{-4} & 60T^{-3} \\ 360T^{-4} & 192T^{-3} & 36T^{-2} \\ 60T^{-3} & 36T^{-2} & 9T^{-1} \end{bmatrix} + \dots, \quad T \to 0. \tag{8.2.92}$$

Substituting expansions (8.2.92) of the elements ψ_{ij}, $i, j = 1,2,3$, into expression (8.2.87) and taking into account that $1 - \cos t = t^2/2 + O(t^4)$, $\sin t = t + O(t^3)$ for small t, we get the following representation of the function $p(t,T)$ for small T and $t \in [0,T]$:

$$p(t,T) = 360^2 T^{-6} \left(\frac{t^2}{T^2} - \frac{t}{T} + \frac{1}{6} \right)^2. \tag{8.2.93}$$

Formula (8.2.93) yields

$$\max_{0 \le t \le T} p(t,T) = p(0,T) = p(T,T) = 3600T^{-6}. \tag{8.2.94}$$

Substituting (8.2.94) into (8.2.88), we obtain

$$r(T) = \frac{T^3}{60}, \quad T \to 0. \tag{8.2.95}$$

Now, let the time of the process T be large. We substitute into equality (8.2.87) for the function $p(t,T)$ the expressions for ψ_{ij}, $i, j = 1,2,3$, calculated by formula (8.2.91), and expand the function $p(t,T)$ in a series of negative powers of T:

$$p(t,T) = \frac{4}{T^2} \left[p_0(t) + \frac{1}{T} p_1(t,T) \right] + O\left(\frac{1}{T^4} \right), \quad T \to \infty. \tag{8.2.96}$$

Here, we have introduced the notation

$$p_0(t) = \cos^2 t + \cos t + \frac{5}{4}, \quad p_1(t,T) = 9(1-\cos t)(\sin T + \frac{1}{2}\sin 2T)$$
$$-8\sin T - \frac{5}{2}\sin 2T + \sin^2 t \sin 2T + (-5+5\cos T - 7\sin^2 T)\sin t \quad (8.2.97)$$
$$-2(1-\cos t)^2(\sin T + \sin 2T) + 2(1-\cos t)(1-\cos T + 3\sin^2 T)\sin t.$$

Using expansions (8.2.96), one can easily find the maximum of the function $p(t,T)$ in T for $T \to \infty$. Obviously, the function $p_0(t)$ of (8.2.97) attains its maximum on the interval $[0,T]$ at $t=0$. Since the contribution of the second term in (8.2.96) is small as $T \to \infty$, we have, up to higher-order infinitesimals,

$$\max_{0 \le t \le T} p(t,T) = p(0,T) = \frac{13}{T^2} - \frac{32}{T^3}\sin T - \frac{10}{T^3}\sin 2T, \quad T \to \infty. \tag{8.2.98}$$

Substituting (8.2.98) into (8.2.88) and expanding the resulting expression in a series of T^{-1}, we obtain

$$r(T) = f(T) + O(T^{-1}), \quad T \to \infty. \tag{8.2.99}$$

Here, we have introduced the notation

$$f(T) = \frac{1}{\sqrt{13}}\left(T + \frac{16}{13}\sin T + \frac{5}{13}\sin 2T\right).$$

We differentiate the function $f(T)$ in T:

$$f'(T) = \frac{20\cos^2 T + 16\cos T + 3}{13\sqrt{13}}.$$

The derivative $f'(T)$ vanishes at $\cos T = -0.5$ and $\cos T = -0.3$. Calculations reveal that the function $f(T)$ has strict local maxima at $T = -2\pi/3 + 2\pi n$ and $T = \arccos(-0.3) + 2\pi n$, and strict local minima at $T = 2\pi/3 + 2\pi n$ and $T = -\arccos(-0.3) + 2\pi n$.

A typical structure of these extrema can be seen in Fig. 8.7, that gives the graph of the function $f(T)$. To make it more clear, a piece of the graph is zoomed in 100 times with respect to the y-axis. Thus, $r(T)$ is not a monotonically increasing function as T varies from 0 to ∞, and equation (8.2.90), in general, has a non-unique solution for some x^0.

Let us describe the procedure for designing the control function $u(t)$. We first solve equation (8.2.90) numerically and find T. To that end, we divide the entire semi-infinite interval of variation of T into three parts: $[0,T_0]$, $[T_0,T_1]$, and $[T_1,\infty)$, where T_0 and T_1 are chosen in such a way that the asymptotic representation (8.2.95) is valid on the interval $[0,T_0]$ for small T, and so is the asymptotic representation (8.2.99) on the semi-infinite interval $[T_1,\infty)$ for large T. We denote $r_i = r(T_i)$, $i = 0,1$. We compare the given quantity $|x^0|$ with r_0 and r_1 and determine T as follows. If $|x^0| \in [0,r_0]$, then, by formula (8.2.95), we have $T = \left(60|x^0|\right)^{1/3}$. If $|x^0| \in [r_0,r_1]$,

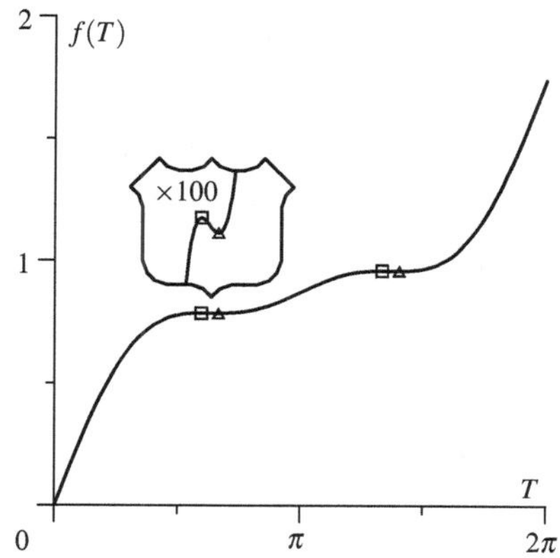

Fig. 8.7 Function $f(T)$

then we find T using the table of values of $r(T)$ that are calculated on the interval $[T_0, T_1]$ (if there are several solutions, we choose the least one). If $|x^0| \in [r_1, \infty)$, we use asymptotic representation (8.2.99). In this case, we search for T in the form

$$T = \sqrt{13}|x^0| + \theta. \tag{8.2.100}$$

Substituting (8.2.100) into (8.2.99), we obtain the equation for θ:

$$F(\theta) = 13\theta + 16\sin(\sqrt{13}|x^0| + \theta) + 5\sin[2(\sqrt{13}|x^0| + \theta)].$$

We solve this equation numerically. If there are several solutions, we find the least value of θ in order to reduce the time of process T.

When the process duration T has been determined for the specified initial vector x^0, we calculate the control $u(t)$ at any time instant t by formula (8.2.85), using analytical expressions (8.2.91) for the functions ψ_{ij}, $i, j = 1, 2, 3$. The control thus obtained we substitute into the right-hand side of system (8.2.76), which is integrated numerically under initial conditions (8.2.78). The control constructed is not time-optimal one, but it is rather simple for the calculation and practical implementation. In Table 8.1, the results obtained for time T are compared with time T^* of the time-optimal process that was determined in [40] for the various initial vectors $x^0 = (x_1^0, x_2^0, x_3^0)$.

The results of simulation are presented in Figs. 8.8 and 8.9 for the case where $x^0 = (1, 0.5, -1)$. In this case, we have $T = 6.0$. Figure 8.8 shows the time history of $u = u(t)$, and Fig. 8.9 depicts the trajectory $x(t)$, i.e., the time histories of $y, \dot{y}$, and z.

Table 8.1 Time T compared with the optimal time T^*

x_1^0	0	0	−5	−4.9015	0.0245	−0.2070
x_2^0	0	0	−4	3.5253	−1.9302	0.5441
x_3^0	1.0471	6.2831	−1	−2.7750	1.5452	−1.0979
T	5.4	24.2	24.5	24.8	10.7	5.7
T^*	3.1415	6.2831	9.8362	8.0612	3.7410	1.0979
T/T^*	1.7	3.9	2.5	3.1	2.9	5.2

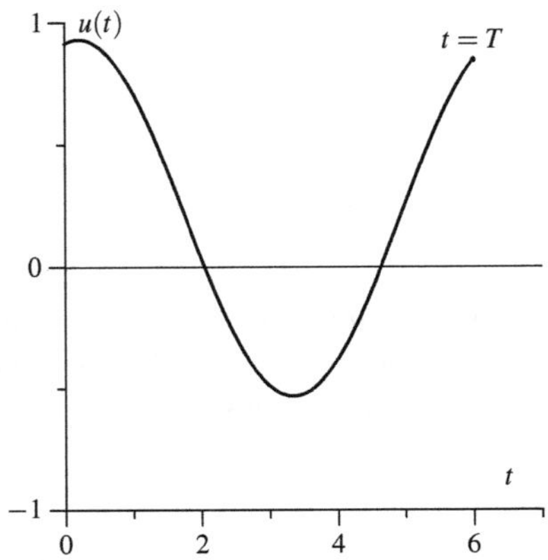

Fig. 8.8 Control $u(t)$

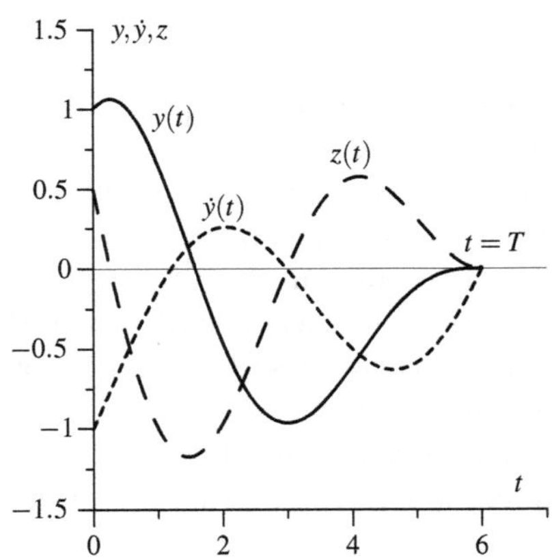

Fig. 8.9 Time histories of $y, \dot{y}$, and z

8.3 Application to electro-mechanical systems

8.3.1 *Model of the electro-mechanical system*

Let us consider a two-mass system controlled by a direct-current electric motor with independent excitation. We take the equations of motion of the system in the form

$$m_1\ddot{\xi}_1 = c(\xi_2 - \xi_1) + F, \quad m_2\ddot{\xi}_2 = c(\xi_1 - \xi_2). \tag{8.3.1}$$

Here, ξ_1 and ξ_2 are the coordinates of the system, m_1 and m_2 are constant inertial coefficients, c is the constant stiffness of the elastic connection, and F is the control (force or torque) produced by the electric motor. Figures 8.10–8.12 show some specific systems described by equations (8.3.1).

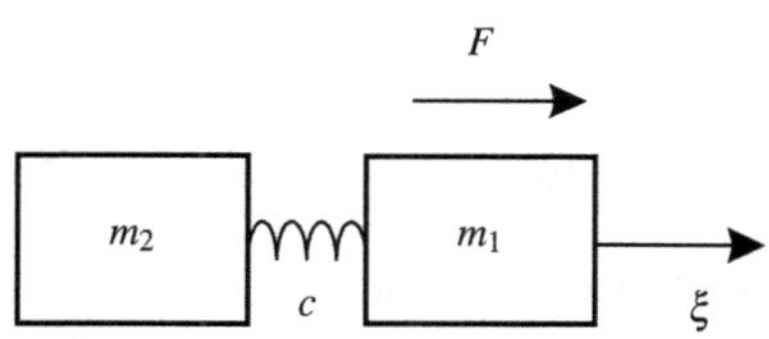

Fig. 8.10 Two-mass system

Figure 8.10 shows a system of two bodies of masses m_1 and m_2 that move along the ξ-axis under the action of force F. Here, ξ_1 and ξ_2 are the coordinates of the bodies, c is the stiffness of the spring, and F is the control force applied to the first body.

Figure 8.11 shows a cart of mass m_1 being moved along the ξ-axis by the force F. A mathematical pendulum of mass m_2 and length l suspended from the cart performs small oscillations.

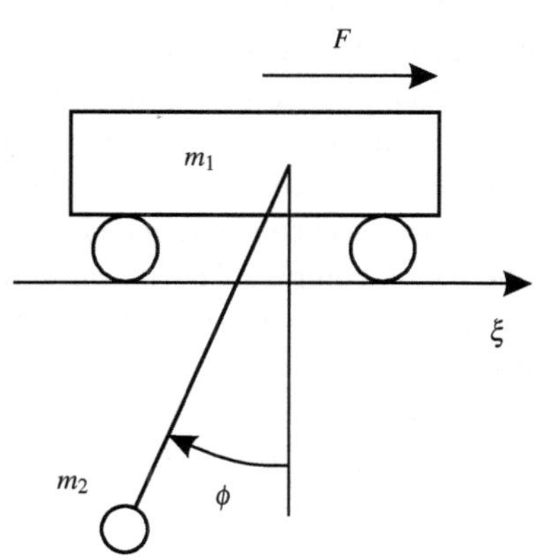

Fig. 8.11 Pendulum on a cart

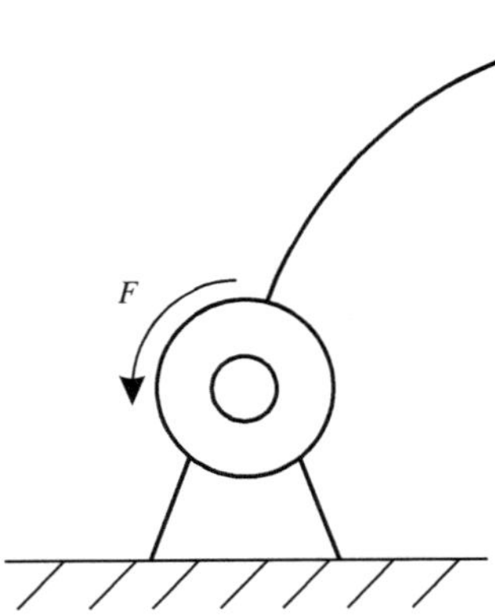

Fig. 8.12 Electric motor and elastic rod with a mass at its end

The equations of motion for the system of Fig. 8.11 have the form:

$$(m_1+m_2)\ddot{\xi}_1 - m_2 l\ddot{\phi} = F, \quad m_2 l^2\ddot{\phi} + m_2 g l\phi = m_2 l\ddot{\xi}_1. \tag{8.3.2}$$

Here, ϕ is the angle of deviation of the pendulum from the vertical direction, g is the acceleration of gravity. The first of equations (8.3.2) is the equation for the change of momentum along the ξ-axis; the second one describes the change of the angular momentum about the axis of the pendulum.

We introduce the notation

$$\xi_2 = \xi_1 - l\phi, \quad c = \frac{m_2 g}{l}. \tag{8.3.3}$$

With notation (8.3.3), equations (8.3.2) take form (8.3.1).

The system shown in Fig. 8.12 is an electric motor. An elastic rod of mass m at its end is attached to the axis of the motor. Neglecting the mass of the rod in comparison with mass m and denoting by c its stiffness, we again arrive at equations (8.3.1), where the variables and constants have the following meanings: m_1 is the moment of inertia of the rotor of the electric motor and the rotating parts of the gear, m_2 is the moment of inertia of the rod with mass m at its end, ξ_1 and ξ_2 are the absolute angles of rotation of the roller of the motor and mass m about the axis of rotation, and F is the moment produced by the electric motor. The equations of motion of the motor with an elastic rod can be reduced to system (8.3.1) also in the case where the mass of the rod is comparable with m, if we restrict our attention with the fundamental (lowest) tone of elastic oscillations. We can regard the system of Fig. 8.12 as the simplest model of an elastic manipulator.

The control F in system (8.3.1) is proportional (or equal) to the moment produced by the electric motor and, hence, is proportional to the current I in the circuit of the rotor, i.e.,

$$F = k_1 I, \quad k_1 > 0, \tag{8.3.4}$$

where k_1 is a constant coefficient. The equation of balance of electric voltages in the circuit of the rotor has the form

$$L\dot{I} + RI + k_2\dot{\xi}_1 = U. \tag{8.3.5}$$

Here, L is the inductance, R is the electric resistance, k_2 is a constant coefficient, and U is the electric voltage. The term $k_2\dot{\xi}_1$ in (8.3.5), equal to the counter-emf, is proportional to the angular velocity of the roller of the motor, which, in turn, is proportional (or equal) to $\dot{\xi}_1$. The first term in the left-hand side of equation (8.3.5) is usually small in comparison with the other terms, and it can be neglected. Then, from (8.3.4) and (8.3.5), we obtain

$$F = \frac{k_1}{R}(U - k_2\dot{\xi}_1) > 0. \tag{8.3.6}$$

Let us consider the constraints imposed on the control and phase coordinates of system (8.3.1). The control voltage is bounded in magnitude by a constant:

$$|U| \le U_0. \tag{8.3.7}$$

The current I and the moment of the motor proportional to I should also be bounded in magnitude. This, by virtue of (8.3.4), leads to the constraint

$$|F| \le F_0, \tag{8.3.8}$$

where F_0 is constant. Furthermore, the angular velocity of rotation of the roller, that is proportional to $\dot{\xi}_1$, is also bounded in such a way that, for a voltage $U = \pm U_0$ of the maximum magnitude and for the maximum angular velocity, the control moment cannot speed up the motor. By virtue of (8.3.6), this constraint can be written in the form

$$|\dot{\xi}_1| \le \frac{U_0}{k_2}. \tag{8.3.9}$$

Taking into account (8.3.6), we represent the set of constraints (8.3.7)–(8.3.9) as follows:

$$|F| \le F_0, \quad |\dot{\xi}_1| \le \frac{U_0}{k_2}, \quad |F + \frac{k_1 k_2}{R}\dot{\xi}_1| \le \frac{k_1}{R}U_0. \tag{8.3.10}$$

We introduce new (dimensionless) variables, using the formulas

$$\begin{gathered} t' = \omega t, \quad x_1 = \frac{m_1\xi_1 + m_2\xi_2}{(m_1+m_2)l_0}, \quad x_2 = \frac{m_1\dot{\xi}_1 + m_2\dot{\xi}_2}{(m_1+m_2)l_0\omega}, \\ x_3 = \frac{m_1(\xi_1 - \xi_2)}{(m_1+m_2)l_0}, \quad x_4 = \frac{m_1(\dot{\xi}_1 - \dot{\xi}_2)}{(m_1+m_2)l_0\omega}, \quad u = \frac{F}{(m_1+m_2)l_0\omega^2}, \\ \omega^2 = \frac{c(m_1+m_2)}{m_1 m_2}, \quad l_0 = \frac{F_0 m_1 m_2}{c(m_1+m_2)^2}. \end{gathered} \tag{8.3.11}$$

Then, equations (8.3.1) take the form that does not contain any parameters:

$$\dot{x}_1 = x_2, \quad \dot{x}_2 = u, \quad \dot{x}_3 = x_4, \quad \dot{x}_4 = -x_3 + u. \tag{8.3.12}$$

Here and below, a dot denotes the derivatives with respect to the dimensionless time t', but the prime will be omitted. Let us make substitution (8.3.11) also in constraints (8.3.10). We obtain

$$|u| \leq 1, \quad |px_2 + \mu px_4| \leq 1, \quad |px_2 + \mu px_4 + qu| \leq 1, \tag{8.3.13}$$

where

$$p = \frac{l_0 \omega k_2}{U_0}, \quad q = \frac{(m_1 + m_2) l_0 \omega^2 R}{k_1 U_0}, \quad \mu = \frac{m_2}{m_1}. \tag{8.3.14}$$

For system (8.3.12) with constraints (8.3.13), we state the problem of constructing an admissible control $u(t)$ that takes the system from the initial state

$$x_1(0) = x_1^0, \quad x_2(0) = x_2^0, \quad x_3(0) = x_3^0, \quad x_4(0) = x_4^0 \tag{8.3.15}$$

to the zero terminal state

$$x_1(T) = 0, \quad x_2(T) = 0, \quad x_3(T) = 0, \quad x_4(T) = 0. \tag{8.3.16}$$

Here, T is the as yet unfixed time of termination of the process.

The problem stated above is a special case of more general problem formulated in Sect.8.1.1. First, we consider a simplified version of the problem described by a second-order system.

8.3.2 Analysis of the simplified model

Let us set $m_2 = 0$ in (8.3.1). We obtain a system with a single degree of freedom that is described, in the dimensionless variables (8.3.11), by the equations

$$\dot{x}_1 = x_2, \quad \dot{x}_2 = u. \tag{8.3.17}$$

In constraints (8.3.13) with $m_2 = 0$, we need, by (8.3.14), to set $\mu = 0$. The number of parameters in the constraints can be reduced with the aid of the change of variables

$$t = p^{-1} t', \quad x_1 = p^{-2} x_1', \quad x_2 = p^{-1} x_2'. \tag{8.3.18}$$

By substituting (8.3.18) into system (8.3.17) and omitting the primes of the new variables, we arrive at the previous system (8.3.17), while constraints (8.3.13) take the form

$$|u| \leq 1, \quad |x_2| \leq 1, \quad |x_2 + qu| \leq 1, \quad q > 0. \tag{8.3.19}$$

Instead of the boundary conditions (8.3.15) and (8.3.16), we have

$$x_1(0) = x_1^0, \quad x_2(0) = x_2^0, \quad x_1(T) = x_2(T) = 0. \tag{8.3.20}$$

We apply the approach of Sect. 8.1.2 to problem (8.3.17), (8.3.19), and (8.3.20). In notation (8.1.1) and (8.1.6), we have for system (8.3.17):

$$A = \begin{bmatrix} 0 & 1 \\ 0 & 0 \end{bmatrix}, \quad B = \begin{bmatrix} 0 \\ 1 \end{bmatrix}, \quad f = \begin{bmatrix} 0 \\ 0 \end{bmatrix}, \quad \Phi(t) = \begin{bmatrix} 1 & t \\ 0 & 1 \end{bmatrix}. \tag{8.3.21}$$

Using (8.3.21), we find the inverse matrix $\Phi^{-1}(t)$, and then the matrices $Q(t)$ and $R(t)$ defined by (8.1.11) and (8.1.13):

$$\Phi^{-1}(t) = \begin{bmatrix} 1 & -t \\ 0 & 1 \end{bmatrix}, \quad Q(t) = \begin{bmatrix} -t \\ 1 \end{bmatrix}, \quad R(t) = \begin{bmatrix} t^3/3 & -t^2/2 \\ -t^2/2 & t \end{bmatrix}. \tag{8.3.22}$$

Here, $t_0 = 0$ in accordance with (8.3.20). Using (8.3.22), we also find the inverse matrix

$$R^{-1}(T) = 2T^{-3} \begin{bmatrix} 6 & 3T \\ 3T & 2T^2 \end{bmatrix}. \tag{8.3.23}$$

Using equalities (8.3.20), (8.3.21), and (8.1.9), we obtain

$$x^1 = 0, \quad x^* = -x^0. \tag{8.3.24}$$

With the aid of relations (8.3.22)–(8.3.24), we represent the phase vector (8.1.16) and control (8.1.17) in the form

$$\begin{aligned} x(t) &= \Phi(t)[E_2 - R(t)R^{-1}(T)]x^0 = X(t,T)x^0, \\ u(t) &= -Q^\top(t)R^{-1}(T)x^0 = \langle w(t,T), x^0 \rangle. \end{aligned} \tag{8.3.25}$$

Here, E_2 is the identity (2×2)-matrix. The elements of the (2×2)-matrix $X(t,T)$ and the two-dimensional vector $w(t,T)$ are equal to

$$\begin{aligned} X_{11}(t,T) &= 1 - 3\tau^2 + 2\tau^3, \quad X_{12}(t,T) = T\tau(1 - 2\tau + \tau^2), \\ X_{21}(t,T) &= \frac{6\tau(\tau-1)}{T}, \quad X_{22}(t,T) = 1 - 4\tau + 3\tau^2, \\ w_1(t,T) &= \frac{6(2\tau-1)}{T^2}, \quad w_2(t,T) = \frac{2(3\tau-2)}{T}, \quad \tau = \frac{t}{T}. \end{aligned} \tag{8.3.26}$$

Each of constraints (8.3.19) can be represented in the form

$$|\alpha x_2 + \beta u| \le 1, \quad 0 \le \alpha \le 1, \quad \beta \ge 0, \tag{8.3.27}$$

where α and β are constant coefficients. We substitute the expressions for x_2 and u from (8.3.25) and (8.3.26) into (8.3.27) and estimate from above the left-hand side of inequality (8.3.27):

$$|\alpha x_2 + \beta u| \le (\alpha|X_{21}| + \beta|w_1|)\,|x_1^0| + (\alpha|X_{22}| + \beta|w_2|)\,|x_2^0|. \tag{8.3.28}$$

Since $t \in [0,T]$, we have $\tau \in [0,1]$. We estimate from above the maxima, for $\tau \in [0,1]$, of the absolute values of the elements (8.3.26) that appear in (8.3.28):

$$|X_{21}| \le \frac{3}{2T}, \quad |X_{22}| \le 1, \quad |w_1| \le \frac{6}{T^2}, \quad |w_2| \le \frac{4}{T}. \tag{8.3.29}$$

Let us substitute the estimates (8.3.29) into (8.3.28):

$$|\alpha x_2 + \beta u| \le \left(\frac{3\alpha}{2T} + \frac{6\beta}{T^2}\right)|x_1^0| + \left(\alpha + \frac{4\beta}{T}\right)|x_2^0|. \tag{8.3.30}$$

Let us consider each of constraints (8.3.19) separately. Comparing inequalities (8.3.19) and (8.3.27), we set $\alpha = 0$ and $\beta = 1$ for the first constraint (8.3.19), $\alpha = 1$ and $\beta = 0$ for the second, and $\alpha = 1$ and $\beta = q$ for the third. Then, we obtain from (8.3.17) and (8.3.30) the following inequalities:

$$\begin{gathered}\frac{6}{T^2}|x_1^0| + \frac{4}{T}|x_2^0| \le 1, \quad \frac{3}{2T}|x_1^0| + |x_2^0| \le 1, \\ \left(\frac{3}{2T} + \frac{6q}{T^2}\right)|x_1^0| + \left(1 + \frac{4q}{T}\right)|x_2^0| \le 1.\end{gathered} \tag{8.3.31}$$

Since $q > 0$ by (8.3.19), the second inequality of (8.3.31) follows from the third one. Therefore, the set of two inequalities

$$\frac{6}{T^2}|x_1^0| + \frac{4}{T}|x_2^0| \le 1, \quad \left(\frac{3}{2T} + \frac{6q}{T^2}\right)|x_1^0| + \left(1 + \frac{4q}{T}\right)|x_2^0| \le 1 \tag{8.3.32}$$

constitute the sufficient conditions for solvability of the control problem (8.3.17), (8.3.19), and (8.3.20). These conditions connect the initial state and the time of the process, and can be regarded as the sufficient conditions for controllability of the system from the given initial state x^0 in the time T.

Let us analyze conditions (8.3.32). Suppose that the initial state x_1^0, x_2^0 is given. If $|x_2^0| > 1$, then both conditions (8.3.32) are not satisfied, which is quite natural since the initial state, in this case, violates the phase constraint $|x_2| \le 1$ imposed in (8.3.19). If $|x_2^0| < 1$, then both conditions (8.3.32) are definitely satisfied for sufficiently large T. The minimum time T^* for which both inequalities (8.3.32) hold is of interest. Solving the quadratic inequalities (8.3.32) in T^{-1}, we get

$$\begin{gathered}T \ge T^* = \max\{\frac{1}{z_1}, \frac{1}{z_2}\}, \quad |x_2^0| \le 1, \\ z_1 = \frac{(6a_1 + 4a_2^2)^{1/2} - 2a_2}{6a_1}, \quad a_i = |x_i^0|, \quad i = 1,2, \\ z_2 = \frac{[(3a_1 + 8qa_2^2)^{1/2} + 96qa_1(1 - a_2)]^{1/2} - 3a_1 - 8qa_2}{24qa_1}.\end{gathered} \tag{8.3.33}$$

Thus, if $|x_2^0| < 1$ and $T \geq T^*$, the proposed control method ensures bringing system (8.3.17) to the given state, and constraints (8.3.19) hold. Fixing any $T \geq T^*$, we find the desired control $u(t)$ and the phase trajectory $x(t)$ from the explicit formulas (8.3.25) and (8.3.26). This solves the problem stated.

Let us compare the constructed solution with the solution obtained in [35] (see Chapter 9) for the time-optimal control problem for system (8.3.17) with constraints (8.3.19) and boundary condition (8.3.20). We confine ourselves to the case of zero initial velocity $x_2^0 = 0$ and, for definiteness, take $q = 1/2$ in (8.3.19). We then obtain from (8.3.33)

$$T^*(a) = \max\{(6a)^{1/2}, \frac{12a}{(9a^2+48a)^{1/2}-3a}\} =$$
$$= \begin{cases} (6a)^{1/2} & \text{for } a \in [0, 2/3], \quad a = |x_1^0|, \\ \dfrac{12a}{(9a^2+48a)^{1/2}-3a} & \text{for } a \in [2/3, \infty). \end{cases} \tag{8.3.34}$$

From (8.3.34), we find the asymptotic behavior of the dependence $T^*(a)$ as $a \to \infty$:

$$T^*(a) = \frac{3a}{2} + O(1), \quad a \to \infty. \tag{8.3.35}$$

We note that the chosen value of the parameter $q = 1/2$ corresponds, by (8.3.19), to the set of admissible values of the variables (x_2, u) in the form of the hexagon shown in Fig. 8.13. The dependence of the optimal time of process T^0 on $a = |x_1^0|$ is given by the relationships [35]:

$$T^0(a) = \begin{cases} 2a^{1/2} & \text{for } a \in [0, 1/4], \\ \dfrac{1}{2} + v - \dfrac{\log[2(1-v)]}{2}, & v \geq \dfrac{1}{2}, \ a \geq \dfrac{1}{4}, \end{cases} \tag{8.3.36}$$
$$a = \frac{3}{8} - \frac{v(1-v)}{2} - \frac{\log[2(1-v)]}{2}.$$

For $a \geq 1/4$, the dependence $T^0(a)$ is given by (8.3.36) in the parametric form, where $v \geq 1/2$ is a parameter. Let us find the asymptotic behavior of the dependence $T^0(a)$ as $a \to \infty$. To that end, we let $v \to 1$. From (8.3.36), we obtain

$$T^0(a) = a + \frac{9}{8} + O(e^{-2a}), \quad a \to \infty. \tag{8.3.37}$$

Comparing dependences (8.3.34) and (8.3.36), we see that, for small a, the time $T^*(a)$ exceeds the optimal time $T^0(a)$ by about 22%. For large a, we see from (8.3.35) and (8.3.37) that the time $T^*(a)$ exceeds by 50% the optimal time. The dependences $T^*(a)$ and $T^0(a)$ are shown in Fig. 8.14. Thus, the proposed method of control brings the system to the given state in the time not greatly different from the optimal one.

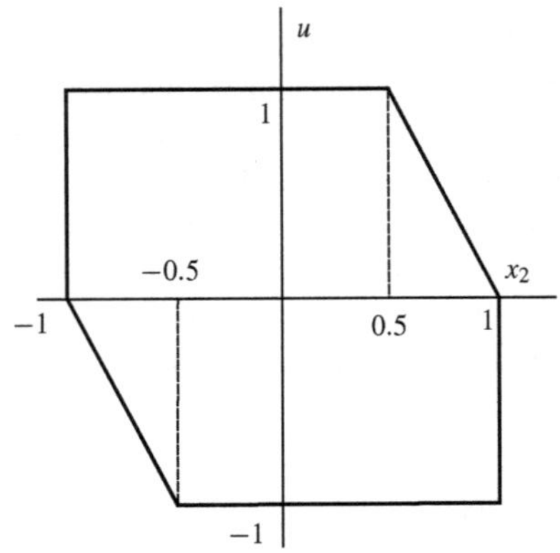

Fig. 8.13 Set of admissible values of the variables (x_2, u)

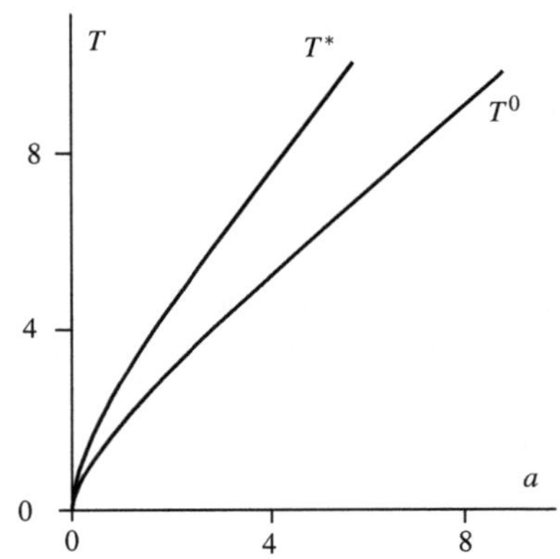

Fig. 8.14 Dependence of times T^* and T^0 on parameter a

8.3.3 Control of the electro-mechanical system of the fourth order

We apply the proposed approach to system (8.3.12) subject to constraints (8.3.13) and the boundary conditions (8.3.15) and (8.3.16). In notation (8.1.1), we have for system (8.3.12)

$$A = \begin{bmatrix} 0 & 1 & 0 & 0 \\ 0 & 0 & 0 & 0 \\ 0 & 0 & 0 & 1 \\ 0 & 0 & -1 & 0 \end{bmatrix}, \quad B = \begin{bmatrix} 0 \\ 1 \\ 0 \\ 1 \end{bmatrix}, \quad f = \begin{bmatrix} 0 \\ 0 \\ 0 \\ 0 \end{bmatrix}. \tag{8.3.38}$$

We write, for the system under consideration, the fundamental matrix (8.1.6) and its inverse:

$$\Phi(t)=\begin{bmatrix}1 & t & 0 & 0\\ 0 & 1 & 0 & 0\\ 0 & 0 & \cos t & \sin t\\ 0 & 0 & -\sin t & \cos t\end{bmatrix},\quad \Phi^{-1}(t)=\begin{bmatrix}1 & -t & 0 & 0\\ 0 & 1 & 0 & 0\\ 0 & 0 & \cos t & -\sin t\\ 0 & 0 & \sin t & \cos t\end{bmatrix}. \tag{8.3.39}$$

Using relationships (8.3.38) and (8.3.39), let us set up the matrix $Q(t)$ from (8.1.11) as follows:

$$Q(t)=(-t,1,-\sin t,\cos t)^{\top}. \tag{8.3.40}$$

We set up the matrix $R(t)$ from (8.1.13), using (8.3.40) and taking $t_0=0$:

$$R(t)=\begin{bmatrix}\frac{t^3}{3} & -\frac{t^2}{2} & \sin t-t\cos t & 1-\cos t-t\sin t\\ -\frac{t^2}{2} & t & \cos t-1 & \sin t\\ \sin t-t\cos t & \cos t-1 & \frac{2t-\sin 2t}{4} & \frac{\cos 2t-1}{4}\\ 1-\cos t-t\sin t & \sin t & \frac{\cos 2t-1}{4} & \frac{2t+\sin 2t}{4}\end{bmatrix}. \tag{8.3.41}$$

To simplify the calculations, as in Sect. 8.2.2, we put $T=2\pi k,\ k=1,2,\dots$. Then, by (8.3.41), the matrix $R(T)$ and its inverse take the forms

$$R(T)=\begin{bmatrix}\frac{T^3}{3} & -\frac{T^2}{2} & -T & 0\\ -\frac{T^2}{2} & T & 0 & 0\\ -T & 0 & \frac{T}{2} & 0\\ 0 & 0 & 0 & \frac{T}{2}\end{bmatrix},$$
$$R^{-1}(T)=\frac{1}{T^2-24}\begin{bmatrix}\frac{12}{T} & 6 & \frac{24}{T} & 0\\ 6 & \frac{4(T^2-6)}{T} & 12 & 0\\ \frac{24}{T} & 12 & 2T & 0\\ 0 & 0 & 0 & \frac{2(T^2-24)}{T}\end{bmatrix}. \tag{8.3.42}$$

Using equations (8.3.16), (8.3.38), and (8.1.9), we obtain, similarly to (8.3.24), that

$$x^1=0,\quad x^*=-x^0. \tag{8.3.43}$$

With the aid of (8.3.43), we represent the phase vector (8.1.16) and control (8.1.17) in the form similar to (8.3.25):

$$x(t) = X(t,T)x^0, \quad u(t) = \langle w(t,T), x^0 \rangle. \tag{8.3.44}$$

Here, $X(t,T)$ is the (4×4)-matrix

$$X(t,T) = \Phi(t) - \Phi_1(t)R^{-1}(T), \quad \Phi_1(t) = \Phi(t)R(t), \tag{8.3.45}$$

and $w(t,T)$ is the four-dimensional vector

$$w(t,T) = -Q^\top(t)R^{-1}(T). \tag{8.3.46}$$

Let us analyze constraints (8.3.13). First of all, we note that since the coefficients p and q are, by (8.3.14), positive, to meet these constraints, it is sufficient that the following two inequalities hold:

$$|u| \leq 1, \quad p|x_2 + \mu x_4| + q|u| \leq 1. \tag{8.3.47}$$

Let us substitute relationships (8.3.44) into inequalities (8.3.47). We get

$$\sum_{i=1}^{4} |w_i||x_i^0| \leq 1, \tag{8.3.48}$$

$$p|\sum_{i=1}^{4} (X_{2i} + \mu X_{4i})x_i^0| + q\sum_{i=1}^{4} |w_i||x_i^0| \leq 1. \tag{8.3.49}$$

We calculate the components of the vector $w(t,T)$ from (8.3.46), using equalities (8.3.40) and (8.3.42). We have

$$\begin{gathered} w_1 = \frac{12t - 6T + 24\sin t}{T(T^2-24)}, \quad w_2 = \frac{6tT - 4T^2 + 24 + 12T\sin t}{T(T^2-24)}, \\ w_3 = \frac{24t - 12T + 2T^2\sin t}{T(T^2-24)}, \quad w_4 = \frac{-2\cos t}{T}. \end{gathered} \tag{8.3.50}$$

To calculate the elements of the matrix $X(t,T)$ of (8.3.45), we first multiply the matrices $\Phi(t)$ from (8.3.39) and $R(t)$ from (8.3.41). We obtain

$$\Phi_1(t) = \begin{bmatrix} -\frac{t^3}{6} & \frac{t^2}{2} & \sin t - t & 1 - \cos t \\ -\frac{t^2}{2} & t & \cos t - 1 & \sin t \\ \sin t - t & 1 - \cos t & \frac{t\cos t - \sin t}{2} & \frac{t\sin t}{2} \\ \cos t - 1 & \sin t & -\frac{t\sin t}{2} & \frac{t\cos t + \sin t}{2} \end{bmatrix}. \tag{8.3.51}$$

We now calculate those elements of the matrix $X(t,T)$ that appear in constraints (8.3.49). To that end, we substitute into equality (8.3.45) for $X(t,T)$ relationships (8.3.39) for $\Phi(t)$, (8.3.51) for $\Phi_1(t)$, and (8.3.42) for $R^{-1}(T)$. We obtain

$$
\begin{aligned}
X_{21} &= \frac{6t(t-T)+24(1-\cos t)}{T(T^2-24)},\\
X_{22} &= \frac{T(3t^2-4tT+T^2)+12(2t-T)-12T\cos t}{T(T^2-24)},\\
X_{23} &= \frac{12t(t-T)+2T^2(1-\cos t)}{T(T^2-24)}, \quad X_{24} = \frac{-2\sin t}{T},\\
X_{41} &= \frac{12(1-\cos t)+6(2t-T)\sin t}{T(T^2-24)},\\
X_{42} &= \frac{6T(1-\cos t)-4(T^2-6)\sin t+6Tt\sin t}{T(T^2-24)},\\
X_{43} &= \frac{24(1-\cos t)+T(12+tT-T^2)\sin t}{T(T^2-24)},\\
X_{44} &= \frac{(T-t)\cos t-\sin t}{T}.
\end{aligned} \tag{8.3.52}
$$

Inequalities (8.3.48) and (8.3.49) must hold for all $t \in [0,T]$. We first estimate from the above the maximum values, for $t \in [0,T]$, of the quantities $|w_i|$ in (8.3.48) and (8.3.49). We have

$$
\begin{aligned}
&|12t-6T+24\sin t| \le 6T+24, \quad t\in[0,T],\\
&|24t-12T+2T^2\sin t| \le 12T+2T^2,\\
&6tT-4T^2+24+12T\sin t \le 2T^2+12T+24,\\
&6tT-4T^2+24+12T\sin t \ge -4T^2-12T+24.
\end{aligned} \tag{8.3.53}
$$

Since $T=2\pi k,\ k\ge 1$, we have $T^2>24$. Then, it follows from the last two of inequalities (8.3.53) that

$$
\begin{aligned}
&|6tT-4T^2+24+12T\sin t|\\
&\le \max\{2T^2+12T+24, 4T^2+12T-24\} = 4T^2+12T-24.
\end{aligned} \tag{8.3.54}
$$

Substituting estimates (8.3.53) and (8.3.54) into (8.3.50), we obtain

$$
|w_i(t,T)| \le A_i, \ \ i=1,2,3,4, \ \ t\in[0,T]. \tag{8.3.55}
$$

Here,

$$
A_1 = \frac{6(T+4)}{T(T^2-24)}, \ A_2 = \frac{4(T^2+3T-6)}{T(T^2-24)}, \ A_3 = \frac{2(T+6)}{T^2-24}, \ A_4 = \frac{2}{T}. \tag{8.3.56}
$$

Turning to inequality (8.3.49), we note that the elements X_{22}, X_{43}, and X_{44} in (8.3.52) do not approach zero as $T\to\infty$. To get more precise estimate of the principal terms in inequality (8.3.49) as $T\to\infty$, we estimate the first term in that inequality as follows:

$$|\sum_{i=1}^{4}(X_{2i}+\mu X_{4i})x_i^0| \le B_1|x_1^0|+B_2|x_2^0|+B_3\left(x_3^{0\,2}+x_4^{0\,2}\right)^{1/2}. \tag{8.3.57}$$

Here, we introduced the notation

$$B_1 = |X_{21}+\mu X_{41}|, \quad B_2 = |X_{22}+\mu X_{42}|,$$
$$B_3 = \left[(X_{23}+\mu X_{43})^2+(X_{24}+\mu X_{44})^2\right]^{1/2}. \tag{8.3.58}$$

Let us estimate the quantities B_i, $i=1,2,3$. Taking into account the obvious estimates

$$|2t-T| \le T, \quad -\frac{T^2}{4} \le t(t-T) \le 0, \quad t\in[0,T], \tag{8.3.59}$$

we obtain from (8.3.52) that

$$T(T^2-24)(X_{21}+\mu X_{41}) \le 6t(t-T)+48+24\mu+6\mu T \le 24(2+\mu)+6\mu T,$$

$$T(T^2-24)(X_{21}+\mu X_{41}) \ge 6t(t-T)-6\mu T \ge -\frac{3T^2}{2}-6\mu T.$$

This implies the following estimate for the quantity B_1 in (8.3.58):

$$B_1 \le D_1 = \frac{\max\{3T^2/2, 24(2+\mu)\}+6\mu T}{T(T^2-24)}. \tag{8.3.60}$$

Then, by virtue of (8.3.52), we obtain

$$T(T^2-24)(X_{22}+\mu X_{42}) = T(3t^2-4tT+T^2)+12(2t-T)$$
$$+6\mu T-6T(2+\mu)\cos t+2\mu T(3t-2T)\sin t+24\mu\sin t. \tag{8.3.61}$$

One can easily see that

$$|3t^2-4tT+T^2| \le T^2, \quad |3t-2T| \le 2T, \quad t\in[0,T]. \tag{8.3.62}$$

By (8.3.61), (8.3.62), and (8.3.59), we obtain the following estimate for the quantity B_2 in (8.3.58):

$$B_2 \le D_2 = \frac{T^3+4\mu T^2+12(2+\mu)T+24\mu}{T(T^2-24)}. \tag{8.3.63}$$

To single out the principal terms as $T\to\infty$, we make, using formulas (8.3.52), the following transformation:

$$X_{23}+\mu X_{43} = \frac{\mu(t-T)\sin t}{T}+X_3',$$
$$X_3' = \frac{12t(t-T)+(2T^2+24\mu)(1-\cos t)+12\mu(2t-T)\sin t}{T(T^2-24)}, \tag{8.3.64}$$

$$X_{24}+\mu X_{44}=\frac{\mu(T-t)\cos t}{T}+X_4',\quad X_4'=-\frac{(2+\mu)\sin t}{T}.$$

Substituting (8.3.64) into expression (8.3.58) for B_3, we obtain that

$$\begin{aligned} B_3^2 &= \frac{\mu^2(T-t)^2}{T^2}+\frac{2\mu(T-t)(X_4'\cos t-X_3'\sin t)}{T}+X_3'^2+X_4'^2 \\ &\le \mu^2+2\mu\left(X_3'^2+X_4'^2\right)^{1/2}+X_3'^2+X_4'^2. \end{aligned} \tag{8.3.65}$$

Using (8.3.59), we get for the quantity X_3' in (8.3.64) the estimates

$$T(T^2-24)X_3'\le 12t(t-T)+4(T^2+12\mu)+12\mu T\le 4T^2+12\mu T+48\mu,$$

$$T(T^2-24)X_3'\ge 12t(t-T)-12\mu T\ge -3T^2-12\mu T.$$

These estimates imply the inequality

$$|X_3'|\le\frac{4T^2+12\mu T+48\mu}{T(T^2-24)}. \tag{8.3.66}$$

It follows from (8.3.64) that

$$|X_4'|\le\frac{2+\mu}{T}. \tag{8.3.67}$$

Substituting inequalities (8.3.66) and (8.3.67) into (8.3.65), we get

$$|B_3|\le D_3=\mu+\left[\frac{(4T^2+12\mu T+48\mu)^2}{T^2(T^2-24)^2}+\frac{(2+\mu)^2}{T^2}\right]^{1/2}. \tag{8.3.68}$$

Now, let us analyze inequalities (8.3.48) and (8.3.49) that are sufficient conditions for solvability of the control problem in time $T=2\pi k$, $k=1,2,\dots$. Substituting estimates (8.3.55) into (8.3.48), we obtain the condition

$$\sum_{i=1}^{4}A_i|x_i^0|\le 1. \tag{8.3.69}$$

Substituting (8.3.55) and (8.3.57) into (8.3.49) and using estimates (8.3.60), (8.3.63), and (8.3.68) for B_i, we get

$$p\left[D_1|x_1^0|+D_2|x_2^0|+D_3\left(x_3^{02}+x_4^{02}\right)^{1/2}\right]+q\sum_{i=1}^{4}A_i|x_i^0|\le 1. \tag{8.3.70}$$

Conditions (8.3.69) and (8.3.70) are sufficient conditions for controllability of system (8.3.12) in the finite time $T=2\pi k$, $k=1,2,\dots$. In other words, if these conditions, for some initial state x^0 and time $T=2\pi k$ are satisfied, then system (8.3.12) can be brought from the initial state (8.3.15) to the given terminal state (8.3.16) in time T. The control law $u(t)$ and the phase trajectory $x(t)$ of the system are given by relations (8.3.44), in which the matrix $X(t,T)$ and the vector $w(t,T)$ are defined by equalities (8.3.52) and (8.3.50), respectively. Thus, all the quantities sought are determined in an explicit analytical form, once we have found the time of the process T.

It remains to choose the time $T = 2\pi k$, $k = 1,2,\dots$, in such a way that, for the given initial state x^0, inequalities (8.3.69) and (8.3.70) hold. To do this, we shall assign to k successively the values $k = 1,2,\dots$, calculate the coefficients A_i, $i = 1,2,3,4$, from formulas (8.3.56) and D_j, $j = 1,2,3$, from formulas (8.3.60), (8.3.63), and (8.3.68), and then verify inequalities (8.3.69) and (8.3.70). We note that $A_i \to 0$, $i = 1,2,3,4$; $D_1 \to 0$, $D_2 \to 1$, $D_3 \to \mu$ as $T \to \infty$. Consequently, inequality (8.3.69) always holds for sufficiently large T. Inequality (8.3.70) also holds for sufficiently large T, if

$$p\left[|x_2^0| + \mu\left(x_3^{02} + x_4^{02}\right)^{1/2}\right] \le 1. \tag{8.3.71}$$

Thus, inequality (8.3.71) is a sufficient condition for solvability of the control problem stated, i.e., for bringing the system under consideration to the terminal state in finite time.

If, in system (8.3.12), there is only one constraint on the control $|u| \le 1$, whereas the remaining (phase and mixed) constraints (8.3.13) are absent, then time T must be chosen in such a way as to satisfy only one condition (8.3.69). In this case, the control problem is always solvable in finite time.

We note that although the control law $u(t)$ is represented in the open-loop form (8.3.44), it can also be used for the feedback correction. To that end, one should determine, at intervals, the current phase vector x and to treat it as the initial vector x^0 in (8.3.44), recalculating each time the duration of the process T, in accordance with the algorithm described above. Since we have explicit relationships, this recalculation is not difficult.

Let us give an example of the numerical implementation of the described algorithm for control of system (8.3.12) under constraints (8.3.13). The dimensionless parameters (8.3.14) are taken as follows:

$$p = 0.1, \quad q = 0.5, \quad \mu = 0.5. \tag{8.3.72}$$

The initial data (8.3.15) are taken in the form

$$x_1^0 = -5, \quad x_2^0 = 0, \quad x_3^0 = -5, \quad x_4^0 = 0. \tag{8.3.73}$$

One can easily verify that parameters (8.3.72) and (8.3.73) satisfy condition (8.3.71) of controllability in finite time.

As a result of the numerical implementation of the control algorithm described in Sect. 8.3.3, we first find the minimum integer k for which conditions (8.3.69) and (8.3.70) hold, and then construct the control $u(t)$ and the phase trajectory $x(t)$. In this case, we have $k = 3$ and $T = 6\pi$.

In Fig. 8.15, curves 1–3 show the time histories of the quantities u, $p(x_2+\mu x_4)$, and $p(x_2+\mu x_4)+qu$, respectively, that appear in constraints (8.3.13). One can see that these constraints hold everywhere. In Fig. 8.16, the projections of the four-dimensional phase trajectory $x(t)$ on the (x_1,x_2)- and the (x_3,x_4)-planes are depicted by the curves 1 and 2, respectively.

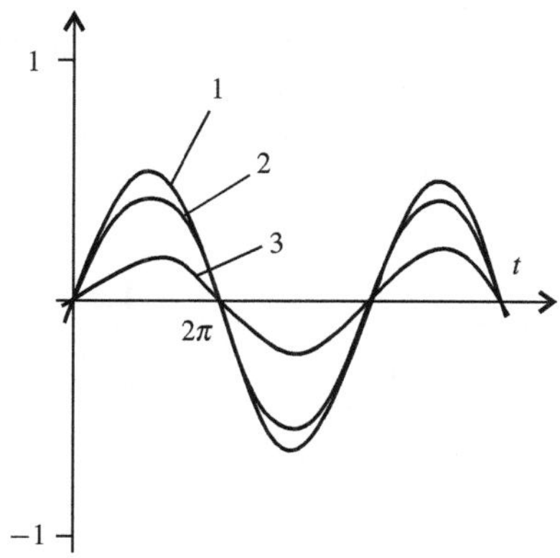

Fig. 8.15 Time histories of u, $p(x_2+\mu x_4)$, and $p(x_2+\mu x_4)+qu$

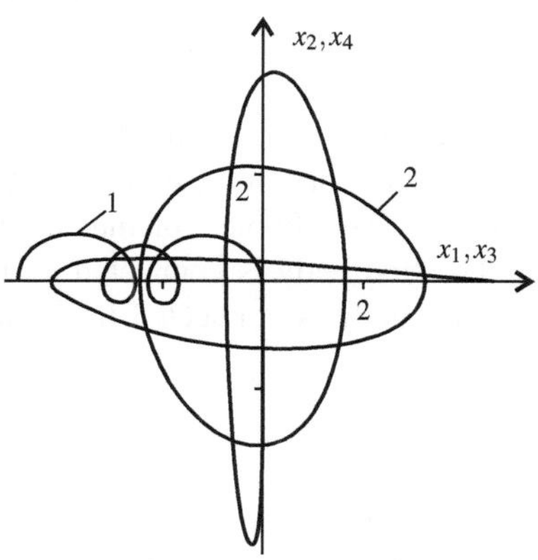

Fig. 8.16 Projections of the phase trajectory

8.3.4 Active dynamical damper

We consider the problem of damping oscillations of a load attached to the end of an elastic beam by means of an active dynamical damper with a moving mass. The control variable is the interaction force between the damper and the load. Systems of this type are used, for example, in a spacecraft (SC), where measuring devices are mounted, by means of a long rod, on a platform (P) at a significant distance from the main body of the spacecraft. High accuracy in positioning and stabilizing the measuring instruments is required in order to perform measurements; hence, it is of great importance to damp any oscillations of the rod, and this should be taken into consideration in spacecraft design. One way to solve this problem is to use a controlled damper located on the platform itself. The damper consists of a guide 1 perpendicular to the axis of the rod 2, and a movable mass 3 that can be displaced along the guide by an electric drive. This scheme is suitable for damping transverse oscillations of the rod (see Fig. 8.17).

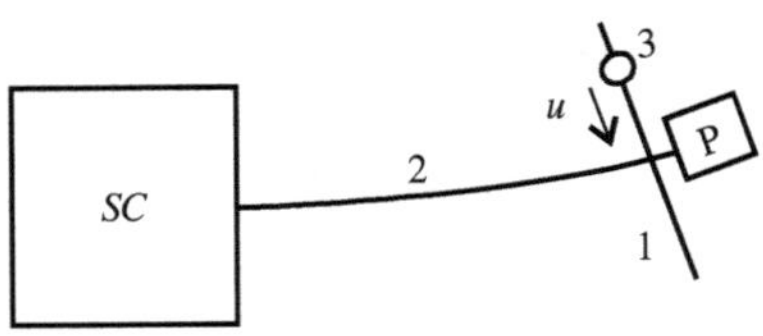

Fig. 8.17 Rod with a load and an active dynamical damper

A particular feature of this problem is the presence of two natural constraints on the different variables of the system. One, due to the restricted possibilities of the drive, is imposed on the control force; the other, due to the bounded path of the damper mass (the damper guide is limited in size), is imposed on the displacement of the mass relative to the platform.

Under certain simplifying assumptions [3], the following two-mass mechanical controlled system containing oscillatory link (see Fig. 8.18) may serve as a model for the structures just described. Two bodies, of masses m_1 and m_2, move along a horizontal straight line. The first body is connected to a fixed base by a spring of stiffness $c > 0$. The second body is connected to the first one by a drive that generates force u.

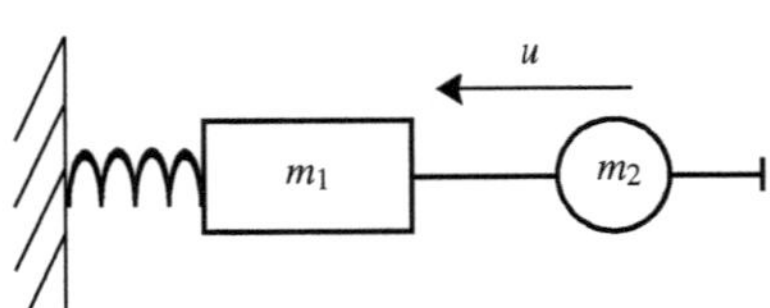

Fig. 8.18 Simplified model

The equations of motion of the system are

$$m_1\ddot{y} + cy = -u, \quad m_2\ddot{z} = u. \tag{8.3.74}$$

Here, y and z are the coordinates of the first and second bodies, respectively, on the straight line. The following constraints are imposed on the control force u and the displacement $z(t)-y(t)$ of the second body relative to the first one:

$$|u(t)| \leq a, \quad a > 0, \tag{8.3.75}$$

$$|z(t)-y(t)| \leq d, \quad d > 0. \tag{8.3.76}$$

It is required to construct control $u(t)$ that meets constraint (8.3.75) and brings system (8.3.74) from the given initial state

$$y(0) = y^0,\ \dot{y}(0) = \dot{y}^0,\ z(0) = z^0,\ \dot{z}(0) = \dot{z}^0 \tag{8.3.77}$$

to the state of rest

$$y(T) = z(T) = 0, \quad \dot{y}(T) = \dot{z}(T) = 0. \tag{8.3.78}$$

In addition, the coordinates $y(t)$ and $z(t)$ must satisfy condition (8.3.76) throughout the whole motion. The time of the process termination T is not fixed.

We introduce the new variables

$$x_1 = \frac{c}{a}y,\ x_3 = -\frac{m_2 c}{m_1 a}z,\ t' = \sqrt{\frac{c}{m_1}}t,\ u' = -au. \tag{8.3.79}$$

In terms of variables (8.3.79), system (8.3.74) and constraint (8.3.75) become (here and below, we denote by dots derivatives with respect to the new time t', the primes of t' and u' are omitted)

$$\ddot{x}_1 + x_1 = u, \quad \ddot{x}_3 = u, \tag{8.3.80}$$

$$|u| \leq 1, \tag{8.3.81}$$

i.e., just the same as in (8.2.36). Constraint (8.3.76) takes the form

$$\left|\frac{m_1}{m_2}x_3 + x_1\right| \leq \frac{cd}{a}.$$

We put $\dot{x}_1 = x_2$, $\dot{x}_3 = x_4$ and denote the phase vector of system (8.3.80) by $x = (x_1, x_2, x_3, x_4)$. We introduce a constant vector $p = (1, 0, m_1/m_2, 0)$ and rewrite the last inequality as follows:

$$|\langle p, x(t)\rangle| \leq \frac{cd}{a}. \tag{8.3.82}$$

After the change of variables (8.3.79), conditions (8.3.77) and (8.3.78) take the form

$$x_i(0) = x_i^0,\ x_i(T) = 0,\ i = 1,\ldots,4. \tag{8.3.83}$$

Here, x_i^0 are given constants, and $T > 0$ is the as yet unknown time when the process terminates.

The problem stated reduces to constructing a control that satisfies the constraint $|u| \leq 1$ and brings system (8.3.80) from the given initial state (8.3.83) to the coordinate origin without violation constraint (8.3.82) during the motion. It is the phase constraint (8.3.82) that essentially distinguishes this problem from those considered in Sect. 8.2. Besides, here we will use other estimation technique than before.

The solution presented below obtained in [7].

We rewrite system (8.3.80) in the vector form

$$\dot{x} = Ax + bu, \tag{8.3.84}$$

$$A = \begin{bmatrix} 0 & 1 & 0 & 0 \\ -1 & 0 & 0 & 0 \\ 0 & 0 & 0 & 1 \\ 0 & 0 & 0 & 0 \end{bmatrix}, \qquad b = \begin{bmatrix} 0 \\ 1 \\ 0 \\ 1 \end{bmatrix}.$$

The initial and terminal states (8.3.83) become

$$x(0) = x^0, \quad x(T) = 0. \tag{8.3.85}$$

As in the case of the pendulum with a suspension point controlled by acceleration, the fundamental matrix of the homogeneous system and the inverse matrix have form (8.2.39), and the matrix $Q(t)$ in (8.1.11) is the four-dimensional column vector given by (8.2.40).

The expression for the control function $u(t)$ that brings system (8.3.84) from the initial state x^0 to the coordinate origin can be written as follows:

$$u(t) = \langle V(t,T), x^0 \rangle, \quad V(t,T) = -R^{-1}(T)Q(t). \tag{8.3.86}$$

Here, the matrix $R(T)$ is given by (8.2.54).

We will show that, if the duration T of the process is sufficiently large, the control $u(t)$ meets constraint (8.3.81). To that end, we estimate the function $u(t)$ as follows:

$$|u(t)| \le \sum_{i=1}^{4} |V_i(t,T)x_i^0| \le |V(t,T)|_\infty \, |x^0|_1. \tag{8.3.87}$$

Here, by $|\cdot|_\infty$ and $|\cdot|_1$ we denote the norms in the spaces R^4_∞ and R^4_1, respectively, which have the following form for an arbitrary vector q:

$$|q|_\infty = \max_{1 \le i \le 4} |q_i|, \quad |q|_1 = \sum_{i=1}^{4} |q_i|.$$

We introduce the auxiliary function

$$v(T) = \max_{0 \le t \le T} |V(t,T)|_\infty \tag{8.3.88}$$

and rewrite estimate (8.3.87) as follows:

$$\max_{0 \le t \le T} |u(t)| \le v(T) \, |x^0|_1. \tag{8.3.89}$$

We propose two ways of determining the duration T of the motion for which constraint (8.3.81) holds. The first is based on the analytical estimation of the function $v(T)$, and the second on the numerical construction of the function.

Similarly to what has been done in Sect. 8.2, in order to simplify calculations, we choose the time of the termination of motion in the form $T = 2\pi k$, where k is an integer. In this case, the matrix $R(T)$ is given by (8.2.42) and its inverse matrix is

$$R^{-1}(T) = \frac{1}{\triangle}\begin{bmatrix} 2T & 0 & \frac{24}{T} & 12 \\ 0 & \frac{2\triangle}{T} & 0 & 0 \\ \frac{24}{T} & 0 & \frac{12}{T} & 6 \\ 12 & 0 & 6 & \frac{4(T^2-6)}{T} \end{bmatrix}, \quad \triangle = T^2 - 24. \tag{8.3.90}$$

We write down the components of the vector-valued function $V(t,T)$ using expressions (8.3.90) and (8.2.40) for the matrix $R^{-1}(T)$ and vector $Q(t)$, respectively. Then, taking into account the inequality $T \geq 2\pi$, we estimate the components $V_i(t,T)$ as follows:

$$|V_1(t,T)| = \frac{|2T^2\sin t + 24t - 12T|}{T\triangle} \leq \frac{2T+12}{\triangle} \leq \frac{4T}{\triangle},$$

$$|V_2(t,T)| = \frac{|2\cos t|}{T} \leq \frac{2}{T} \leq \frac{4T}{\triangle},$$

$$|V_3(t,T)| = \frac{|-24\sin t - 12t + 6T|}{T\triangle} \leq \frac{6T+24}{T\triangle} \leq \frac{4T}{\triangle},$$

$$|V_4(t,T)| = \frac{|-12T\sin t - 6Tt + 4T^2 - 24|}{T\triangle} \leq \frac{4T}{\triangle}.$$

These estimates and definition (8.3.88) of the function $v(T)$ imply that $v(T) \leq 4T/\triangle$. Hence, using (8.3.89), we obtain the following estimate for the control function $u(t)$:

$$\max_{0\leq t\leq T} |u(t)| \leq \frac{4T}{T^2-24}|x^0|_1. \tag{8.3.91}$$

Since $T = 2\pi k,\ k = 1,2\ldots$, we have, for sufficiently large k, that

$$\frac{4T}{T^2-24} \leq \frac{1}{|x^0|_1}. \tag{8.3.92}$$

Inequalities (8.3.91) and (8.3.92) ensure constraint (8.3.81).

We also propose another method of choosing the process duration T that ensures constraint (8.3.81) for the control function (8.3.86). To that end, we construct the function $v(T)$ numerically, using relations (8.2.54), (8.3.86) and (8.3.88). The function $v(T)$ is completely defined by the matrix A and vector b of system (8.3.84), therefore, it suffices to construct this function once for the given system.

Figure 8.19 shows the graph of the function $v(T)$. As we expected, $v(T)$ is a decreasing function, and the longer the time of motion of the system to its terminal state, the lower the maximum value of the control function u. As the time of the process termination one can choose any time that satisfies inequality

$$v(T) \leq \frac{1}{|x^0|_1}. \tag{8.3.93}$$

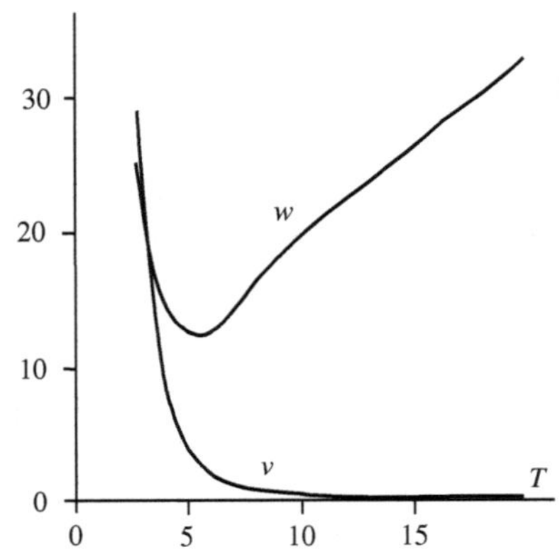

Fig. 8.19 Functions $v(T)$ and $w(T)$

We now describe the method of choosing the process duration T that ensures constraint (8.3.82). With the above notation, the solution of system (8.3.84) that starts at the time instant $t = 0$ at the point x^0 has the form

$$x(t) = \Phi(t)\left(x^0 + \int_0^t Q(\tau)u(\tau)d\tau\right).$$

Substituting into this formula expression (8.3.86) for the control function $u(t)$ and using relations (8.2.54) that define the matrix $R(t)$, we obtain

$$\begin{aligned} x(t) &= \Phi(t)\left(x^0 - \int_0^t Q(\tau)\left[Q^\top(\tau)R^{-1}(T)x^0\right]d\tau\right) = \\ &= \Phi(t)\left(x^0 - \left[\int_0^t Q(\tau)Q^\top(\tau)d\tau\right]R^{-1}(T)x^0\right) = \\ &= \Phi(t)\left(x^0 - R(t)R^{-1}(T)x^0\right) = W(t,T)x^0, \\ W(t,T) &= \Phi(t)\left[R(T) - R(t)\right]R^{-1}(T). \end{aligned} \tag{8.3.94}$$

As in the case of constraint (8.3.81), we propose two methods of determining the values of T that ensure constraint (8.3.82).

We first choose the time of the process termination to be $T = 2\pi k$, $k = 1,2\ldots$, and estimate the norm of the vector $x(t)$ in terms of the norms of the matrices on the right-hand side of (8.3.94). The quantity $\|\Phi(t)\|^2$ is equal (see Sect. 3.2.4) to the maximal eigenvalue $\phi(t)$ of the matrix

$$\Phi^\top(t)\Phi(t) = \begin{bmatrix} 1 & 0 & 0 & 0 \\ 0 & 1 & 0 & 0 \\ 0 & 0 & t^2+1 & t \\ 0 & 0 & t & 1 \end{bmatrix}, \quad t \in [0,T].$$

It is not difficult to calculate that $\phi(t) \le t^2+2$, whence we obtain

$$\|\Phi(t)\| \le (T^2+2)^{1/2}. \tag{8.3.95}$$

The matrix $R^{-1}(T))$ is symmetric and positive-definite; consequently, its eigenvalues are positive, and the maximal of them equals the norm of the matrix. In addition [64], the sum of all these eigenvalues is equal to the trace of $R^{-1}(T)$ (by definition, the trace $\mathrm{tr}\, Z$ of a square matrix Z is the sum of the elements on the main diagonal of Z). It follows from expression (8.3.90) for the matrix $R^{-1}(T)$ that the number $2/T$ is one of the eigenvalues of this matrix, and

$$\frac{\mathrm{tr}\, R^{-1}(T)}{4} = \frac{2T^2-15}{T(T^2-24)} > \frac{2}{T}.$$

Consequently, $2/T$ is not the maximal eigenvalue, and

$$\|R^{-1}(T)\| \le \mathrm{tr}\, R^{-1}(T) - \frac{2}{T} = \frac{6(T^2-2)}{T(T^2-24)}, \quad T = 2\pi k. \tag{8.3.96}$$

Since system (8.3.84) is controllable, the matrices $R(t), R(T)$, and

$$R(T) - R(t) = \int_t^T Q(\tau)Q^\top(\tau)d\tau.$$

are symmetric and positive definite. Besides,

$$R(T) - R(t) < R(T), \quad 0 < t \le T, \tag{8.3.97}$$

(the inequality $X < Y$ for the symmetric matrices X and Y means that the matrix $Y - X$ is positive definite). Since the norm of a symmetric positive definite matrix is equal to the maximal of all its eigenvalues, and inequality (8.3.97) implies [64] the corresponding inequality for the eigenvalues of matrices $R(T) - R(t)$ and $R(T)$, we have $\|R(T) - R(t)\| \le \|R(T)\|$.

To estimate $\|R(T)\|$, we employ reasoning similar to that used above for the matrix $R^{-1}(T)$. It follows from expression (8.2.42) for $R(T)$ that the number $T/2$ is a non-maximum eigenvalue of the matrix $R(T)$, therefore,

$$\|R(T)\| \le \operatorname{tr} R(T) - \frac{T}{2} = \frac{T(2T^2+9)}{6}, \quad T = 2\pi k.$$

By (8.3.94)–(8.3.96), equality $|p| = (m_1^2+m_2^2)^{1/2}/m_2$, and the last relation, we have

$$|\langle p, x(t)\rangle| \le |p| \|\Phi(t)\| \|R(T)-R(t)\| \|R^{-1}(T)\| |x^0| \le$$
$$\le \frac{(m_1^2+m_2^2)^{1/2}(2T^2+9)(T^2-2)(T^2+2)^{1/2}}{m_2(T^2-24)} |x^0|.$$

Consequently, we can choose as the time of the process termination those values of $T = 2\pi k$ that satisfy the inequality

$$\frac{(2T^2+9)(T^2-2)(T^2+2)^{1/2}}{T^2-24} \le \frac{m_2 cd}{a(m_1^2+m_2^2)^{1/2}|x^0|}. \tag{8.3.98}$$

We propose one more method of choosing the process duration T that ensures constraint (8.3.82). It follows from (8.3.94) that

$$|\langle p, x(t)\rangle| \le |\langle p, W(t,T)x^0\rangle| \le w(T)|x^0|, \quad w(T) = \max_{0\le t\le T} |W^\top(t,T)p|.$$

We construct the function $w(T)$ numerically. This function is completely defined by the matrix A and vectors b and p, so it suffices to construct this function once for the given system (8.3.84) and given vector p in constraint (8.3.82).

Figure 8.19 shows graph of the function $w(T)$ for $p = (1,0,10,0)$. One can see, that with the increase of the process duration T, the function $w(t)$ first decreases and then increases. As the duration of the process, one can take any T for which the value of the function $w(t)$ satisfies the inequality

$$w(T) \le \frac{cd}{a|x^0|}. \tag{8.3.99}$$

Thus, we propose the following procedure for constructing the control function $u(t)$. First, given the initial state vector x^0 we choose the time T of the process duration. The value of T may be chosen in the form $T = 2\pi k$, where the natural number k must ensure conditions (8.3.92) and (8.3.98). Another way to determine T is to find the functions $v(T)$ and $w(T)$ numerically and then to find such a value of T that guarantees inequalities (8.3.93) and (8.3.99).

After the time T of the process duration has been determined, we calculate the control function $u(t)$ analytically using (8.3.86). As we have already mentioned, the expression for the inverse matrix $R^{-1}(T)$, obtained by using the computer program of symbolic calculations, turns out to be quite cumbersome. For illustration, the element in the upper left corner of this matrix is presented in Sect.8.2.3 [see formula (8.2.61)].

In Fig. 8.20, we present the results of the numerical simulation of the dynamics of system (8.3.84). The system is brought from the initial state $x^0 = (0.5, -0.5, 0.5, 0.5)$ to the coordinate origin. The time of the process duration is taken as $T = 10$. The solid curve represents the projection of the phase trajectory on the plane $x_1, x_2 = \dot{x}_1$, while the thin curve represents its projection on the plane $x_3, x_4 = \dot{x}_3$.

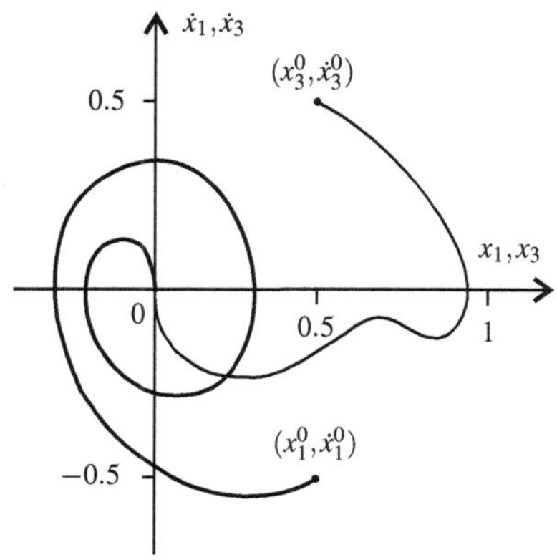

Fig. 8.20 Projections of the phase trajectory

The solid curves in Fig. 8.21 are graphs of the control function $u(t)$ and the quantity

$$|\langle p, x(t)\rangle| = \left|\frac{m_1}{m_2}x_3 + x_1\right|$$

that appears in constraint (8.3.82), as a function of time, for the case where $m_1/m_2 = 10$. For comparison, the dashed curves represent the same functions for the process duration taken as $T = 5$.

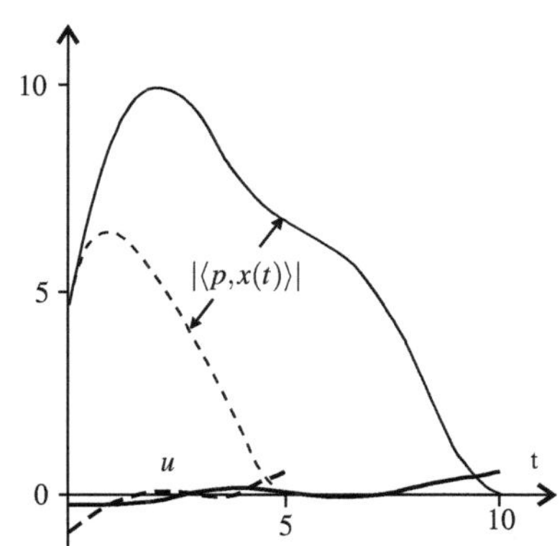

Fig. 8.21 Control function $u(t)$ and quantity $|\langle p, x(t)\rangle|$

One can see that, for the case of $T = 5$, the quantity $|\langle p, x(t)\rangle|$ is smaller, whereas the maximum modulus of the control function $u(t)$ is greater, than for $T = 10$.

Chapter 9
Optimal control problems under complex constraints

The simplest dynamical system with one degree of freedom

$$m\ddot{\xi} = F \tag{9.0.1}$$

often serves as a model in control theory. Here, ξ is a generalized coordinate, $\dot{\xi}$ is a generalized velocity, m is a constant inertial characteristic (the mass or moment of inertia), F is the control (the force or the moment of the force), and dots denote derivatives with respect to the time t. On the one hand, this model is used to work out and demonstrate schemes and methods for solving control problems. On the other, a system with one degree of freedom has been used as an element in certain schemes for decomposing non-linear systems with many degrees of freedom into simpler subsystems (see Chapters 2 and 3).

In this chapter, solutions of a series problems concerning the steering system (9.0.1) to the origin of the phase plane under various constraints are presented. These results have been obtained in papers [35, 48, 49, 38, 37].

In a number of practical control problems, the so-called mixed constraints are imposed at every time instant on the current control and state variables. For example, if the electric drive is present in the system, then the restrictions imposed on the control torque of the motor, angular velocity of the shaft, and other mechanical and electric parameters are often imposed. As known, the existence of the mixed constraints in the optimal control problems causes significant difficulties even for linear systems.

A time-optimal control problem in the presence of mixed constraints is analyzed in Sect. 9.1. The results are applied to the control of an electric motor [35, 37].

In Sect. 9.2, the case, where system (9.0.1) is acted upon by a control force with a bounded rate of change, is considered. The time-optimal open-loop control of the system is constructed. The feedback optimal control is given in a closed form [48, 49].

In Sect. 9.3, the case, where system (9.0.1) controlled by a force of bounded magnitude, is considered. It is assumed that the magnitude of the force may increase gradually at a finite rate and that the force is switched off instantaneously. Under these restrictions, which simulate real servo-systems, a control is constructed that steers the system to the origin and has the simplest possible structure [38].

9.1 Time-optimal control problem under mixed and phase constraints

9.1.1 Problem statement

Let us introduce the notation

$$x = \xi, \qquad v = \dot{\xi}, \qquad w = \frac{F}{m} \tag{9.1.1}$$

and rewrite (9.0.1) in the form

$$\dot{x} = v, \qquad \dot{v} = w. \tag{9.1.2}$$

We will consider control system of the second order (9.1.2), where x and v are the phase coordinates (coordinate and velocity), and w is the control (acceleration). The constraints

$$v_1 \leq v \leq v_2, \qquad v_1 < 0, \qquad v_2 > 0, \tag{9.1.3}$$

where v_1 and v_2 are given constants, and the mixed constraint

$$f_1(v) \leq w \leq f_2(v), \qquad v \in [v_1, v_2] \tag{9.1.4}$$

are imposed on the control and phase coordinates. Here, $f_1(v)$ and $f_2(v)$ are piecewise continuous functions given on the interval $[v_1, v_2]$ and such that

$$f_1(v) < 0, \qquad f_2(v) > 0, \qquad v \in (v_1, v_2). \tag{9.1.5}$$

Functions $f_1(v)$ and $f_2(v)$ may vanish only on the boundaries of the interval $[v_1, v_2]$. The domain in the plane (v, w) restricted by conditions (9.1.3) and (9.1.4) is shown in Fig. 9.1.

Let us state the problem of determining the feedback control $w(x, v)$ bringing system (9.1.2) under constraints (9.1.3) and (9.1.4) from any (in the domain $v_1 \leq v \leq v_2$) initial phase state

$$x(t_0) = x_0, \qquad v(t_0) = v_0 \tag{9.1.6}$$

to the state

$$x(T) = 0, \qquad v(T) = 0 \tag{9.1.7}$$

in a minimal time ($T \to \min$). If we replace control w by a new control u in accordance with formula

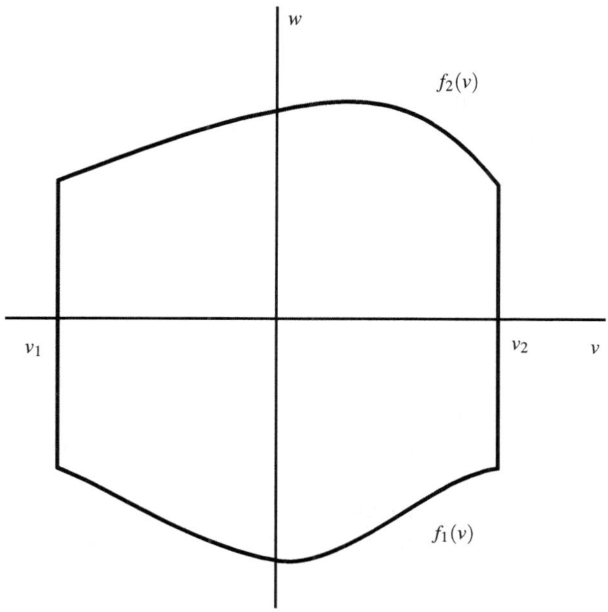

Fig. 9.1 Domain of admissible values (v,w)

$$w = f_3(v)u + f_4(v), \quad f_3(v) = \frac{f_2(v) - f_1(v)}{2}, \quad f_4(v) = \frac{f_1(v) + f_2(v)}{2},$$

then system (9.1.2) with constraints (9.1.3) and (9.1.4) takes the form

$$\dot{x} = v, \qquad \dot{v} = f_3(v)u + f_4(v), \qquad |u| \leq 1, \qquad v_1 \leq v \leq v_2. \tag{9.1.8}$$

Here, the mixed constraint is absent. If we omit the phase constraint $v_1 \leq v \leq v_2$, then the obtained time-optimal control problem for system (9.1.8) with boundary conditions (9.1.6) and (9.1.7) becomes the particular case of a more general time-optimal control problem for which the regular synthesis is constructed [23]. The presence of the phase constraint leads to some modification of this synthesis. Below, the time-optimal feedback control for the formulated time-optimal control problem in the initial form (9.1.2)–(9.1.7) is proposed and the direct proof of its optimality without using the maximum principle is given.

9.1.2 Time-optimal control under constraints imposed on the velocity and acceleration

Supposing $w = f_i(v)$, $i = 1,2$, we obtain from (9.1.2)

$$\frac{dx}{dv} = \frac{v}{f_i(v)}, \qquad i = 1,2. \tag{9.1.9}$$

Integrating (9.1.9), we find equations of the phase trajectories corresponding to the controls $w = f_i(v)$ in the form (C_i is a constant)

$$x = X_i(v) + C_i, \qquad i = 1,2, \tag{9.1.10}$$

where

$$X_i(v) = \int_0^v \frac{vdv}{f_i(v)}, \qquad i = 1,2. \tag{9.1.11}$$

Due to (9.1.5), functions (9.1.11) have the following properties. Function $X_1(v)$ increases monotonically for $v < 0$ and decreases monotonically for $v > 0$, and $X_2(v)$, vice versa, decreases monotonically for $v < 0$ and increases monotonically for $v > 0$. Function $X_1(v)$ has a zero maximum at $v = 0$, and function $X_2(v)$ has a zero minimum at $v = 0$. According to (9.1.2), the motion along trajectories (9.1.10) with $i = 1$ occurs in the direction of decreasing of v, and along trajectories (9.1.10) with $i = 2$—in the direction of increasing of v. If $f_i(v_j) = 0$ for some $i, j = 1,2$ and integral $X_i(v_j)$ in (9.1.11) diverges, then the corresponding phase trajectory (9.1.10) has the horizontal asymptote $v = v_j$. Otherwise, i.e., if either $f_i(v_j) \neq 0$ or $f_i(v_j) = 0$, but the integral $X_i(v_j)$ converges, the curve $x = X_i(v_j)$ intersects the straight line $v = v_j$ at the point $x = X_i(v_j)$.

First of all, we describe the proposed feedback control, and after that, we prove its optimality. Let us define the switching curve in the phase plane (x, v) by the equalities

$$x = X(v) = \begin{cases} X_2(v), & v \in [v_1, 0], \\ X_1(v), & v \in [0, v_2]. \end{cases} \tag{9.1.12}$$

Due to the properties of the functions $X_i(v)$, $i = 1,2$, function $X(v)$ decreases monotonically on the interval $[v_1, v_2]$ with $X(0) = 0$. The switching curve (9.1.12) has an inflection point at the origin of coordinates. This curve is shown in Figs. 9.2 and 9.3 by the thick line. It divides the strip $v_1 < v < v_2$ defined by the phase constraints (9.1.3) into two domains: D_1 [at the right top of curve (9.1.12)] and D_2 (to the lower right of this curve), see Figs. 9.2 and 9.3. Figure 9.2 corresponds to the case where both integrals $X_1(v_1)$ and $X_2(v_2)$ diverge, and Fig. 9.3—to the case where all $X_i(v_j)$ are bounded, $i, j = 1,2$. We set in the open domains D_i

$$\begin{aligned} w &= f_1(v) \quad \text{for} \quad x > X(v), \quad v \in (v_1, v_2) \quad (\text{in } D_1); \\ w &= f_2(v) \quad \text{for} \quad x < X(v), \quad v \in (v_1, v_2) \quad (\text{in } D_2). \end{aligned} \tag{9.1.13}$$

Let us define the control on the boundaries of the domains D_1 and D_2. We set

$$\begin{aligned} w &= f_1(v) \quad \text{for} \quad x = X_1(v), \quad v \in [0, v_2]; \\ w &= f_2(v) \quad \text{for} \quad x = X_2(v), \quad v \in [v_1, 0] \end{aligned} \tag{9.1.14}$$

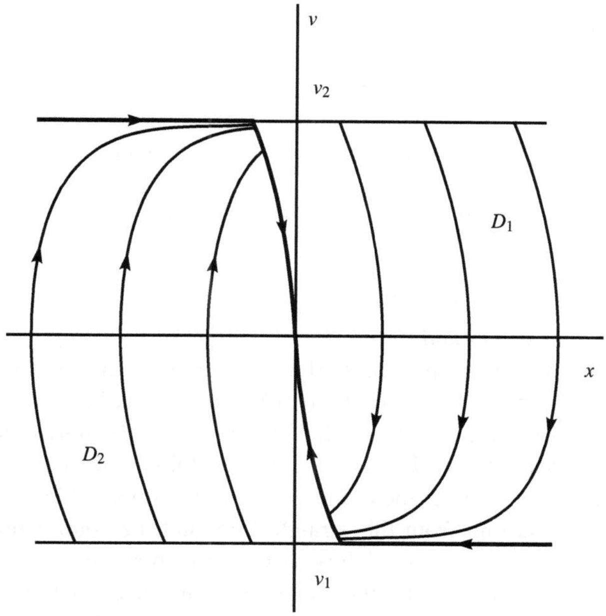

Fig. 9.2 Switching curve and phase trajectories [$X_1(v_1)$ and $X_2(v_2)$ diverge]

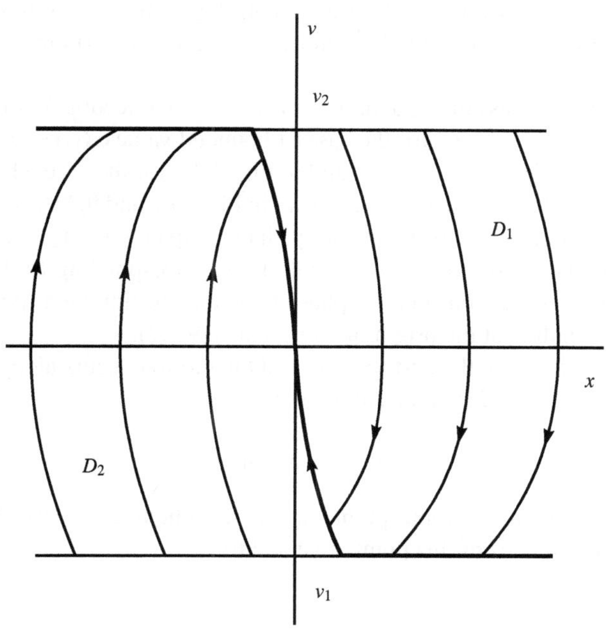

Fig. 9.3 Switching curve and phase trajectories [all $X_i(v_j)$ are bounded]

on the switching curve (9.1.12) and

$$\begin{aligned} &w=0 \quad \text{for} \quad v=v_1, \quad x>X_2(v_1);\\ &w=f_2(v_1) \quad \text{for} \quad v=v_1, \quad x<X_2(v_1);\\ &w=0 \quad \text{for} \quad v=v_2, \quad x<X_1(v_2);\\ &w=f_1(v_2) \quad \text{for} \quad v=v_2, \quad x>X_1(v_2) \end{aligned} \tag{9.1.15}$$

on the straight lines $v=v_1$ and $v=v_2$.

Relations (9.1.15) have the sense, if the magnitudes $X_2(v_1)$ and $X_1(v_2)$ are bounded. If any of them $[X_2(v_1)$ or $X_1(v_2)]$ is unbounded, i.e., the corresponding integral (9.1.11) diverges, then the switching curve has the horizontal asymptote $v=v_1$ or $v=v_2$, respectively. In this case, the control is undefined on this straight line ($v=v_1$ or $v=v_2$). Under the circumstances, it is impossible to reach the domain $v_1<v<v_2$ starting from this straight line, and the control problem is unsolvable, if the initial point (x_0,v_0) lies on the corresponding straight line. If the magnitudes $X_1(v_1)$ or $X_2(v_2)$ are unbounded, then it means that the phase trajectories starting inside the domains D_1 and D_2, respectively, do not reach the corresponding straight lines v_1 and v_2 and, therefore, intersect the switching curve. If the initial point (x_0,v_0) lies on one of these straight lines, then, according to (9.1.15), the phase state moves along this line under the control $w=0$ until it reaches the switching curve.

The phase trajectories corresponding to the constructed feedback control (9.1.13)–(9.1.15) are shown in Fig. 9.2 for the case of bounded values $X_1(v_2)$ and $X_2(v_1)$ but unbounded values $X_1(v_1)$ and $X_2(v_2)$, and in Fig. 9.3—for the case where all values $X_i(v_j)$ for $i,j=1,2$ are bounded. The arrows in Figs. 9.2 and 9.3 show the direction of motion. The trajectories cover compactly all the strip $v_1\le v\le v_2$. The every phase trajectory consists of no more than three sections corresponding to three different controls. To be definite, consider the phase trajectory beginning at point (9.1.6) in domain D_2 and ending at the origin of coordinates (9.1.7).

We have $w=f_2(v)$ on the first section, and the motion occurs along the curve of family (9.1.10) with $i=2$, namely, along the curve

$$x=X_2(v)-X_2(v_0)+x_0 \tag{9.1.16}$$

from the point (x_0,v_0) to some point on the straight line $v=v_2$. According to (9.1.15), the coordinates of this point are equal to

$$x=x_2=X_2(v_2)-X_2(v_0)+x_0, \qquad v=v_2. \tag{9.1.17}$$

We have $w=0$ on the second section, and the motion occurs along the straight line $v=v_2$ from point (9.1.17) to the point

$$x=X_1(v_2), \qquad v=v_2 \tag{9.1.18}$$

on the switching curve $x = X(v)$. We have $w = f_1(v)$ on the third section, and the motion occurs along the switching curve $x = X_1(v)$ from point (9.1.18) to the origin of coordinates.

The second section is absent, if $X_2(v_2)$ and $X_1(v_2)$ are unbounded, and also if point (9.1.17) lies to the right of point (9.1.18), i.e., for $x_2 > X_1(v_2)$. In these cases, the transition from the first to the third section occurs at the intersection point of the curves (9.1.16) and $x = X_1(v_2)$. If the initial point lies on the straight line $v = v_2$, $X_1(v_2)$ is bounded and $x_0 < X_1(v_2)$, then the first section is absent. Finally, if the initial point lies on the switching curve, then the first and second sections are absent. The phase trajectories in the domain D_1 have a similary structure.

Compare the motion times along two trajectories for the constructed feedback control starting at the points (x_0', v_0) and (x_0'', v_0) lying on the straight line $v = v_0$ in the domain D_2. If $x_0' < x_0''$, then it follows from the structure of the phase trajectories that the first trajectory starting at $x_0 = x_0'$ contains all those sections that belong to the second one (with the origin at $x_0 = x_0''$). Therefore, the entire time of motion along the first trajectory is greater than that for the second one. This property of monotone dependence of the motion time on the abscissa x_0 of the initial point will be used below.

Let us turn to the optimality proof of the proposed feedback control. To this end, together with some phase trajectory of the constructed synthesis, which will be called an original one, we consider an arbitrary tentative trajectory satisfying constraints (9.1.3) and (9.1.4). The tentative trajectory starts at the time instant $t = t_0$ at the point (x_0, v_0), which is the same as the starting point for the original trajectory, and ends at the moment $t = T_*$ at the origin of coordinates. To prove the optimality of the constructed feedback control, it is sufficient to show that $T_* > T$.

First of all, we note that the tentative trajectory may have self-intersections, i.e., contain closed loops. The motion time along each of the loops is positive. If one deletes all the loops from the tentative trajectory joining the beginnings and the ends for each of the loops, then the obtained new tentative trajectory is admissible, and the time of motion along it will be less that the motion time for the tentative trajectory with self-intersections. Hence, without loss of generality, it is sufficient to consider further only tentative trajectories without self-intersections.

First, we assume that the point (x_0, v_0) lies on the switching curve $x = X_1(v)$. According to (9.1.2), the time of motion along the original and tentative trajectories are equal to

$$T - t_0 = \int_{v_0}^{0} \frac{dv}{f_1(v)}, \qquad T_* - t_0 = \int_{v_0}^{0} \frac{dv}{w}, \tag{9.1.19}$$

respectively.

If the velocity v strictly decreases along the tentative trajectory (as it takes place along the original trajectory), then $dv < 0$ and $f_1(v) \le w < 0$ due to (9.1.4), and, hence, it follows from (9.1.19) that $T \le T_*$. If v changes non-monotonically along the tentative trajectory, then the second integral in (9.1.19) should be regarded as an integral along a curve. If there are sections of the tentative trajectory where v

increases, then $dv > 0$ and $w > 0$ on these sections. This gives an additional positive input to the second integral in (9.1.19). Hence, we have $T \le T_*$ in all cases.

Thus, if the phase point (initial or current) reaches the switching curve, then the further optimal motion occurs along this curve. Therefore, one may not consider further tentative trajectories intersecting the switching curve and assume that the final section of the tentative trajectory entering the origin of coordinates lies on the switching curve.

Now, let the initial point (x_0, v_0) lies in the domain D_2 with $v_0 \ge 0$. First, we assume that the tentative trajectory does not intersect the axis x for $t \ge t_0$. Then, the coordinate x increases monotonically both on the original and tentative trajectories. The time of motion along these trajectories can be presented in the form

$$T - t_0 = \int_{x_0}^{0} \frac{dx}{v(x)}, \qquad T_* - t_0 = \int_{x_0}^{0} \frac{dx}{v_*(x)}, \tag{9.1.20}$$

in accordance with (9.1.2).

The dependence $v(x)$ for the original trajectory, according to the facts stated above, consists of no more than three sections: increasing from v_0 to v_2, motion with maximal possible velocity $v = v_2$, and decreasing from v_2 to 0 (some of these sections may be absent). The dependence $v_*(x)$ for the tentative trajectory satisfies the inequalities

$$\frac{f_2(v_*)}{v_*} \le \frac{dv_*}{dx} \le \frac{f_1(v_*)}{v_*}, \qquad v_* \ge 0, \tag{9.1.21}$$

in accordance with (9.1.4) and (9.1.9). The equal signs in (9.1.21) are attained at the original trajectory. From (9.1.21), constraint $v_* \le v_2$, and boundary conditions $v_*(x_0) = v_0$ and $v_*(0) = 0$ that are common for the original and tentative trajectories, it follows that the dependence $v_*(x)$ (for every fixed v) increases no faster and decreases no slower than $v(x)$. In other words, the graph of the function $v_*(x)$ lies no higher than the graph of dependence $v(x)$, i.e., $v_*(x) \le v(x)$ for $x \in [x_0, 0]$. Then, it follows from (9.1.20) that $T \le T_*$.

Now, let us assume that the tentative trajectory starting at the point (x_0, v_0) of the domain D_2 with $v_0 \ge 0$ intersects the axis x for some $t_1 > t_0$. We have $x_1 > x_0$ in the intersection point, where the velocity changes its sign, and x starts to decrease. Nevertheless, the tentative trajectory must eventually reach the domain $v > 0$. In this case, either the self-intersection of the tentative trajectory may occur, the possibility of which was excluded above, or the tentative trajectory comes to the point with the coordinates (x_2, v_0), where $x_2 < x_0$, and after that the trajectory remains in the domain $v_0 \ge 0$. The time t_2^* of motion along the tentative trajectory from the point (x_2, v_0) to the point $(0,0)$, in accordance to the facts proven above, is no less than the corresponding time t_2 for the original trajectory with the same starting point (x_2, v_0). Due to the monotone dependence of the time of motion along the original trajectory on the coordinates of the initial point, we have $t_2 > T - t_0$ because $x_2 < x_0$. Hence, we have $T_* - t_0 > t_2^* > t_2 > T - t_0$. Thus, $T < T_*$ in the case considered.

Thus, if the phase point comes to the axis x with $x < 0$, then the further optimal motion occurs along the original trajectory passing throw the given point of the axis x. Therefore, we conclude that the tentative trajectory in the half-plane $v > 0$ coincides with the original one.

Finally, consider the case, where the initial point (x_2, v_0) lies in domain D_2 and $v_0 < 0$. Let us present both the tentative and original trajectories in the form of two sections lying in the half-planes $v < 0$ and $v > 0$, respectively. Let us compare the times of motions along the first sections ($v < 0$) similarly to the consideration concerning the initial point on the switching curve [see (9.1.19)]. We obtain that the tentative trajectory reaches the axis x not before the original one. The estimates of the form (9.1.21) allow one to establish that the point $x = x_1^*$, where the tentative trajectory intersects the axis x, lies not to the right of the corresponding point $x = x_1$ of the intersection of the axis x and the original trajectory, i.e., $x_1^* \leq x_1$. Due to the monotone dependence of the time of motion along the original trajectory on the abscissa of the initial point proven above, we conclude that the duration of the second section of the tentative trajectory is not less than that for the second section of the original trajectory. Hence, it is proved that $T \leq T_*$ also in this case. Thus, the time-optimality of the constructed feedback control given by (9.1.13)–(9.1.15) is completely proved.

9.1.3 Problem of control of an electric motor

As an example, consider a simple model of a DC (direct current) motor with an independent excitation. The torque M produced by the motor is proportional to the armature circuit current I

$$M = k^M I, \tag{9.1.22}$$

where k^M is a constant coefficient. Equation of balance of the electric voltages in the armature circuit has the from

$$L\dot{I} + RI + k^E \omega = u, \qquad k^E > 0. \tag{9.1.23}$$

Here, L and R are the inductance and resistance of the armature winding, u is the control voltage, ω is the angular velocity of the armature rotation, $k^E \omega$ is the counter-emf, and k^E is a constant coefficient. The first term on the left-hand side of (9.1.23) is usually small compared with the remaining terms and can be omitted. Then, having eliminated I from (9.1.22) by using (9.1.23), we obtain

$$M = k^M R^{-1}(u - k^E \omega). \tag{9.1.24}$$

The magnitude of the control voltage is bounded

$$|u| \leq u_0, \tag{9.1.25}$$

where u_0 is a constant. It follows from (9.1.24) and (9.1.25) that torque M is bounded by the inequalities

$$-k^M R^{-1}(u_0 + k^E \omega) \leq M \leq k^M R^{-1}(u_0 - k^E \omega). \tag{9.1.26}$$

Additionally to (9.1.26), the restrictions on the magnitude of the electric current I and torque M are often imposed; the later are associated with the requirements related to the reduction gear. Both types of the constraints are reduced, according to (9.1.22), to the restriction of the form $|M| \leq M_0$, where M_0 is a constant. We will bind the absolute value of the angular velocity by $\omega_0 = u_0 \left(k^E\right)^{-1}$. Under this restriction, the left- and right-hand sides of inequality (9.1.26) have the opposite signs, i.e., the moment M for $|\omega| < \omega_0$ may be both accelerating and decelerating. As a result, we come to the following set of the constraints

$$\begin{gathered} -k(\omega_0 + \omega) \leq M \leq k(\omega_0 - \omega), \quad |M| \leq M_0, \quad |\omega| \leq \omega_0, \\ k = k^M R^{-1} k^E, \quad \omega_0 = u_0 \left(k^E\right)^{-1}. \end{gathered} \tag{9.1.27}$$

The equations of rotation of the rotor of the electric drive have the form

$$\dot{\alpha} = \omega, \qquad J\dot{\omega} = M + M_1. \tag{9.1.28}$$

Here, α is the angle of rotation of the rotor, J is the moment of inertia of the rotor and other parts of the reduction gear, M_1 is the moment of all external forces except the electromagnetic torque M, which can be considered as the control in system (9.1.28). Constraints (9.1.27), among which there are the mixed constraint and also constraints on the control and phase coordinate, are imposed on system (9.1.28). Let us introduce the non-dimensional variables t', x, v, w, and parameter κ by relations

$$t = J\omega_0 M_0^{-1} t', \quad \alpha = J\omega_0^2 M_0^{-1} x, \quad \omega = \omega_0 v, \quad M = M_0 w, \quad \kappa = k\omega_0 M_0^{-1}. \tag{9.1.29}$$

Then, equations (9.1.28) with $M_1 = 0$ and restrictions (9.1.27) are transformed to the form

$$\dot{x} = v, \quad \dot{v} = w, \qquad |w| \leq 1, \quad |v| \leq 1, \quad -\kappa(1+v) \leq w \leq \kappa(1-v), \tag{9.1.30}$$

where dot denotes derivatives with respect to the non-dimensional t' from (9.1.29). Below, the prime at the variable t' is omitted. Equations (9.1.30) coincide with (9.1.2), and constraints (9.1.30) are reduced to the form of (9.1.3) and (9.1.4), where

$$f_1(v) = -f_2(-v) = \begin{cases} -\kappa(1+v) & \text{for} \quad -1 \leq v \leq v_c, \\ -1 \quad \text{for} \quad v_c \leq v \leq 1, \end{cases} \tag{9.1.31}$$

$$v_c = \kappa^{-1} - 1, \quad v_1 = -1, \quad v_2 = 1.$$

Here, the functions f_1 are f_2 are piecewise-linear and have the break points at $v = \pm v_c$, respectively.

If $\kappa \le 1/2$, then the break points lie outside the interval $(-1, 1)$. In this case, the functions f_1 and f_2 are linear on the interval $[-1, 1]$, the constraint $|w| \le 1$ is satisfied automatically and becomes non-essential. Then, system (9.1.30) is reduced to the form of the linear controlled system with the new control w_1:

$$\dot{x} = v, \quad \dot{v} = -\kappa v + w_1, \qquad |w_1| \le \kappa, \quad |v| \le 1.$$

Here, the mixed constraint is absent.

The points $\pm v_c$ lie inside the interval $(-1, 1)$ for $\kappa > 1/2$. The points v_c and $-v_c$ lie in the intervals $(0, 1)$ and $(-1, 0)$ for $\kappa \in (1/2, 1)$, respectively, and in the intervals $(-1, 0)$ and $(0, 1)$ for $\kappa > 1$, respectively.

The domain in the plane (v, w) allowed by constrains (9.1.30) is the hexagon in the general case, possessing the central symmetry about the origin of coordinates. It is shown in Fig. 9.4 for the case, where $\kappa = 2$.

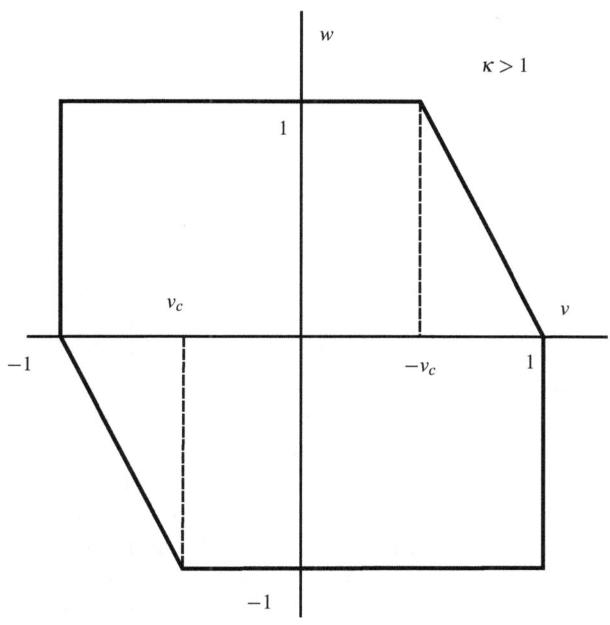

Fig. 9.4 Domain of admissible values (v, w)

Let us consider the time-optimal control problem for the described model of the electric drive. In the non-dimensional variables (9.1.29), this model is given by relations (9.1.30) that are the particular case of relations (9.1.2)–(9.1.4), where notation (9.1.31) is used. Conditions (9.1.5) for functions f_1 and f_2 from (9.1.31) hold true. The boundary conditions still have the form of (9.1.6) and (9.1.7). We will specify the general solution of the time-optimal feedback control problem obtained for system (9.1.2)–(9.1.7) in the context of system (9.1.30).

Let us substitute (9.1.31) into (9.1.11) and find the corresponding integrals, considering separately the cases $\kappa \le 1/2$, $\kappa \in [1/2, 1]$, and $\kappa \ge 1$. We obtain

$$X_1(v) = -\kappa^{-1}[v - \log(1+v)] \quad \left(\kappa \le \frac{1}{2}\right),$$

$$X_1(v) = \begin{cases} -\kappa^{-1}[v - \log(1+v)] & \text{for} \quad v \in [-1, v_c], \\ \frac{1}{2}(1 - v^2 - \kappa^{-2}) - \kappa^{-1}\log\kappa & \text{for} \quad v \in [v_c, 1], \end{cases}$$

$$\left(\frac{1}{2} \le \kappa \le 1, \quad 0 \le v_c \le 1\right),$$

$$X_1(v) = \begin{cases} -\kappa^{-1}[v - \log(1+v)] + \frac{1}{2}(\kappa^{-2} - 1) + \kappa^{-1}\log\kappa & \text{for} \quad v \in [-1, v_c], \\ -\frac{1}{2}v^2 \quad \text{for} \quad v \in [v_c, 1], \end{cases}$$

$$(\kappa \ge 1, \quad -1 \le v_c \le 0),$$

$$X_2(v) = -X_1(-v), \quad v_c = \kappa^{-1} - 1. \tag{9.1.32}$$

The switching curve $x = X(v)$, given by (9.1.12), divides the strip $|v| \le 1$ into the domains D_1, where $x > X(v)$, and D_2, where $x < X(v)$. The feedback time-optimal control is defined by (9.1.13)–(9.1.15), where $v_1 = -1$ and $v_2 = 1$, and functions f_i and X_i are defined by (9.1.31) and (9.1.32), respectively, $i = 1, 2$. Note that the switching curve and the entire field of optimal trajectories possess the central symmetry about the origin of coordinates due to (9.1.32). The field of optimal trajectories for the case $\kappa = 2$, $v_c = -1/2$ is shown in Fig. 9.2.

Since the magnitudes of $X_1(1)$ and $X_2(-1)$ are bounded [see (9.1.32)], the time-optimal feedback control is defined in the entire strip $|v| \le 1$ including its boundaries, in accordance with the remark in Sect. 9.1.2. On the other hand, it follows from (9.1.32) that the magnitudes of $X_1(-1)$ and $X_2(1)$ are not bounded. Therefore, according to Sect. 9.1.2, the phase trajectories starting inside the strip, i.e., with $|v| < 1$, do not reach the boundary of the strip and intersect the switching curve (see Fig. 9.2). Hence, all these trajectories consist of no more than two sections: the middle section described in Sect. 9.1.2 is absent. If the initial point lies on the boundary of the strip, then the phase point moves along the boundary under the control $w = 0$ up to the switching curve, if $v = -1$ and $x > X_2(-1)$ or if $v = 1$ and $x < X_1(1)$. In other cases, i.e., if $v = -1$ and $x < X_2(-1)$ or if $v = 1$ and $x > X_1(1)$, the phase point comes inside the strip (see Fig. 9.2).

Let us find the motion time along the optimal phase trajectory starting at $t = 0$ at the point $(x_0, 0)$ of the x-axis, where $x_0 < 0$. According to the constructed synthesis and remarks made above, the optimal trajectory starts in the domain D_2 and consists of the two sections. The first section is the arc of curve (9.1.16), i.e., of the curve

$$x = X_2(v) + x_0 \tag{9.1.33}$$

from the point $(x_0, 0)$ to the point (x_m, v_m), where curve (9.1.33) intersects the switching curve $x = X_1(v)$. Therefore, we have the following relations for determining the coordinates (x_m, v_m):

$$x_m = X_1(v_m) = X_2(v_m) + x_0. \tag{9.1.34}$$

The second section of the phase trajectory is the arc of the curve $x = X_1(v)$ from the point (x_m, v_m) to the origin of coordinates. The velocity v increases from 0 to v_m on the first section, where $w = f_2(v)$, and decreases from v_m to 0 on the second section, where $w = f_1(v)$. We present the total motion time in the form

$$T = \int_0^{v_m} \frac{dv}{f_2(v)} + \int_{v_m}^{0} \frac{dv}{f_1(v)}. \tag{9.1.35}$$

To be definite, we restrict our consideration by the case $\kappa > 1$ and denote

$$\xi = -x_0 > 0, \quad q = -v_c = 1 - \kappa^{-1} > 0, \quad \kappa > 1. \tag{9.1.36}$$

Let us determine integrals (9.1.35) substituting into them expressions (9.1.31) and using notation (9.1.36). We obtain

$$\begin{aligned} &T = 2v_m \quad \text{for} \quad v_m \le q, \\ &T = q + v_m - (1-q)\log\left|\frac{1-v_m}{1-q}\right| \quad \text{for} \quad v_m \ge q. \end{aligned} \tag{9.1.37}$$

Substituting (9.1.32) with $\kappa > 1$ into (9.1.34) and using notation (9.1.36), we obtain the relations connecting v_m and ξ

$$\begin{aligned} &\xi = v_m^2 \quad \text{for} \quad v_m \le q, \\ &\xi = \frac{q^2 + v_m^2}{2} + (1-q)\left\{(q - v_m) - \log\left|\frac{1-v_m}{1-q}\right|\right\} \quad \text{for} \quad v_m \ge q. \end{aligned} \tag{9.1.38}$$

Relations (9.1.37) and (9.1.38) determine the dependence $T(\xi)$ in the parametric form. Assigning the values from 0 to 1 to the parameter v_m, we obtain the required dependence $T(\xi)$ for every fixed q. Note that (9.1.37) and (9.1.38) yield

$$T(\xi) = 2\xi^{1/2} \quad \text{for} \quad \xi \le q^2. \tag{9.1.39}$$

Let us find the asymptotic behavior of the function $T(\xi)$ for $\xi \to \infty$. To this end, suppose $v_m \to 1$ and obtain from (9.1.37) and (9.1.38)

$$T = -(1-q)\log(1-v_m) + q + 1 + (1-q)\log(1-q) + O(1-v_m),$$

$$\xi = -(1-q)\log(1-v_m) + 2q - \frac{q^2+1}{2} + (1-q)\log(1-q) + O(1-v_m). \tag{9.1.40}$$

From (9.1.40), we find the desired asymptotic behavior

$$T(\xi) = \xi + \frac{3}{2} - q + \frac{q^2}{2} + O\Big(\exp[-\xi(1-q)^{-1}]\Big), \quad \xi \to \infty. \tag{9.1.41}$$

Let us consider separately the limiting case $\kappa \to \infty$. According to (9.1.36), we have here $v_c = -1$ and $q = 1$, and constraints (9.1.30) determine the square $|v| \leq 1$, $|w| \leq 1$ in the plane (v, w). In this case, according to (9.1.32),

$$X_1(v) = -X_2(v) = -\frac{v^2}{2} \quad \text{when} \quad v \in [-1, 1]. \tag{9.1.42}$$

All phase trajectories consist of the arcs of parabolas defined by (9.1.10) and (9.1.42), and sections of the straight lines $v = \pm 1$, as shown in Fig. 9.3.

Let us find the total time T of motion from the initial point $(x_0, 0)$, where $x_0 = -\xi < 0$, to the origin. If $v_m < 1$, then relation (9.1.39) holds; if $v_m = 1$, then the motion time is the sum of the motion times along the parabolic arcs and along the straight line $v = 1$. In the non-dimensional variables, the motion time along each of the parabolas is equal to 1, and the time of motion along the straight line $v = 1$ is equal to the length of the segment of this straight line equal to $\xi - 1$. As a result, we obtain for $q = 1$

$$T = 2\xi^{1/2}, \quad \text{if} \quad \xi \leq 1,$$

$$T = \xi + 1, \quad \text{if} \quad \xi \geq 1.$$

Results of numerical calculations of the dependence $T(\xi)$ for various q are presented in Fig. 9.5.

Thus, the complete solution of the problem stated above is obtained.

9.2 Time-optimal control under constraints imposed on the rate of change of the acceleration

9.2.1 Statement of the problem

We consider a system with a single degree of freedom described by equation (9.0.1). When formulating optimal control problems, it is usually assumed that the absolute magnitude of the force F is bounded by a constant F_0, that is, $|F| \leq F_0$. In the case

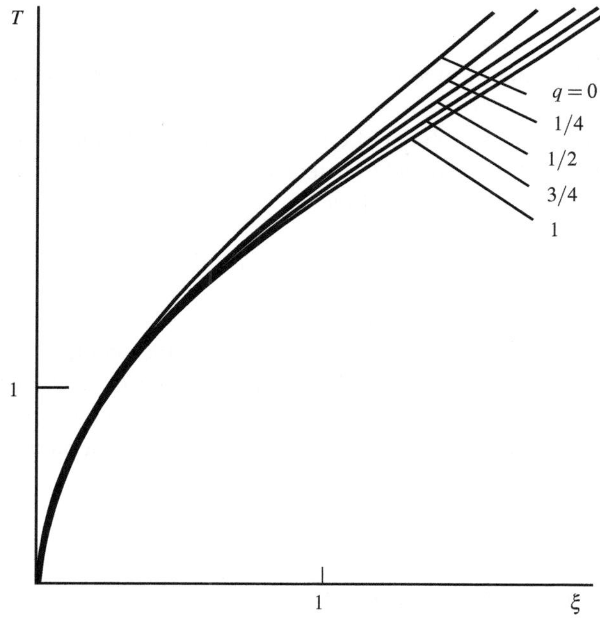

Fig. 9.5 Dependence $T(\xi)$ for various q

of a time-optimal control problem, it is well-known that this constraint leads to the bang-bang form of the optimal control. In this case, the force $F(t)$ takes limiting values $\pm F_0$ and instantaneously switches from one of these values to the other. Such a control is not always practicable, for example, if an electric drive is used to realize the control.

In this section, we assume that there is a more realistic constraint on the rate of change of the control force of the form

$$|\dot{F}| \leq v_0, \tag{9.2.1}$$

where $v_0 > 0$ is a specified constant. We shall also assume that the bound on the absolute magnitude of the force is not attained and $|F(t)| < F_0$ always.

Making the change of variables

$$\xi = \frac{v_0}{m}x, \quad \dot{\xi} = \frac{v_0}{m}y, \quad F = v_0 z,$$

we reduce (9.0.1) and constraint (9.2.1) to the form

$$\dot{x} = y, \quad \dot{y} = z, \quad \dot{z} = u, \quad |u| \leq 1. \tag{9.2.2}$$

Now, the variables x, y, and z are phase coordinates and u plays the role of a bounded control.

The initial conditions for system (9.2.2) are specified in the form

$$x(0) = x_0, \quad y(0) = y_0, \quad z(0) = z_0, \tag{9.2.3}$$

where the initial instant of time is assumed to be equal to zero without any loss in generality.

We now formulate the problem of constructing the control $u(t)$ that satisfies the constraint $|u(t)| \leq 1$ for $t \geq 0$ and transfers system (9.2.2) from an arbitrary initial state (9.2.3) to the specified terminal set

$$x(T) = 0, \quad y(T) = 0 \tag{9.2.4}$$

for arbitrary $z(T)$ in the shortest time T.

In addition to determining the open-loop control, the problem of the feedback time-optimal control for system (9.2.2) will also be solved. This control $u(x,y,z)$, which is expressed as a function of the current (or initial) phase coordinates x, y, and z, ensures that system (9.2.2) is brought to the specified terminal set (9.2.4) in the shortest time.

9.2.2 Open-loop optimal control

We now apply the maximum principle to the time-optimal control problem (9.2.2)–(9.2.4). We set up the Hamiltonian function

$$H = p_x y + p_y z + p_z u \tag{9.2.5}$$

and write down the adjoint equations

$$\dot{p}_x = 0, \quad \dot{p}_y = -p_x, \quad \dot{p}_z = -p_y. \tag{9.2.6}$$

Here, p_x, p_y, and p_z are the adjoint variables. System (9.2.6) is integrated subject to the transversality condition $p_z(T) = 0$ that corresponds to the condition that $z(T)$ is not fixed, and we obtain

$$p_x = c_x, \quad p_y = c_y + c_x \tau, \quad p_z = c_y \tau + \frac{1}{2} c_x \tau^2. \tag{9.2.7}$$

Here, $\tau = T - t$ is the time measured from the end of the process (the "inverse" time), c_x and c_y are arbitrary constants. The condition for the Hamiltonian (9.2.5) to be a maximum with respect to u subject to the constraint $|u| \leq 1$ from (9.2.2) gives

$$u(t) = \operatorname{sign} p_z(t).$$

It follows from formula (9.2.7) for p_z that the function $p_z(t)$ changes sign not more than once when $t \leq T$ and $\tau \geq 0$. Consequently, the optimal control $u(t) = \pm 1$ has not more than one switching when $t \leq T$.

We denote the lengths of the two possible segments of constancy of the control $u(t)$ by θ_1 and θ_2 and the value of $u(t)$ in the first of these segments by $\sigma = \pm 1$. The optimal control can then be represented in the form

$$u(t) = \begin{cases} \sigma & \text{when} \quad t \in (0, \theta_1), \\ -\sigma & \text{when} \quad t \in (\theta_1, T), \quad \theta_1 + \theta_2 = T. \end{cases} \tag{9.2.8}$$

We now substitute control (9.2.8) into system (9.2.2) and integrate it subject to the initial conditions (9.2.3). We obtain

$$\begin{aligned}
& x(t) = x_0 + y_0 t + \frac{1}{2} z_0 t^2 + \frac{1}{6} \sigma t^3, \\
& y(t) = y_0 + z_0 t + \frac{1}{2} \sigma t^2, \quad z(t) = z_0 + \sigma t \quad \text{when} \quad t \in (0, \theta_1); \\
& x(t) = x_0 + y_0 \theta_1 + \frac{1}{2} z_0 \theta_1^2 + \frac{1}{6} \sigma \theta_1^3 + \left(y_0 + z_0 \theta_1 + \frac{1}{2} \sigma \theta_1^2\right)(t - \theta_1) \\
& \qquad + \frac{1}{2}(z_0 + \sigma \theta_1)(t - \theta_1)^2 - \frac{1}{6} \sigma (t - \theta_1)^3, \\
& y(t) = y_0 + z_0 \theta_1 + \frac{1}{2} \sigma \theta_1^2 + (z_0 + \sigma \theta_1)(t - \theta_1) - \frac{1}{2} \sigma (t - \theta_1)^2, \\
& z(t) = z_0 + \sigma \theta_1 - \sigma (t - \theta_1) \quad \text{when} \quad t \in (\theta_1, T).
\end{aligned} \tag{9.2.9}$$

Substituting solution (9.2.9) into conditions (9.2.4), we obtain two relations and, on solving these for x_0 and y_0, we obtain

$$\begin{aligned}
x_0 &= \frac{1}{2} z_0 T^2 + \frac{1}{3} \sigma (\theta_1^3 + 3\theta_1^2 \theta_2 - \theta_2^3), \\
y_0 &= -z_0 T - \frac{1}{2} \sigma (\theta_1^2 + 2\theta_1 \theta_2 - \theta_2^2).
\end{aligned} \tag{9.2.10}$$

Introduce the following notation

$$\begin{aligned}
& \xi = z_0^{-3} x_0, \quad \eta = z_0^{-1} |z_0|^{-1} y_0, \quad \zeta = \operatorname{sign} z_0, \\
& s = |z_0|^{-1} T, \quad \lambda = \theta_2 T^{-1} \qquad (z_0 \neq 0), \\
& X(\lambda) = \frac{1}{3}(1 - 3\lambda^2 + \lambda^3), \quad Y(\lambda) = \lambda^2 - \frac{1}{2}.
\end{aligned} \tag{9.2.11}$$

Relations (9.2.10) then take the form

$$\zeta \left(\xi s^{-3} - \frac{1}{2} s^{-1}\right) = \sigma X(\lambda), \quad \zeta \left(\eta s^{-2} + s^{-1}\right) = \sigma Y(\lambda). \tag{9.2.12}$$

When $z_0 = 0$, relations (9.2.10) give

$$x_0 T^{-3} = \sigma X(\lambda), \quad y_0 T^{-2} = \sigma Y(\lambda). \tag{9.2.13}$$

When the parameter λ changes from 0 to 1, a point with coordinates $X(\lambda)$, $Y(\lambda)$ moves along the arc of the curve that joins points A_1 and A_2 with coordinates $(1/3,-1/2)$ and $(-1/3,1/2)$, respectively. When $\lambda \in [0,1]$ and $\sigma = \pm 1$, points with coordinates $\sigma X(\lambda)$, $\sigma Y(\lambda)$ form a closed curve Γ that is symmetric about the origin of coordinates and has corner points A_1 and A_2, see Fig. 9.6. The curve Γ bounds a convex domain containing the origin of the system of coordinates.

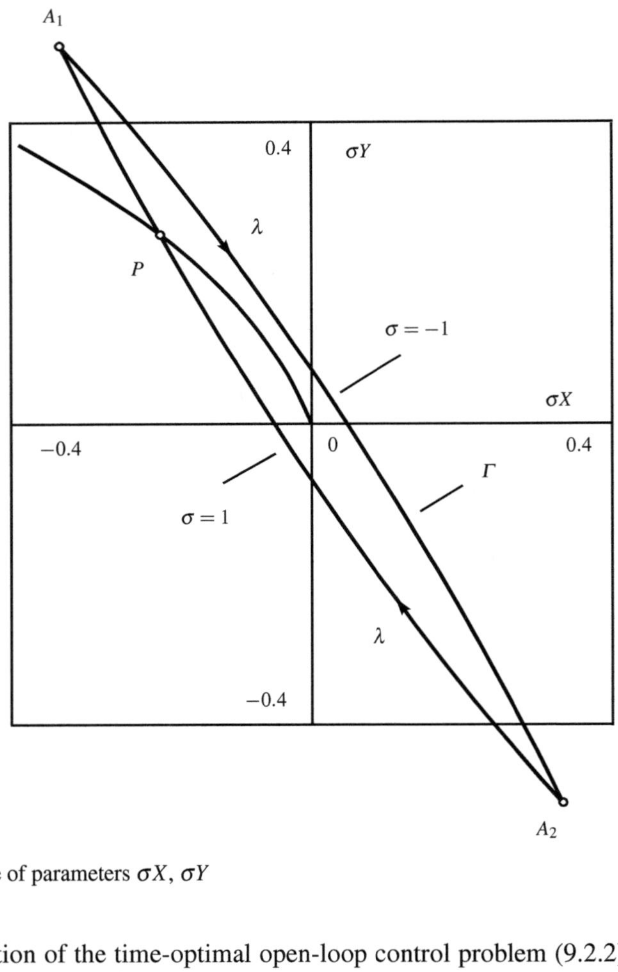

Fig. 9.6 Plane of parameters σX, σY

The solution of the time-optimal open-loop control problem (9.2.2)–(9.2.4) can then be represented as follows.

We initially assume that $z_0 \neq 0$ and determine ξ, η, and ζ, in accordance with (9.2.11), using the specified initial data x_0, y_0, and z_0 from (9.2.3). The left-hand sides of relations (9.2.12) specify the coordinates of a certain point P that depends on the parameter $s \in [0,\infty)$. As s changes from ∞ to 0, P moves along a smooth semi-infinite curve from the origin of the system of coordinates (when $s \to \infty$) to infinity (when $s \to 0$). This point falls at least once on the closed curve Γ that encircles the origin of the system of coordinates. The least value of $s = s_*$ for which $P \in \Gamma$ is found numerically. According to (9.2.11), the optimal time is equal to $T = |z_0|s_*$. The position of the point P on the curve Γ when $s = s_*$ determines the values of the parameters $\sigma = \pm 1$ and $\lambda \in [0,1]$. By virtue of (9.2.11), the lengths of the segments of constancy of the control are equal to $\theta_1 = (1-\lambda)T$ and $\theta_2 = \lambda T$.

When $z_0 = 0$, we consider equalities (9.2.13) instead of (9.2.12). The left-hand sides of these equalities specify the coordinates of the point P that depends on the parameter T. When T changes from ∞ to 0, the point P moves along a semicubic parabola from the origin of the system of coordinates (when $T \to \infty$) to infinity (when $T \to 0$). The least value of the parameter T for which $P \in \Gamma$ is the optimal time. The values of the parameters σ, λ, θ_1, and θ_2 are determined from the position of the point P on Γ as in the case, where $z_0 \neq 0$.

When the quantities σ, θ_1, and θ_2 have been determined, the optimal control $u(t)$ and the corresponding optimal trajectory are specified by equalities (9.2.8) and (9.2.9). The proposed algorithm completely determines the solution of the time-optimal open-loop control problem. According to the construction, this solution is unique.

As an example, we present the results of the determination of the optimal control for the initial data

$$x_0 = -72 + 27\sqrt{3} \approx -25.2, \; y_0 = 3, \; z_0 = 1.$$

In this case, one obtains

$$T = s = 6, \; \sigma = 1, \; \theta_1 = 6 - 3\sqrt{3} \approx 0.80, \; \theta_2 = 3\sqrt{3} \approx 5.20.$$

The corresponding trajectory of the point P when T changes from ∞ to 0 is shown in Fig. 9.6.

9.2.3 Feedback optimal control

In order to construct the feedback optimal control, it suffices to find the switching surfaces in the phase space (x,y,z) on which the sign of the control $u = \pm 1$ changes. On these surfaces, the length of one of the segments of constancy vanishes, that is, $\theta_1 = 0$ or $\theta_2 = 0$. From (9.2.11), we have here $\lambda = 0$ or $\lambda = 1$. According to (9.2.11), values of X and Y equal to $\pm 1/3$ and $\mp 1/2$ correspond to these values of λ, respectively. From (9.2.12), we obtain the conditions

$$\zeta\left(\xi s^{-3} - \frac{1}{2}s^{-1}\right) = \pm\frac{1}{3}\sigma, \quad \zeta\left(\eta s^{-2} + s^{-1}\right) = \mp\frac{1}{2}\sigma \tag{9.2.14}$$

that are satisfied in the (ξ,η)-plane on the switching curves when $z_0 \neq 0$. However, relations (9.2.14) are insufficient for determining the switching curves: for this, we require a direct analysis of relations (9.2.12) that will be carried out below.

Note that, in the feedback control, the initial conditions x_0, y_0, and z_0 can be treated as the current values of the phase coordinates x, y, and z. Let us consider relations (9.2.11) as formulas for the change of variables

$$\xi = z^{-3}x, \quad \eta = z^{-1}|z|^{-1}y, \quad \zeta = \operatorname{sign} z \tag{9.2.15}$$

in phase space. This change of variables that introduces the self-similar variables ξ and η enables one, when $z \neq 0$, to reduce the dimension of the phase space by one and to construct the feedback optimal control in the (ξ,η)-plane.

We will first consider separately the case, where $z = 0$. Similarly to (9.2.14), we obtain the conditions

$$xT^{-3} = \pm\frac{1}{3}\sigma, \quad yT^{-2} = \mp\frac{1}{2}\sigma \tag{9.2.16}$$

from (9.2.13). These conditions are satisfied at the intersection of the switching surfaces with the plane $z = 0$. When $z = 0$, conditions (9.2.16) define two halves of the semicubic parabolas that form the switching curve in the plane $z = 0$. The parabolas are described by the equation

$$\gamma(x,y) \equiv 3x + 2y|2y|^{1/2} = 0. \tag{9.2.17}$$

An analysis of the signs of σ on the branches of the switching curve (9.2.17) enables one to determine the signs of the controls on the different sides of the switching curve. As a result, we obtain the feedback optimal control for $z = 0$ in the form

$$u(x,y,0) = \begin{cases} -\operatorname{sign}\gamma(x,y) & \text{when} \quad \gamma \neq 0, \\ \operatorname{sign} x = -\operatorname{sign} y & \text{when} \quad \gamma = 0. \end{cases} \tag{9.2.18}$$

When $z \neq 0$, the change of variables (9.2.15) transforms the first two equations of (9.2.2) to the form

$$\dot{\xi} = |z|^{-1}(\eta - 3u\zeta\xi), \quad \dot{\eta} = |z|^{-1}(1 - 2u\zeta\eta). \tag{9.2.19}$$

By dividing the first equation of (9.2.19) by the second, we obtain the linear differential equation for $\xi(\eta)$

$$\frac{d\xi}{d\eta} = \frac{\eta - 3\alpha\xi}{1 - 2\alpha\eta}, \quad \alpha = u\zeta = \pm 1. \tag{9.2.20}$$

The parameter α retains a constant value along the optimal trajectories that do not intersect the plane $z = 0$. By integrating (9.2.20) in the case of constant α, we find the general solution of equation (9.2.20):

$$\xi = \Phi(\eta, \alpha, A) \equiv \alpha\eta - \frac{1}{3} + A|1 - 2\alpha\eta|^{3/2}, \qquad (9.2.21)$$

where A is an arbitrary constant. Note that the second equation of (9.2.19) enables one to determine the direction of motion along the optimal trajectories. If $\alpha = 1$, the motion occurs in the direction of the increase in η when $\eta < 1/2$ and in the direction of the decrease in η when $\eta > 1/2$. If, however, $\alpha = -1$, the motion occurs in the direction of the decrease in η when $\eta < -1/2$ and the increase in η when $\eta > -1/2$.

We will now construct the feedback optimal control. As was shown above, in order to do this, it is sufficient to establish the sign of the control $u = \sigma$ at the initial instant of time $t = 0$ as a function of the initial data x_0, y_0, and z_0. On changing to self-similar variables and returning to relations (9.2.12), the feedback problem can be formulated as follows: it is required to find the value of $\sigma = \pm 1$ that corresponds to the solution of relations (9.2.12) (for fixed ξ, η, and ζ, where $\zeta = \pm 1$) with the least s, where $s > 0$ and $\lambda \in [0, 1]$.

We will now briefly describe the solution algorithm and subsequently explain its most important features.

First, we note that relations (9.2.12) retain their form, if ζ and σ change signs simultaneously. Consequently, if ζ is replaced by $-\zeta$, the required quantity σ also changes sign. It is therefore sufficient to construct the solution in the case, where $\zeta = 1$ for arbitrary ξ and η, and then for the case, where $\zeta = -1$, simply to change the sign in the resulting dependence $\sigma(\xi, \eta)$.

Without loss of generality, we therefore put $\zeta = 1$ and eliminate λ using the second of equations (9.2.12). We obtain

$$\lambda = \left[\frac{1}{2} + \sigma\left(\eta s^{-2} + s^{-1}\right)\right]^{1/2}, \qquad \sigma = \pm 1. \qquad (9.2.22)$$

Since $\lambda \in [0, 1]$, then, for fixed $\sigma = \pm 1$ and η, the ranges of variation of s in which λ is real and $\lambda \leq 1$ are determined from (9.2.22). We substitute λ from (9.2.22) into the first equation of (9.2.12) and find the dependences of ξ on s, η, and $\sigma = \pm 1$. For fixed η, we shall denote these dependences by $\xi^+(s)$ and $\xi^-(s)$ for $\sigma = \pm 1$, respectively. Subject to the condition $\lambda \in [0, 1]$, they define two curves in the (s, ξ)-plane, each of which consists, generally speaking, of a finite number of arcs. We investigate these curves and find their domains of definition and their extrema in the whole range of variation in the argument s and the parameter η. After that, we analyze their position with respect to each other. A line $\xi = \text{const}$ is then mentally drawn in the (s, ξ)-plane, and the minimum value of the abscissa $s > 0$ is found for which this line intersects one of the above-mentioned curves. The value of $\sigma = \pm 1$ that corresponds to that curve with which this intersection occurs determines the required control $u = \sigma$ for the data ξ, η, and $\zeta = 1$, and the value of s that corresponds to this point of intersection is equal to the normalized optimal time: $s = T|z|^{-1}$.

We will now describe these operations in greater detail taking account of the fact that the following constructions are true only when $s > 0$. From (9.2.12) and

(9.2.22), we have

$$\xi^{\pm}(s) = \mp\frac{1}{6}s^3 - \frac{1}{2}s^2 - s\eta \pm \frac{1}{3}\Big(\frac{1}{2}s^2 \pm \eta \pm s\Big)^{3/2}. \qquad (9.2.23)$$

If $s \to +\infty$, then $\xi^{\pm}(s) \approx \pm(-1+1/\sqrt{2})s^3/6 \to \mp\infty$.

Let us consider the function $\xi^+(s)$. In the case, where $\sigma = 1$, the condition $\lambda \leq 1$ selects the set $s \in (0, s_2] \cup [s_1, +\infty)$, where $s_{1,2} = 1 \pm \sqrt{1+2\eta}$. The expression for $\xi^+(s)$ is determined if $s \in [s_5, +\infty)$, where $s_5 = -1+\sqrt{1-2\eta}$. The derivative $d\xi^+/ds$ vanishes at the points s_5 and $s_7 = -1+\sqrt{2(1-2\eta)}$, if s_5 and s_7 exist and $s_5 \leq s_7$. Furthermore, $d^2\xi^+/ds^2 < 0$ when $s = s_7$, that is, s_7 is a maximum point. It can be shown that, if s_1, s_2, s_5, and s_7 exist, then $s_2 \leq s_7 \leq s_1$ and $s_2 \geq s_5$.

If $\eta \geq 0$, then $s_2 \leq 0$ and $d\xi^+/ds < 0$ when $s \geq s_1$, that is, the function $\xi^+(s)$ is defined when $s \in [s_1, +\infty)$ and decreases from $\xi^+(s_1)$ to $-\infty$. If $-1/2 \leq \eta < 0$, then $s_5 > 0$, that is, the function $\xi^+(s)$ is defined when $s \in [s_5, s_2] \cup [s_1, +\infty)$. It has a null derivative when $s = s_5$, increases in the interval $[s_5, s_2]$ and decreases from $\xi^+(s_1)$ to $-\infty$ when $s \in [s_1, +\infty)$. If $\eta < -1/2$, then $s_5 > 0$ and the value of s_2 is undefined. Then, the function $\xi^+(s)$ is defined when $s \in [s_5, +\infty)$, $d\xi^+/ds = 0$ when $s = s_5$, and $\xi^+(s)$ increases up to its maximum at the point $s = s_7$; after that it decreases from $\xi^+(s_7)$ to $-\infty$.

We now consider the function $\xi^-(s)$. We require that $\lambda \leq 1$ in (9.2.22) and obtain that $s \in [s_5, +\infty)$, where $s_5 = -1+\sqrt{1-2\eta}$. If $s \in (0, s_2] \cup [s_1, +\infty)$, then the function $\xi^-(s)$ from (9.2.23) is defined. Its derivative vanishes at the points s_1, s_2, and $s_3 = 1-\sqrt{2(1+2\eta)}$, if they exist and $s_3 \geq s_1$. In addition, $d^2\xi^-/ds^2 > 0$ when $s = s_3$, that is, $s = s_3$ is the point of a minimum.

If $\eta \geq 0$, then $s_2 \leq 0$, and the function $\xi^-(s)$ is defined when $s \in [s_1, +\infty)$, and $d\xi^-/ds = 0$ when $s = s_1$. The function $\xi^-(s)$ decreases from $\xi^-(s_1)$ up to the point of a minimum $s = s_3$; after that it increases from $\xi^-(s_3)$ up to $+\infty$. If $-1/2 \leq \eta < 0$, then $s_5 > 0$, that is, the dependence $\xi^-(s)$ is defined when $s \in [s_5, s_2] \cup [s_1, +\infty)$. The function $\xi^-(s)$ increases as s varies from $s = s_5$ to $s = s_2$ ($d\xi^-/ds = 0$ when $s = s_2$) and increases when $s \in [s_1, s_3]$; we have $d\xi^-/ds = 0$ when $s = s_1$ and $s = s_3$. Then, $\xi^-(s)$ increases from $\xi^-(s_3)$ to $+\infty$. If $\eta < -1/2$, then $s_5 > 0$, the values of s_1, s_2, and s_3 are not defined, and $d\xi^-/ds > 0$ when $s \geq s_5$, that is, the function $\xi^-(s)$ is defined when $s \in [s_5, +\infty)$ and increases over the whole of this interval up to $+\infty$.

We now make two remarks concerning the mutual position of the pair of curves from the two families investigated, for the same value of the parameter η.

First, we find the point of intersection of the curves $\xi^+(s)$ and $\xi^-(s)$; that requires solution of the equation

$$1 - \Big(\frac{1}{2} - \frac{\eta}{s^2} - \frac{1}{s}\Big)^{3/2} = \Big(\frac{1}{2} + \frac{\eta}{s^2} + \frac{1}{s}\Big)^{3/2}. \qquad (9.2.24)$$

We square both sides of (9.2.24), reduce similar terms and then again square both sides of the equation to obtain the equation in s:

$$\left[\left(\frac{\eta}{s^2}+\frac{1}{s}\right)^2+2\right]\left[\left(\frac{\eta}{s^2}+\frac{1}{s}\right)^2-\frac{1}{4}\right]^2=0. \qquad (9.2.25)$$

An analysis of the roots of (9.2.25) shows that only s_1, s_2, and s_5 are roots of (9.2.24), and, moreover, they are positive for at least a single value of η. We shall denote coincident values $\xi^+ = \xi^-$ at the above-mentioned points by $\xi^\pm$.

Secondly, we establish that $\xi^\pm(s_5) > \xi^\pm(s_1)$, if and only if $-\sqrt{3}/4 < \eta \leq 0$.

As a result, it turns out to be convenient to pick out four ranges of values of the parameter η that correspond to different mutual positions of the curves $\xi^+(s)$ and $\xi^-(s)$. Thus, we will determine the required control for all ξ and η with the exception of $\xi^\pm(s_1)$, $\xi^\pm(s_2)$, and $\xi^\pm(s_5)$.

When $\eta \geq 0$ for any $\xi < \xi^\pm(s_1)$, the minimum permissible abscissa s is reached on the curve $\xi^+(s)$. When $\xi > \xi^\pm(s_1)$, the same result holds for $\xi^-(s)$.

When $-\sqrt{3}/4 < \eta < 0$, the closed isolated curve for $s_5 \leq s \leq s_2$ is added to the curves $\xi^+(s)$ and $\xi^-(s)$ that have the same properties as considered above. The curve $\xi^-(s)$ lies above the curve $\xi^+(s)$, and $\xi^\pm(s_5) < \xi^\pm(s_2)$. Moreover, $\xi^\pm(s_1) < \xi^\pm(s_5)$, that is, $\xi^\pm(s_1)$ lies below the lowest point of the closed isolated curve. Consequently, the required control is defined in the same way as in the preceding case.

When $-1/2 \leq \eta \leq -\sqrt{3}/4$, the inequality $\xi(s_1) > \xi(s_5)$ is satisfied and, for any $\xi < \xi^\pm(s_5)$, the minimum permissible abscissa s is attained on the curve $\xi^+(s)$. When $\xi > \xi^\pm(s_5)$, the same assertion holds for $\xi^-(s)$.

The close isolated curve disappears when $\eta < -1/2$, and the required control is specified as in the preceding case.

We now determine the control on the curves $\xi^\pm(s_1(\eta))$, $\xi^\pm(s_2(\eta))$, and $\xi^\pm(s_5(\eta))$ in the (ξ,η)-plane. We recall that the dependences of s_1, s_2, and s_5 on η have been presented above. By (9.2.22), we have $\lambda = 0$ when $\sigma = -1$ on the curve $\xi^\pm(s_1(\eta))$, that is, the time interval in which it is necessary to take $u = 1$ is equal to zero. Consequently, it is necessary to take $u = -1$ on the curve $\xi^\pm(s_1(\eta))$ and it is a switching curve when $\eta > -\sqrt{3}/4$. Similarly, on the curve $\xi^\pm(s_2(\eta))$, one must use $u = -1$ when $-1/2 \leq \eta < 0$ but this curve will not be a switching curve. It is easy to show using the same method that we have $u = 1$ when $\eta < 0$ on the curve $\xi^\pm(s_5(\eta))$. This curve serves as a switching curve.

We now completely present the feedback optimal control. To be specific, we shall take $z > 0$ and $\zeta = 1$. The switching curve in the (ξ,η)-plane is defined by the equalities

$$\xi = f(\eta) = \begin{cases} \Phi(\eta, 1, \frac{1}{3}), & \eta \leq \eta^*, \\ \Phi(\eta, -1, -\frac{1}{3}), & \eta > \eta^*; \quad \eta^* = -\frac{\sqrt{3}}{4}, \end{cases} \qquad (9.2.26)$$

where the notation (9.2.21) is used. The switching curve is continuous and has a kink at the point K with the coordinates $\xi^* = 1/12$, $\eta^* = -\sqrt{3}/4$. This curve is represented by the solid line in Figs. 9.7 and 9.8. Since the scale in Fig. 9.7 is smaller than that in Fig. 9.8, the points K and R shown in Fig. 9.8 are practically indistinguishable in Fig. 9.7 and are therefore not labelled (the coordinates of the point R are $\xi = 1/6$, $\eta = -1/2$). On the other hand, the scale used in Fig. 9.7 enables us to depict all the characteristic phase trajectories, the important part of which is missing in Fig. 9.8. The rest of the notation employed in Figs. 9.7 and 9.8 is identical. The branches of the switching curve corresponding to $\eta < \eta^*$ and $\eta > \eta^*$ are denoted by the letters M and N, respectively. In the (ξ,η)-plane, we have

$$u = \begin{cases} 1 & \text{when} \quad \xi < f(\eta), \\ 1 & \text{when} \quad \xi = \Phi(\eta,1,1/3), \quad \eta \le 0, \\ -1 & \text{at the remaining points of the } (\xi,\eta)\text{-plane.} \end{cases} \tag{9.2.27}$$

Hence, $u = 1$ to the left of and below the switching curve (9.2.26), on its segment KM to the right of and below point K and also on the arc of the curve $\xi = \Phi(\eta,1,1/3)$ that joins the origin of the system of coordinates and the point K, see Fig. 9.8, where this arc is a part of the switching curve. In the remaining part of the (ξ,η)-plane, we have $u = -1$.

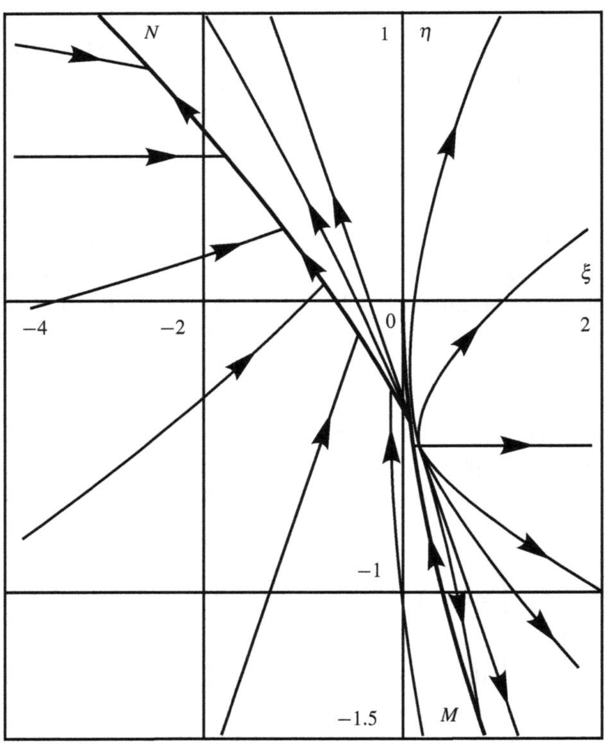

Fig. 9.7 Switching curve and optimal trajectories

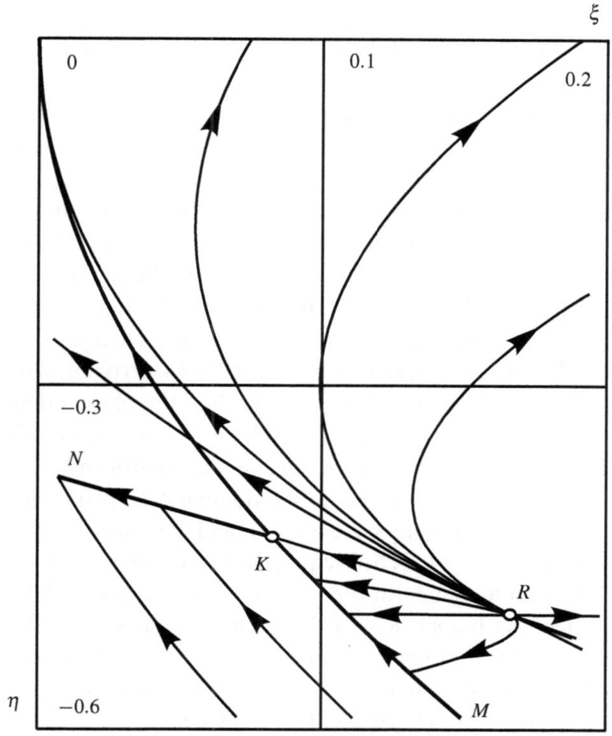

Fig. 9.8 Switching curve and optimal trajectories around the origin of the system of coordinates

When $z < 0$ and $\zeta = -1$, the switching curve remains the same and one simply has to interchange the positions of the set of points (ξ, η), where $u = 1$ and $u = -1$ in relations (9.2.27). Thus, the synthesis of the optimal control $u(x, y, z)$ is completely determined by relations (9.2.15), (9.2.17), (9.2.18), (9.2.21), (9.2.26), and (9.2.27) for all x, y, and z.

We now describe the set of optimal trajectories that, in the variables ξ and η, consist of arcs of the curves (9.2.21). Suppose that the initial point (x, y, z) is specified and, to be specific, we shall assume that $z > 0$. According to formulas (9.2.15), we find ξ, η, and $\zeta = 1$.

If a point (ξ, η) lies on the curve $\xi = \Phi(\eta, 1, 1/3)$, where $\eta \leq 0$, then motion occurs along this curve $MK0$ with a control $u = 1$ until it reaches the origin of the system of coordinates (see Fig. 9.8).

All the remaining optimal trajectories also arrive at the origin of the system of coordinates along this curve. An exception is the segment $R0$ of the curve $\xi = \Phi(\eta, -1, 1/3)$ when $\eta \in [-1/2, 0]$: this segment is a phase trajectory for $u = -1$ that begins at the point R with the coordinates $(1/6, -1/2)$ and reaches the origin of the system of coordinates. Phase trajectories are denoted by thin lines in Figs. 9.7 and 9.8, and arrows indicate the direction of the motion.

If the initial point lies in the domain

$$\eta \leq 0, \quad \Phi\left(\eta, 1, \frac{1}{3}\right) < \xi < \Phi\left(\eta, -1, \frac{1}{3}\right), \tag{9.2.28}$$

then the optimal trajectory consists of a segment with $u = -1$ until it reaches the curve $\xi = \Phi(\eta, 1, 1/3)$ and of the subsequent motion along this curve with $u = 1$ (see Fig. 9.8).

If the initial point lies in the domain $\xi < f(\eta)$, the motion initially occurs with $u = 1$ until it intersects the curve $\Phi = \xi(\eta, -1, -1/3)$ that is the part KN of the switching curve (9.2.26) (see Figs. 9.7 and 9.8); then the phase point moves with $u = -1$ along this curve that departs to infinity. By (9.2.11), we have $z = 0$ at an infinitely distant point of the (ξ, η)-plane. At infinity, z changes its sign, and then we have $z < 0$, $\zeta = -1$. The phase trajectory continues, arriving with $u = -1$ from infinity along the curve $\xi = \Phi(\eta, 1, 1/3)$; after that, the phase point moves along this curve to the origin of the system of coordinates. Note that motion through an infinitely distant point occurs without a change in the control and takes a finite time.

It remains to consider initial points in the domain $\xi > f(\eta)$ but outside of the domain (9.2.28). Here, we initially have $u = -1$, and the trajectory $\xi = \Phi(\eta, -1, A)$ departs to infinity with $A > -1/3$. Subsequently, the motion occurs along the curves $\xi = \Phi(\eta, 1, -A)$ with a change in the sign of A. These curves lie in the domain $\xi < f(\eta)$ and meet the branch KN of the switching curve $\xi = \Phi(\eta, -1, -1/3)$. Then the trajectory with $u = 1$ departs to infinity along this curve, where the sign of z changes again. Later, the motion occurs with $u = 1$ along the curve $\xi = \Phi(\eta, 1, 1/3)$ until it reaches the origin of the system of coordinates.

Note that certain phase trajectories contain segments of the lines $\xi = \pm\eta - 1/3$ and $\eta = \pm(2\alpha)^{-1}$ that correspond to the values $A = 0$ and $A = \infty$ in (9.2.21), respectively. On departing to infinity along these lines, the variable x (in the case of the straight line with $A = 0$) or the variable y (in the case of the straight line with $A = \infty$) simultaneously vanishes together with z, as is easily shown using (9.2.15). In other respects, these lines are treated in the same manner as the remaining trajectories (9.2.21).

Hence, for any initial point (x, y, z), the motion is completely described by the trajectories of Figs. 9.7 and 9.8 and contains not more than two segments, where the control is constant. The sign of z cannot change more than twice.

We now present the results of the investigation of the normalized optimal time s as a function of ξ and η. The dependence of s on ξ for different fixed values of η is studied, where s_1, s_2, and s_5 are again considered to be the functions of η that were introduced above. When $\eta \geq 0$, the function $s(\xi, \eta)$ decreases as ξ increases, if $\xi < \xi^{\pm}(s_1)$, and has a discontinuity, if $\xi = \xi^{\pm}(s_1)$. It increases on passing from $\xi < \xi^{\pm}(s_1)$ to $\xi > \xi^{\pm}(s_1)$ and when ξ increases from $\xi = \xi^{\pm}(s_1)$ to $+\infty$.

When $-\sqrt{3}/4 < \eta < 0$, the function $s(\xi, \eta)$ decreases as ξ increases, if $\xi < \xi^{\pm}(s_1)$, and has a discontinuity, if $\xi = \xi^{\pm}(s_1)$. It increases on passing from $\xi < \xi^{\pm}(s_1)$ to $\xi > \xi^{\pm}(s_1)$. There is a further discontinuity when $\xi = \xi^{\pm}(s_5)$. The function $s(\xi, \eta)$ decreases on passing from $\xi < \xi^{\pm}(s_5)$ to $\xi > \xi^{\pm}(s_5)$ but increases when $\xi^{\pm}(s_5) \leq \xi \leq \xi^{\pm}(s_2)$. There is another discontinuity when $\xi = \xi^{\pm}(s_2)$. The function $s(\xi,\eta)$ also increases on passing from $\xi < \xi^{\pm}(s_2)$ to $\xi > \xi^{\pm}(s_2)$ and also when ξ increases from $\xi = \xi^{\pm}(s_2)$ to $+\infty$.

When $-1/2 \le \eta \le -\sqrt{3}/4$, the function $s(\xi,\eta)$ decreases as ξ increases, if $\xi < \xi^{\pm}(s_5)$, and has a discontinuity at $\xi = \xi^{\pm}(s_5)$. It decreases on passing from $\xi < \xi^{\pm}(s_5)$ to $\xi > \xi^{\pm}(s_5)$ but increases when $\xi^{\pm}(s_5) \le \xi \le \xi^{\pm}(s_2)$. The next discontinuity occurs when $\xi = \xi^{\pm}(s_2)$. The function $s(\xi,\eta)$ increases on passing from $\xi < \xi^{\pm}(s_2)$ to $\xi > \xi^{\pm}(s_2)$ and when ξ increases from $\xi = \xi^{\pm}(s_2)$ to $+\infty$.

When $\eta < -1/2$, the function $s(\xi,\eta)$ decreases as ξ increases when $\xi < \xi^{\pm}(s_5)$ and has a discontinuity at $\xi = \xi^{\pm}(s_5)$. It decreases on passing from $\xi < \xi^{\pm}(s_5)$ to $\xi > \xi^{\pm}(s_5)$ and increases as ξ increases from $\xi = \xi^{\pm}(s_5)$ to $+\infty$.

The computation results for the normalized optimal time $s(\xi,\eta)$ are presented in Figs. 9.9 and 9.10. In Fig. 9.9, the thin lines are level lines of the function $s(\xi,\eta)$, and the bold lines are the lines of discontinuity of this function. The rest of the notation is the same as in Fig. 9.8. A three-dimensional graph of the function $s(\xi,\eta)$ is shown in Fig. 9.10, where the darker is the background, the smaller is the corresponding value.

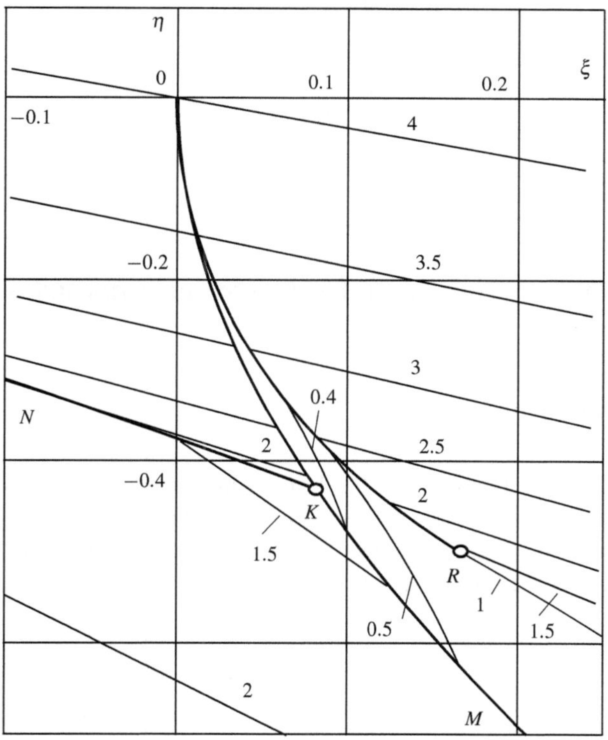

Fig. 9.9 Level lines of the function $s(\xi,\eta)$

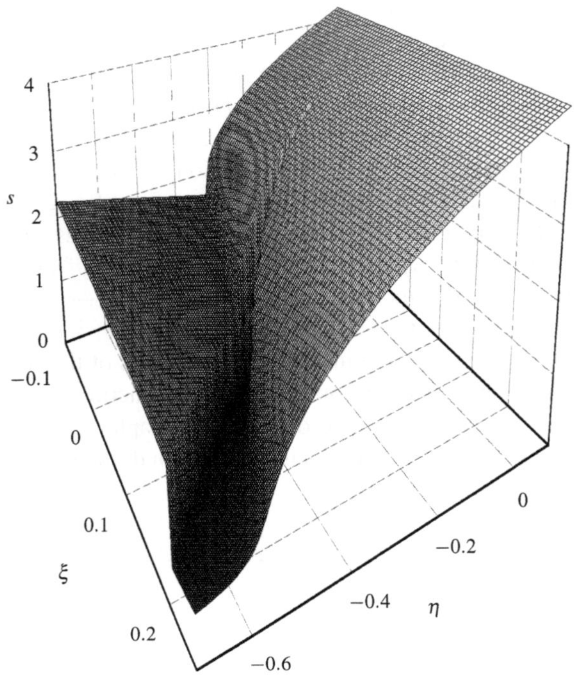

Fig. 9.10 Function $s(\xi,\eta)$

9.3 Time-optimal control under constraints imposed on the acceleration and its rate of change

9.3.1 Problem statement

When formulating problem in Sect. 9.2.1, it was assumed that the constraint $|F| \le F_0$ is not attained. Let us now cancel this assumption. We will consider system (9.0.1) under constraints imposed on the magnitude of the control force and the rate of its change, i.e., under the following restrictions

$$|F| \le F_0, \tag{9.3.1}$$

$$|\dot{F}| \le v_0, \tag{9.3.2}$$

where F_0 and v_0 are constants.

Let us analyze the problem of bringing system (9.0.1) to the origin of the phase plane, that is, to the state $\xi = \dot{\xi} = 0$.

It is well-known that if restriction (9.3.2) is absent, then the time-optimal control is a bang-bang control with at most one switching point. The typical time-dependence of this control is shown in Fig. 9.11.

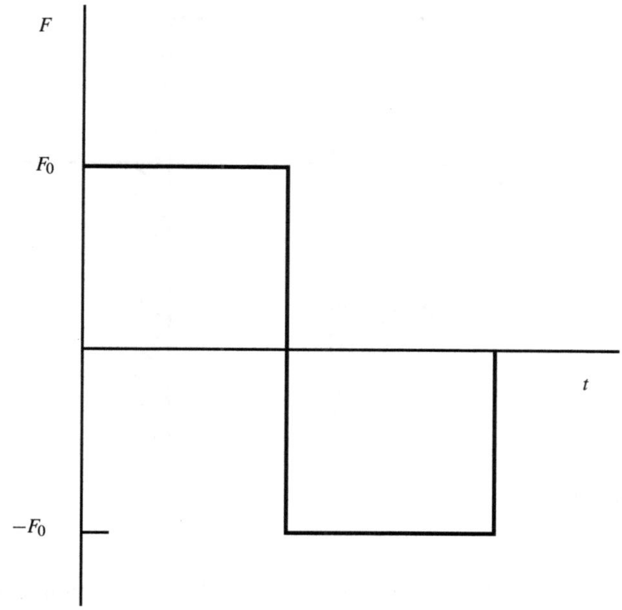

Fig. 9.11 Bang-bang control

On the other hand, the solution of the time-optimal control problem in the absence of constraint (9.3.1) is given in Sect. 9.2. As to the time-optimal control problem for system (9.0.1) taking both restrictions (9.3.1) and (9.3.2) into account, we have not come upon any solution in the literature.

The control problem for system (9.0.1) will be treated in the following formulation. It is assumed that the control force is bounded as in (9.3.1), while condition (9.3.2) will only be satisfied when the magnitude of the force increases, that is, when $d|F|/dt > 0$. At the same time, the force may be switched off instantaneously. These restrictions may be written as a system of inequalities

$$|F| \leq F_0;$$
$$\dot{F} \leq v_0, \quad \text{if} \quad F \geq 0, \qquad (9.3.3)$$
$$\dot{F} \geq -v_0, \quad \text{if} \quad F \leq 0.$$

The domain defined by inequalities (9.3.3) in the $(F, \dot{F})$-plane is shown in Fig. 9.12.

Conditions (9.3.3) simulate the following situation: the control force may be increased only gradually, at a finite rate, but it can be switched off instantaneously.

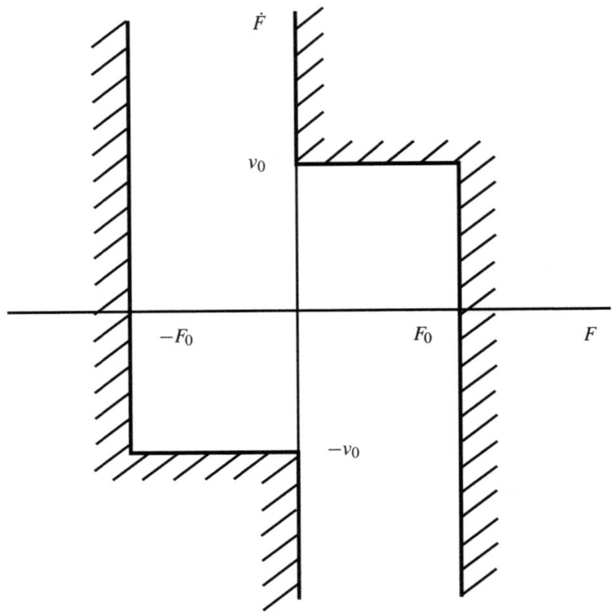

Fig. 9.12 Domain of restrictions

This is not infrequently the case in practice, since the deceleration is often implemented by means other than the acceleration.

We introduce the following non-dimensional variables

$$t' = v_0 F_0^{-1} t, \quad x = m v_0^2 F_0^{-3} \xi,$$
$$y = m v_0 F_0^{-2} \dot{\xi}, \quad z = F_0^{-1} F, \quad u = v_0^{-1} \dot{F}. \tag{9.3.4}$$

Equation (9.0.1) and constraints (9.3.3) take the following form in terms of these new variables

$$\dot{x} = y, \quad \dot{y} = z, \quad \dot{z} = u, \tag{9.3.5}$$

$$|z| \le 1;$$
$$u \le 1, \quad \text{if} \quad z \ge 0, \tag{9.3.6}$$
$$u \ge -1, \quad \text{if} \quad z \le 0.$$

Here and below, dots denote differentiation with respect to the new (non-dimensional) time. The prime indicating non-dimensional time will be omitted from now on.

The initial conditions for system (9.3.5) are

$$x(0) = x_0, \qquad y(0) = y_0, \qquad z(0) = 0, \tag{9.3.7}$$

and the terminal state is

$$x(T) = 0, \qquad y(T) = 0. \tag{9.3.8}$$

Note that the terminal value of $z(T)$ may always be made equal to zero by adjusting the jump of the force $z(t)$ at time $t = T$ that is allowed by conditions (9.3.6). We may therefore assume throughout that $z(T) = 0$.

We now formulate the following problem.

It is required to find a control $u(t)$ and the corresponding trajectory, that is, functions $x(t), y(t)$, and $z(t)$ satisfying (9.3.5), constraints (9.3.6), initial conditions (9.3.7), and terminal conditions (9.3.8) at some (non-fixed) time $T > 0$.

Henceforth, we will construct a solution that solves the problem and has the simplest structure satisfying constraints (9.3.6). This control is presumably time-optimal.

9.3.2 Possible modes of control

The possible time histories of the non-dimensional force $z(t)$ are shown in Fig. 9.13. The figure shows the intervals in which the force increases or decreases gradually, here $\dot{z} = \pm 1$, and the intervals over which the force is constant, $z = \pm 1$. These modes have the following properties.

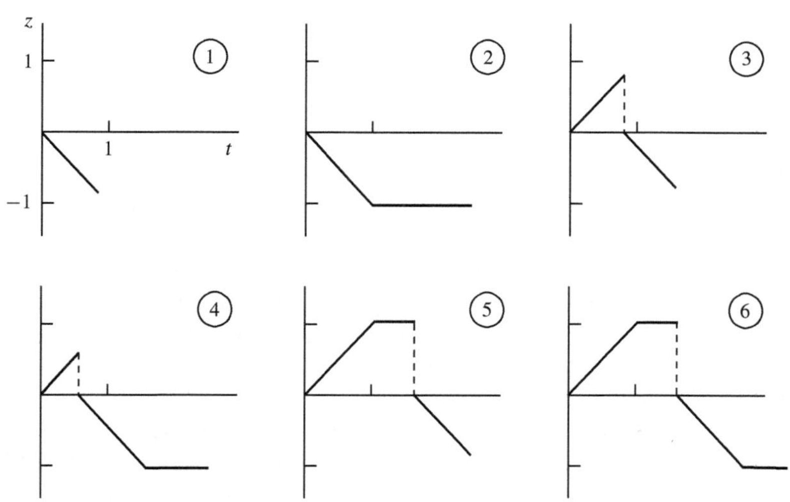

Fig. 9.13 Modes of function $z(t)$

1°. At the beginning of the process, we have $z(0) = 0$, in accordance with initial conditions (9.3.7).

2°. Modes 1–6 of Fig. 9.13 satisfy conditions (9.3.6).

3°. Just before the end of the process, $z(t) < 0$ as $t \to T$. This condition is assumed for definiteness and does not affect the generality of our solution, since, besides the forms illustrated in Fig. 9.13, we might have similarly considered their mirror images in the t axis, namely, $z'(t) = -z(t)$.

4°. The modes shown in Fig. 9.13 have at most one jump and one change in the sign of the force $z(t)$.

5°. Mode 6 of Fig. 9.13 is a direct extension of the bang-bang form of Fig. 9.11 to the case of gradual increase of the magnitude of the force, that is, to the case of constraints (9.3.6).

6°. Modes 1–5 of Fig. 9.13 are special cases of 6. Indeed, in 5 the bound $z = -1$ is not achieved; in 4 the bound $z = 1$ is not achieved; in 3 neither bound $z = \pm 1$ is achieved; in 1 and 2 there is no jump in the function $z(t)$, the bound $z = -1$ being achieved in 2 but not in 1.

As will be shown below, using modes 1–6 for $z(t)$ as well as their mirror-image laws $z'(t) = -z(t)$, one can steer system (9.3.5) from any initial state (9.3.7) to the terminal state (9.3.8).

We now introduce a domain D in the (x,y)-plane defined by the inequalities

$$D = \left\{ (x,y) : \begin{array}{ll} x < -\varphi(-y), & \text{if} \quad y \le 0, \\ x \le \varphi(y), & \text{if} \quad y > 0. \end{array} \right. \tag{9.3.9}$$

Here, function $\varphi(y)$ is defined as follows:

$$\varphi(y) = \begin{cases} -\dfrac{(2y)^{3/2}}{3}, & \text{if} \quad 0 \le y \le \dfrac{1}{2}, \\ \dfrac{1}{24} - \dfrac{y}{2} - \dfrac{y^2}{2}, & \text{if} \quad y \ge \dfrac{1}{2}. \end{cases} \tag{9.3.10}$$

It is not difficult to verify that these relations define $\varphi(y)$ as a smooth function, decreasing monotonically from 0 to $-\infty$ over the non-negative real line $y \in [0, \infty)$. At the point $y = 1/2$, we have $\varphi(y) = -1/3$ and $\varphi'(y) = -1$.

The curves Γ and Γ' defined for $y \ge 0$ and $y \le 0$ by the formulas $x = \varphi(y)$ and $x = -\varphi(-y)$, respectively, are shown in Fig. 9.14 (thicker curves). The curves are symmetrical to one another about the origin. We also show on these curves the points $A = (-1/3, 1/2)$ and $A' = (1/3, -1/2)$ at which the sections defined by formulas (9.3.10) meet smoothly.

The curves Γ and Γ' form the boundary of the domain D; according to (9.3.9), Γ that lies in the second quadrant of the (x,y)-plane and belongs to D, while Γ' that lies in the fourth quadrant is not contained in D. The union of the domain D with the domain D' symmetric to it with respect to the origin gives the whole (x,y)-plane punctured at the origin O. By (9.3.7), O is the terminal point and is, therefore, of no interest as an initial point: if $x = y = 0$ at time $t = 0$, the control is needless.

Below, we will construct a control and trajectories, that is, functions $u(t)$, $x(t)$, $y(t)$, and $z(t)$, for all initial points $(x_0, y_0) \in D$. If $(x_0, y_0) \in D'$, the required solution will be given by functions $\{-u(t), -x(t), -y(t), -z(t)\}$, where the set $\{u(t), x(t), y(t), z(t)\}$ is the solution for the initial point $(-x_0, -y_0) \in D$ symmetric

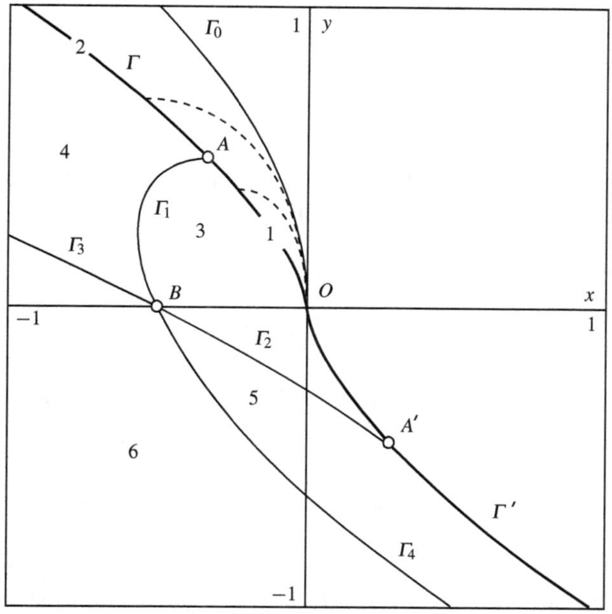

Fig. 9.14 Division of (x,y)-plane into domains D_i

to $(x_0,y_0) \in D'$. Then, equations (9.3.5) and constraints (9.3.6) will be satisfied, the trajectories starting from $(x_0,y_0) \in D'$ will be symmetric to the trajectories coming from the point $(-x_0,-y_0) \in D$ and will also lead to the origin, moreover, in the same time.

Thus, it will suffice to solve the control problem as formulated for an initial point $(x_0,y_0) \in D$. This will be done with the help of the modes 1–6 shown in Fig. 9.13.

9.3.3 Construction of the trajectories

We will construct appropriate trajectories for each of modes 1–6 in Fig. 9.13 and determine the domains D_i, $i = 1,\dots,6$, of initial data x_0 and y_0 in the domain D from which the mode in question steers the system to the terminal state $x(T) = y(T) = 0$. The sets D_i are indicated in Fig. 9.14 by the corresponding digits $i = 1,\dots,6$.

Mode 1

By Fig. 9.13, we have $u = -1$ for $t \in [0,T]$. Integrating (9.3.5) for initial data (9.3.7), we obtain

$$u = -1, \quad z = -t, \quad y = y_0 - \frac{1}{2}t^2,$$
$$x = x_0 + y_0 t - \frac{1}{6}t^3. \tag{9.3.11}$$

Let us set $t = T$ in formulas (9.3.11) and substitute the results into the terminal conditions (9.3.8). Eliminating T, we obtain

$$T = (2y_0)^{1/2} < 1, \tag{9.3.12}$$

$$x_0 = -\frac{1}{3}(2y_0)^{3/2}. \tag{9.3.13}$$

The inequality $T < 1$ follows from the fact that the bound $z = -1$ is not achieved in mode 1; see Fig. 9.13. Relations (9.3.12) and (9.3.13) imply the inequalities

$$-\frac{1}{3} < x_0 < 0, \quad 0 < y_0 < \frac{1}{2}. \tag{9.3.14}$$

Thus, mode 1 defined by (9.3.11) is implemented, if the initial point (x_0, y_0) lies on the arc of the curve defined by (9.3.13) and inequalities (9.3.14). Consequently, the set D_1 is the arc of the curve Γ, see (9.3.10), indicated by the numeral 1 in Fig. 9.14, enclosed between the points O and $A = (-1/3, 1/2)$. All phase trajectories beginning on this arc will ultimately lead to the origin if mode 1 is applied. The trajectories are defined by (9.3.11), and the duration of the motion by (9.3.12). It is easily verified that all these trajectories lie in the domain between the curve Γ and the parabola Γ_0 defined by the formulas

$$\Gamma_0 : x = \varphi_0(y) = -\frac{1}{2}y^2, \quad y \geq 0. \tag{9.3.15}$$

This parabola Γ_0 is at the same time the switching curve and a phase trajectory leading to the origin for the time-optimal problem, if constraints (9.3.6) are replaced by the simple restriction $|z| \leq 1$.

Mode 2

We have

$$u = -1, \quad z = -1, \quad \text{if} \quad 0 \leq t < 1,$$
$$u = 0, \quad z = -1, \quad \text{if} \quad 1 < t < T. \tag{9.3.16}$$

Motion along the first part of the trajectory ($t < 1$) is defined by relationships (9.3.11). We conclude from (9.3.11) that at $t = 1$ we have:

$$y(1) = y_0 - \frac{1}{2}, \quad x(1) = x_0 + y_0 - \frac{1}{6}. \tag{9.3.17}$$

Integrating (9.3.5), taking (9.3.16) and initial data (9.3.17) in the second part of the motion ($t > 1$) into account, we obtain

$$y(t) = y(1) - (t-1), \quad x(t) = x(1) + y(1)(t-1) - \frac{1}{2}(t-1)^2. \tag{9.3.18}$$

Let us substitute expressions (9.3.18) into terminal conditions (9.3.8) and eliminate T. We obtain

$$T = y(1) + 1 > 1, \quad x(1) = -\frac{[y(1)]^2}{2}. \tag{9.3.19}$$

Thus, the point $(x(1), y(1))$ lies on the parabola Γ_0 of (9.3.15), and, according to (9.3.16), motion along the second part of the trajectory (for $t \in [1, T]$) takes place along this parabola until the origin is reached. Substituting (9.3.17) into (9.3.19), we obtain the conditions

$$x_0 = -\frac{1}{2}y_0^2 - \frac{1}{2}y_0 + \frac{1}{24}, \quad y_0 \geq \frac{1}{2}. \tag{9.3.20}$$

Formulas (9.3.20) define the set D_2 of initial data for which mode 2 ensures that the system will reach the origin. This set D_2 is the part of the curve Γ [see (9.3.10)] from the point $A = (-1/3, 1/2)$ inclusive to infinity. All trajectories starting in that set are enclosed between Γ and Γ_0 with their second parts (for $t > 1$) lying on the parabola Γ_0. Typical trajectories for modes 1 and 2 are shown in Fig. 9.14 by dashed curves.

Thus, if the initial point (x_0, y_0) lies on the curve Γ, our problem is solved by controls 1 and 2 with mode 1 applying, if (x_0, y_0) is between O and A, and mode 2, if it is to the left of A in Fig. 9.14.

We now consider modes 3–6 of Fig. 9.13 letting θ denote the time at which the function $z(t)$, $\theta \in (0, T)$, experiences a jump. It is not difficult to see that the functions $z(t)$ for $t > \theta$ for all modes 3–6 of Fig. 9.13 are identical with $z(t)$ for $t > 0$ for one of modes 1 or 2: for modes 3 and 5 the relevant mode is 1, and for modes 4 and 6 it is 2. Hence, the segments of trajectories for modes 3–6 for $t > \theta$ coincide with trajectories for one of modes 1 or 2. Consequently, the point $(x(\theta), y(\theta))$ for modes 3–6 must belong to the sets of initial data for the appropriate modes 1 or 2, namely,

$$\begin{aligned} &\{x(\theta), y(\theta)\} \in D_1 \quad \text{for modes 3 and 5;} \\ &\{x(\theta), y(\theta)\} \in D_2 \quad \text{for modes 4 and 6.} \end{aligned} \tag{9.3.21}$$

To compute the numbers $x(\theta)$ and $y(\theta)$, we note that, apart from sign, the time history of $z(t)$ for $t < \theta$ in cases 3 and 4 is identical with mode 1, and in cases 5 and 6—with mode 2. Therefore, changing signs when needed and setting $t = \theta$, we conclude from (9.3.11) that for modes 3 and 4

$$y(\theta) = y_0 + \frac{1}{2}\theta^2, \quad x(\theta) = x_0 + y_0\theta + \frac{1}{6}\theta^3, \quad \theta < 1. \tag{9.3.22}$$

Using formulas (9.3.17) and (9.3.18) and proceeding in analogous fashion, we obtain the results for modes 5 and 6

$$y(\theta) = y_0 + \frac{1}{2} + (\theta - 1) = y_0 + \theta - \frac{1}{2},$$

$$x(\theta) = x_0 + y_0 + \frac{1}{6} + \left(y_0 + \frac{1}{2}\right)(\theta - 1) + \frac{1}{2}(\theta - 1)^2 \tag{9.3.23}$$

$$= x_0 + y_0\theta + \frac{1}{2}\theta^2 - \frac{1}{2}\theta + \frac{1}{6}, \quad \theta \geq 1.$$

Let us determine the domains D_i in the (x,y)-plane containing the initial data x_0 and y_0 for the appropriate modes, $i = 3,4,5,6$. To do this, we use formulas (9.3.21)–(9.3.23) and the previously presented definitions of D_1 and D_2.

Mode 3

Substituting expressions (9.3.22) for $x(\theta)$ and $y(\theta)$ in place of x_0 and y_0 into equation (9.3.13) and inequalities (9.3.14) defining the set D_1, we obtain

$$x_0 = -y_0\theta - \frac{1}{6}\theta^3 - \frac{1}{3}(2y_0 + \theta)^{2/3},$$

$$0 < 2y_0 + \theta^2 < 1, \quad 0 < \theta < 1. \tag{9.3.24}$$

Let us determine the boundaries of the set D_3 defined parametrically by formulas (9.3.24). To do this, it will suffice to consider four cases corresponding to equality in each of the four inequalities (9.3.24).

We first assume that $2y_0 + \theta^2 = 0$. Substituting the value of θ found from this equality into (9.3.24), we obtain

$$x_0 = \frac{1}{3}(-2y_0)^{3/2}, \quad -\frac{1}{2} < y_0 < 0. \tag{9.3.25}$$

By (9.3.9) and (9.3.10), formulas (9.3.25) define a segment of the curve Γ' from the origin to the point $A' = (1/3, -1/2)$, see Fig. 9.14.

Putting $2y_0 + \theta^2 = 1$ and substituting the value of θ thus determined into (9.3.24), we obtain

$$x_0 = -\frac{1}{3} - y_0(1 - 2y_0)^{1/2} - \frac{1}{6}(1 - 2y_0)^{3/2}, \quad 0 < y_0 < \frac{1}{2}. \tag{9.3.26}$$

Formulas (9.3.26) define an arc of a curve Γ_1 in the (x,y)-plane joining the points $A = (-1/3, 1/2)$ and $B = (-1/2, 0)$. This curve is shown in Fig. 9.14.

Setting $\theta = 0$, we obtain from (9.3.24)

$$x_0 = -\frac{1}{3}(2y_0)^{3/2}, \quad 0 < y_0 < \frac{1}{2}.$$

By (9.3.10), this segment of the boundary of D_3 coincides with the set D_1, that is, with the arc OA of the curve Γ.

Finally, setting $\theta = 1$, we obtain from (9.3.24)

$$x_0 = -\frac{1}{6} - y_0 - \frac{1}{3}(2y_0+1)^{3/2}, \quad -\frac{1}{2} < y_0 < 0. \tag{9.3.27}$$

These formulas define an arc of a curve in the (x,y)-plane joining the points B and A'. This curve Γ_2 touches the curve Γ' at the point A', see Fig. 9.14.

Thus, the set D_3 is a curvilinear quadrilateral $OABA'$ bounded by arcs of the curves Γ (from O to A), Γ_1, Γ_2, and Γ' (from A' to O).

Mode 4

Substituting expressions (9.3.22) for $x(\theta)$ and $y(\theta)$ in place of x_0 and y_0 into relations (9.3.20) defining the set D_2, we obtain

$$\begin{gathered} x_0 = \frac{1}{24} - \frac{y_0^2}{2} - \frac{y_0\theta^2}{2} - y_0\theta - \frac{1}{2}y_0 - \frac{\theta^4}{8} - \frac{\theta^3}{6} - \frac{\theta^2}{4}, \\ 2y_0 + \theta^2 \geq 1, \quad 0 < \theta < 1. \end{gathered} \tag{9.3.28}$$

The boundaries of the set D_4 will be found by replacing the inequality sign in each of the three inequalities of (9.3.28) in turn by an equality sign.

We first assume that $2y_0 + \theta^2 = 1$ and eliminate θ: $\theta = (1-2y_0)^{1/2}$ from the given equality. Substituting the resulting value of θ into (9.3.28) and simplifying, we obtain the relation defining the arc Γ_1.

Setting $\theta = 0$ in (9.3.28), we obtain, as is easily verified, relations (9.3.20) that define the set D_2, that is, the arc of the curve Γ from the point A to infinity.

Setting $\theta = 1$ in (9.3.28), we have

$$x_0 = -\frac{y_0^2}{2} - 2y_0 - \frac{1}{2}, \quad y_0 \geq 0. \tag{9.3.29}$$

The curve Γ_3 defined by these relations begins at the point $B = (-1/2, 0)$ and goes off to infinity, see Fig. 9.14.

As a result, the set D_4 is bounded by the set D_2, the curve Γ_1, along which it borders on D_3, and the curve Γ_3.

Mode 5

Substituting expressions (9.3.23) for $x(\theta)$ and $y(\theta)$ in place of x_0 and y_0 into equation (9.3.13) and inequalities (9.3.14) defining the set D_1, we obtain

$$x_0 = -y_0\theta - \frac{\theta^2}{2} + \frac{1}{2}\theta - \frac{1}{6} - \frac{1}{3}(2y_0 + 2\theta - 1)^{3/2},$$
$$\frac{1}{2} < y_0 + \theta < 1, \quad \theta \geq 1. \tag{9.3.30}$$

We now determine the boundaries of the set D_5 reasoning similarly to the previous cases and replacing each of the three inequalities (9.3.30) in turn by equalities.

Setting $y_0 + \theta = 1/2$, we find that $\theta = 1/2 - y_0$. Substituting this value of θ into equality (9.3.30), we obtain

$$x_0 = \frac{y_0^2}{2} - \frac{1}{2}y_0 - \frac{1}{24}, \quad y_0 \leq -\frac{1}{2}.$$

By (9.3.10), these formulas define the arc of the curve Γ' from the point $A' = (1/3, -1/2)$ to infinity; this arc is symmetric to D_2 about the origin.

Settings $y_0 + \theta = 1$, we obtain $\theta = 1 - y_0$. Substituting this value into equality (9.3.30), we obtain

$$x_0 = y_0^2 - \frac{1}{2}y_0 - \frac{1}{2}, \quad y_0 \leq 0. \tag{9.3.31}$$

Formulas (9.3.31) define a curve Γ_4 beginning at $B = (-1/2, 0)$ and going off to infinity, see Fig. 9.14.

Setting $\theta = 1$ in (9.3.30), we obtain formulas (9.3.27) defining the curve Γ_2.

Thus, the set D_5 is bounded by an arc of the curve Γ_2, along which it borders on the set D_3, the curve Γ_4, and the arc of the curve Γ' from A' to infinity.

Mode 6

Substituting expressions (9.3.23) for $x(\theta)$ and $y(\theta)$ in place of x_0 and y_0 into formulas (9.3.20) defining the set D_2, we obtain

$$x_0 = -\frac{y_0^2}{2} - 2y_0\theta - \theta^2 + \frac{1}{2}\theta, \quad y_0 + \theta \geq 1, \quad \theta \geq 1. \tag{9.3.32}$$

Replacing the first of inequalities (9.3.32) by an equality, we obtain $\theta = 1 - y_0$. Substituting this expression into equality (9.3.32), we obtain relations (9.3.31) defining the curve Γ_4.

Setting $\theta = 1$ in (9.3.32), we obtain relations (9.3.29) defining the curve Γ_3.

Thus, the set D_6 borders on the sets D_4 and D_5 along the curves Γ_3 and Γ_4, respectively, and lies below and to the left of these curves that have a common point $B = (-1/2, 0)$.

Note that the curves Γ_2 and Γ_3 have a common tangent at the point B, and the same is true of Γ_1 and Γ_4.

The solution of the control problem as formulated may be described as follows. Given an initial state (9.3.7) in the domain D of the (x, y)-plane, we determine to which of the domains D_i, $i = 1,2,\dots,6$, it belongs. The boundaries between the domains are given by the curves Γ, Γ', Γ_1, Γ_2, Γ_3, and Γ_4 defined by formulas (9.3.9), (9.3.10), (9.3.26), (9.3.27), (9.3.29), and (9.3.31). The boundary between D_1 and D_2 is the point $A = (-1/3, 1/2)$.

1°. If $(x_0, y_0) \in D_1$, then we define $u \equiv -1$ for $t > 0$. The system reaches the given state $x = y = 0$ in a time $T < 1$.

2°. If $(x_0, y_0) \in D_2$, we put $u = -1$ for $t \in (0,1)$ and $u = 0$ for $t \geq 1$. The system reaches the terminal state in a time $T \geq 1$.

3°. If $(x_0, y_0) \in D_3$, then $u = 1$ for $t \in (0, \theta)$, where the time $\theta < 1$ is defined by the condition $\{x(\theta), y(\theta)\} \in D_1$. At time $t = \theta$, we equate z to zero by a jump that is admitted by restrictions (9.3.6). At $t > \theta$, we define $u = -1$ up to the end of the process. The trajectory for $t > \theta$ is the same as for mode 1.

4°. If $(x_0, y_0) \in D_4$, then $u = 1$ for $t \in (0, \theta)$, where the time $\theta < 1$ is defined by the condition $\{x(\theta), y(\theta)\} \in D_2$. At time $t = \theta$, we equate z to zero by a jump. We then define $u = -1$ for $t \in (\theta, \theta + 1)$ and $u = 0$ for $t \in (\theta + 1, T)$.

5°. If $(x_0, y_0) \in D_5$, then $u = 1$ for $t \in (0,1)$ and $u = 0$ for $t \in (1, \theta)$, where the time $\theta > 1$ is defined by the condition $\{x(\theta), y(\theta)\} \in D_1$. At time θ, we equate z to zero by a jump. We then define $u = -1$ for $t \in (\theta, T)$ up to the end of the process.

6°. If $(x_0, y_0) \in D_6$, then $u = 1$ for $t \in (0,1)$ and $u = 0$ for $t \in (1, \theta)$, where the time $\theta > 1$ is defined by the condition $\{x(\theta), y(\theta)\} \in D_2$. At time θ, we equate z to zero by a jump. We then define $u = -1$ for $t \in (\theta, \theta + 1)$ and $u = 0$ for $t \in (\theta + 1, T)$.

Note that $T < 1$ in case 1, $T > 1$ in cases 2, 4, and 5, and $T > 2$ in case 6.

All trajectories beginning in the domain D lie in the domain bounded by the curves Γ_0 and Γ' (to the left of and below those curves; see Fig. 9.14). They reach the origin O either touching the curve Γ_0 (for modes 1, 3, and 5) or coinciding with Γ_0 over its last part (for modes 2, 4, and 6; see the curves in Fig. 9.14).

If the initial point (x_0, y_0) is in the domain D' symmetric to D about the origin, the control is taken equal in magnitude and opposite in sign to the control corresponding to the point $(-x_0, -y_0) \in D$.

The solution we have constructed was obtained for initial data (9.3.7), which presume that $z(0) = 0$. The general case of initial data

$$x(0) = x_0, \qquad y(0) = y_0, \qquad z(0) = z_0$$

is reduced to that considered above, if at time $t = 0$ we change z by a jump equating it to zero that is admitted by restrictions (9.3.6). After that, one can use the solution constructed for the initial data (9.3.7). Thus, terminal conditions (9.3.8) will be satisfied. In that case, however, the property of time-optimality is not to be expected.

Chapter 10
Time-optimal swing-up and damping feedback controls of a nonlinear pendulum

A pendulum is a well-know example of a nonlinear mechanical system that is often regarded as a benchmark for control algorithms. In a number of papers, various feedback controls have been proposed that bring the pendulum to the upper unstable or lower stable equilibrium position. These controls are called swing-up and damping controls, respectively. Time-optimal controls have been also considered, but the obtained solutions were not complete.

In this chapter, the time-optimal controls both for the swing-up and damping control problems are obtained. The solution is based on the maximum principle and involves analytical investigations and extensive numerical computation for a wide range of parameters. As a result, the switching and dispersal curves are obtained that bound the domains in the phase plane corresponding to different values of the optimal bang-bang control.

Optimal trajectories can intersect the switching curves but not the dispersal curves. The latter curves have the following property: two different optimal trajectories start from each point of the dispersal curve.

The switching and dispersal curves are obtained for various values of the maximal admissible control torque. These curves completely determine the feedback optimal control.

Fine details of the structure of these curves as well as of the field of optimal trajectories are analyzed. The structure depends essentially on the magnitude of the control torque. In particular, numerical results show how the breaks of the switching curves (in the case of the damping control) are formed at the transition from high torques, where these curves are smooth, to low torques corresponding to the switching curves with breaks.

The chapter is based on papers [46, 47, 104, 105, 106],

10.1 Optimal control structure

10.1.1 Statement of the problem

Consider a pendulum that can rotate about a horizontal axis O and is controlled by a torque M applied to the pendulum (Fig. 10.1). Denote by φ the angle between the pendulum and the vertical axis, by m the mass of the pendulum, by J its moment of inertia about the axis O, by l the distance from the axis O to the center of mass of the pendulum, and by g the acceleration due to gravity.

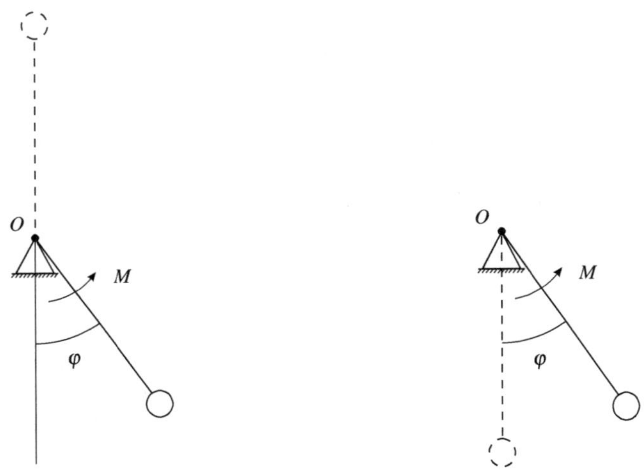

Fig. 10.1 Swing-up and damping control problems

The equation of motion of the pendulum is

$$J\ddot{\varphi} + mgl\sin\varphi = M, \tag{10.1.1}$$

where the dots denote derivatives with respect to time t. Let the control torque be restricted by the constraint

$$|M| \leq M_0, \tag{10.1.2}$$

where M_0 is a given constant.

Denote

$$\omega = \left(\frac{mgl}{J}\right)^{1/2}, \qquad t' = \omega t. \tag{10.1.3}$$

Here, ω is the natural frequency of small oscillations of the pendulum, and t' is a non-dimensional time.

Let us introduce non-dimensional variables

$$x_1 = \varphi, \qquad x_2 = \frac{d\varphi}{dt'}, \qquad u = \frac{M}{M_0}, \qquad k = \frac{M_0}{mgl} \tag{10.1.4}$$

and rewrite (10.1.1) as follows:

$$\dot{x}_1 = x_2, \qquad \dot{x}_2 = -\sin x_1 + ku. \tag{10.1.5}$$

Here and below, we denote by dots derivatives with respect to non-dimensional time t'. We will omit $'$ after t'.

Constraint (10.1.2) can be rewritten as follows:

$$|u(t)| \leq 1. \tag{10.1.6}$$

The initial conditions for system (10.1.5) are arbitrary

$$x_1(0) = x_1^0, \qquad x_2(0) = x_2^0, \tag{10.1.7}$$

and the terminal coordinates correspond to the upper unstable or lower stable equilibrium position

$$\begin{aligned} x_1(T) &= \pi + 2\pi n, \qquad x_2(T) = 0 \qquad \text{(swing-up control)}, \\ x_1(T) &= 2\pi n, \qquad x_2(T) = 0 \qquad \text{(damping control)}, \end{aligned} \tag{10.1.8}$$

where n is an arbitrary integer.

We will find the control, both in an open-loop and feedback form, that satisfies constraint (10.1.6) for all $t \in [0, T]$ and brings system (10.1.5) from any initial state (10.1.7) to the terminal state (10.1.8) in a minimal possible time T.

10.1.2 Phase cylinder

It is known that the presence of nonlinearity in the motion equation of the pendulum results in a periodic structure (cylindrical property) in the angle of the synthesis pattern. An infinite set of terminal points in the state space corresponds to the upper or lower equilibrium position.

The cylindrical property of the state space results in specific features of the synthesis. The main specific feature is the presence of a dispersal curve (the terminology of [69]) on the cylinder such that two optimal trajectories with the same motion time are originated from every its point.

For a very large control torque, we can omit the nonlinear term in the motion equation of the pendulum. The synthesis pattern in this case consists of parabolic switching curves passing through terminal points and dispersal curves situated between them, see Fig. 10.2. The equation of the dispersal curves in the case of a very large control torque can be obtained in an explicit form [63].

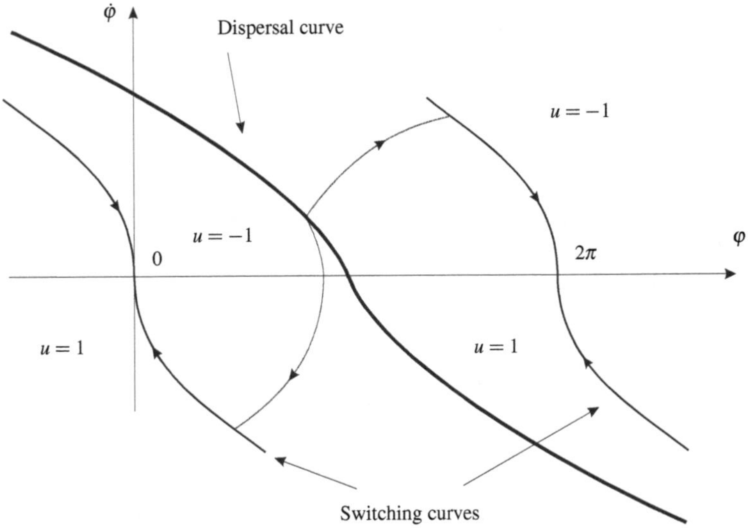

Fig. 10.2 Time-optimal feedback control for a very large control torque: $k \gg 1$

The question arises of how the synthesis pattern changes when the maximum admissible control torque gradually decreases beginning with a certain sufficiently large value. The answer is given below, where time-optimal feedback controls are designed for steering the nonlinear pendulum to the upper unstable or lower stable equilibrium position. The solution is given for various values of the maximum possible control torque.

10.1.3 Maximum principle

Following the maximum principle, we introduce the Hamiltonian for system (10.1.5)

$$H = p_1 x_2 + p_2(-\sin x_1 + ku). \tag{10.1.9}$$

Here, p_1 and p_2 are the adjoint variables that satisfy the equations

$$\dot{p}_1 = p_2 \cos x_1, \qquad \dot{p}_2 = -p_1. \tag{10.1.10}$$

The optimal control satisfying constraint (10.1.6) is determined by the condition

$$u = \operatorname{sign} p_2. \tag{10.1.11}$$

It follows from (10.1.8), (10.1.9), and (10.1.11) that, at the terminal time instant $t = T$, we have

$$H_T = k|p_2(T)| \geq 0.$$

This inequality is one of the necessary optimality conditions [see (1.2.22)]. It follows from the results of Chapter 7 of [83] that there are no singular controls in our optimal control problem.

It follows from (10.1.11) that the optimal control takes the values $u = \pm 1$, and, in order to obtain the optimal feedback control, it is sufficient to find the switching and dispersal curves in the (x_1, x_2)-plane that bound the domains, where $u = +1$ and $u = -1$.

Note that the switching curves consist of points, where the control $u(t)$ changes its sign along the optimal trajectory. The dispersal curves consist of such points, where the optimal control can be taken equal to either $+1$ or -1, and two optimal trajectories starting from each of these points reach terminal states (10.1.8) (with the same or different n) in the same time.

10.1.4 Numerical algorithm

To construct the complete field of optimal phase trajectories, we integrate numerically the system of equations (10.1.5), (10.1.10), and (10.1.11) backward with respect to time (starting from some instant T). The conditions for the phase coordinates x_1 and x_2 at the time instant T are given by (10.1.8); the conditions for the adjoint variables p_1 and p_2 are arbitrary. It is convenient to divide all optimal trajectories into families identified by the integer parameter n, see (10.1.8). Within each family, an individual trajectory is identified by two parameters, $p_1(T)$ and $p_2(T)$.

The maximum principle implies that at least one of the values $p_1(T)$ and $p_2(T)$ is nonzero. Therefore, we can normalize the adjoint variables so that $p_1^2(T) + p_2^2(T) = 1$. Thus, the family of trajectories corresponding to a fixed n depends on one parameter. For example, we can take $p_1(T) = \cos\alpha$ and $p_2(T) = \sin\alpha$. By changing α in the interval $\alpha \in [0, 2\pi)$, we obtain all extremal trajectories $(x_1(t), x_2(t))$ of system (10.1.5) for a given n.

On each phase trajectory, we are interested in the points at which the adjoint variable $p_2(t)$ changes in sign. We will refer to the points of this type as 1-points. In addition, we are interested in the projections of the points of intersection of the trajectory under consideration with other trajectories in the augmented phase space (x_1, x_2, t) onto the (x_1, x_2)-plane. From all such projection points, we select only one point corresponding to the maximum time instant t. We find such a point for every trajectory. We will refer to the points of the latter type as 2-points. The intersecting trajectories may belong to either the same or different families, i.e., they may correspond to the same or different n. Note that some points can be both 1-points and 2-points simultaneously.

According to (10.1.11), 1-points are switching points for the optimal control. From 2-points, the system can arrive at the desired terminal point along different trajectories corresponding to different controls.

Continuous loci of 1-points form the *switching curves*, whereas continuous loci of 2-points form the *dispersal curves*. Points that are 1-points and 2-points simultaneously are points of connection of switching and dispersal curves. Both switching curves and dispersal curves divide the domains in the phase plane (x_1, x_2) corresponding to $u = -1$ and $u = +1$. In terms of the field of optimal trajectories, the difference between the switching curves and the dispersal curves is that the optimal trajectories can only depart from the dispersal curves but cannot arrive at these curves. No optimal trajectories can coincide with the dispersal curves. At the same time, the phase state (x_1, x_2) of the system can arrive at a switching curve, move along this curve, or depart from it.

The field of phase trajectories will be constructed using the maximum principle that provides only the necessary optimality conditions. Nevertheless, the obtained phase portrait provides the optimal feedback control under the condition that an optimal control exists in the class of piecewise continuous functions $u(t)$ for each pair of the initial and final conditions (10.1.7) and (10.1.8). This follows from the fact that the trajectory arriving at the terminal point from an initial point that does not belong to the dispersal curve is unique. If the initial point lies on the dispersal curve, then there exist two extremal curves that start from this point and arrive at the terminal state, but the motion time for both trajectories is the same.

The software for the numerical calculations presented below in this chapter was developed in Borland Delphi 5.1 environment. A widespread procedure RKF45 (see [60, 56, 57, 111]) was used for the numerical construction of smooth parts of the trajectories. A reliable PASCAL version of the aforementioned procedure can be found in [55]. We do not give a detailed description of the numerical algorithm due to its complexity, especially for the unit forming the dispersal curves.

10.2 Swing-up control

10.2.1 Literature overview

The problem of stabilization of a pendulum at the upper equilibrium position has been a matter of considerable scientific interest. In [114, 74, 92], a pendulum is stabilized due to the vertical motion of its base. In [66], a pendulum with a fixed suspension point is stabilized by means of a rotating flywheel.

In a number of papers, controls have been proposed that bring the planar pendulum on a cart to the upper unstable equilibrium position. Global stabilization of this model has been studied, for instance, in [112, 18, 81]. Time-optimal control has been studied in [97, 119]. These papers are focused on computing exact switching times for an open-loop control starting from the down equilibrium.

However, the problem of time-optimal swing-up feedback control for the nonlinear pendulum has not been solved.

10.2.2 Special trajectories

Let us analyze special trajectories that reach the terminal states $(\pi+2\pi n, 0)$ with constant control $u=+1$ or $u=-1$. We will present detailed construction of the special trajectory for $x_1=\pi$ $(n=0)$, $x_2=0$, and $u=-1$. All other special trajectories can be constructed in a similar way. We will set terminal values for the adjoint variables so that $p_1(T)<0$ and $p_2(T)=0$ and prove that $p_2(t)<0$ for $t<T$. In this case, condition (10.1.11) will be satisfied. Without loss of generality, we can normalize the adjoint variables so that

$$p_1(T)=-1, \qquad p_2(T)=0. \tag{10.2.1}$$

We substitute $u=-1$ into equations (10.1.5) and find the first integral of these equations

$$\frac{x_2^2}{2}-\cos x_1+kx_1=C_1.$$

To find the constant C_1, we insert the terminal conditions $x_1=\pi$ and $x_2=0$ into this integral. As a result we obtain

$$\frac{x_2^2}{2}=1+\cos x_1+k(\pi-x_1). \tag{10.2.2}$$

For the trajectory that reaches the terminal state with $u=-1$, we have, according to (10.1.5), $x_2>0$ and $x_1<\pi$ for small positive $T-t$, i.e., at the end of motion. Introduce the change of variable

$$\pi-x_1=y \tag{10.2.3}$$

to represent (10.2.2) as follows:

$$x_2=R(y)=\left[2\left(1-\cos y+ky\right)\right]^{1/2}. \tag{10.2.4}$$

This function is shown in Fig. 10.3 for various values of k. Its behavior for small and large y is determined by the expansions

$$\begin{aligned} R^2(y) &= 2ky+y^2+O\left(y^4\right) \qquad \text{as} \qquad y\to 0,\\ R^2(y) &= 2ky+O(1) \qquad \text{as} \qquad y\to\infty. \end{aligned} \tag{10.2.5}$$

It follows from (10.1.5) and (10.2.3) that

$$\frac{dy}{dt}=-R(y)<0.$$

Therefore, as t changes form T to $-\infty$, the variable y grows monotonically from 0 to ∞. Hence, y can be taken as an independent variable along the trajectory under consideration.

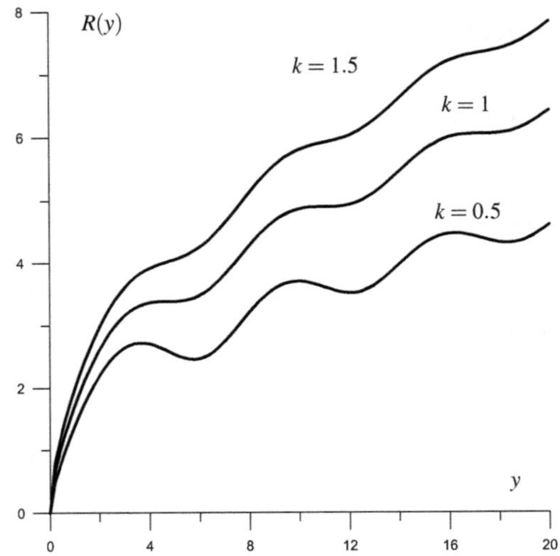

Fig. 10.3 Function $R(y)$ for different values of k

Let us analyze the behavior of the adjoint variable $p_2(y)$ for this trajectory. Note that, according to (10.1.11), the sign of p_2 determines the control.

We rewrite the adjoint equations (10.1.10) using y as an independent variable and taking into account (10.1.5) and (10.2.4):

$$\frac{dp_1}{dy} = \frac{p_2 \cos y}{R(y)}, \qquad \frac{dp_2}{dy} = \frac{p_1}{R(y)}. \tag{10.2.6}$$

Eliminating p_1 from these equations, we come to the following equation for p_2:

$$\frac{d}{dy}\left(R\frac{dp_2}{dy}\right) = \frac{p_2 \cos y}{R}. \tag{10.2.7}$$

Let us multiply (10.2.7) by R and transform it taking into account the relation

$$R\frac{dR}{dy} = \sin y + k \tag{10.2.8}$$

following from (10.2.4). We have

$$\frac{d}{dy}\left(R^2\frac{dp_2}{dy}\right) = \frac{d}{dy}\left[(\sin y + k)p_2\right].$$

Integrating this equation, we obtain

$$R^2\frac{dp_2}{dy} - (\sin y + k)p_2 = C_2, \tag{10.2.9}$$

where C_2 is a constant.

At the terminal time instant $t = T$, we have, according to (10.1.8), (10.2.1), and (10.2.3),

$$y = 0, \qquad p_2 = 0.$$

In addition, from (10.2.1), (10.2.5), and (10.2.6), it follows that

$$R^2 \frac{dp_2}{dy} = Rp_1 \to 0 \qquad \text{as} \qquad y \to 0.$$

Substituting these data into (10.2.9), we find $C_2 = 0$, and (10.2.9) becomes

$$R^2 \frac{dp_2}{dy} - (\sin y + k) p_2 = 0. \tag{10.2.10}$$

Note that the general solution of this homogeneous equation can be expressed as follows:

$$p_2 = CR(y), \tag{10.2.11}$$

where C is a constant; this can be easily verified by using relation (10.2.8).

Finally, we obtain from (10.2.6), (10.2.8), and (10.2.11) the expression for p_1:

$$p_1 = R\frac{dp_2}{dy} = CR\frac{dR}{dy} = C(\sin y + k). \tag{10.2.12}$$

Therefore, in order to satisfy the terminal condition (10.2.1) for p_1, we should set $C = -1/k$ in (10.2.11) and (10.2.12).

Thus, we see that along the phase trajectory corresponding to the control $u = -1$, the adjoint variable $p_2(t)$ is negative for all $y > 0$ and, therefore, for all $t < T$.

Hence, this trajectory with $u = -1$ satisfies the necessary optimality conditions. Numerical analysis shows that the special trajectory considered above is optimal and coincides with the switching curve. This is the case also for the trajectory with $u = 1$ that is symmetric to the previous one about the point $(\pi, 0)$. The special trajectories leading to the point $(\pi + 2\pi n, 0)$ can be obtained by shifting the special trajectories leading to the point $(\pi, 0)$ by $2\pi n$.

10.2.3 Numerical results

Figures 10.4 and 10.5 present the field of optimal trajectories constructed according to the procedure described in Sect. 10.1.4. The switching curves and dispersal curves are designated by less thick and more thick lines, respectively. The optimal trajectories are depicted by thin lines. The arrows specify the direction of the growth of time along the trajectories. The figures present a part of the phase plane confined by the switching curves passing through the points $(-\pi, 0)$ and $(\pi, 0)$. The complete phase portrait can be obtained by a translation of this segment to the left and right by the quantity $2\pi n$, $n = \pm 1, \pm 2, \ldots$.

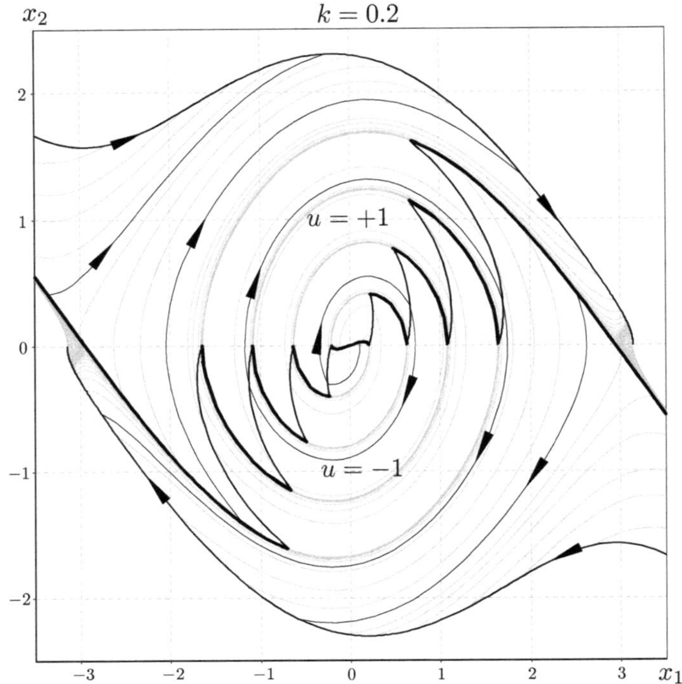

Fig. 10.4 Time-optimal swing-up feedback control for $k = 0.2$

Let us describe the properties of the synthesis patterns obtained. In Figs. 10.4 and 10.5, all switching curves touch the axis $x_2 = 0$ by one of their ends, and each of the dispersal curves divides the optimal trajectories from different families (corresponding to different n). The specified properties are related to a special choice of the values $k = 0.2$, 0.25, 0.35, 0.5, 0.85, 1, and 2 and may be absent for the other values.

The figures show that, when k increases, the number of switching curves and dispersal curves decreases gradually. After passing the threshold value $k \approx 0.80$, only special trajectories (investigated in Sect. 10.2.2) and only the smooth dispersal curve passing between them remain. The specified mechanism of transformation of the phase portrait corresponding to the optimal feedback control is depicted on a larger scale in Fig. 10.6 (special trajectories are not shown).

Let us describe the properties of these phase portraits. Note that, in Fig. 10.6, the switching curves do not touch the axis $x_2 = 0$, and the dispersal curve passing through the point $(0,0)$ has three clearly distinguishable smooth legs. This is explained by the fact that its middle leg separates the optimal trajectories that belong to two different families, and any of the extreme legs separates optimal trajectories of the same family. A similar pattern takes place in a domain close to the origin and in other ranges of k when the number of switching curves and dispersal curves changes (for example, see Fig. 10.7, where the threshold value is $k \approx 0.44$).

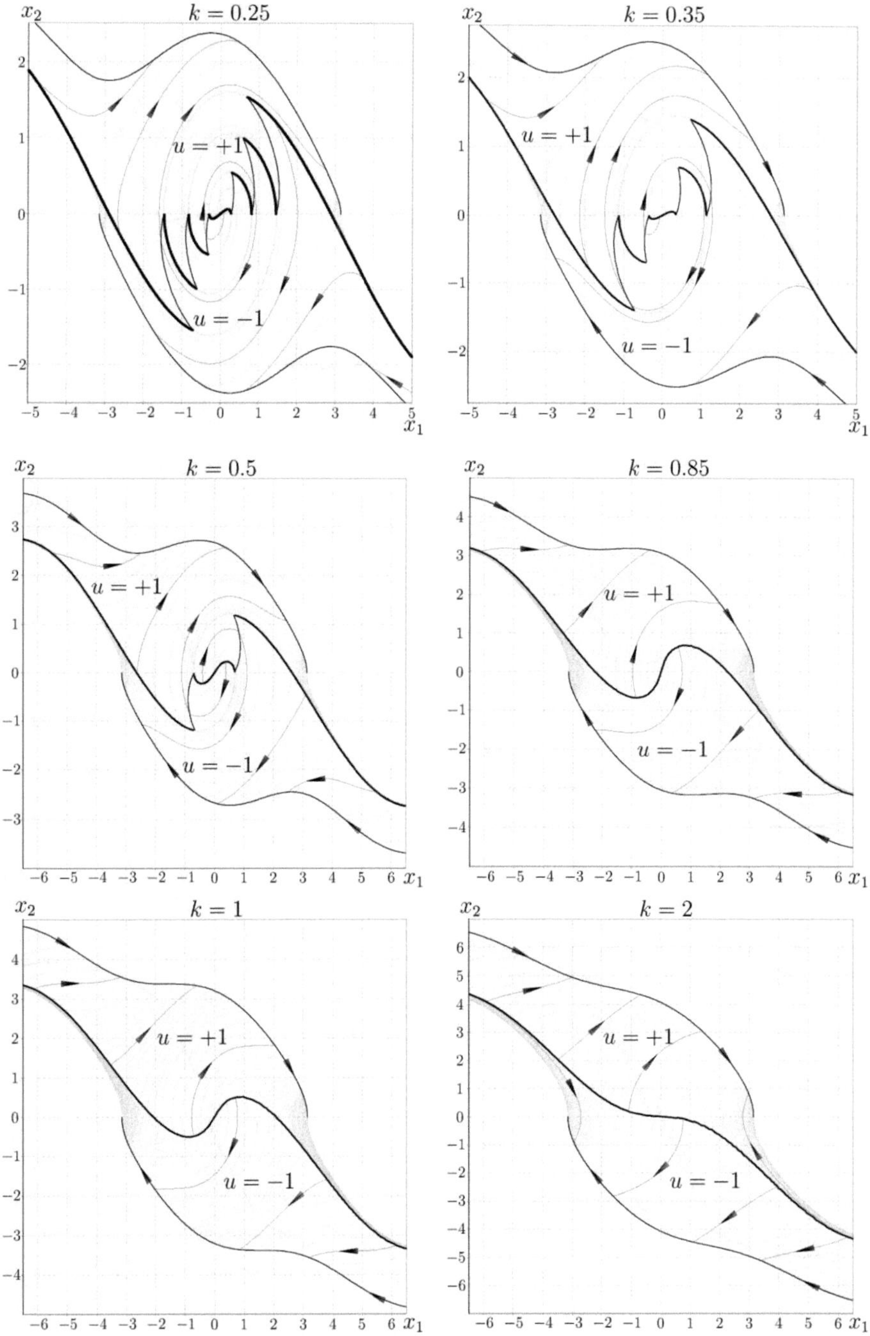

Fig. 10.5 Time-optimal swing-up feedback control for $k = 0.25, 0.35, 0.5, 0.85, 1$, and 2

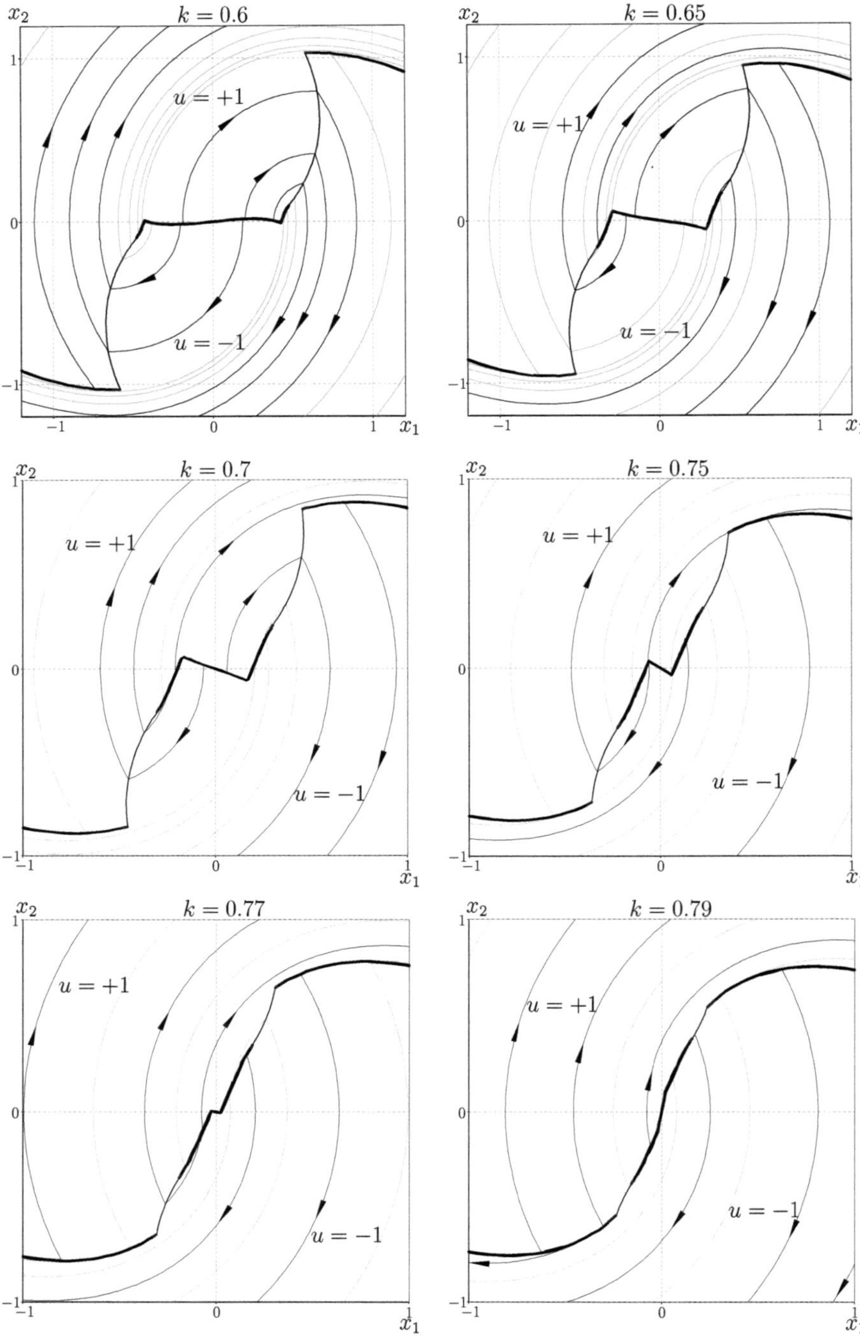

Fig. 10.6 Transition to a unique (smooth) dispersal curve when k increases, $k = 0.6$, 0.65, 0.7, 0.75, 0.77, and 0.79

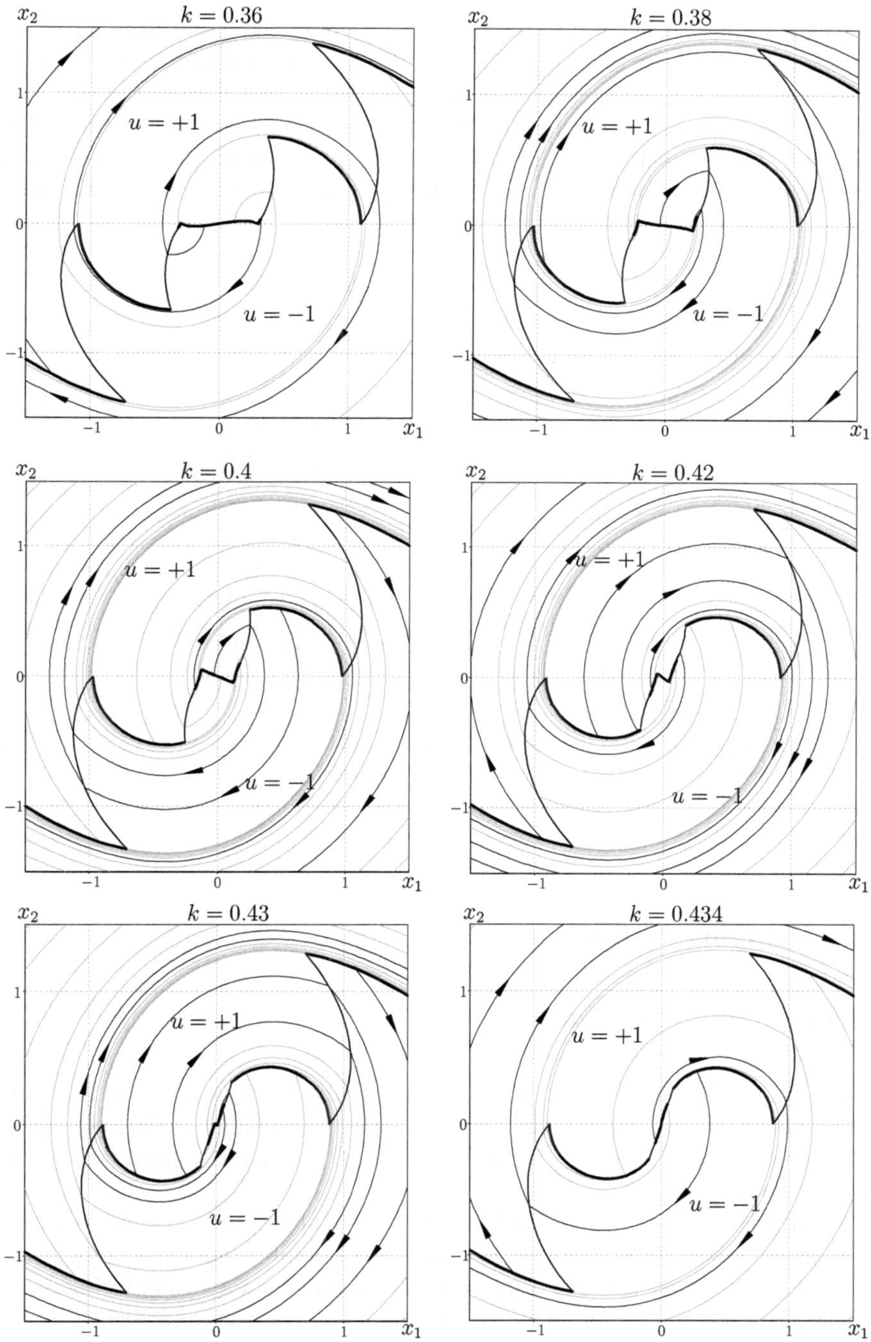

Fig. 10.7 Transformation of the phase portrait when k increases, $k = 0.36$, 0.38, 0.4, 0.42, 0.43, and 0.434

We can suppose that, when k decreases, the switching curves and dispersal curves [except for those that pass through the point $(0,0)$ or go to infinity] lie between the coordinate axes and the special trajectories in the first and third quadrants. Note that these curves run through the specified domains of the phase plane *more or less uniformly*. It is this behavior of the switching curves and dispersal curves that was obtained as a result of numerical computations. Figure 10.8 presents this behavior.

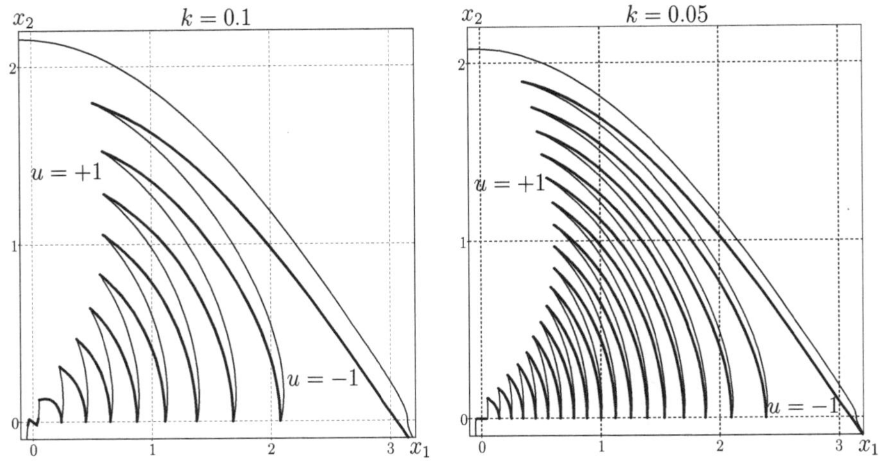

Fig. 10.8 Time-optimal swing-up feedback control in the first quadrant of the phase plane for $k = 0.1$ and 0.05

10.3 Damping control

10.3.1 Literature overview

In [20, 65, 83, 82, 58], for sufficiently large values of the control torque, the open-loop problem and the optimal synthesis of steering the pendulum to the lower stable equilibrium position were investigated both on the whole phase plane and on the phase cylinder.

Paper [65] analyzes control problem (10.1.5) and (10.1.6) for an equation that is more general than (10.1.5). However, in that paper, a time-optimal steering of any point of the phase plane to the origin was considered, and the cylindrical property of the phase space was not taken into account. The presence of so-called FLAG domains in the state space is the most essential feature found in [65]. The results of [65] were used later in [20], where problem (10.1.5)–(10.1.8) was considered for even n in (10.1.8), i.e., the cylindrical property of the state space arising in the problems of controlling a satellite was taken into account.

Remark 10.1. Here, we use the FLAG term introduced in [20] that is generated by capital letters of the names of the authors of [65].

In [83, 82, 58], nothing is said about the existence of FLAG domains, and they are not shown in the synthesis patterns. In other words, the control problem is considered under the large control torque and in the neighborhood of the terminal state. In [58], the size of such neighborhood is determined by the auxiliary constraint imposed on the motion time along the optimal trajectory. This constraint, in its turn, is forced to be related with the maximal admissible control torque.

In [17], the problem of optimal control of the mathematical pendulum that also describes a controlled revolution of a satellite in the plane of circular orbit [16] was considered. Numerical and qualitative investigations of the open-loop problem of rotation of the pendulum that is placed initially at the lower stable position by 360° about the suspension point were presented. The dependences of the optimal time and the number of switchings on the torque were constructed.

In [75], a controlled mechanical system in the form of a pendulum with a suspension point on the axis of a wheel that can roll on a horizontal plane without slip was considered. The control torque applied to the wheel was bounded. On the phase cylinder, a time-optimal control for damping oscillations of the pendulum was constructed. An algorithm for constructing an open-loop control that steers the system from the lower equilibrium position to the unstable upper one with damping of the velocity of the suspension point at the end of motion was given. The problem addressed in [75] is essentially different from the problem that is solved in this chapter.

Note that the solution in a neighborhood of each of terminal points is close to the optimal synthesis for a linear oscillator (see Example 2 in Sect. 1.4). The switching curve for a linear oscillator consists of an infinite number of semicircles, whose radius is equal to the maximum admissible control torque. Therefore, the more stringent constraints on control, the smaller the radii of the specified semicircles and the more frequent the breaks on the switching curve. As k decreases from large values ($k \gg 1$, see Fig. 10.2) to small ones ($k \ll 1$, see Fig. 10.9), the switching curves transform from smooth ones to the curves with breakes. The question arises: at which values of k the breaks first appear?

The calculations of the authors of this book have shown that the very first of the specified breaks are generated as a result of transformation of the boundaries of the FLAG domains that are confined by curves consisting of arcs of switching curves and dispersal curves. Appearing of the FLAG domains is closely associated with generation of additional switching curves corresponding to the optimal trajectories with two switchings of control. Such bifurcation has been analyzed in details in [101], where the bifurcation value of the maximal admissible control torque has been obtained. If $k \approx 1.04$, then one of such switching curves (infinitely small in size) arises at the point with $x_1 \approx -51.7$ and $x_2 \approx 10.4$ of the phase plane. An optimal trajectory with two switchings starting from this point goes the terminal point $(0,0)$. Its first interval with constant control $u = +1$ is assumed infinitely small. As can be seen from these data, for the pendulum it is required to perform approximately eight complete revolutions until its phase point gets to the terminal point $(0,0)$ from the specified point.

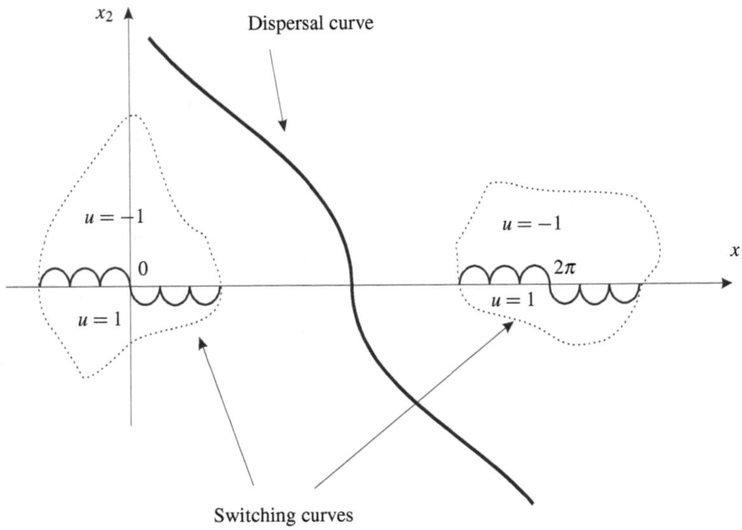

Fig. 10.9 Switching curves in the neighborhoods of the terminal points: $k \ll 1$

The further generation of breaks on the main switching curves (when the control torque decreases) occurs according to a similary pattern, and the process is shown in figures in Sect. 10.3.3 for various constraints imposed on the control torque.

10.3.2 Special trajectories

Let us analyze the special trajectories that lead to the terminal states $(2\pi n, 0)$ for the constant control $u = +1$ or $u = -1$. Consider the special trajectory for $x_1 = 0$ $(n = 0)$, $x_2 = 0$, and $u = +1$. All other special trajectories can be analyzed in a similar way.

For the trajectory that reaches the terminal state with the control $u = +1$, by (10.1.5), we have $x_2 < 0$ and $x_1 > 0$ for small positive $T - t$, i.e., at the end of the motion. In this case, the solution of (10.1.5) can be represented in the form

$$x_2 = -R(x_1) = -\left[2\left(-1 + \cos x_1 + kx_1\right)\right]^{1/2}. \tag{10.3.1}$$

The special trajectory (or its bounded final part, see Remarks 10.2–10.4 below) described by (10.3.1) satisfies the necessary optimality conditions. This fact can be proved similar to the case of swing-up control, see Sect. 10.2.2.

Remark 10.2. The function $R(x_1)$ is defined on the whole positive semi-axis if $k \geq k^*$, where k^* is determined by the relations

$$k^* = \sin z, \qquad z = \tan\frac{z}{2}, \qquad \frac{\pi}{2} < z < \pi. \tag{10.3.2}$$

By calculations, we obtain $k^* \approx 0.7246$ and $z \approx 2.3311$. Figure 10.10 shows the dependence $R^2(x_1)$ for various values of k.

Remark 10.3. The special trajectory with $u = 1$ and $k \geq k^*$ is unbounded and described by (10.3.1) for $0 \leq x_1 < \infty$. We found that, if $k \geq k_{\text{opt}}$, $k_{\text{opt}} \approx 0.9$, then the whole special trajectory is optimal. If $k^* \leq k < k_{\text{opt}}$, then only some final part of the considered special trajectory is optimal.

Remark 10.4. The special trajectory with $u = 1$ and $k < k^*$ is bounded. Its final part is described by (10.3.1), where $0 < x_1 \leq x_1^*$. Here, x_1^* is the minimal positive root of the following equation

$$R(x_1^*) = 0.$$

Numerical analysis shows that this final part of the special trajectory is optimal.

The same situation takes place for the trajectory for $u = -1$ that is situated symmetrically to the considered one relative to the point $(0,0)$. The special trajectories that arrive at the points $(2\pi n, 0)$ can be obtained by shifting special trajectories arriving at the point $(0,0)$ by $2\pi n$.

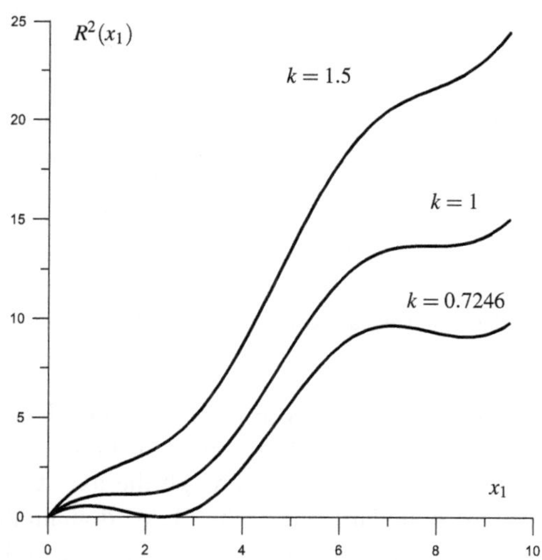

Fig. 10.10 Function $R^2(x_1)$ for different values of k

10.3.3 Numerical results

Figures 10.11–10.14 present the field of optimal trajectories constructed according to the procedure described in Sect. 10.1.4. Switching curves and dispersal curves are designated by less thick or more thick lines, respectively. Optimal trajectories are depicted by thin lines. Arrows specify the direction of time growth along trajectories. The figures present a segment of the phase plane confined by the dispersal curves passing through the points $(-\pi, 0)$ and $(\pi, 0)$. The complete phase portrait can be obtained by translation of this segment to the right and left by $2\pi n$, $n = \pm 1, \pm 2, \ldots$.

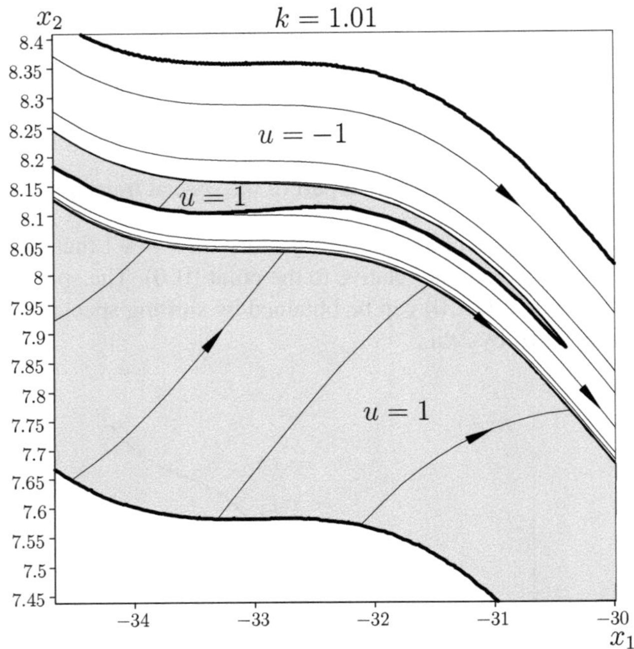

Fig. 10.11 Time-optimal damping feedback control for $k = 1.01$

Let us describe the properties of synthesis patterns obtained. Figures 10.11–10.13 ($k = 1.01$, 1, 0.85, 0.75, 0.65, 0.64, and 0.62) show the process of generation of the first break on the switching curve passing through the axis $x_2 = 0$. In Fig. 10.11 ($k = 1.01$), we can see the boundary part of the FLAG domain that does not touch the main switching curve [the special trajectory passing trough the point $(0,0)$]. In Fig. 10.12 ($k = 1$, 0.85, and 0.8), similary areas are denoted by a rectangle as well as shown separately with a larger scale. The comparison of synthesis patterns for $k = 1.01$ and $k = 1$ allows one to make the conclusion that the location of the right boundary of the FLAG domain is sensitive to the parameter k. For $k = 0.85$, the FLAG domain and the switching curve merge generating a sufficiently long "slot" that is much shorter for $k = 0.8$; and for $k = 0.75$, 0.65, and 0.64, it transforms gradually into a sharp "tooth" that is turned so that its peak touches the axis $x_2 = 0$ for $k = 0.62$. Meanwhile, the dispersal curve that generates the initial bottom boundary of the FLAG domain disappears completely.

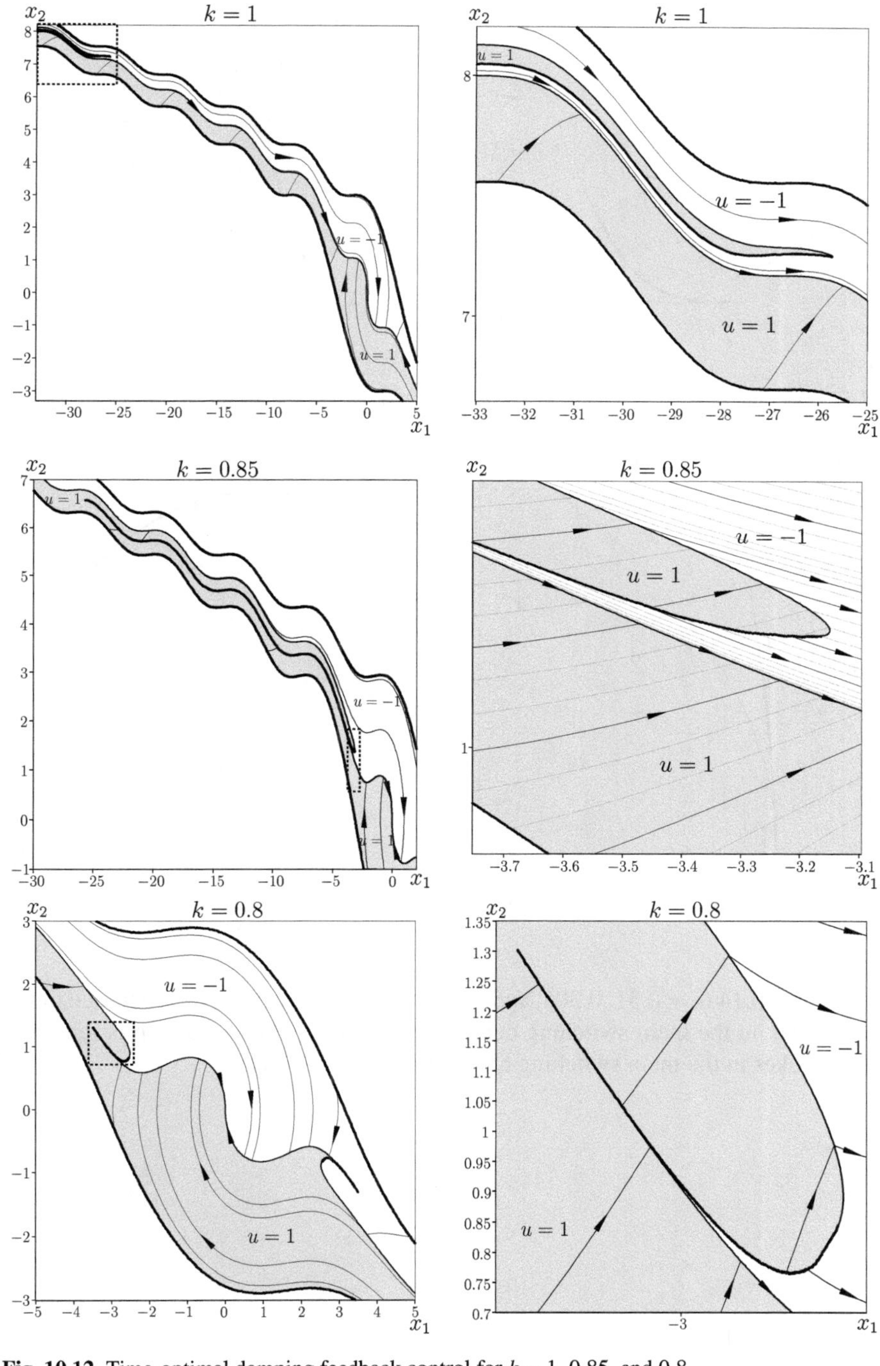

Fig. 10.12 Time-optimal damping feedback control for $k = 1, 0.85$, and 0.8

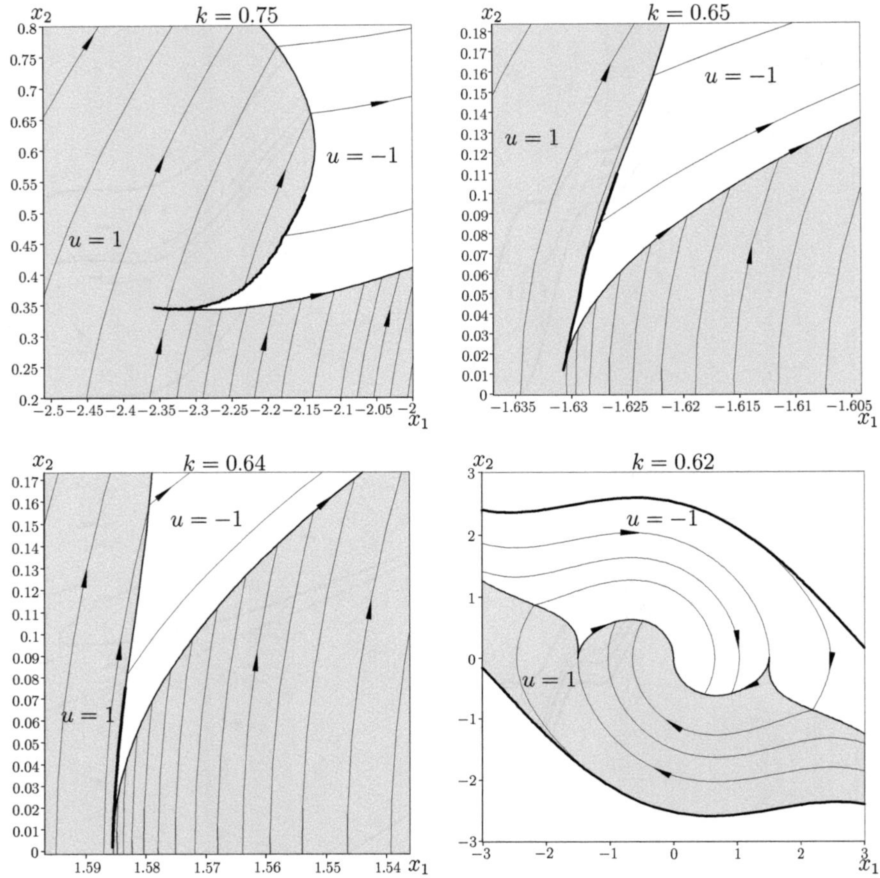

Fig. 10.13 Time-optimal damping feedback control for $k = 0.75, 0.65, 0.64$, and 0.62

Figure 10.14 ($k = 0.31, 0.305$, and 0.3) illustrates the process of generation of the third break in the main switching curve. This means that the situation is repeated: new breakes in the main switching curve emerge as k decreases.

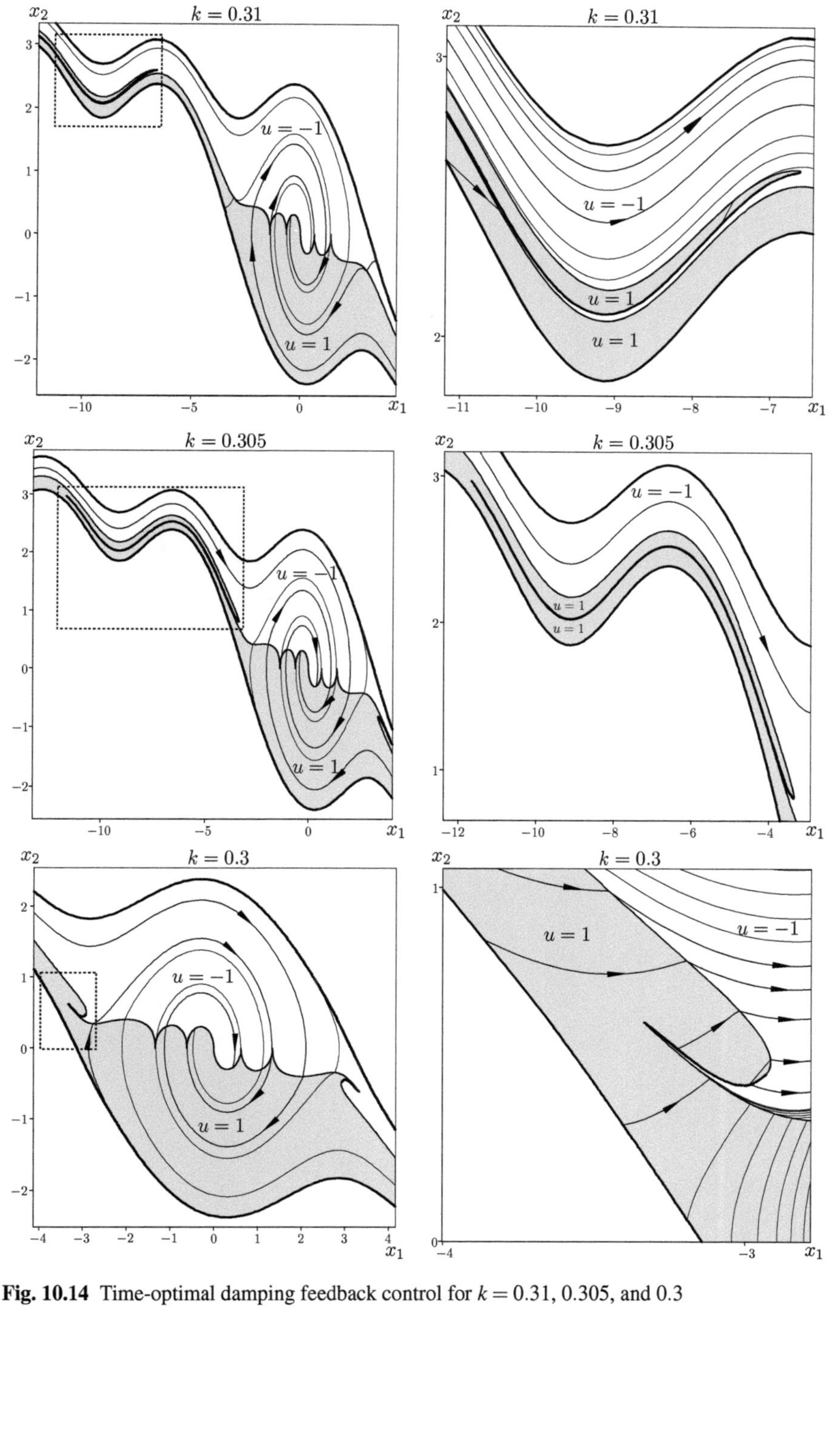

Fig. 10.14 Time-optimal damping feedback control for $k = 0.31$, 0.305, and 0.3

References

1. Agmon, S.: On kernels, eigenvalues and eigenfunctions of operators related to elliptic problems. Communs Pure and Appl Math **18**(4), 627–663 (1965)
2. Akulenko, L.D., Bolotnik, N.N.: Synthesis of optimal control of transport motions of manipulation robots. Mechanics of Solids **21**(4), 18–26 (1986)
3. Akulenko, L.D., Bolotnik, N.N., Kumakshev, S.A., Chernov, A.A.: Active damping of vibrations of large-sized load-carrying structures by moving internal masses. J. of Computer and Systems Sciences International **39**(1), 128–138 (2000)
4. Ananievski, I.M.: The control of a mechanical system with unknown parameters by a bounded force. J. Appl. Maths Mechs **61**(1), 49–58 (1997)
5. Ananievski, I.M.: Constrained control of a mechanical system in conditions of uncertainty. Doklady Mathematics **57**(2), 328–330 (1998)
6. Ananievski, I.M.: Control of a two-mass system with unknown parameters. J. of Computer and Systems Sciences International **37**(2), 232–242 (1998)
7. Ananievski, I.M.: Control of a fourth-order linear system with mixed constraints. J. Appl. Maths Mechs **64**(6), 863–870 (2000)
8. Ananievski, I.M.: Limited control of a rheonomous mechanical system under uncertainty. J. Appl. Maths Mechs **65**(5), 785–796 (2001)
9. Ananievski, I.M.: Continuous feedback control of mechanical systems subjected to disturbances. Doklady Mathematics **66**(1), 143–146 (2002)
10. Ananievski, I.M.: Continuous feedback control of perturbed mechanical systems. J. Appl. Maths Mechs **67**(2), 163–178 (2003)
11. Ananievski, I.M.: Synthesis of continuous control for a mechanical system with constraints dependent on time. J. of Computer and Systems Sciences International **42**(6), 852–862 (2003)
12. Ananievski, I.M.: Synthesis of bounded control for nonlinear uncertain mechanical systems. In: IUTAM Symposium on Dynamics and Control of Nonlinear Systems with Uncertainty, pp. 277–286. Nanjing, China (2006)
13. Ananievski, I.M.: Synthesis of continuous control for a mechanical system with an unknown inertia matrix. J. of Computer and Systems Sciences International **45**(3), 356–367 (2006)
14. Ananievski, I.M., Dobrynina, I.S., Chernousko, F.L.: Use of the decomposition method in controlling a mechanical system. J. of Computer and Systems Sciences International **34**(5), 138–149 (1996)
15. Ananievski, I.M., Reshmin, S.A.: The method of decomposition in the problem of tracking the trajectories of mechanical systems. J. of Computer and Systems Sciences International **41**(5), 695–702 (2002)
16. Anchev, A.A.: Equilibrium attitude transitions of a three-rotor gyrostat in a circular orbit. AIAA J. **11**(4), 467–472 (1973)
17. Anchev, A.A., Melikyan, A.A.: Optimal reorientation of a satellite on a circular orbit. Mechanics of Solids **15**(6), 28–32 (1980)
18. Angeli, D.: Almost global stabilization of the inverted pendulum via continuous state feedback. Automatica **37**, 1103–1108 (2001)
19. Athans, M., Falb, P.L.: Optimal control: an introduction to the theory and its applications. McGraw-Hill (1966)
20. Beletskii, V.V.: On optimal steering of an artificial Earth's satellite in a gravitationally stable position. Kosm. Issl. **9**(3), 366–375 (1971)
21. Bellman, R.E.: Dynamic programming. Princeton University Press, New Jersey (1957)
22. Bogoliubov, N.N., Mitropolsky, Y.A.: Asymptotic methods in the theory of nonlinear oscillations. Gordon and Breach Sci. Publ., New York (1961)
23. Boltyanskii, V.G.: The mathematical methods of optimal control. Nauka, Moscow (1969)
24. Bryson, A.E., Ho, Y.E.: Applied optimal control: optimization, estimation, and control. Blaisdell Pub. Co., Waltham Mass (1969)

25. Butkovsky, A.G.: Theory of optimal control of distributed parameter systems. Elsevier Publ. Co., New York (1969)
26. Chernousko, F.L.: On the construction of a bounded control in oscillatory systems. J. Appl. Maths Mechs **52**(4), 426–433 (1988)
27. Chernousko, F.L.: Decomposition and suboptimal control in dynamical systems. J. Appl. Maths Mechs **54**(6), 727–734 (1990)
28. Chernousko, F.L.: Control synthesis for a system with non-linear resistance. J. Appl. Maths Mechs **55**(6), 759–771 (1991)
29. Chernousko, F.L.: Decomposition and synthesis of control in dynamical systems. Soviet J. of Computer and Systems Sciences **29**(5), 126–144 (1991)
30. Chernousko, F.L.: Bounded controls in distributed-parameter systems. J. Appl. Maths Mechs **56**(5), 707–723 (1992)
31. Chernousko, F.L.: Control synthesis for a non-linear dynamical system. J. Appl. Maths Mechs **56**(2), 157–166 (1992)
32. Chernousko, F.L.: The decomposition of controlled dynamic systems. In: A.B. Kurzhanski (ed.) Advances in nonlinear dynamics and control, pp. 1–40. Birkhäuser, Boston Basel Berlin (1993)
33. Chernousko, F.L.: State estimation for dynamic systems. CRC Press, Boca Raton (1994)
34. Chernousko, F.L.: Decomposition and synthesis of control in nonlinear dynamical systems. Proceedings of the Steklov Institute of Mathematics **211**, 414–428 (1995)
35. Chernousko, F.L.: The time-optimal control problem under mixed constraints. Izv. Ross. Akad. Nauk. Teoriya i Sistemy Upravleniya (4), 103–113 (1995)
36. Chernousko, F.L.: Control of elastic systems by bounded distributed forces. Applied Mathematics and Computation **78**(2), 103–110 (1996)
37. Chernousko, F.L.: Control of a system with one degree of freedom under complex restrictions. J. Appl. Maths Mechs **63**(5), 669–676 (1999)
38. Chernousko, F.L.: Control of systems with one degree of freedom under constraints on the controlling force and the rate of its change. Doklady Mathematics **60**(2), 309–311 (1999)
39. Chernousko, F.L.: Control of oscillations in systems with many degrees of freedom. In: N. Van Dao, E.J. Kreuzer (eds.) Proc. IUTAM Symposium on Recent Developments in Nonlinear Oscillations of Mechanical Systems, Hanoi, 1999, pp. 45–54. Kluwer Academic Publishers, Dordrecht, Netherlands (2000)
40. Chernousko, F.L., Akulenko, L.D., Sokolov, B.N.: Control of Oscillations. Nauka, Moscow (1980)
41. Chernousko, F.L., Bolotnik, N.N., Gradetsky, V.G.: Manipulation robots: dynamics, control, and optimization. CRC Press, Boca Raton (1994)
42. Chernousko, F.L., Reshmin, S.A.: Decomposition and synthesis of control in a nonlinear dynamic system. In: Proc. International Conference on Informatics and Control (ICI&C'97), vol. 1, pp. xlv–lii. St. Petersburg (1997)
43. Chernousko, F.L., Reshmin, S.A.: Decomposition of control for nonlinear lagrangian systems. In: Proc. 4th IFAC Nonlinear Control Systems Design Symposium (NOLCOS'98), vol. 1, pp. 209–214. Enschede, The Netherlands (1998)
44. Chernousko, F.L., Reshmin, S.A.: Decomposition of control for robotic manipulators. In: Proc. 4th ECPD International Conference on Advanced Robotics, Intelligent Automation and Active Systems, pp. 184–189. Moscow (1998)
45. Chernousko, F.L., Reshmin, S.A.: Method of decomposition and its applications to uncertain dynamical systems. In: Proc. 16th IFAC World Congress, on CD. Prague, Czech Republic (2005)
46. Chernousko, F.L., Reshmin, S.A.: Time-optimal swing-up and damping feedback controls of a nonlinear pendulum. In: Proc. of the ECCOMAS Thematic Conference on Multibody Dynamics, on CD. Milan, Italy (2007)
47. Chernousko, F.L., Reshmin, S.A.: Time-optimal swing-up feedback control of a pendulum. Nonlinear Dynamics **47**(1–3), 65–73 (2007)
48. Chernousko, F.L., Shmatkov, A.M.: Synthesis of time-optimal control in a third-order system. Doklady Mathematics **55**(3), 467–470 (1997)

49. Chernousko, F.L., Shmatkov, A.M.: Time-optimal control in a third-order system. J. Appl. Maths Mechs **61**(5), 699–707 (1997)
50. Chilikin, M.G., Klyuchev, B.I., Sandler, A.S.: Theory of automatic electric drive. Energy, Moscow (1979)
51. Corless, M., Leitmann, G.: Adaptive control of systems containing uncertain functions and unknown functions with uncertain bounds. J. of Optimization Theory and Applications **42**(1), 155–168 (1983)
52. Corless, M., Leitmann, G.: Adaptive controllers for a class of uncertain systems. Annales Foundation de Broglie **9**, 65–95 (1984)
53. Dobrynina, I.S., Chernousko, F.L.: Constrained control of a fourth-order linear system. J. of Computer and Systems Sciences International **32**(41), 108–115 (1994)
54. Dobrynina, I.S., Karpov, I.I., Chernousko, F.L.: Computer modelling of the control of the motion of a system of connected rigid bodies. Izv. Ross. Akad. Nauk. Tekhn. Kibernetika (1), 167–180 (1994)
55. Faronov, V.V.: Programming for personal computers in Turbo-Pascal environment. Mosk. Gos. Tekh. Univ., Moscow (1991)
56. Fehlberg, E.: Low-Order Classical Runge-Kutta Formulas with Stepsize Control. NASA Technical Report R-315 (1969)
57. Fehlberg, E.: Klassische Runge-Kutta-Formeln vierter und niedrigerer Ordnung mit Schrittweiten-Kontrolle und ihre Anwendung auf Wärmeleitungsprobleme. Computing **6**, 61–71 (1970)
58. Filimonov, Yu.M.: On an optimal control problem for a mathematical pendulum. Differ. Uravn. **1**(8), 1007–1015 (1965)
59. Fomin, V.N., Fradkov, A.L., Yakubovich, V.A.: Adaptive control of dynamical systems. Nauka, Moscow (1981)
60. Forsythe, G.E., Malcolm, M.A., Moler, C.B.: Computer methods for mathematical computations. Prentice-Hall, Englewood Cliffs, N.J. (1977)
61. Fradkov, A.L., Miroshnik, I.V., Nikiforov, V.O.: Nonlinear and adaptive control of complex systems. Kluwer Academic Publishers, Dordrecht (1999)
62. Fradkov, A.L., Pogromsky, A.Yu.: Introduction to control of oscillations and chaos. World scientific series on nonlinear science. World Scientific Publishing Co. Pte. Ltd., Singapore (1998)
63. Friedland, B., Sarachik, P.: Indifference regions in optimum attitude control. IEEE Trans. Autom. Control. **9**(2), 180–181 (1964)
64. Gantmaher, F.R.: Matrix theory, 3rd edn. Nauka, Moscow (1967)
65. Garcia Almuzara, J.L., Flügge-Lots, I.: Minimum time control of a nonlinear system. J. Differen. Equations **4**(1), 12–39 (1968)
66. Grishin, A.A., Lenskii, A.V., Okhotsimsky, D.E., Panin, D.A., Formal'skii, A.M.: A control synthesis for a unstable object: an inverted pendulum. J. of Computer and Systems Sciences International (5), 14–24 (2002)
67. Il'in, V.A.: On the uniform convergence of eigenfunction expansions throughout a closed domain. Mat. Sbor. **45**(87), 195–232 (1958)
68. Il'in, V.A.: Spectral theory of differential operators. Nauka, Moscow (1991)
69. Isaacs, R.: Differential Games. Wiley, New York (1965)
70. Isidori, A.: Nonlinear Control Systems, 3 edn. Springer-Verlag, New York (1995)
71. Isidori, A.: Nonlinear Control Systems, vol. II. Springer-Verlag, New York (1999)
72. Kalman, R.E.: General theory of control systems. In: Proc. I Int. IFAC Congress, vol. 2. Butterworths, London (1960)
73. Kalman, R.E., Falb, P.L., Arbib, M.A.: Topics in Mathematical System Theory. McGraw-Hill (1969)
74. Kapitsa, P.L.: Dynamic stability of a pendulum with oscillating suspension point. Zh. Éksp. Teor. Fiz. **21**(5), 588–598 (1951)
75. Kayumov, O.R.: Optimal control of an elliptical pendulum. Mechanics of Solids **20**(4), 35–41 (1985)

76. Klimov, D.M., Rudenko, V.M.: Methods of computer algebra in problems of mechanics. Nauka, Moscow (1989)
77. Krasovskii, N.N.: Some problems of motion stability theory. Fizmatgiz, Moscow (1959)
78. Krasovskii, N.N.: The theory of control of motion. Nauka, Moscow (1968)
79. Krasovskii, N.N.: Game-theoretic problems on the encounter of motions. Nauka, Moscow (1970)
80. Krasovskii, N.N., Subbotin, A.I.: Positional differential games. Nauka, Moscow (1974)
81. Kuipers, B., Ramamoorthy, S.: Qualitative modeling and heterogeneous control of global system behavior. In: Hybrid systems computation and control, pp. 294–307. Stanford, California (2002)
82. Lee, E.B., Markus, L.: On necessary and sufficient optimality conditions in time optimality for second-order nonlinear systems. In: Proc. 2nd International IFAC Congress. Basel, Switzerland (1963)
83. Lee, E.B., Markus, L.: Foundations of optimal control theory. John Wiley & Sons, Inc., New York, London, Sydney (1967)
84. Leitmann, G.: Introduction to optimal control. McGraw-Hill, Maidenhead (1966)
85. Leitmann, G.: Deterministic control of uncertain systems. Acta Astronautica **7**, 1457–1461 (1980)
86. Lions, J.L.: Optimal control of systems governed by partial differential equations. Springer-Verlag, New York (1971)
87. Lions, J.L.: Exact controllability, stabilization and perturbations for distributed systems. SIAM Review **30**(1), 1–68 (1988)
88. Lyapunov, A.M.: The general problem of the stability of motion. Republished Princeton University Press (1947)
89. Moiseev, N.N.: Elements of the theory of optimal systems. Nauka, Moscow (1974)
90. Nijmeijer, H., Rodriguez-Angeles, A.: Synchronization of mechanical systems. World scientific series on nonlinear science. World Scientific Publishing Co. Pte. Ltd. (2003)
91. Nijmeijer, H., van der Schaft, A.J.: Nonlinear Dynamic Control Systems. Springer-Verlag, New York (1990)
92. Ovseyevich, A.I.: The stability of an inverted pendulum when there are rapid random oscillations of the suspension point. J. Appl. Maths Mechs **70**, 762–768 (2006)
93. Pontryagin, L.S., Boltyanskii, V.G., Gamkrelidze, R.V., Mishchenko, E.F.: The mathematical theory of optimal processes. Gordon and Breach, New York (1986)
94. Pyatnitskii, Ye.S.: Synthesis of the control of manipulating robots using the principle of decomposition. Izv. Akad. Nauk SSSR. Tekhn. Kibernetika (3), 92–99 (1987)
95. Pyatnitskiy, Ye.S.: A decoupling principle in control of mechanical systems. Reports of the Russian Academy of Sciences **300**(2), 300–303 (1988)
96. Pyatnitskiy, Ye.S.: Criteria of complete controllability for classes of mechanical systems with bounded control. J. Appl. Maths Mechs **60**(5), 785–796 (1996)
97. Åström, K.J., Furuta, K.: Swinging up a pendulum by energy control. Automatica **36**(2), 287–295 (2000)
98. Reshmin, S.A.: Synthesis of control of a manipulator with two links. J. of Computer and Systems Sciences International **36**(2), 299–303 (1997)
99. Reshmin, S.A.: Bounded control of a linear system of the third order. J. of Computer and Systems Sciences International **35**(1), 18–22 (1998)
100. Reshmin, S.A.: Decomposition method in the problem of controlling an inverted double pendulum with the use of one control moment. J. of Computer and Systems Sciences International **44**(6), 861–877 (2005)
101. Reshmin, S.A.: Finding the principal bifurcation value of the maximum control torque in the problem of optimal control synthesis for a pendulum. J. of Computer and Systems Sciences International **47**(2), 163–178 (2008)
102. Reshmin, S.A., Chernousko, F.L.: Control synthesis in a nonlinear dynamic system based on a decomposition. J. Appl. Maths Mechs **62**(1), 115–122 (1998)

103. Reshmin, S.A., Chernousko, F.L.: Method of decomposition for the control of nonlinear dynamical systems. In: Proc. of the Fifth EUROMECH Nonlinear Oscillations Conference (ENOC-2005), on CD, pp. 790–799. Eindhoven, Netherlands (2005)
104. Reshmin, S.A., Chernousko, F.L.: Time-optimal control of an inverted pendulum in the feedback form. J. of Computer and Systems Sciences International **45**(3), 383–394 (2006)
105. Reshmin, S.A., Chernousko, F.L.: Time-optimal control synthesis in problems of swing-up and damping of oscillations of a nonlinear pendulum. In: Proc. of 9th International Chetaev Conference on Analytical Mechanics, Stability, and Motion Control, vol. 3, pp. 179–196. Irkutsk, Russia (2007)
106. Reshmin, S.A., Chernousko, F.L.: Time-optimal feedback control of a nonlinear pendulum. J. of Computer and Systems Sciences International **46**(1), 9–18 (2007)
107. Salle, J.L., Lefschetz, S.: Stability by Liapunov's direct method with applications. Academic Press, New York, London (1961)
108. Schiehlen, W.: Multibody systems handbook. Springer Verlag, Berlin (1990)
109. Schiehlen, W.: Advanced multibody system dynamics — simulation and software tools. Kluwer Academic Publishers, Dordrecht (1993)
110. Seeley, R.: Interpolation in L^p with boundary conditions. Stud Math **44**(1), 47–60 (1972)
111. Shampine, L.E., Watts, H.A., Davenport, S.: Solving non-stiff ordinary differential equations — the state of the art. Sandia Laboratories Report SAND75-0182, Albuquerque, New Mexico; SIAM Review **18**(3), 376–411 (1976)
112. Shiriaev, A., Egeland, O., Ludvigsen, H.: Global stabilization of unstable equlibrium point of pendulum. In: Proc. of IEEE conference on decision and control, pp. 4584–4585. Tampa, Florida (1998)
113. Sirazetdinov, T.K.: Optimization of systems with distributed parameters. Nauka, Moscow (1977)
114. Stephenson, A.: On a new type of dynamical stability. Memoirs and Proceedings of the Manchester Literary and Philosophical Society **52**(8), Pt. 2, 1–10 (1908)
115. Utkin, V., Guldner, J., Shi, J.: Sliding Mode Control in Electromechanical Systems. Taylor & Francis, London (1999)
116. Utkin, V.I.: Sliding modes in optimization and control. Springer-Verlag, New York (1992)
117. Utkin, V.I., Orlov, Y.V.: The theory of infinitely-dimensional control systems in sliding regimes. Nauka, Moscow (1990)
118. Van der Schaft, A.: L_2-gain and passivity techniques in nonlinear control. Springer, London (2000)
119. Xu, Y., Iwase, M., Furuta, K.: Time optimal swing-up control of a single pendulum. Transactions of the ASME **123**, 518–527 (2001)
120. Yefimov, G.B., Pogorelov, D.Y.: Some algorithm for automate synthesis of motion equations of rigid bodies system. Preprint of the Institute of Applied Mathematics RAS (84) (1993)
121. Yefimov, G.B., Pogorelov, D.Y.: "Universal mechanism"—software tool for modelling dynamics of rigid bodies systems. Preprint of the Institute of Applied Mathematics RAS (77) (1993)
122. Yegorov, A.I.: Optimal control of thermal and diffusive processes. Nauka, Moscow (1978)
123. Yemelyanov, S.V.: Automatic control systems with variable structure. Nauka, Moscow (1967)

Index

MIX
Papier aus verantwortungsvollen Quellen
Paper from responsible sources
FSC® C105338
FSC
www.fsc.org

If you have any concerns about our products,
you can contact us on
ProductSafety@springernature.com

In case Publisher is established outside the EU,
the EU authorized representative is:
Springer Nature Customer Service Center GmbH
Europaplatz 3, 69115 Heidelberg, Germany

Printed by Libri Plureos GmbH
in Hamburg, Germany